Studies in Computational Intelligence

Volume 793

Series editor

Janusz Kacprzyk, Polish Academy of Sciences, Warsaw, Poland
e-mail: kacprzyk@ibspan.waw.pl

The series "Studies in Computational Intelligence" (SCI) publishes new developments and advances in the various areas of computational intelligence—quickly and with a high quality. The intent is to cover the theory, applications, and design methods of computational intelligence, as embedded in the fields of engineering, computer science, physics and life sciences, as well as the methodologies behind them. The series contains monographs, lecture notes and edited volumes in computational intelligence spanning the areas of neural networks, connectionist systems, genetic algorithms, evolutionary computation, artificial intelligence, cellular automata, self-organizing systems, soft computing, fuzzy systems, and hybrid intelligent systems. Of particular value to both the contributors and the readership are the short publication timeframe and the world-wide distribution, which enable both wide and rapid dissemination of research output.

More information about this series at http://www.springer.com/series/7092

Krassimir Georgiev · Michail Todorov
Ivan Georgiev
Editors

Advanced Computing in Industrial Mathematics

12th Annual Meeting of the Bulgarian Section of SIAM December 20–22, 2017, Sofia, Bulgaria Revised Selected Papers

Editors
Krassimir Georgiev
Institute of Information
and Communication Technology
Bulgarian Academy of Sciences
Sofia, Bulgaria

Michail Todorov
Faculty of Applied Mathematics
and Informatics
Technical University of Sofia
Sofia, Bulgaria

Ivan Georgiev
Institute of Mathematics and Informatics
and Institute of Information and
Communication Technologies
Bulgarian Academy of Sciences
Sofia, Bulgaria

ISSN 1860-949X ISSN 1860-9503 (electronic)
Studies in Computational Intelligence
ISBN 978-3-030-07329-9 ISBN 978-3-319-97277-0 (eBook)
https://doi.org/10.1007/978-3-319-97277-0

Softcover re-print of the Hardcover 1st edition 2019

This Springer imprint is published by the registered company Springer Nature Switzerland AG
The registered company address is: Gewerbestrasse 11, 6330 Cham, Switzerland

Preface

The 12th Annual Meeting of the Bulgarian Section of the Society for Industrial and Applied Mathematics (BGSIAM) was held in Sofia, December 20–22, 2017. The Section was formed in 2007 with the purpose to promote and support the application of mathematics to science, engineering, and technology in Bulgaria.

The goals of BGSIAM follow and creatively develop the general goals of SIAM:

- To advance the application of mathematics and computational science to engineering, industry, science, and society;
- To promote research that will lead to effective new mathematical and computational methods and techniques for science, engineering, industry, and society;
- To provide media for the exchange of information and ideas among mathematicians, engineers, and scientists.

During the BGSIAM'17 conference, a wide range of problems concerning recent achievements in the field of industrial and applied mathematics were presented and discussed. The meeting provided a forum for exchange of ideas between scientists, who develop and study mathematical methods and algorithms, and researchers, who apply them for solving real-life problems.

Among the topics of interest are high-performance computing, numerical methods and algorithms, analysis of partial differential equations and their applications, mathematical biology, control and uncertain systems, stochastic models, molecular dynamics, neural networks, genetic algorithms, metaheuristics for optimization problems, generalized nets, and Big Data.

The invited speakers were:

- *Krassimir Atanassov (Bulgarian Academy of Sciences), Generalized Nets—Theory and Applications*
- *Peter Minev (University of Alberta, Canada), High-order Artificial Compressibility for the Navier–Stokes Equations*
- *Maya Neytcheva (Uppsala University, Sweden), Enhanced degree of parallelism when solving optimal control problems constrained by evolution equations*

- *Zahari Zlatev (Aarhus University, Denmark), Application of repeated Richardson Extrapolation in the treatment of some chemical modules of environmental pollution models.*

We would like to thank all the referees for the constructive remarks and criticism, which furthered considerable improvements of the quality of the papers in this book.

Sofia, Bulgaria

Krassimir Georgiev
Michail Todorov
Ivan Georgiev

Contents

Method for Indoor Localization of Mobile Devices Based on AoA and Kalman Filtering

A. Alexandrov and V. Monov

Abstract The mobile devices are a widely used tools for indoor localization. By different reasons the localization of mobile devices in closed areas has not yet fully developed because of missing of a reliable positioning technology. The paper presents a hybrid method for improving the accuracy of indoor positioning approach for Bluetooth Low Energy (BLE) mobile devices based on optimized combination of Angle of Arrival (AoA) and Receive Signal Strength (RSS) technologies. We propose a hybrid optimization method for indoor positioning, realized by two stage data fusion process using Extended Kalman filtering approach and Fraser-Potter equation. The test results show that the proposed method can achieve sensitively better accuracy in a real environment compared to existing indoor localization methods and techniques.

1 Introduction

Indoor Positioning is a challenge topic in public areas, which are used by large number of people. The problem of the mobile devices localization in closed areas and buildings has become more difficult because of the big complexity and scale of the public space. Nowadays there is a lot of researches in the area of implementation of evolutionary algorithms and AI neural networks in the localization process [1, 2]. The positioning and the localization of assets in indoor spaces is useful for several reasons. Loss and theft of equipment take a large expense of the budget. When it is possible to have the position of a device in real time, a system could be developed that locates the assets through the public area. The main goal of this paper is to develop a method and algorithm for an optimized indoor positioning system for localization of mobile devices (MD) and assets in indoor public areas.

A. Alexandrov (✉)
IICT-BAS, Sofia, Bulgaria
e-mail: akalexandrov@iit.bas.bg

V. Monov
e-mail: vmonov@iit.bas.bg

K. Georgiev et al. (eds.), *Advanced Computing in Industrial Mathematics*,
Studies in Computational Intelligence 793,
https://doi.org/10.1007/978-3-319-97277-0_1

2 Related Work

A large number of solutions, such as angle of arrival (AOA) [3, 4], received signal strength (RSS) [5–7], time of arrival (ToA) [8], time difference of arrival (TDOA) [9], and etc., have been proposed to attain mobile devices localization by measuring the received radio signal between the mobile device and the localization sensor node.

2.1 Typical AoA Based Method

In the AoA based techniques [3, 4, 8], the localization sensors using directional antennas have the capability of localization the Radio Frequency (RF) signal angle of arrival. For this purpose, some techniques like RF signal angle diversity are used in order to determine the directionality of the receivers antenna system. AoA method can fix the 2-D coordinate of mobile device (MD) with two angle values measured from two localization sensors (LS) to the MD. As it is shown in Fig. 1, LS1 and LS2 represent two LSs which coordinates are already known and MD is supposed to be the mobile device. θ_1 is the measured angle between LS1 and MD and θ_2 is the angle between LS2 and MD.

In this case we have

$$\begin{bmatrix} \tan\theta_1 \\ \tan\theta_2 \end{bmatrix} = \begin{bmatrix} (y - y_1)/(x - x_1) \\ (y - y_2)/(x - x_2) \end{bmatrix} \tag{1}$$

By solving (1), we can obtain the coordinate of MD as follows,

$$\begin{bmatrix} x \\ y \end{bmatrix} = \begin{bmatrix} \frac{y_2 - y_1 + x_1 \tan\theta_1 - x_2 \tan\theta_2}{\tan\theta_1 - \tan\theta_2} \\ \frac{y_2/\tan\theta_2 - y_1 \tan\theta_1 + x_1 - x_2}{1/\tan\theta_2 - 1/\tan\theta_1} \end{bmatrix} \tag{2}$$

The measurement angle error has some deviations which is caused by the measuring equipment and environmental noise. Therefore, more than one possible coordinates of MD should be calculated by using any two LS, and these calculated position points will vary around the real coordinate of MD. Varieties of methods have been

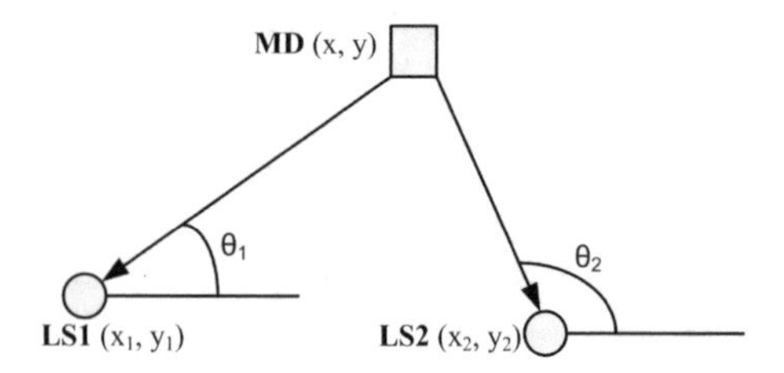

Fig. 1 AoA localization principle

proposed to restrain the influence caused by environment noise. According to the distribution character of these varying points, the simplest method is taking the mean of these point coordinates.

2.2 RSS Method

The RSS based techniques, use a distance measurement based on the attenuation introduced by the propagation of the signal from the MD to the LS. In [5, 7, 10] is proposed an empirical mathematical model to calculate the distance according to signal propagation:

$$p(R) = p(R_0) - 10n \log \frac{R}{R_0} - \left\{ \begin{matrix} nW \times WAF & (nW < C) \\ C \times WAF & (nW \geq C) \end{matrix} \right\} \tag{3}$$

In the presented above formula (3), R denotes the distance between the MD and the LS, R_0 is an already known distance acting as reference, $p(R)$ and $p(R_0)$ represent the signal strength received at R and R_0 respectively, nW is the number of obstructions between the MD and the LS, WAF is the average attenuation coefficient of the wall, C is the maximum number of attenuation barriers between the MD and the LS, and n is the routing attenuation coefficient which could be empirically determined. Based on the represented RSS technology, a few methods have been proposed to estimate the position of the MD. For example, the fingerprint based solution [7] for target positioning is typical application of RSS technology. In general, we can divide the fingerprint methodology into two phases: offline detection and sampling and online matching. For the sampling phase realization, a database is created offline to store the RF signal parameters including the geographical positions and the corresponding signal levels. In the online matching phase, the corresponding RF signals collected for the MD are compared against the already stored in the database records. By this way, it will be able to calculate the MD coordinates, as long as any RF signal record in the database is matched.

2.3 ToA Method

ToA localization algorithm as shown in Fig. 2 is based on ToA circumference equation, through the different combination of intersecting lines between corresponding circles, and generates different positioning equations [1].

The geometry model represents the signal transmit from MD to LSs. The measured distance is d_i :

$$d_i = \sqrt{(x_i - x_t)^2 + (y_i - y_t)^2}, \; i = 1, 2, 3 \tag{4}$$

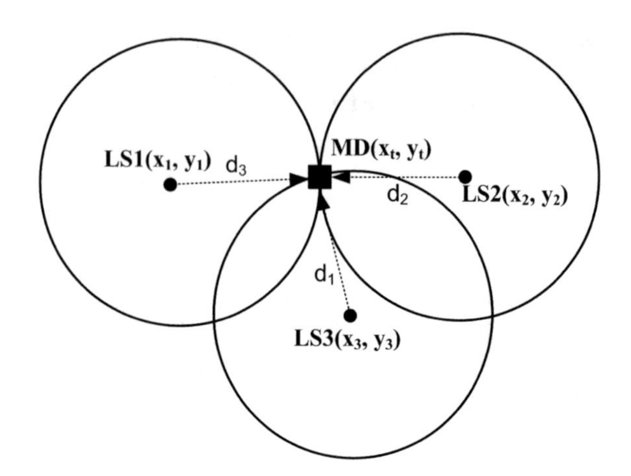

Fig. 2 ToA localization principle

(x_i, y_i) is the LSs coordinates, (x_t, y_t) is coordinate of the MD. Then the MD must be located on the circumference of a circle with radius d_i centered on the LS. When there are three LS coordinates, ToA measurement equation can be expressed as follow:

$$(x_2 - x_1)x_t + (y_2 - y_1)y_t = \frac{1}{2}[(x_2^2 + y_2^2) - (x_1^2 + y_1^2) + (d_1^2 - d_2^2)] \tag{5}$$

$$(x_3 - x_2)x_t + (y_3 - y_2)y_t = \frac{1}{2}[(x_3^2 + y_3^2) - (x_2^2 + y_2^2) + (d_2^2 - d_3^2)] \tag{6}$$

By these equations, we can calculate the coordinates of the MD:

$$x_t = \frac{(y_2 - y_1)D_3 - (y_3 - y_2)D_1}{[(x_3 - x_2)(y_2 - y_1) - (x_2 - x_1)(y_3 - y_2)]} \tag{7}$$

$$y_t = \frac{(x_2 - x_1)D_3 - (x_3 - x_2)D_1}{[(y_3 - y_2)(x_2 - x_1) - (y_2 - y_1)(x_3 - x_2)]} \tag{8}$$

where:

$$D_1 = \frac{1}{2}[(x_2^2 + y_2^2) - (x_1^2 + y_1^2) + (d_1^2 - d_2^2)] \tag{9}$$

$$D_3 = \frac{1}{2}[(x_3^2 + y_3^2) - (x_2^2 + y_2^2) + (d_2^2 - d_3^2)] \tag{10}$$

2.3.1 Time Difference-of-Arrival (TDoA)

This technology [9] uses two different kind RF signals - one transmitted from LS and another from referent base station. It is common practice to use equal frequency RF signals with different polarization. The time offset between the received two radio signals is used to calculate the MD's position. The calculation is based on the following equation:

$$\frac{R_1}{c_1} - \frac{R_2}{c_2} = t_1 - t_2 \tag{11}$$

where in vacuum $c_1 = c_2 = c$. The velocity of the RF signals can vary depend of the medium.

In (11), c_1 represents the velocity of the first RF signal, c_2 is the velocity of second RF signal, t_1 and t_2 are the times for these two signals traveling from MD to LS and from the MD to the referent station, and R_1 is the distance between the MD and the LS and R_2 is the distance between the MD and the referent station. A large number of works have explored TDoA-based methods. For instance [9, 11–13].

2.4 *Kalman Filter Observation and Transition Models*

The equations of the Kalman filter approach [14, 15] can be specified in two main categories: time update equations and measurement update equations. Generally, the target of the time update equations is to predict the current state and error estimations and at the same time to obtain a priori estimates for the next time period step. The target of the measurement update equations is the feedback, i.e. the realization of a new measurement based on a priori estimate and to obtain an improved a posteriori estimate. We can describe the time update equations as predictor equations, at the same time the measurement update equations can be specified as corrector equations (see Fig. 3).

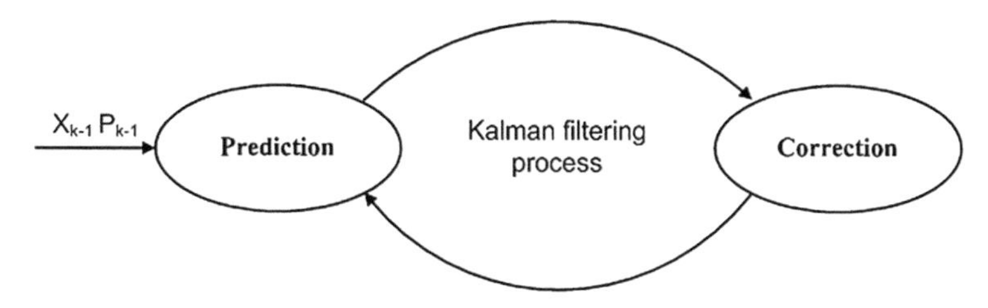

Fig. 3 Kalman filtering approach

The prediction phase starts with initial estimation of $x_{k-1}\hat{}$ and covariance vector P_{k-1} and proceed with

$$\hat{x_k} = A\hat{x}_{k-1} + Bu_k + Cw_k \quad (12)$$

$$z_k = H_k x_k + D_k v_k \quad (13)$$

where $\hat{x_k}$ is the estimated value, A is the transition state matrix of the process, B is the input matrix, C is the noise state transition matrix, u_k is the known input value, w_k is the noise, z_k is the observation vector, v_k is a variable describing observation noise, and H_k is the matrix of the observed value z_k and D_k is a matrix describing the contribution of noise to the observation. The measurement correction adjusts the projected estimate by an actual measurement at that time. In our case, we will be focused mainly on the measurement update algorithm. The current paper doesn't focus on details in the mathematical side of the Kalman filter measured updated equations. Details can be found in [13, 14, 16–18]. Based on the above cited papers we accept that the final extended Kalman filter measurement update equations are formulated as follows:

$$G_k = \frac{P_k H_k}{H_k P_k H_K^t + R_k} \quad (14)$$

$$\hat{x_k} = \hat{x_k} + G_k(z_k - H_k\hat{x_k}) \quad (15)$$

$$P_k = (1 - G_k H_k)P_k \quad (16)$$

In formula (14) G_k is the so called Kalman gain, P_k is an error covariance vector, H_k is a matrix of the observed value vector z_k and R_k is the covariance matrix. The initial task during the measurement update is to compute the Kalman gain G_k. The next phase is to actually measure the process to receive and then to calculate a posteriori state estimate by adding the correction based on the measurement and estimation as in (15). The last phase of the process is to obtain an a posteriori error covariance estimate via (16). After each measurement update pair, the process is repeated with the previous a posteriori estimates. In the present equation of the Kalman filter, each of the measurement error covariance matrices R_k can be measured before the execution of the Kalman filter data fusion process. In the case of the measurement error covariance R_k, especially this makes sense because there is a need to measure the process while operating the filter. We should be able to take some off-line sample measurements to determine the variance of the measurement error. In both cases, we have a good basis for choosing the parameters. Very often superior filter performance can be obtained by modification of the filter parameter R_k. The modification is usually performed off-line, frequently with the help of another

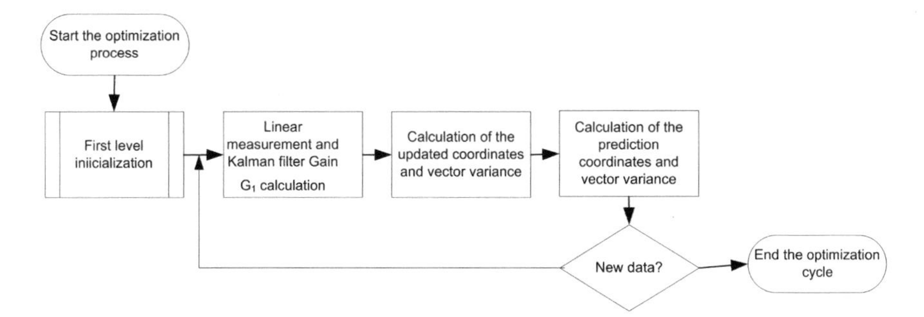

Fig. 4 Kalman filtering process flow diagram

Kalman filter. Finally, we can note that in cases where R_k is constant, both the estimation error covariance P_k and the Kalman gain G_k will start to stabilize very fast and then remain constant. If this is the case, the R_k parameter can be precomputed by running the filter off-line.

Extended Kalman localization of the flow diagram as shown in Fig. 4, mainly includes the state variables and the error variance matrix initialization. The state equation and measurement equation are linearized, computing the Kalman gain, update the status variables, predictions and the rest four steps.

3 Hybrid Indoor Localization Method Based on RSS and AoA

In the proposed localization method, we will be focused mainly on Angle of Arrival and RSS techniques as one of the reliable and low cost methods for indoor target localization. The proposed technique is based on ranging, whereby angle approximations are obtained [8] including some syntheses of interaction information [19, 20]. In this context, geometric approaches will be used to calculate the position of the mobile device of the target as an intersection of position lines obtained from the position-related parameters. Since the RF signal measurements in real systems have a deviation, especially in indoor areas, some optimization and error reducing techniques are used to improve the accuracy of the indoor positioning. The main factors who reduce the mobile device positioning accuracy are:

- the distance between the localization sensor and the mobile device of the target.
- the level of the RF noise and the signals reflected from indoor walls and metal equipment surfaces.

The second influencing factor can be sensitively reduced by using Bayesian filtering like Kalman filter and related statistical methods as Fraser-Potter smoothing. For the AoA implementation we use custom designed grid of directed antennas oriented at different azimuth and evaluation positions. Instead of the use of two different

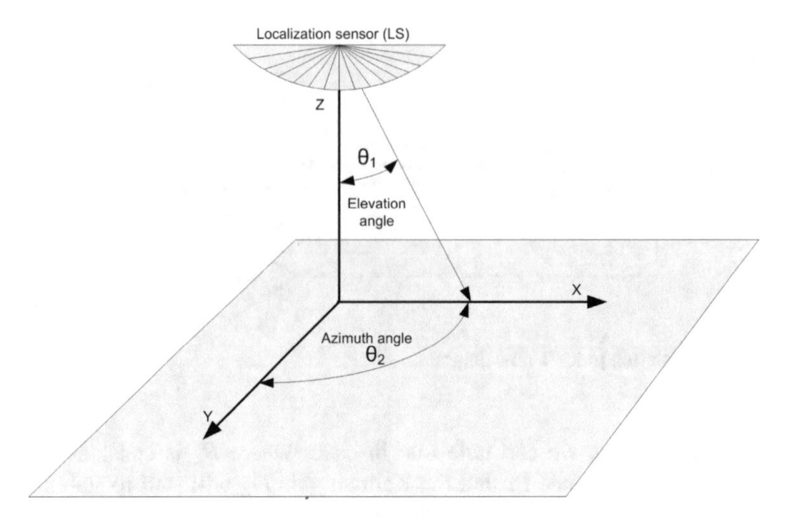

Fig. 5 Angular parameters for position measurement

localization sensors situated on different places to measure the angle of arrival as described in Sect. 2.1, we use only one localization sensor. The received signal angle information can be calculated because the antenna array geometry is known. The proposed AoA principle doesn't need of time synchronization between localization sensors. If there is a need for 2D localization only a single localization sensor is enough to find the target coordinates. The case of 3D localization needs a minimum of two sensor localization nodes to calculate the mobile device x, y and z coordinates. In the current paper, we are focused only on the 2D indoor localization problem as shown on Fig. 5.

Implementation of 2D position localization using RSS and AoA

In Fig. 6, the 2D tracking plane is shown. When a localization sensor detects a mobile device, the system can return the x, y coordinates of the target. Usually, the height ($h = 1.2$ m) component is fixed for all the measurements.

The presented approach relies on precise range determination, using combination of directional grid antennas, RSS and Angle of Arrival techniques for the device position calculation. The measured data pass a process of optimization, based on the extended Kalman filter approach and the Fraser-Potter equation. In this case, each reference sensor node (with known position) sends a ranging request to the BLE mobile devices in the area. Then the mobile device replies the sensor request, which is received by the grid of directional antennas. Based on the antenna directional diagram and an embedded in the receiver RSSI (Received Signal Strength Indication) capability could be calculated the azimuth, the evaluation and the strength of the received RF signal. Then the standard deviation of the measured data is calculated and it starts the Kalman filter based local optimization process for the last 6 measurements. In the current case we used an experimental test system. The localization sensor node with integrated data fusion algorithm realizes a number (in our case 6 which depends

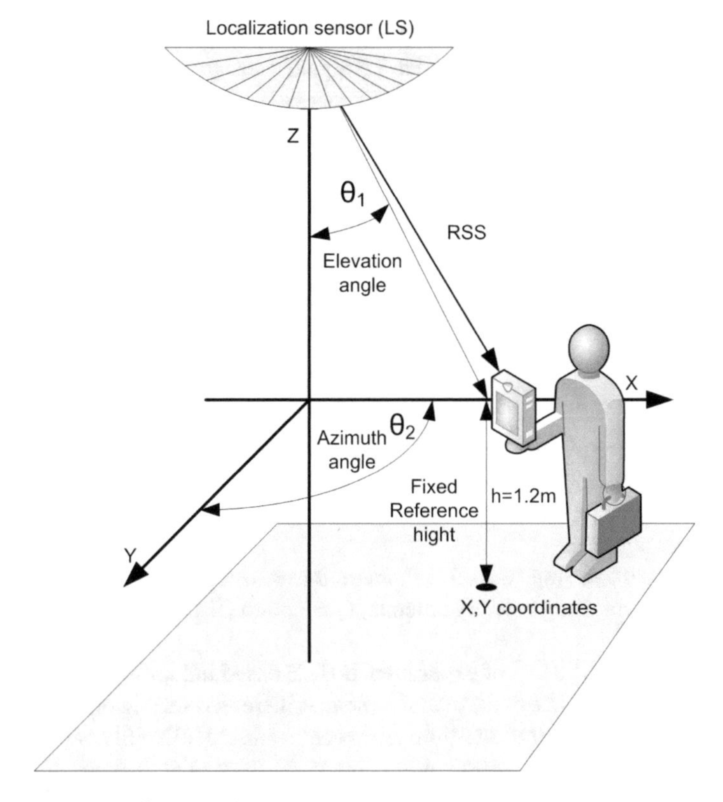

Fig. 6 Tracking plane for 2D positioning

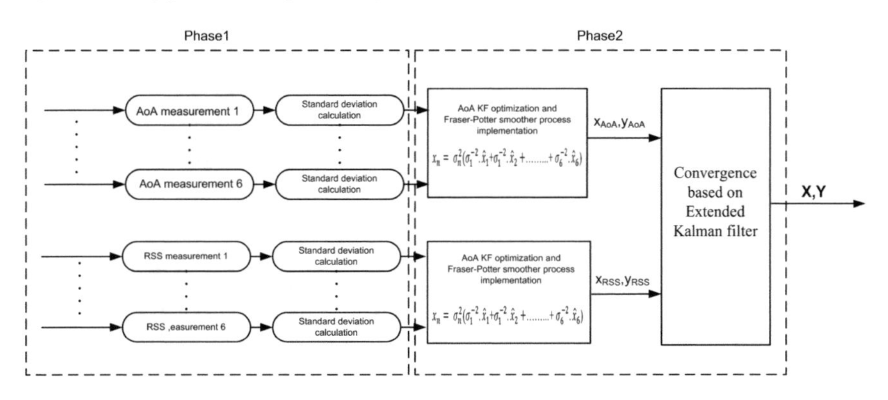

Fig. 7 Hybrid AoA/RSS localization method diagram

from the processor memory capacity) consistent measurements with 100 ms period. The implemented algorithm of the node fuses the results using Kalman filtering method. Based on a decision criterion (Fig. 7), the node sends the fused data of the localization measurement to the control center.

As shown on Fig. 7 the proposed indoor localization process can be divided into two main phases. In Phase 1 the RF signal measurement and target position calculation are executed. The measuring process starts simultaneously trough two independent channels - AoA and RSS measurement channels. During the first phase, some RF signal requests are sent to the mobile device and the RF response is received by one or more localization sensors (LS) with fixed coordinates. In Phase-2 the process of measured data optimization starts by two level Kalman filter optimization. At the first level, a KF based optimization process is executed for each measurement channel AoA and RS, respectively. The local optimization process is combined with smoothing procedure based on Fraser-Potter equations

$$x_n = \sigma_n^2(\sigma_1^{-2}\hat{x}_1 + \sigma_2^{-2}\hat{x}_2 + \cdots + \sigma_6^{-2}\hat{x}_6) \tag{17}$$

where $\sigma_n^2 = \sigma_1^{-2} + \sigma_2^{-2} + \cdots + \sigma_6^{-2}$ is the variance of the combined estimate and x_n represents the combined measurement. Finally, the calculated and optimized data from each measurement channel are optimized at the second level by Extended Kalman filter approach.

The proposed two stage data fusion method was tested by experimental localization indoor system using rotating antenna, Qualcomm BLE tags and regular mobile devices (Fig. 8).

The results from the test of experimental BLE based indoor localization (Fig. 9) show that the performance of the overall system is increased sensitively because of the reduced communication traffic to the control center and the reliability of the measured data from the localization sensor node compared to regular Kalman filtering. At the same time the decentralized Kalman filter algorithm effectively reduces and measurement noises.

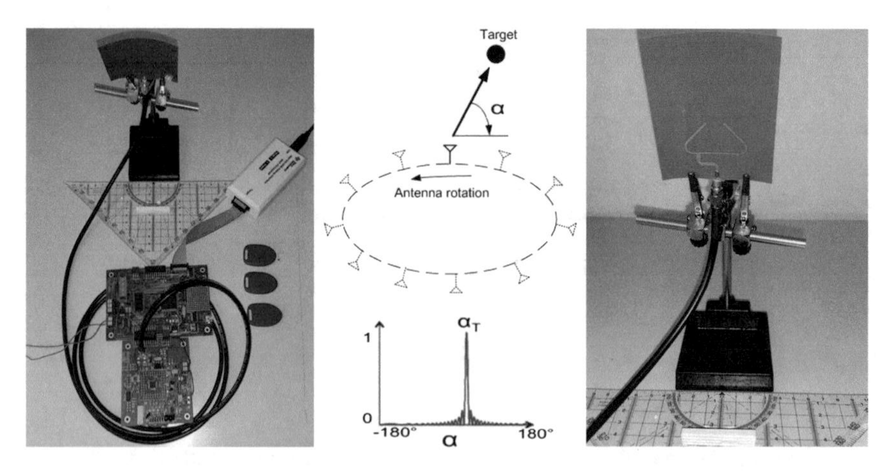

Fig. 8 Experimental AoA/RSS localization method system

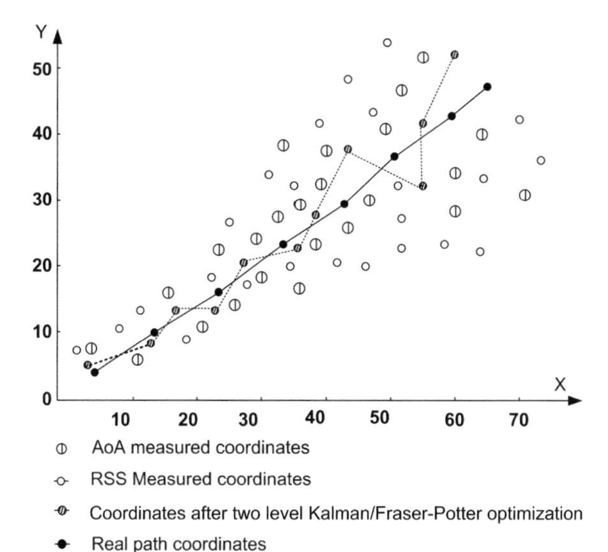

Fig. 9 Kalman filter/Fraser-Potter optimization experimental test results

4 Conclusions and Future Work

In this paper, a new hybrid method for improving indoor localization was presented. The method utilizes two independent channel position calculation processes based on RSS and AoA, respectively. Two level optimization is implemented. At the first level Kalman Filter and Fraser-Potter equation statistical approach is realized. The second level employs Extended Kalman filter which works as data fusion device between the channels. It additionally reduces the noise and improves the reliability of the measured data. The performance of the proposed method was tested in real environment by an experimental indoor localization system. The presented method allows the information from a consistent number of measurements to be combined and integrated in real time. Our approach gives the flexibility and reliability of the process of the mobile devices indoor localization. The experimental results show that the implementation of the proposed hybrid method improves substantially the accuracy in typical office building applications. The future research directions include the noise characteristics analysis of the received signal strength and choosing the most appropriate filtering initial parameters and correction coefficients.

References

1. Balabanov, T., Zankinski, I., Dobrinkova, N.: Time Series Prediction by Artificial Neural Networks and Differential Evolution in Distributed Environment, pp. 198–205. Large-Scale Scientific Computing, Springer, Berlin Heidelberg (2012)

2. Balabanov, T., Keremedchiev, D., Goranov, I.: Web distributed computing for evolutionary training of artificial neural networks. In: Proceedings of of the International Conference InfoTech-2016, Publishing House of Technical University—Sofia (2016). ISSN:1314-1023, 210–216
3. Dakkak, M., Nakib, A., Daachi, B., Siarry, P., Lemoine, J.: Indoor localization method based on RTT and AOA using coordinates clustering. Comput. Netw. **55**(8), 17941803 (2011)
4. Chen, B.Y., Chiu, C.C., Tu, L.C.: Mixing and combining with AOA and TOA for the enhanced accuracy of mobile location. In: 5th European Personal Mobile Communications Conference. (Conf. Publ. No. 492). IET, pp. 276–280 (2003)
5. Luo, R.C., Yih, C.-C., Su, K.L.: Multisensor fusion and integration: approaches, applications, and future research directions. IEEE Sens. J. **2**(2), 107119 (2002)
6. Llinas, J., Bowman, C., Rogova, G., Steinberg, A., Waltz, E., White, F.: Revisiting the JDL Data Fusion Model II. Technical Report, DTIC Document (2004)
7. Wang, L., Zawodniok, M.: RSSI-based localization in cellular networks. In: Proceedings of the IEEE 37th Conference on Local Computer Networks Workshops, pp. 820826 (2012)
8. Shen, J., Molisch, A.F., Salmi, J.: Accurate passive location estimation using TOA measurements. IEEE Trans. Wirel. Commun. **11**(6), 21822192 (2012)
9. Saloranta, J., Abreu, G.: Solving the fast moving vehicle localization problem via TDOA algorithms. In: Proceedings of the 8thWorkshop on PositioningNavigation and Communication (WPNC 11), pp. 127130, IEEE, Dresden, Germany, April 2011
10. Hall, D.L., Llinas, J.: An introduction to multisensor data fusion. Proc. IEEE **85**(1), 623 (1997)
11. Al Nuaimi, K., Kamel, H.: A survey of indoor positioning systems and algorithms. In: 2011 International Conference on Innovations in Information Technology (IIT) (pp. 185–190). IEEE (2011)
12. Mautz, R.: Indoor Positioning Technologies. Doctoral dissertation, Habilitationsschrift ETH Zurich (2012)
13. Venkatraman, S., Caffery, Jr., J.: Hybrid TOA/AOA techniques for mobile location in non-line-of-sight environments. In: WCNC Wireless Communications and Networking Conference, 2004. 2004 IEEE. 1: 274–278 (2004)
14. Welch, G., Bishop, G.: An Introduction to the Kalman Filter. Department of Computer Science University of North Carolina, UNC-Chapel Hill, TR 95-041, 24 Jul 2006
15. Julier, S.J., Uhlmann, J.K.: A new extension of the Kalman filter to nonlinear systems. In: Proceedings of the International Symposium on Aerospace/Defense Sensing, Simulation and Controls, vol. 3 (1997)
16. Blasch, E.P., Plano, S.: JDL level 5 fusion model user refinement issues and applications in group tracking. In: Proceedings of the Signal Processing, Sensor Fusion, and Target Recognition XI, pp. 270279, Apr 2002
17. Brown, C., Durrant-Whyte, H., Leonard, J., Rao, B., Steer, B.: Distributed data fusion using Kalman filtering: a robotics application. In: Abidi, M.A., Gonzalez, R.C. (eds.) Data, Fusion in Robotics and Machine Intelligence, pp. 267309 (1992)
18. Durrant-Whyte, H.F., Stevens, M.: Data fusion in decentralized sensing networks. In: Proceedings of the 4th International Conference on Information Fusion, pp. 302307, Montreal, Canada (2001)
19. Tashev, T.D., Hristov, H.R.: Modeling and synthesis of information interactions. J. Probl. Tech. Cybern. Robot. Sofia, Academic Publishing House "Prof. Marin Drinov", No 52, pp. 75–80 (2001)
20. Hristov, H.R., Tashev, T.D.: Computer-aided synthesis for interconnection process model. J. Probl. Tech. Cybern. Robot. Sofia, Academic Publishing House "Prof. Marin Drinov", No 51, pp. 20–25 (2001)

Cross-Validated Sequentially Constructed Multiple Regression

Slav Angelov and Eugenia Stoimenova

Abstract In this paper will be briefly presented a technique that we call *Cross-validated sequentially constructed multiple regression*. It is applied to a multiple regression model with at least 2 variables. This technique combines sequentially some of the model variables into components using a leave-one-out cross-validation procedure while taking into account the correlations between the model variables. The uncombined model variables along with the obtained components are then used to estimate a regression model. The newly obtained model is with lower multi-collinearity, and it tends to give better out-of-sample error while recalculated with additional observations. The proposed method is tested on a real accounting data concerning the Bulgarian gas utilities.

1 Introduction

Let us observe the multiple regression as a tool for prediction. A well-known fact is that under the Gaus-Markov conditions the least squares estimator is the best linear unbiased estimator for multiple regression, see for example [12, p. 41]. Let us assume that the G-M conditions hold, the input data are suitable for linear model, and that the coefficients are estimated by the least squares estimator. Another important thing is that we are assuming that we have independent model variables. However, usually in practice, we have correlations between the variables, and as a result, the model has multicollinearity problem. Multicollinearity causes some of the coefficients in the model to be estimated with high variance [9] or even to be biased because of the presence of suboptimal solutions. This leads to unstable model, and the final result is

S. Angelov (✉)
Department of Informatics, New Bulgarian University, 21 Montevideo str., Sofia, Bulgaria
e-mail: slav_angelov@abv.bg

E. Stoimenova
Institute of Mathematics and Informatics, Bulgarian Academy of Sciences (BAS),
Acad. G. Bontchev str. bl. 8, Sofia, Bulgaria
e-mail: jeni@math.bas.bg

K. Georgiev et al. (eds.), *Advanced Computing in Industrial Mathematics*,
Studies in Computational Intelligence 793,
https://doi.org/10.1007/978-3-319-97277-0_2

poor predictions. Moreover, if there is a presence of multicollinearity, and the model is recalculated by a data set with outliers than the adverse effects will be increased. If we examine the outliers as a separate problem [6], we should say that they can be influential to the model or not. Usually, even if they are not too influential they have a higher impact over the model than the other observations, and their individual or group effect causes the model to be biased which again may lead to poor prediction results for observations out of the learning set. Additionally, poor prediction result may be caused by overfitting the model - there is a small number of observations per model variable, and as a result, the estimated model is misleading [2]. Under our initial assumptions, we can summarize that to obtain a linear model with better prediction performance than a multiple regression we need to take into consideration all of the mentioned factors.

To simplify the understanding of the aim of this paper, we will define one task. Let us assume that over a chosen learning set we have estimated a regression model with the desired model statistics and prediction results checked over some test sets. We want to assure that this model will remain with as nice diagnostics and prediction performance as possible after adding more observations to its estimation. The method proposed in this paper is an instrument for such a handling.

2 Cross-Validated Sequentially Constructed Multiple Regression

We start this section with some notions, then we present the core of the method which incorporates two model variables.

We are observing a multiple regression model:

$$Y = \alpha_0 + \alpha_1.X_1 + \alpha_2.X_2 + \cdots + \alpha_k.X_k + \xi\ , \tag{1}$$

where Y is the predicted variable, α_i is the coefficient in front of predictor X_i for $i = 1, \ldots, k$, α_0 is the model intercept term, and ξ is the error term.

A standard approach to measure the model's quality of fit is to split the primary data set into a test set and a learning set. We estimate the model over the learning set and then test it over the test set using a function over the model residuals. Such a function is the *Root mean squared error* (RMSE), Eq. 2 , for more information see [3].

$$RMSE(A) = \sqrt{\frac{\sum_{i=1}^{n} \varepsilon^2(i)}{n}}, \tag{2}$$

where A is the test set, $\varepsilon(i)$ is the error of the i-th observation, and n is the number of observations in set A.

When the number of the observations in the primary data set is small it is not very convenient to split the set into two parts, as a result we have to estimate the model's quality of fit by the *RMSE* measure over the primary set (the full data set). However, the RMSE estimated over the learning set leads to a misleading results because it is in general biased downwards, see [7, p. 292]. To obtain a more realistic estimate of how good is the model in that case we can use the *Root mean squared error after cross-validation* (RMSECV) measure, see Eq. 3.

$$RMSECV = \sqrt{\frac{\sum_{i=1}^{n} \varepsilon_i^2}{n}}, \tag{3}$$

where n is the number of the observations in the primary data set, ei is the error for the i-th observation which is obtained from the regression model estimated over the primary data set without the i-th observation. RMSECV consists of out of sample errors which makes it more realistic measure for the model fit. RMSECV measure is based on a Leave-one-out cross-validation (LOOCV). This type of cross-validation is the one with the smallest variance of the error compared to the other types of cross-validation, to see more about the topic check [8, 10].

The technique that we are proposing merges some of the model variables into components reducing in that manner the total number of variables. The way we merge two chosen variables into one component is the core of the method. Each component is forged in a procedure that involves two model variables (or other already obtained components). Let us assume for simplicity that we want to combine variables X_1 and X_2 into one component Z. Moreover, the estimate a_1 of the coefficient in front of X_1 is bigger in absolute value than the estimate a_2 of the coefficient in front of X_2. Then $Z(k) = X_1 + k * X_2$, where we are searching for the optimal $k \in (k_0 - \mu, k_0 + \mu)$, $k_0 = \frac{a_2}{a_1}$, μ is a small number usually less than 1, while minimizing the RMSECV function for the multiple regression with the rest of the model variables (except the mentioned two) and the new component $Z(k)$, see Eq. 4. For $k = k_0$ we will attain the initial regression model (after substituting $Z(k_0)$ with X_1 and X_2), k_0 is in the interval $[-1,1]$ by the definition of $Z(k)$, μ is usually small because we do not expect estimates much more different than the regression model's ones.

$$\min_k RMSECV(k) = \min_k \sqrt{\frac{\sum_{i=1}^{n} \varepsilon_i^2(k)}{n}}, \quad k \in (k_0 - \mu, k_0 + \mu),\ \mu > 0, \tag{4}$$

where $\varepsilon_i(k) = Y(i) - f_i(k)$ is the error from the i-th observation from the model $f_i(k)$. $f_i(k)$ is the multiple regression model with input variables $Z(k), X_3, X_4, ..., X_n$ estimated with the full data set except the i-th observation. The RMSECV(k) function is continuous and positive. Moreover, it is bounded from above, the simplest way to see this, without going into details, is to note that RMSECV(k) is constructed from errors derived from regression models, and these errors are expected to be as small as possible. Thus, searching for the global minimum of the RMSECV(k) function is a valid operation.

2.1 The Proposed Method's Framework for More Than Two Variables

The proposed method has the following algorithm:
We have a multiple regression model with n variables.

1. We choose the model variable which has the worst estimate based on its absolute t-value. Then we find the most correlated with it model variable, and we combine them into one component;
2. The regression model is estimated with the new component instead of the two chosen variables from Step 1 and all of the rest model variables;
3. Step 1 is repeated with the n−1 model variables;
4. Step 2 is repeated with the obtained component from Step 3 and so on.

The procedure ends when we have achieved absolute t-values over a chosen threshold for all of the derived model variables or when the model variables are reduced to a predefined percentage from their initial number. We suggest that the procedure should stop when all of the model variables are with absolute t-values over 4.5, or the number of the derived model variables is around 50% of the initial number of variables.

The obtained components are then used instead of the model variables from which they are made from.

2.2 The Idea Behind the Approach

First, let us note that we are merging the variable with the highest estimation error for its coefficient with a variable that is most highly correlated with it. Thus, we are reducing the level of multicollinearity for the model, and as a consequence, we hope to achieve better-estimated coefficients for the model with the new component.

Second, while merging variables into components we reduce the total number of model variables, as a result we have smaller amount of model coefficients to compute while the model is updated with additional observations. Smaller number of variables along with bigger number of observations is a preposition for a better estimated model.

Third, each component is produced while minimizing RMSECV. RMSECV can be regarded as the RMSE over the learning set while performing LOOCV. Its goal is to see how well a regression model is capable of predicting out of sample observations. While minimizing RMSECV(k), we are searching for a component $Z(k)$ which allows the model that uses it to be as robust to changes in the learning set as possible concerning the out of sample prediction performance. To make this statement more clear we will present one valuable property of the LOOCV errors which shows the connection between them and their related errors from a multiple regression.

Theorem 1 *The out-of-sample residuals ε_i, $i = 1, ..., n$ used in the RMSECV measure for Model 1 can be computed by the following formula:*

$$\varepsilon_i = \frac{\hat{\xi}_i}{1 - h_i}, \; i = 1, ..., n, \tag{5}$$

where $\hat{\xi}_i$ is the error of the i-th observation from Model 1, n is the number of observations, and $h_{i,i}$ is the i-th element from the diagonal of the hat matrix $H = X(X^tX)^{-1}X^t$, *where X is the design matrix of Model 1, X^t is its transposed matrix.*

Proof This proof will follow a similar construction as the Theorem which is for the support vector regression case [4]. It is essential to show that for $i \in [1, ..., n]$:

$$\hat{\alpha} - \hat{\alpha}_i = \frac{\hat{\xi(i)}}{1 - h_{i,i}}(X^tX)^{-1}x_i^t, \tag{6}$$

where $\hat{\alpha}$ is the vector of the estimates of the coefficients of Model 1, i.e., the regression with the full data set; $\hat{\alpha}_i$ is the vector of the estimates of the coefficients of the regression model without the i-th observation; X_i is the design matrix of the model without the i-th observation, i.e., it is X without the i-th row. To prove (6) we need the following:

$$[(X^tX)^{-1} + \left(\frac{1}{1 - x_i^t(X^tX)^{-1}x_i}\right)(X^tX)^{-1}x_ix_i^t(X^tX)^{-1}]((X^tX) - x_ix_i^t) = I, \tag{7}$$

where I is the unit matrix, x_i is the vector with the values for the i-th observation. Equation 7 can be easily verified if we note that $x_i^t(X^tX)^{-1}x_i \neq 1$ is a number. From (7) by right multiplication with $((X^tX) - x_ix_i^t)^{-1}$ we have:

$$(X^tX - x_ix_i^t)^{-1} = (X^tX)^{-1} + \frac{(X^tX)^{-1}x_ix_i^t(X^tX)^{-1}}{1 - x_i^t(X^tX)^{-1}x_i}. \tag{8}$$

Let us note that (8) can be directly obtained by the Bartlett's matrix inversion formula [1]. To continue, we use one trick from the matrix algebra, $X^tX = \sum_{i=1}^{n} x_ix_i^t$. From it we have:

$$X_i^tX_i = X^tX - x_ix_i^t \Longrightarrow (X_i^tX_i)^{-1} = (X^tX - x_ix_i^t)^{-1}. \tag{9}$$

From (8) and (9), and a right multiplication with $X_i^tY_i$, we have:

$$(X_i^tX_i)^{-1}X_i^tY_i = \left((X^tX)^{-1} + \frac{(X^tX)^{-1}x_ix_i^t(X^tX)^{-1}}{1 - x_i^t(X^tX)^{-1}x_i}\right)X_i^tY_i. \tag{10}$$

We know that the coefficients of a regression model can be obtained with the formula $\hat{\alpha} = (X^tX)^{-1}X^tY$, see [12, p. 30]. Additionally, we have $X_i^tY_i = X^tY - x_iy_i$. Then (10) becomes:

$$\hat{\alpha}_i = \left((X^tX)^{-1} + \frac{(X^tX)^{-1}x_ix_i^t(X^tX)^{-1}}{1 - x_i^t(X^tX)^{-1}x_i} \right)(X^tY - x_iy_i). \tag{11}$$

We know from the theory that the *hat* matrix of a regression model is $H = X(X^tX)^{-1}X^t$, and its diagonal contains the leverage of each of the observations - $h_{i,i} = x_i^t(X^tX)^{-1}x_i$ is the i-th diagonal element which corresponds to the leverage of the i-th observation, see [13]. Expanding the brackets in (11) we have:

$$\begin{aligned} \hat{\alpha}_i = (X^tX)^{-1}X^tY - (X^tX)^{-1}x_iy_i + \frac{(X^tX)^{-1}x_ix_i^t(X^tX)^{-1}}{1 - h_{i,i}}X^tY \\ - \frac{(X^tX)^{-1}x_ix_i^t(X^tX)^{-1}}{1 - h_{i,i}}x_iy_i = \hat{\alpha} + (X^tX)^{-1}x_i\left(-y_i + \frac{x^t\hat{\alpha}}{1 - h_{i,i}} - \frac{h_{i,i}y_i}{1 - h_{i,i}}\right) = \\ = \hat{\alpha} + (X^tX)^{-1}x_i\left(\frac{x^t\hat{\alpha} - y_i}{1 - h_{i,i}}\right) \end{aligned} \tag{12}$$

The residual for the i-th observation is $\hat{\xi}_i = y_i - x^t\hat{\alpha}$, then:

$$\hat{\alpha}_i = \hat{\alpha} - \frac{\hat{\xi}_i}{1 - h_{i,i}}(X^tX)^{-1}x_i. \tag{13}$$

We know that $\varepsilon_i = y_i - x_i^t\hat{\alpha}_i$ and $\hat{\xi}_i = y_i - x_i^t\hat{\alpha}$. We multiply (13) by x_i^t:

$$\begin{aligned} x_i^t\hat{\alpha}_i = x_i^t\hat{\alpha} - \frac{\hat{\xi}_i}{1 - h_{i,i}}x_i^t(X^tX)^{-1}x_i \Leftrightarrow y_i - \varepsilon_i = y_i - \hat{\xi}_i - \frac{\hat{\xi}_ih_{i,i}}{1 - h_{i,i}} \Leftrightarrow \\ \Leftrightarrow \varepsilon_i = \frac{\hat{\xi}_i}{1 - h_{i,i}},\ i = 1, ..., n. \end{aligned} \tag{14}$$

Corollary 1 *We can compute the RMSECV by running only one regression making the computation almost as fast as the computation of a single regression model.*

From the theorem we now see that the out-of-sample residuals ε_i, $i = 1, ..., n$ depend from the model errors $\hat{\xi}_i$, $i = 1, .., n$ and from their corresponding leverage values $h_{i,i}$, $i = 1, ..., n$. The leverage $h_{i,i} \in [0, 1]$ of observation i shows how much the values of the input variables for the i-th observation will influence the model to obtain an estimate of the i-th observation close to the real output value y_i of the i-th observation, for more information see [5, 14]. For the proposed method, while making a component Z we are searching for the optimum component $Z(k)$ that will form a design matrix which will balance between its leverage values and the obtained model errors in order to lower the out-of-sample errors. With other words, the *Cross-validated sequentially constructed multiple regression* procedure is trying to make a model that is more robust to the input data with respect to smaller out-of-sample errors by replacing some of the model variables with components.

3 An Example

The proposed step by step procedure for obtaining components is tested on a real example concerning accounting and macroeconomic information from the firms in the Bulgarian gas distribution sector in the period 2007–14. The goal is the predicting of the *Return on assets* (ROA) financial ratio for the next observed period t using the input data from the current period $t-1$. The full data set consists of 116 observations (three of these observations are omitted to improve the model). The model has seven variables one of which is a macroeconomic one and the others are financial ratios. The description of the input will be skipped because we want to highlight the proposed method, not the example.

$$ROA(t) = \alpha_0 + \alpha_1 * X_1(t-1) + \alpha_2 * X_2(t-1) + \alpha_3 * X_3(t-1) + \alpha_4 * X_4(t-1) \\ + \alpha_5 * X_5(t-1) + \alpha_6 * X_6(t-1) + \alpha_7 * X_7(t-1). \quad (15)$$

The new method will be demonstrated against a multiple regression by several steps. First, the proposed method is applied to a training set A_{2010} which contains the observations from years 2007–2010 (45 observations) to derive the needed components. The derived components by the method are three (43% of the primary variables):

$$\begin{aligned} C_1 &= X_1, \\ C_2 &= X_2 + 0.024 * (-0.93 * (X_3 + 0.26 * X_7) + X_5), \\ C_3 &= -0.66 * X_4 + X_6. \end{aligned} \quad (16)$$

Only these components will be used further in the example. Let us note that the components are derived from a subset over which the multiple regression has some insignificant coefficients. The aim is to demonstrate that components that are estimated over not so perfect data set can still have robust prediction results while further used. In the perfect case, it would be preferable to estimate the components from a subset over which a multiple regression model has significant model coefficients and desirable out-of-sample performance.

Second, we start to update the models that uses the three components C_1, C_2, C_3 and the multiple regression that uses the primary model variables $X_1, ..., X_7$ by adding the observations from year 2011 to the learning set A_{2010}. We compared them using the RMSE measure on the left out-of-sample observations, in this case they are from years 2012, 2013, 2014 (49 observations). Third, we add year 2012 to the learning set A_{2011} that already contains year 2011, update the models, and test them on years 2013, 2014. Finally, we add three a little bit more influential observations to the learning set A_{2012} that contains year 2012, and test the models on years 2013, 2014.

We present three tables that correspond to the three cases that we are analysing. Each table containers model statistics for the regressions with the components and the primary variables. The model statistics list contains some standard statistics about a multiple regression - the model coefficients, model R^2, adjusted R^2 and the model standard error. These summary statistics are obtained using the *lm()* function in the R language [11].

The result of the comparison between the recalculated models using the learning set A_{2011}, can be seen in Table 1. We see that the three components are well estimated, based on the t-values, while more data is added.

If we observe the out-of-sample error for the model in Table 1, we have $RMSE_{mult.\,reg.} = 0.01564$ for the seven-variables model and $RMSE_{new\,method} = 0.01557$ for the three-component model. The results are similar, but let us again note that the models are with significant difference in the number of the observations and the inner structure of the three components is not recalculated (they are derived from 43 observations, that is 38% of the full data set).

Let us continue adding more observations. We add year 2012 to the learning set. The coefficients in front of the components will be recalculated using the additional data. We again compare with a multiple regression over the updated training set. The results can be seen in Table 2. We see that this time the coefficients from the multiple regression with seven variables are well estimated. The coefficients from the three-component model are even better estimated compared to their corresponding estimates in Table 1, but it is not clear if this improvement will be enough to obtain better results than the improved seven-variable model. When we test the two models over a test set which contains years 2013 and 2014, we see that the $RMSE_{mult.\,reg.} = 0.01615$ and the $RMSE_{new\,method} = 0.01624$. This result is even expected while observing the

Table 1 The new method and a multiple regression for ROA, learning set 2007–2011

The new method				Multiple regression			
Coefficient	Estimate	t-value	p-value	Coefficient	Estimate	t-value	p-value
Intercept	0.03527	19.783	2e−16	Intercept	0.03527	19.523	2e−16
C_1	0.70213	12.082	2e−16	X_1	0.64911	8.907	2.55e−12
C_2	−0.77059	−7.664	1.84e−10	X_2	−0.65439	−5.043	5.13e−06
C_3	0.05170	9.922	2.87e−14	X_3	0.01297	2.916	0.00509
				X_4	−0.02737	−3.236	0.00204
				X_5	−0.02829	−2.209	0.03129
				X_6	0.05294	4.725	1.59e−05
				X_7	0.00788	2.294	0.02554
R^2:	0.8123	adj. R^2:	0.8029	R^2:	0.8201	adj. R^2:	0.7976
RSE[a]:	0.01426			RSE[a]:	0.01445		

[a]Residual standard error

Table 2 The new method and a multiple regression for ROA, learning set 2007–2012

The new method				Multiple regression			
Coefficient	Estimate	t-value	p-value	Coefficient	Estimate	t-value	p-value
Intercept	0.02993	19.473	2e−16	Intercept	0.02993	19.575	2e−16
C_1	0.66454	15.093	2e−16	X_1	0.59633	10.536	2e−16
C_2	−0.76598	−9.546	6.64e−15	X_2	−0.60584	−5.536	4.12e−07
C_3	0.04896	10.448	2e−16	X_3	0.01102	2.942	0.004303
				X_4	−0.02710	−3.539	0.000684
				X_5	−0.02736	−3.958	0.000167
				X_6	0.04739	5.914	8.65e−08
				X_7	0.00886	3.234	0.001802
R^2:	0.8314	adj. R^2:	0.8251	R^2:	0.8413	adj. R^2:	0.8269
RSE[a]:	0.01417			RSE[a]:	0.0141		

[a]Residual standard error

expected model error (RSE). The RSE of the multiple regression with seven variables is slightly lower than the RSE of the model with the three components, but still similar.

We have mentioned at the beginning of the section that three of the observations are omitted due to their influence on the model performance. Let us add these three observations to the learning set which containers all the observations up to year 2012. The slight deterioration of the two models can be seen in Table 3. This time the RSE of the three-variable model is better. While testing over the out-of-sample years 2013 and 2014, we see that $RMSE_{mult.\,reg.} = 0.01911$ and $RMSE_{new\,method} = 0.01883$. Now, the out-of-sample error of the model with the components is slightly lower, and this is achieved while adding problematic observations. This indicates a possible bigger robustness of the models using components obtained by the proposed methodology.

Table 3 The new method and a multiple regression for ROA, learning set 2007–2012 with 3 added observations

The new method				Multiple regression			
Coefficient	Estimate	t-value	p-value	Coefficient	Estimate	t-value	p-value
Intercept	0.03058	18.962	2e−16	Intercept	0.03058	18.871	2e−16
C_1	0.68674	14.963	2e−16	X_1	0.62775	10.545	2e−16
C_2	−0.76650	−9.051	5.70e−14	X_2	−0.63424	−5.453	5.64e−07
C_3	0.04670	9.551	5.78e−15	X_3	0.01194	2.998	0.003648
				X_4	−0.02560	−3.141	0.002379
				X_5	−0.026830	−3.642	0.000485
				X_6	0.04655	5.452	5.64e−07
				X_7	0.00770	2.657	0.009562
R^2:	0.8157	adj. R^2:	0.809	R^2:	0.823	adj. R^2:	0.8071
RSE[a]:	0.01496			RSE[a]:	0.01503		

[a] Residual standard error

4 Summary

We have presented in this paper a technique that aims to make a model more robust to new observations into its training set with respect to the out-of-sample performance. The tests of the technique for the real example show that it gives similar to the multiple regression performance, no matter that the components are only 3, and are constructed from only 38% of the data set (not the most informative 38% of the data set). There are some theoretical issues that are still not covered, but the achieved results stimulate further research on the topic.

Acknowledgements The authors acknowledge funding by the Bulgarian fund for scientific investigations Project DN 12/5.

References

1. Bartlett, M.: An inverse matrix adjustment arising in discriminant analysis. Ann. Math. Stat. **22**(1), 107–111 (1951)
2. Babyak, M.: What you see may not be what you get: a brief, nontechnical introduction to overfitting in regression-type models. Psychosom. Med. **66**, 411–421 (2004)
3. Chai, T., Draxler, R.R.: Root mean square error (RMSE) or mean absolute error (MAE)? Arguments against avoiding RMSE in the literature. Geosci. Model Dev. **7**, 1247–1250 (2014)
4. Cawley, G., Talbot, N.: Fast exact leave-one-out cross-validation of sparse least-squares support vector machines. Neural Netw. **17**, 1467–1475 (2004)
5. Cook, R., Weisberg, S.: Residuals and Influence in Regression. Monographs on Statistics and Applied Probability. Chapman & Hall, New York (1982)
6. Choi, S.-W.: The effect of outliers on regression analysis: regime type and foreign direct investment. Q. J. Polit. Sci. **4**(2), 153–165 (2009)
7. Davison, A., Hinkley, D.: Bootstrap Methods and their Application. Cambridge Series in Statistical and Probabilistic Mathematics. Cambridge University Press, Cambridge, UK (1997)
8. Hawkins, D., Basak, S., Mills, D.: Assessing model fit by cross-validation. J. Chem. Inf. Comput. Sci. **43**(2), 579–586 (2003)
9. Kroll, C., Song, P.: Impact of multicollinearity on small sample hydrologic regression models. Water Resour. Res. **49**(6), 3756–3769 (2013)
10. Mevik, B., Cederkvist, H.: Mean squared error of prediction (MSEP) estimates for principal component regression (PCR) and partial least squares regression (PLSR). J. Chemometr. **18**, 422–429 (2004)
11. R Core Team.: R: A Language and Environment for Statistical Computing. R Foundation for Statistical Computing, Vienna, Austria. URL https://www.R-project.org/ (2017)
12. Sen, A., Srivastava, M.: Regression Analysis: Theory, Methods, and Applications. Springer, New York Inc., New York (1990)
13. Scanlon, E.: Residuals and Influence in Regression. Insurance Mathematics and Economics (1996)
14. Weisberg, S.: Applied Linear Regression. Wiley, New York (1985)

How to Assess Multi-population Genetic Algorithms Performance Using Intuitionistic Fuzzy Logic

Maria Angelova and Tania Pencheva

Abstract In this investigation a step-wise "cross-evaluation" procedure has been implemented aiming to assess the quality of multi-population genetic algorithms (MpGA) performance. Three MpGA, searching for an optimal solution applying main genetic operators selection, crossover and mutation in different order, have been here applied in such a challenging object as parameter identification of a fermentation process model. The performance quality of standard MpGA algorithm, denoted as MpGA_SCM (coming from selection, crossover, mutation), and two modifications, respectively MpGA_MCS (mutation, crossover, selection) and MpGA_CMS (crossover, mutation, selection) have been investigated for the purposes of parameter identification of *S. cerevisiae* fed-batch cultivation. As an alternative to conventional criteria for assessing the quality of algorithms performance, here an intuitionistic fuzzy logic (IFL) is going to be implemented. Also, this is the first time when two modifications of standard MpGA_SCM, in which the selection operator is performed as the last one, after crossover and mutation, are going to be evaluated. The performance of three MpGA is going to be assessed applying a step-wise procedure implementing IFL. As a result, MpGA_SCM has been approved as a leader between three considered here MpGA. The leadership between MpGA_CMS and MpGA_MCS depends on the researcher choice between a bit slower, but more highly evaluated MpGA_CMS towards faster one, but a bit less highly evaluated MpGA_MCS.

1 Introduction

Biotechnological processes, and particularly fermentation processes (FP), are in the core of food and pharmaceutical industries, genetic engineering, etc. The complex structure of FP, since they combine the dynamics of biological and non-biological

---

M. Angelova (✉) · T. Pencheva
Institute of Biophysics and Biomedical Engineering,
Bulgarian Academy of Sciences , 105 Acad. G. Bonchev Str., 1113 Sofia, Bulgaria
e-mail: maria.angelova@biomed.bas.bg

T. Pencheva
e-mail: tania.pencheva@biomed.bas.b

K. Georgiev et al. (eds.), *Advanced Computing in Industrial Mathematics*,
Studies in Computational Intelligence 793,
https://doi.org/10.1007/978-3-319-97277-0_3

processes, makes their modeling, optimization and further high quality control a real challenge for the investigators. Failure of conventional optimization methods [1] to lead to a satisfied solution for model parameter identification of non-linear FP provokes the idea some stochastic algorithms to be applied. As a quite promising stochastic global optimization technique, genetic algorithms (GA) [2] are here applied as proved as successful in various difficult to be solved tasks. GA are among the methods based on biological evolution and inspired by Darwins theory of "survival of the fittest". Properties like hard problems solving, noise tolerance, easy to interface and hybridize make GA suitable and more workable for parameter identification of fermentation process models [1, 3–5].

Standard multi-population genetic algorithm [2] imitates the processes occurred in the nature and searches a global optimal solution using three main genetic operators in a sequence selection, crossover and mutation and as such denoted as MpGA_SCM (coming from selection, crossover, mutation). Since the basic idea of GA is to imitate the mechanics of natural selection and genetics, one can make an analogy with the processes occurring in the nature. So, it could be said that the probability mutation to come first and then crossover is comparable to that both processes to occur in a reverse order; or selection to be performed after crossover and mutation, no matter of their order. Following this idea eight modifications of MpGA_SCM, differ in the execution sequence of main genetic operators, have been developed aiming to improve model accuracy and algorithm convergence time for the purposes of parameter identification of a fed-batch cultivation of *S. cerevisiae* [5].

From a mathematical and engineering point of view, the objective function value and the algorithm convergence time are the most representative criteria that might be used to assess the quality of algorithms performance. But from a biotechnological point of view it might not be sufficient. Intuitionistic fuzzy logic (IFL) is an alternative in assessing the quality of different algorithms for various purposes. The aim of this investigation is IFL to be applied to assess the quality of performance of three different kinds of MpGA. A step-wise "cross-evaluation" procedure implementing IFL [6] is going to be applied aiming to assess MpGA performance for the purposes of parameter identification of *S. cerevisiae* fed-batch cultivation. Up to now the procedure has been successfully applied to assess the performance of different modifications of simple GA (SGA), different modifications of MpGA, standard SGA towards standard MpGA, as well as to assess the performance of SGA and MpGA at different values of GGAP, proven as the most sensitive genetic algorithm parameter [6].

Two unexplored modifications of standard MpGA are going to be evaluated in this investigation, in which the selection operator is performed as the last one, after crossover and mutation. As such, MpGA_CMS (crossover, mutation, selection) and MpGA_MCS (mutation, crossover, selection) are in the investigation focus since they have not been evaluated by IFL so far. Each one of them is going to be compared to the standard one MpGA, namely MpGA_SCM, implementing IFL.

2 Background

2.1 Mathematical Model of S. cerevisiae Fed-Batch Cultivation

Experimental data of *S. cerevisiae* fed-batch cultivation, obtained in the *Institute of Technical Chemistry University of Hannover, Germany*, consist of on-line measurements of substrate (glucose) and dissolved oxygen and off-line measurements of biomass and ethanol. The detailed description of the process conditions and experimental data might be found in [1].

Mathematical model of *S. cerevisiae* fed-batch cultivation is commonly described by the following system of non-linear differential equations, according to the mass balance [1]:

$$\frac{dX}{dt} = \mu X - \frac{F}{V}X \tag{1}$$

$$\frac{dS}{dt} = -q_S X + \frac{F}{V}\left(S_{in} - S\right) \tag{2}$$

$$\frac{dE}{dt} = q_E X - \frac{F}{V}E \tag{3}$$

$$\frac{dO_2}{dt} = -q_{O_2} X + k_L^{O_2} a \left(O_2^* - O_2\right) \tag{4}$$

$$\frac{dV}{dt} = F \tag{5}$$

where X, S, E, and O_2, are concentrations, respectively of biomass [g/l], substrate [g/l], ethanol [g/l] and dissolved oxygen [%]; O_2^* – dissolved oxygen saturation concentration, [%]; F – feeding rate, [l/h]; V – volume of bioreactor, [l]; $k_L^{O_2}a$ – volumetric oxygen transfer coefficient, [1/h]; S_{in} – glucose concentration in the feeding solution, [g/l]; μ, q_S, q_E, q_{O_2} – specific growth/utilization rates of biomass, substrate, ethanol and dissolved oxygen, [1/h].

According to functional state modeling approach [1], specific rates in Eqs. (1)–(5) are as follows:

$$\begin{aligned} \mu &= \mu_{2S}\frac{S}{S+k_S} + \mu_{2E}\frac{E}{E+k_E}, \quad q_S = \frac{\mu_{2S}}{Y_{SX}}\frac{S}{S+k_S} \\ q_E &= -\frac{\mu_{2E}}{Y_{EX}}\frac{E}{E+k_E}, \quad q_{O_2} = q_E Y_{OE} + q_S Y_{OS} \end{aligned} \tag{6}$$

where μ_{2S}, μ_{2E} are the maximum growth rates of substrate and ethanol, [1/h]; k_S, k_E – saturation constants of substrate and ethanol, [g/l]; Y_{ij} – yield coefficients, [g/g].

All functions in the model (Eqs. 1–6) are continuous and differentiable, as well as all model parameters fulfil the non-zero division requirement.

As an optimization criterion, mean square deviation between the model output and the experimental data obtained during the cultivation has been used:

$$J = \sum (Y - Y^*)^2 \rightarrow min, \tag{7}$$

where Y is the experimental data, Y^* – model predicted data, $Y = [X, S, E, O_2]$.

2.2 *Multi-population Genetic Algorithms*

According to [2, 7], standard MpGA, namely MpGA_SCM, works with many populations, called subpopulations, evolved independently from each other. In the beginning MpGA starts with a choice through a selection of chromosomes (coded parameter set) representing better possible solutions according to their own objective function values. After that, crossover proceeds to form a new offspring. Mutation might be then applied with a determinate probability. After a certain number of generations a number of individuals are distributed between the subpopulation (migration). GA is terminated when a certain criterion is fulfilled, in this case number of generations set to 100.

Two modifications of MpGA, namely MpGA_CMS and MpGA_MCS, are chosen to be in the focus of this investigation, in which the selection operator is performed as the last one, after crossover and mutation, no matter of their order. Following the mathematical model (Eqs. 1–6) of *S. cerevisiae* fed-batch cultivation, altogether nine model parameters have to be estimated, applying different modifications of MpGA. Parameter identification of the model has been performed using *Genetic Algorithm Toolbox* [8] in *Matlab 7* environment, on PC Intel Pentium 4 (2.4 GHz) platform running Windows XP.

2.3 *Intuitionistic Fuzzy Estimations*

In intuitionistic fuzzy logic (IFL) [9, 10], if p is a variable, then its truth-value is represented by the ordered couple

$$V(p) = \langle M(p), N(p) \rangle \tag{8}$$

where $M(p)$ and $N(p)$ are respectively degrees of validity and of non-validity of p, so that $M(p), N(p), M(p) + N(p) \in [0, 1]$. These values can be obtained applying different formula depending on the problem considered. In [9, 10], the relation $\leq$ is defined between the intuitionistic fuzzy pairs $\langle a, b \rangle$ and $\langle c, d \rangle$ by

$$\langle a, b\rangle \leq \langle c, d\rangle \text{ if and only if } a \leq c \text{ and } b \geq d. \tag{9}$$

For the purpose of this investigation the degrees of validity and non-validity can be obtained, e.g., by the following formula:

$$M(p) = \frac{m}{u}, \quad N(p) = 1 - \frac{n}{u}, \tag{10}$$

where m is the lower boundary of the "narrow" range; u – the upper boundary of the "broad" range; n – the upper boundary of the "narrow" range. The "broad" range of the model parameters are based on referent data, while the "narrow" one is based on the preliminary algorithms evaluations.

If there is a database collected having elements with the form $< p, M(p), N(p) >$, different new values for the variables can be obtained. In case of three records in the database, the following new values can be obtained:

$$\begin{aligned} V_{strong\_opt} = \big\langle &M_1(p) + M_2(p) + M_3(p) - M_1(p)M_2(p) - M_1(p)M_3(p) \\ &- M_2(p)M_3(p) + M_1(p)M_2(p)M_3(p), N_1(p)N_2(p)N_3(p)\big\rangle, \end{aligned} \tag{11}$$

$$V_{opt} = \langle \max(M_1(p), M_2(p), M_3(p)), \min(N_1(p), N_2(p), N_3(p))\rangle, \tag{12}$$

$$V_{aver} = \langle (M_1(p) + M_2(p) + M_3(p))/3, (N_1(p) + N_2(p) + N_3(p))/3\rangle, \tag{13}$$

$$V_{pes} = \langle \min(M_1(p), M_2(p), M_3(p)), \max(N_1(p), N_2(p), N_3(p))\rangle, \tag{14}$$

$$\begin{aligned} V_{strong\_pes} = \big\langle &M_1(p)M_2(p)M_3(p), N_1(p) + N_2(p) + N_3(p) - N_1(p)N_2(p) \\ &- N_1(p)N_3(p) - N_2(p)N_3(p) + N_1(p)N_2(p)N_3(p))\big\rangle, \end{aligned} \tag{15}$$

Therefore, for each p:
$V_{strong\_pes}(p) \leq V_{pes}(p) \leq V_{aver}(p) \leq V_{opt}(p) \leq V_{strong\_opt}(p)$.

2.4 Procedure for Genetic Algorithms Quality Assessment Implementing IFL

Implementation of IFL requires the construction of the degrees of validity and non-validity to be performed in two different intervals of model parameters variation. One interval, here called "broad" range, might be based on referent data, known from the literature (here based on [11]). The other interval of model parameters, here called "narrow" range, could be defined based on some criterion for shrinking the range - e.g. model parameter evaluations obtained just from previously algorithms runs, or, e.g., after applying the procedure for purposeful model parameters genesis developed by authors [12].

The procedure for assessment of algorithm quality performance (AAQP) applying IFL, originally proposed in [6], consists of six steps as follows, not omitting any of them and without cycles:

Step 1. *Performance of N runs of each of the algorithms, in two different ranges of model parameters - broad and narrow*

Step 2. *Determination of the average values of the objective function, algorithms convergence time and each of the model parameters for each one of the ranges and each one of the investigated algorithms*

Step 3. *Determination of degrees of validity and non-validity for each of the algorithms, according to Eq.* (10)

Step 4. *Determination of strong optimistic, optimistic, average, pessimistic and strong pessimistic values in case of three objects, according to Eqs.* (11–15), *for each one of the model parameters*

Step 5. *Assignment of the mentioned in* ***Step 4*** *values to each of the model parameters for each of the ranges for each of the algorithms*

Step 6. *Assessment of the quality of each algorithm, based on the values obtained in* ***Step 5***.

3 MpGA Quality Assessment Applying IFL

Following the mathematical model (Eqs. 1–6) of *S. cerevisiae* fed-batch cultivation, 9 model parameters have to be estimated, applying consequently MpGA_SCM, MpGA_CMS and MpGA_MCS. After that the AAQP procedure is going to be applied for quality assessment of the MpGA.

The quality of MpGA performance is going to be assessed before and after the application of the procedure for purposeful model parameter genesis (PMPG) [12]. Thirty runs have been executed for each of the considered here three MpGA. Then, the minimum and the maximum of the objective function are determined and the discrimination number Δ is assigned according to PMPG. After that, the procedure proceeds with the determination of top, middle and low level of performance for each one of the investigated algorithms. The best results hit the interval $[minJ;\ minJ + \Delta - \varepsilon]$, forming the top level (TL) with low boundary (LB) and up boundary (UB). Those results, classified in the middle level (ML), have an objective function varying in the interval $[minJ + \Delta;\ minJ + 2\Delta - \varepsilon]$. The worst solutions for the objective function fall in the interval $[minJ + 2\Delta;\ maxJ]$, forming the low level (LL). Table 1 presents objective functions and level of performance of three considered here kinds of MpGA.

As seen in Table 1, there are no results classified in the middle level. All results obtained with considered here three MpGA hit the top or respectively the low level of performance. For each of the levels, constructed in such a way, the minimum, maximum and average values of each model parameter have been determined. Table 2 presents these values only for the top levels, according to Table 1.

Table 1 Levels of performance of considered MpGA

MpGA	Objective function		Levels of performance		Average convergence time, s
MpGA_SCM	*min J*	0.0221	TL_LB	0.0221	98.96
			TL_UB	0.0221	
	max J	0.0222	LL_LB	0.0222	
			LL_UB	0.0222	
MpGA_CMS	*min J*	0.0221	TL_LB	0.0221	281.56
			TL_UB	0.0221	
	max J	0.0222	LL_LB	0.0222	
			LL_UB	0.0222	
MpGA_MCS	*min J*	0.0221	TL_LB	0.0221	272.22
			TL_UB	0.0221	
	max J	0.0222	LL_LB	0.0222	
			LL_UB	0.0222	

The new boundaries for so-called "narrow" range of the model parameters are constructed in a way that the new minimum and maximum are respectively lower and higher, but close to the average values of the estimated parameters of the top level. Table 3 presents previously used "broad" boundaries for each model parameter according to [11], as well as new boundaries proposed based on the PMPG when applying considered here kinds of MpGA. Additionally, Table 3 consists of intuitionistic fuzzy estimations, obtained based on Eq. (10) as described above.

A "cross-evaluation" between three MpGA, namely MpGA_SCM, MpGA_CMS and MpGA_MCS, is going to be presented. The implementation of AAQP in this case is based on the presented in Table 3 previously used "broad" boundaries for each model parameter according to [11], as well as on the new boundaries advised after PMPG application.

Thus, following AAQP procedure, *strong optimistic*, *optimistic*, *average*, *pessimistic* and *strong pessimistic* prognoses for the performance of considered here MpGA_SCM, MpGA_CMS and MpGA_MCS have been constructed, based on intuitionistic fuzzy estimations Eq. (10) and formula Eq. (11–15). Table 4 presents low (LB) and up boundaries (UB) of the new values, obtained after AAQP procedure for each one of the identified parameters.

Investigated here three kinds of MpGA have been again applied for parameter identification of *S. cerevisiae* fed-batch cultivation involving newly proposed according to Table 3 boundaries. Thirty runs have been performed in order reliable results to be obtained. Table 5 presents the average values of the objective function, convergence time and model parameters when MpGA_SCM, MpGA_CMS and MpGA_MCS have been applied, shown, respectively, as "broad" (before PMPG application) and "narrow" (after PMPG application) ranges.

Table 2 Model parameters values for the top levels of considered MpGA

MpGA		μ_{2S}	μ_{2E}	k_S	k_E	Y_{SX}	Y_{EX}	$k_L^{O_2}a$	Y_{OS}	Y_{OE}
MpGA_SCM	*min*	0.9002	0.1161	0.1313	0.7979	.3987	1.5408	61.129	473.491	228.867
	max	0.9378	0.1438	0.1500	0.8000	0.4182	1.9290	118.423	921.284	809.902
	avrg	0.9155	0.1283	0.1453	0.7998	0.4097	1.7212	92.78	656.597	508.501
MpGA_CMS	*min*	0.9133	0.1172	0.143	0.7938	0.40367	1.57109	76.658	604.327	95.6817
	max	0.9374	0.1375	0.15	0.8	0.41895	1.87371	126.219	995.808	752.84
	avrg	0.92668	0.12687	0.1469	0.79887	0.41167	1.71194	93.6483	743.065	383.282
MpGA_MCS	*min*	0.9	0.1246	0.1444	0.7908	0.40178	1.65481	54.2122	435.152	272.442
	max	0.9382	0.1349	0.15	0.8	0.41386	1.81389	90.714	717.838	768.38
	avrg	0.91602	0.12922	0.14826	0.7981	0.40837	1.7352	72.8336	574.481	509.136

Table 3 Model parameters boundaries for considered MpGA

MpGA			μ_{2S}	μ_{2E}	k_S	k_E	Y_{SX}	Y_{EX}	$k_L^{O_2}a$	Y_{OS}	Y_{OE}
MpGA_SCM	Previously used	LB	0.9	0.05	0.08	0.5	0.3	1	0.001	0.001	0.001
		UB	1	0.15	0.15	0.8	10	10	300	1000	1000
	Advisable after	LB	0.90	0.12	0.14	0.70	0.35	1.5	80	650	220
	PMPG application	UB	0.92	0.15	0.15	0.80	0.45	2	100	800	820
	Degrees of validity of$p$	$M_1(p)$	0.90	0.80	0.93	0.88	0.04	0.15	0.27	0.65	0.22
	Degrees of non-validity of p	$N_1(p)$	0.08	0.00	0.00	0.00	0.96	0.80	0.67	0.20	0.18
MpGA_CMS	Previously used	LB	0.9	0.05	0.08	0.5	0.3	1	0.001	0.001	0.001
		UB	1	0.15	0.15	0.8	10	10	300	1000	1000
	Advisable after	LB	0.91	0.11	0.14	0.75	0.4	1.5	70	600	90
	PMPG application	UB	0.94	0.14	0.15	0.80	0.42	1.9	130	1000	760
	Degrees of validity of p	$M_2(p)$	0.91	0.73	0.93	0.94	0.04	0.15	0.23	0.60	0.09
	Degrees of non-validity of p	$N_2(p)$	0.06	0.07	0.00	0.00	0.96	0.81	0.57	0.00	0.24
MpGA_MCS	Previously used	LB	0.9	0.05	0.08	0.5	0.3	1	0.001	0.001	0.001
		UB	1	0.15	0.15	0.8	10	10	300	1000	1000
	Advisable after	LB	0.90	0.12	0.14	0.75	0.4	1.6	50	420	260
	PMPG application	UB	0.94	0.14	0.15	0.8	0.42	1.9	95	720	770
	Degrees of validity of p	$M_3(p)$	0.90	0.80	0.93	0.94	0.04	0.16	0.17	0.42	0.26
	Degrees of non-validity of p	$N_3(p)$	0.06	0.07	0.00	0.00	0.96	0.81	0.68	0.28	0.23

Table 4 Prognoses for MpGA performance

	μ_{2S}		μ_{2E}		k_S		k_E		Y_{SX}		Y_{EX}		$k_L^{O_2}a$		Y_{OS}		Y_{OE}	
	LB	UB	LB	UB	LB	UB	LB	UB	LB	UB	LB	UB	LB	UB	LB	UB	LB	UB
V_{strong_opt}	1.00	1.00	0.15	0.15	0.15	0.15	0.80	0.80	1.11	1.24	3.93	4.75	159.44	222.56	918.80	1000.00	474.75	990.06
V_{opt}	0.91	0.94	0.12	0.15	0.14	0.15	0.75	0.80	0.40	0.45	1.60	2.00	80.00	130.00	650.00	1000.00	260.00	820.00
V_{aver}	0.90	0.93	0.12	0.14	0.14	0.15	0.73	0.80	0.38	0.43	1.53	1.93	66.67	108.33	556.67	840.00	190.00	783.33
V_{pes}	0.90	0.92	0.11	0.14	0.14	0.15	0.70	0.80	0.35	0.42	1.50	1.90	50.00	95.00	420.00	720.00	90.00	760.00
V_{strong_pes}	0.74	0.81	0.07	0.13	0.12	0.15	0.62	0.80	0.00	0.00	0.04	0.07	3.11	13.72	163.80	576.00	5.15	479.86

Table 5 Results from model parameter identification before and after PMPG

Parameters	MpGA_SCM		MpGA_CMS		MpGA_MCS	
	Broad range	Narrow range	Broad range	Narrow range	Broad range	Narrow range
J	0.0221	0.0220	0.0221	0.0221	0.0221	0.0221
CPU time	98.96	86.52	296.45	261.34	270.95	245.42
μ_{2S}	0.91	0.90	0.92	0.91	0.93	0.90
μ_{2E}	0.12	0.14	0.12	0.13	0.13	0.13
k_S	0.15	0.15	0.15	0.15	0.15	0.15
k_E	0.80	0.80	0.80	0.80	0.80	0.80
Y_{SX}	0.41	0.40	0.42	0.41	0.40	0.41
Y_{EX}	1.62	1.93	1.57	1.81	1.81	1.77
$k_L^{O_2}$	96.34	88.73	76.66	97.14	62.59	84.41
Y_{OS}	768.61	696.56	604.33	770.73	500.75	673.20
Y_{OE}	809.9	291.42	601.03	500.37	454.40	462.89

It is worth to note that in all three considered here kinds of MpGA running the algorithms in "narrow" boundaries leads to expecting decrease of the convergence time, moreover, in all cases there is no loss of model accuracy. The reduction of convergence time in case of MpGA_SCM is about 1.14 times towards "broad" range, in case of MpGA_CMS – 1.13 times and in case of MpGA_MCS – 1.10 times. In addition, the results obtained in the "narrow" range hit the top level of performance, thus showing good effectiveness of PMPG procedure for all considered here kinds of MpGA. But if one compares only the convergence time, the fastest among the investigated algorithms MpGA_SCM in "narrow" range leads to decrease of the convergence time in about 3.13 times towards the slowest one MpGA_CMS in "broad" range.

Table 6 lists the number and kind of estimations assigned to each of the parameters concerning Table 5 for the considered here three kinds of MpGA before and after PMPG application, respectively in "broad" and "narrow" range of model parameters variations.

Table 6 Model parameter estimations before and after PMPG

	MpGA_SCM		MpGA_CMS		MpGA_MCS	
	Broad range	Narrow range	Broad range	Narrow range	Broad range	Narrow range
strong_opt	3	2	3	3	2	2
opt	6	6	3	6	5	6
aver	0	1	3	0	0	1
pes	0	0	0	0	2	0
strong_pes	0	0	0	0	0	0

As seen from Table 6, there are no any *strong pessimistic* prognoses, while *pessimistic* prognoses appear only in the case of MpGA_MCS in "broad" range. This fact verifies the high degree of effectiveness of considered here three kinds of MpGA. There are two pairs with identical behaviour: MpGA_SCM in "broad" range and MpGA_CMS in "narrow" range, and MpGA_SCM and MpGA_MCS, both in "narrow" range. The first pair is with 3 *strong optimistic* prognoses and 6 *optimistic* prognoses, which evaluations distinguish both algorithms as undisputed leaders. Very close to them is the other pair – MpGA_SCM and MpGA_MCS, both in "narrow" range, with 2 *strong optimistic* prognoses, followed by 6 *optimistic* and 1 *average* prognoses. In all of these four distinguished as the most reliable cases the value of the objective function is the lowest one achieved, that means they are with the highest degree of accuracy. So it is again "a question of convergence time" the best algorithm to be distinguished. As such, the unquestionable leader is MpGA_SCM. When one compares MpGA_SCM in "broad" and "narrow" range, the performance of MpGA_SCM in "broad" range is evaluated with one more *strong optimistic* prognosis towards one more *average* prognosis of MpGA_SCM in "narrow" range. So, it is a question of a researcher decision to choose between a bit slower, but more highly evaluated MpGA_SCM in "broad" range towards the fastest one, but a bit less highly evaluated MpGA_SCM in "narrow" range.

As a conclusion of this investigation, MpGA_SCM has been logically distinguished as a leader among three considered here MpGA, based on the proved advantages towards MpGA_MCS, shown in [6], and also towards MpGA_CMS, presented in this investigation. But as seen from results presented here, the decision about "the winner in the fight" between MpGA_CMS and MpGA_MCS is again in the hands of the researcher: to choose between a bit slower, but more highly evaluated MpGA_CMS in "narrow" range towards faster one, but a bit less highly evaluated MpGA_MCS in "narrow" range.

Presented here assessment of three kinds of MpGA approve once again the procedure for assessment of algorithm quality performance as a very effective tool for evaluation of different kinds of algorithms.

4 Conclusion

In this investigation intuitionistic fuzzy logic has been implemented in order to assess the quality performance of three different MpGA algorithms – standard MpGA_SCM and two unexplored until now modifications, namely MpGA_CMS and MpGA_MCS. These algorithms have been investigated for the purposes of parameter identification of *S. cerevisiae* fed-batch cultivation. IFL has been implemented firstly to obtain intuitionistic fuzzy estimations of model parameters and further to construct *strong optimistic*, *optimistic*, *average*, *pessimistic* and *strong pessimistic* prognoses. Results for three considered here kinds of MpGA have been compared and MpGA_SCM in "narrow" range has been distinguished as the fastest, although not the most reliable one, while MpGA_SCM in "broad" range is the most reliable, although not the fastest one. The choice between more highly evaluated, but a bit slower MpGA_CMS in “narrow” range towards a bit less highly evaluated, but faster one MpGA_MCS in “narrow” range is in the researchers hands and might depend on some additional specific considerations.

Acknowledgements The work is partially supported by National Scientific Fund of Bulgaria, grants DM07/1 “Development of New Modified and Hybrid Metaheuristic Algorithms" and DN02/10 “New Instruments for Knowledge Discovery from Data, and Their Modelling".

References

1. Pencheva, T., Roeva, O., Hristozov, I.: Functional State Approach to Fermentation Processes Modeling. Prof. M. Drinov Acad. Publ. House, Sofia (2006)
2. Goldberg, D.E.: Genetic Algorithms in Search. Optimization and Machine Learning. Addison Wesley Longman, London (2006)
3. Roeva, O. (ed.).: Real-World Application of Genetic Algorithms. InTech (2012)
4. Adeyemo, J., Mehablia, A.: Fermentation processes optimization: evolutionary algorithms a review. Sci. Res. Essays **6**(7), 1464–1472 (2011)
5. Angelova, M.: Modified Genetic Algorithms and Intuitionist Fuzzy Logic for Parameter Identification of Fed-batch Cultivation Model, Ph.D. Thesis, Sofia, (2014) (in Bulgarian)
6. Pencheva, T., Angelova, M., Atanassov, K.: Genetic algorithms quality assessment implementing intuitionist fuzzy logic, Chapter 11. In: Vasant, P. (ed.) Handbook of Research on Novel Soft Computing Intelligent Algorithms: Theory and Practical Applications, pp. 327–354. Hershey, Pennsylvania (USA), IGI Global (2013)
7. Gupta, D., Ghafir, S.: An overview of methods maintaining diversity in genetic algorithms. Int. J. Emerg. Technol. Adv. Eng. **2**(5), 56–60 (2012)
8. Chipperfield, A., Fleming, P., Pohlheim, H., Fonseca, C.: Genetic Algorithm Toolbox for Use with MATLAB, Users Guide, Version 1.2., Department of Automatic Control and System Engineering, University of Sheffield, UK (1994)
9. Atanassov, K.: Intuitionist Fuzzy Sets. Springer, Heidelberg (1999)
10. Atanassov, K.: On Intuitionist Fuzzy Sets Theory. Springer, Berlin (2012)
11. Schuegerl, K., Bellgardt, K.-H. (eds.).: Bioreaction Engineering, Modeling and Control. Springer, Berlin Heidelberg New York (2000)
12. Angelova, M., Atanassov, K., Pencheva, T.: Purposeful model parameters genesis in simple genetic algorithms. Comput. Math. Appl. **64**, 221–228 (2012)

Perturbation Analysis of a Nonlinear Matrix Equation Arising in Tree-Like Stochastic Processes

Vera Angelova

Abstract The solution, obtained in the environment of finite precision machine arithmetic must be allays accompanied by an analysis of the conditioning of the problem solved. The perturbation analysis derives measures for the sensitivity of the solution to perturbations in the matrix coefficients. Motivated by these, in order to ascertain the accuracy of an iteratively calculated solution to a nonlinear matrix equation arising in Tree-like stochastic processes, in this paper norm-wise, mixed and component-wise condition numbers, as well as local perturbation bounds are formulated and norm-wise non-local residual bounds are derived using the methods of nonlinear perturbation analysis (Fréchet derivatives, Lyapunov majorants, fixed-point principles). The residual bounds are formulated in terms of the computed approximate solution to the equation and can be used as a stop criteria of the iterations, when solving the considered nonlinear matrix equation by a numerically stable iterative algorithm.

1 Introduction

Consider the nonlinear matrix equation

$$X = \Phi(S), \quad \Phi(S) = C - \sum_{i=1}^{m} A_i X^{-1} D_i, \tag{1}$$

with $S = (A_1, A_2, \ldots, A_m, D_1, D_2, \ldots, D_m, C) \in \Psi = \underbrace{\mathbb{R}^{n\times n} \times \mathbb{R}^{n\times n} \times \ldots \times \mathbb{R}^{n\times n}}_{2m+1}$ — the $(2m+1)$-tuple of the collection of matrix coefficients A_i, D_i, $C \in \mathbb{R}^{n\times n}$ for $i = 1, \ldots, m$ and $X \in \mathbb{R}^{n\times n}$ — a nonsingular solution. This equation is encountered in the analysis of Tree-like stochastic processes [1]. The computation of the

V. Angelova (✉)
Institute of Information and Communication Technologies - BAS,
Akad. G. Bonchev Str. Bl. 2, Sofia, Bulgaria
e-mail: vangelova@iit.bas.bg

K. Georgiev et al. (eds.), *Advanced Computing in Industrial Mathematics*,
Studies in Computational Intelligence 793,
https://doi.org/10.1007/978-3-319-97277-0_4

stationary distribution of the discrete time bi-variate Markov process is reduced to LU factorization of its generator matrix, and finally reduced to the pair of equations

$$N = (-U)^{-1}, \quad U = C + \sum_{i=1}^{n} A_i N D_i, \tag{2}$$

where the matrix N represents the time which is expected to be spent in a node before the first passage to the empty state, and U is the infinitesimal generator of the Markov process in a local node before the first passage to the empty state. For the solution of the pair of equations (2), a computational procedure, consisting in the sequence $(U_n, N_n, : n \geq 1)$: $U_1 = C$; $N_n = \left(-U_n^{-1}\right)$; $U_{n+1} = C + \sum_{i=1}^{n} A_i N_n D_i$ is proposed, as well as necessary conditions to be recurrent and positive recurrent are derived.

In Bini et al. [2] the pair of equations (2) is transformed to the nonlinear matrix equation (1) under the assumptions: $C = B - I$ is the transition matrix within a node, with $B \in \mathbb{R}^{n\times n}$ — a sub-stochastic matrix; the transition matrices A_i (from a node to its left child) and D_i (from the left child to its parent) have non-negative entries; the matrices $I + C + D_j + \sum_{i=1}^{m} A_i$ for $j = 1, \ldots, m$ are stochastic. Recall that a matrix $P = [p_{ij}] \in \mathbb{R}^{n\times n}$ with non-negative elements p_{ij} is stochastic if $\sum_{j=1}^{n} p_{ij} = 1$. For solving equation (1) a fixed-point iteration algorithm is used in [1, 3]. Reducing (1) to a quadratic equation, the cyclic reduction method is applied in [4, 5]. Iteration algorithms applying Newton's scheme are proposed in [6] and then in [2], where a comparative analysis of the effectiveness and the computational complexity of the algorithms is also presented.

Recently, some authors [7–9] have studied the general case, and [10–15] the case $m = 2$ of the introduced in [16] matrix equation, arising in optimal interpolation problem

$$X + \sum_{i=1}^{m} A_i^* X^{-1} A_i = Q, \tag{3}$$

with A_i, Q square matrices, $i = 1, \ldots, m$, Q — Hermitian positive definite matrix and a positive definite solution X required. A^* denotes the complex conjugate transpose of A in the complex case or the transpose of A in the real case. The case $m = 1$ is a problem of a practical importance and have been extensively studied since 1993 [17–24]. Obviously, in terms of the perturbation analysis and ignoring the physical interpretation of (1) and (3) the Eq. (1) generalizes the real case of Eq. (3).

When solving a regular practical problem by a numerically stable algorithm in the environment of machine arithmetic, we often do not avoid some errors of approximation and round-off errors. Then, we only get an approximation of the solution and we would like to know how good was it. The solution, obtained in the environment of the finite precision machine arithmetic must be allays accompanied by an analysis of the conditioning of the problem solved, that is, one needs to know the sensitivity of the solution to perturbations in the matrix coefficients. Motivated by these, in order to ascertain the accuracy of an iteratively calculated solution to Eq. (1), in this paper norm-wise, mixed and component-wise condition numbers, as well as local perturbation bounds are formulated and norm-wise non-local residual bounds are derived using the methods of nonlinear perturbation analysis [25–27] (Fréchet derivatives, Lyapunov majorants, fixed-point principles). The residual bounds are formulated in terms of the computed approximate solution to (1) and can be used as a stop criteria of the iterations, when solving Eq. (1) by a numerically stable iterative algorithm.

The following notations are used later on: $\mathbb{R}^{n\times m}$ is the set of $n \times m$ matrices over the field of real numbers $\mathbb{R}$; $I = I_n$ is the identity $n \times n$ matrix; $A^\top$, stands for the transpose of A; vec(.) is the column-wise vector representation of an $n \times n$ matrix, that is stacking the columns of a matrix in one vector, obtaining the n^2-dimensional vector $\mathrm{vec}(A) = \left[a_1^\top, a_2^\top, \ldots, a_n^\top \right]^\top \in \mathbb{R}^{n^2}$ by arranging column-wise the entries of the matrix $A = [A(\varsigma, j)] \in \mathbb{R}^{n\times n}, A = \left[a_1, a_2, \ldots, a_n \right]$ with $a_j \in \mathbb{R}^n$, where $\mathbb{R}^n = \mathbb{R}^{n\times 1}$. and the mn^2-dimensional vector $\mathrm{vec}(S) = \left[\mathrm{vec}(S_1)^\top, \ldots, \mathrm{vec}(S_m)^\top \right]^\top \in \mathbb{R}^{mn^2}$ for the $m-$tuple $S = \left[S_1, \ldots, S_m \right] \in \mathbb{R}^{n\times nm}$, with $S_i \in \mathbb{R}^{n\times n}$ for $i = \overline{1, m}$; $A \otimes B = [A(\varsigma, j)B]$ is the Kronecker product of the matrices $A = [A(\varsigma, j)]_{\varsigma, j}$ and B; $\|\cdot\|$ is the induced norm in the space of linear operators; $\|\cdot\|_2$ is the Euclidean vector or the spectral matrix norm; $\|\cdot\|_{\mathrm{F}}$ is the Frobenius norm; The notation ':=' stands for 'equal by definition'.

The paper is organized as follows. The definitions of norm-wise, mixed and component-wise condition numbers are recalled and the statement of the problem is given in Sect. 2. Norm-wise, mixed and component-wise condition numbers, as well as local perturbation bounds are formulated in Sect. 3. Non-local residual bounds are derived in Sect. 4. The effectiveness of the residual bound proposed in Sect. 4 is demonstrated in Sect. 5 by a numerical example. Sect. 6 concludes the paper.

2 Preliminaries and Statement of the Problem

In this section some achieved results are summarized and the problem considered is stated. To define the norm-wise, mixed and component-wise condition numbers, the following notations are introduced.

Let $\Upsilon \in \Psi$ be such that for $S \in \Upsilon$ Eq. (1) has a solution $X = \Phi(S) \in \mathbb{R}^{n\times n}$. Suppose that S varies over an open subset Υ^0 of Υ. Denote by $x = \varphi(s)$ the vector representation of the problem (1) $X = \Phi(S)$, where $x := \mathrm{vec}(X) = [x_j] \in \mathbb{R}^{n^2}$, $s := \left[\mathrm{vec}(S_1)^\top \; \mathrm{vec}(S_2)^\top \ldots \mathrm{vec}(S_{2m+1})^\top \right]^\top = [s_j] \in \mathbb{R}^{(2m+1)n^2}$ and $\varphi := \mathrm{vec} \circ \Phi$.

For the forward perturbation analysis, we express the presence of errors of approximation and round-off errors associated with the computation of the solution by a numerically stable iterative algorithm by perturbations δS_l in the matrix coefficients $S_l \in \mathbb{S} := \{A_1, A_2, \ldots, A_m, D_1, D_2, \ldots, D_m, C\} = \{S_1, S_2, \ldots, S_{2m+1}\}$, for $l = \overline{1, 2m+1}$, i.e. $S_l \to S_l + \delta S_l$. Denote by $\delta S := (\delta A_1, \delta A_2, \ldots, \delta D_m, \delta C)$ the collection of perturbations in the data. The perturbed equation, written in the form of equivalent operator equation is

$$F(X+\delta X, S+\delta S) := X+\delta X + \sum_{i=1}^{m}(A_i+\delta A_i)(X+\delta X)^{-1}(D_i+\delta D_i) - C - \delta C = 0. \quad (4)$$

For sufficiently small data perturbations δS the perturbed equation (4) has a solution $\tilde{X} = X + \delta X = \Phi(S+\delta S)$, where the perturbation $\delta X = \Phi(S+\delta S) - \Phi(S)$ is a function of the elements of the data perturbations δS.

Simple to obtain and easy to compute perturbation estimates are the norm-wise absolute and relative condition numbers and the local bounds, based on them. The norm-wise condition numbers are linear or first order homogeneous functions of the perturbations in the data, valid asymptotically, for infinite small data perturbations. They give relevant measure for largest of perturbations, and would be quite pessimistic for smaller or structured perturbations. For the case of data, varying in a specific way, e.i., when some of the elements remain constant, the component-wise perturbation analysis is more convenient to be applied.

Tree kinds of condition numbers-norm-wise, component-wise and mixed, introduced by [26, 28–30] are listed as follows:

Definition 1 The finite quantity

$$K(S) := \lim_{\varepsilon\to 0} \sup\left\{\frac{\|\Phi(S+\delta S)-\Phi(S)\|}{\|\delta S\|} : \delta S \neq 0, \|\delta S\| \leq \varepsilon\right\} \quad (5)$$

is the *absolute norm-wise condition number* of the problem $X = \Phi(S)$.

Definition 2 The quantity $k(S) := K(S)\frac{\|S\|}{\|X\|} = K(S)\frac{\|S\|}{\|\Phi(S)\|}$ is the *relative norm-wise condition number* of the problem $X = \Phi(S)$.

Assume Φ is Fréchet differentiable at S.

Definition 3 If $x \neq 0$, then the quantity

$$m_\infty(\varphi, s) := \frac{\|\varphi'(s)\mathrm{diag}(s_1, s_2, \ldots, s_{(2m+1)n^2})\|_\infty}{\|x\|_\infty} = \frac{\||\varphi'(s)|\,|s|\|_\infty}{\|x\|_\infty} \quad (6)$$
$$\leq k_\infty(s) := \frac{\|\varphi'(s)\|_\infty \|s\|_\infty}{\|x\|_\infty}$$

is the *mixed relative condition number* of the problem $x = \varphi(s)$, where $k_\infty(s)$ is the relative condition number from Definition 2 in the infinity norm $\|.\|_\infty$.

Definition 4 If $x_j \neq 0$, for $j = 1, 2, \ldots, n^2$, then the quantity

$$c(\varphi, s) := \|\mathrm{diag}(1/x_1, 1/x_2, \ldots, 1/x_{n^2})\varphi'(s)\mathrm{diag}(s_1, s_2, \ldots, s_{(2m+1)n^2})\|_\infty \quad (7)$$
$$= \|(|\varphi'(s)|\,|s|)./|x|\|_\infty$$

is the *relative component-wise condition number* of the problem $x = \varphi(s)$.

For the backward perturbation analysis, the perturbation δX, in the computed approximate solution $\tilde{X} = X + \delta X$ to (1) reflects the round-off errors and the errors of approximation due to the implementation of a numerically stable iterative algorithm for solving (1) in finite arithmetic environment with machine precision ε, and $\|\delta X\| \leq \varepsilon g(n)\|X\|$, where $g(n)$ is a low order polynomial in n.

Denote by $R(\tilde{X})$ the residual of (1) with respect to the computed approximate solution $\tilde{X}$

$$R(\tilde{X}) = \tilde{X} + \sum_{i=1}^{m} A_i \tilde{X}^{-1} D_i - C. \tag{8}$$

The goal of the paper is to derive norm-wise, mixed and component-wise condition numbers and to formulate effective norm-wise bounds of the error δX in the approximate computed solution $\tilde{X}$ of Eq. (1) based on the condition numbers, as well as, to derive residual bound for δX.

3 Condition Numbers and Local Perturbation Bounds

Subtracting (1) from (4) and neglecting the terms of second and higher order on δX and δS gives

$$F_X(X,S)(\delta X) = \sum_{S_l \in \mathbb{S}} F_{S_l}(X,S)(\delta S) + O(\Delta^2), \quad \Delta := \|\delta X\|_{\mathrm{F}} + \|\delta S\|_{\mathrm{F}} \to 0. \tag{9}$$

Here

$$F_X(X,S)(H) := H + \sum_{i=1}^{m} A_i X^{-1} H X^{-1} D_i : \mathbb{R}^{n\times n} \to \mathbb{R}^{n\times n} \tag{10}$$

is the partial Fréchet derivative of $F(X,S)$ in X at point (X,S) and $F_{S_l}(X,S) : \mathbb{R}^{n\times n} \to \mathbb{R}^{n\times n}$ are the partial Fréchet derivatives of $F(X,S)$ in S_l, for $S_l \in \mathbb{S}$, $l = \overline{1, 2m+1}$, and S_l stands for A_i, D_i, C, for $i = \overline{1,m}$: $F_{A_i}(X,S)(H) := -HX^{-1}D_i$, $F_{D_i}(X,S)(H) := -A_i X^{-1} H$, and $F_C(X,S)(H) := H$.

Denote $D := \left[D_1^\top, D_2^\top, \ldots, D_m^\top\right]^\top \in \mathbb{R}^{nm\times n}$, $A := \left[A_1, A_2, \ldots, A_m\right] \in \mathbb{R}^{n\times nm}$ and $\hat{X} := I \otimes X \in \mathbb{R}^{nm\times nm}$ — the $m \times m$ block diagonal matrix with $X \in \mathbb{R}^{n\times n}$ on its diagonals.

The matrices of the operators $F_{A_i}(X,S)$ and $F_{D_i}(X,S)$, for $i = \overline{1,m}$, $F_C(X,S)$, and $F_X(X,S)$, respectively are $L_{A_i} = \left(X^{-1}D_i\right)^\top \otimes I_{n^2}$, $L_{D_i} = I_{n^2} \otimes A_i X^{-1}$, $L_C = I_{n^2}$ and

$$L := I_{n^2} + \sum_{i=1}^{m} \left(X^{-1}D_i\right)^{\top} \otimes \left(A_i X^{-1}\right) = I_{(nm)^2} + \left(\hat{X}^{-1}D\right)^{\top} \otimes \left(A\hat{X}^{-1}\right). \quad (11)$$

The operator $F_X(X, S)$ from (10) and his matrix L from (11) are invertible iff the matrix L is not singular, i.e. if for the eigenvalues λ_τ, $\tau = \overline{1, nm}$ is fulfilled

$$\lambda_\tau\left(\hat{X}^{-1}D\right)\lambda_\nu\left(A\hat{X}^{-1}\right) \neq -1, \quad \tau, \nu = 1, \ldots, nm. \quad (12)$$

In what follows we assume that the inequality (12) holds true. It follows from (9) that

$$\delta X = \sum_{S_l \in \mathbb{S}} F_X^{-1} F_{S_l}(X, S)(\delta S_l) + O(\Delta^2), \quad (13)$$

or in vector form, after applying the vec-operation from both sides of (13)

$$\begin{aligned} \mathrm{vec}(\delta X) &= \sum_{S_l \in \mathbb{S}} M_{S_l} \mathrm{vec}(\delta S_l) + O(\Delta^2) & (14)\\ &\approx M\, \mathrm{vec}(\delta S), & (15) \end{aligned}$$

where

$$\begin{aligned} M_{S_l} &:= L^{-1} L_{S_l}, \quad l = \overline{1, 2m+1} & (16)\\ M &:= \left[M_{A_1}\ M_{A_2}\ \ldots\ M_{A_m}\ M_{D_1}\ M_{D_2}\ \ldots\ M_{D_m}\ M_C \right] & (17)\\ &= \left[M_{S_1}\ M_{S_2}\ \ldots M_{S_{2m+1}} \right] \in \mathbb{R}^{n^2 \times (2m+1)n^2}, & \end{aligned}$$

$\mathrm{vec}(\delta X) \in \mathbb{R}^{n^2}$ is the vector representation of the solution $X \in \mathbb{R}^{n \times n}$ and $\mathrm{vec}(\delta S) = \left[\mathrm{vec}(\delta A_1)^{\top} \ldots \mathrm{vec}(\delta A_m)^{\top}\ \mathrm{vec}(\delta D_1)^{\top} \ldots \mathrm{vec}(\delta D_m)^{\top}\ \mathrm{vec}(\delta C)^{\top} \right]^{\top} = \left[\mathrm{vec}(\delta S_1)^{\top}\ \mathrm{vec}(\delta S_2)^{\top} \ldots \mathrm{vec}(\delta S_{2m+1})^{\top} \right]^{\top} \in \mathbb{R}^{(2m+1)n^2}$ is the vector representation of the $(2m+1)$–tuple δS.

3.1 Norm-wise Condition Numbers

Set $\sigma_l := \|S_l\|_{\mathrm{F}}$, $\delta_{S_l} := \|\delta S_l\|_{\mathrm{F}}/\sigma_l$ for $l = \overline{1, 2m+1}$ and $\delta_X := \|\delta X\|_{\mathrm{F}}/\|X\|_{\mathrm{F}}$.

Theorem 1 *Let the coefficient matrices S of Eq. (1) satisfy $S \in \Upsilon^0$. The following norm-wise condition numbers for the solution X of Eq. (1) are valid:*

$$\textit{absolute condition numbers:}\quad K_{S_l} := \|M_{S_l}\|_2,\quad S_l \in \mathbb{S},\quad l = \overline{1, 2m+1} \tag{18}$$

$$\textit{relative condition numbers:}\quad k_{S_l} := \frac{K_{S_l}\|S_l\|_2}{\|X\|_2},\quad S_l \in \mathbb{S},\quad l = \overline{1, 2m+1} \tag{19}$$

overall condition number: (20)

$$k := \lim_{\varepsilon\to 0} \sup \left\{ \frac{\delta_X}{\varepsilon} : \sqrt{\sum_{l=1}^{2m+1} \delta_{S_l}^2} \le \varepsilon \right\} = \frac{\left\| \left[\sigma_1 M_{S_1}\ \sigma_2 M_{S_2}\ \dots\ \sigma_{2m+1} M_{S_{2m+1}} \right] \right\|_2}{\|X\|_{\mathrm{F}}}.$$

Proof The absolute and relative condition numbers (18), (19) follow directly from Definition 1 and Definition 2 and expression (14).

For the proof of the overall condition number k (20) we rewrite expression (14) in the form

$$\mathrm{vec}(\delta X) \approx \left[\sigma_1 M_{S_1}\ \sigma_2 M_{S_2}\ \dots\ \sigma_{2m+1} M_{S_{2m+1}} \right] \begin{bmatrix} \mathrm{vec}(\delta S_1)/\sigma_1 \\ \mathrm{vec}(\delta S_2)/\sigma_2 \\ \dots \\ \mathrm{vec}(S_{2m+1})/\sigma_{2m+1} \end{bmatrix}. \tag{21}$$

Taking the spectral norm of both sides of (21), we obtain $\delta_X \le \frac{1}{\|X\|_{\mathrm{F}}} \left\| \left[\sigma_1 M_{S_1}\ \sigma_2 M_{S_2}\ \dots\ \sigma_{2m+1} M_{S_{2m+1}} \right] \right\|_2 \left\| \begin{bmatrix} \mathrm{vec}(\delta S_1)/\sigma_1 \\ \mathrm{vec}(\delta S_2)/\sigma_2 \\ \dots \\ \mathrm{vec}(S_{2m+1})/\sigma_{2m+1} \end{bmatrix} \right\|_2 = \frac{1}{\|X\|_{\mathrm{F}}} \left\| \left[\sigma_1 M_{S_1}\ \sigma_2 M_{S_2}\ \dots\ \sigma_{2m+1} M_{S_{2m+1}} \right] \right\|_2 \sqrt{\sum_{l=1}^{2m+1} \delta_{S_l}^2}$. Hence (20) holds. □

The relative condition numbers k_{S_l} (19) give measure of the actual error in the computed by a numerically stable algorithm solution of the problem considered. For computing environment with machine precision ε and problem with relative condition number k, such that $\varepsilon k < 1$, the computed solution possesses about $\log_{10}(\varepsilon k)$ true decimal digits, accordingly the heuristic rule, derived in [26].

Using the norm-wise condition numbers from Theorem 1, we formulate the following local estimate $\mathrm{est}(\delta)$ of the perturbations δX in the computed solution of Eq. (1): $\delta_X \le \mathrm{est}(\delta) := \min[\mathrm{est}_2(\delta), \mathrm{est}_3(\delta)]$, where $\mathrm{est}_2(\delta) := k\sqrt{\sum_{l=1}^{2m+1} \delta_{S_l}^2}$; $\mathrm{est}_3(\delta) := \sqrt{\sum_{l=1}^{2m+1} k_{S_l}^2 \delta_{S_l}^2 + \sum_{\nu=1}^{2m+1} \sum_{l=1}^{2m+1} 2\gamma_{l\nu} \delta_{S_l} \delta_{S_\nu}}$, with $\gamma_{l\nu} := \frac{\sigma_l \sigma_\nu}{\|X\|_{\mathrm{F}}^2} \|M_{S_l}^\top M_{S_\nu}\|_2$ for $l, \nu = \overline{1, 2m+1}$ and $\delta := \left[\delta_{S_1} \dots \delta_{S_{2m+1}} \right] \in \mathbb{R}_+^{2m+1}$ -the vector of Frobenius norms of the relative perturbations in the matrix coefficients.

The norm-wise perturbation bounds are relevant measures for largest of perturbations in the data matrices S, or in the solution X, while for smaller or structural perturbations, the norm-wise perturbation bounds would be pessimistic. For problem with data varying in a specific way, e.g., when some of the data elements remain constant, it is convenient to apply the technique of component-wise perturbation analysis.

3.2 Component-Wise Condition Number

The component-wise perturbation bounds estimate the sensitivity of the elements x_j, $j = \overline{1, n^2}$ of the solution X in respect to perturbations in the elements s_j, $j = \overline{1, (2m+1)n^2}$, of the data S. This makes it convenient for problems with significantly different perturbations in the components of the data, the solution or both.

Theorem 2 *Let $S \in \Upsilon^0$. If $x_j \neq 0$ for $j = \overline{1, n^2}$, then, according to Definition 4, the component-wise condition number (7) of Eq. (1) is*

$$c(\varphi, s) := \|(|M|\,|s|)\,./|x|\|_\infty\,, \tag{22}$$

where M is defined in (17) with (16) and the following upper bound is valid

$$c_u(\varphi, s) := \|\mathrm{diag}(\mathrm{vec}(X))^{-1} L^{-1}\|_\infty \beta \geq c(\varphi, s), \tag{23}$$

where

$$\beta := \left\| \mathrm{vec}\left(|C| + 2\sum_{i=1}^{m} |A_i|\,|X^{-1}|\,|D_i| \right) \right\|_\infty . \tag{24}$$

Proof The component-wise condition number (22) defined according to Definitions 4 follows directly from expressions (14), (16) and (17) .

For the proof of the upper bound $c_u(\varphi, s)$ from (23), consider expressions (16) and (17). For the term $\|\,|M|\,|s|\|_\infty$ we have

$$\begin{aligned}
\||M|\,|s|\|_\infty &= \|\,|\left[L^{-1}L_{S_1}\; L^{-1}L_{S_2}\; \ldots\; L^{-1}L_{S_{2m+1}}\right]|\,|s|\|_\infty \\
&\leq \|\,|L^{-1}|\,|\left[L_{S_1}\; L_{S_2}\; \ldots\; L_{S_{2m+1}}\right]|\,|s|\|_\infty \\
&\leq \|L^{-1}\|_\infty \||\left[L_{S_1}\; L_{S_2}\; \ldots\; L_{S_{2m+1}}\right]|\,|s|\|_\infty \\
&\leq \|L^{-1}\|_\infty \|\mathrm{vec}(|C| + 2\sum_{i=1}^{m} |A_i|\,|X^{-1}|\,|D_i|)\|_\infty .
\end{aligned}$$

□

The component-wise perturbation technique require non zero entries of the solution. When the computational problem has non zero solution $X \neq 0$, with some zero components $x_j = 0, j = \overline{1, n^2}$, the use of mixed condition numbers is appropriate.

3.3 Mixed Condition Number

The mixed perturbation bounds are norm-like functions which reflect the changes in the data in a component-wise style. Usually, rounding errors are not introduced in the zero entries of the data matrices. That is the reason to obtain more sharpest estimates by the mixed perturbation bounds.

Theorem 3 *Let* $S \in \Upsilon^0$. *If* $x \neq 0$, *the mixed condition number (6) based on Definition 3 for Eq. (1) is*

$$m_\infty(\varphi, s) := \frac{\| |M| \, |s| \|_\infty}{\|x\|_\infty}. \tag{25}$$

Proof The mixed condition number (25) follows directly from the Definitions 3 and expressions (14), (16) and (17) . □

An effective and simple for computation upper bound of the mixed condition number is $m_u(\varphi, s) := \|\mathrm{vec}(X)\|_\infty^{-1} \|L^{-1}\|_\infty \beta \geq m_\infty(\varphi, s)$ with β given in (24).

The mixed and the component-wise condition numbers from Theorems 2 and 3 can be used to estimate locally in a mixed and a component-wise way the perturbations δX in the solution X of Eq. (1).

Let $|\delta S_1| \leq \varepsilon_1 |S_1|$, $|\delta S_2| \leq \varepsilon_2 |S_2|, \ldots, |\delta S_{2m+1}| \leq \varepsilon_{2m+1} |S_{2m+1}|$ and $\varepsilon := \min\{\varepsilon_1, \varepsilon_2, \varepsilon_3, \ldots, \varepsilon_{2m+1}\}$. Then the following mixed and component-wise perturbation bounds are valid $\|\mathrm{vec}(\delta X)\|_\infty / \|x\|_\infty \leq \varepsilon \, m_\infty(\varphi, s)$, and $\| |\mathrm{vec}(\delta X)| ./ |x| \|_\infty \leq \varepsilon \, c(\varphi, s)$.

4 Residual Bounds

Consider Eq. (8) for the residual of (1). Applying the matrix inversion lemma $\tilde{X}^{-1} = (X + \delta X)^{-1} = X^{-1} - \tilde{X}^{-1} \delta X X^{-1}$ and taking into account (1), a straightforward calculation leads to the following equation of the error δX in the calculated approximate solution $\tilde{X}$

$$\delta X = R(\tilde{X}) + \sum_{i=1}^{m} A_i (\tilde{X} - \delta X)^{-1} \delta X \tilde{X}^{-1} D_i. \tag{26}$$

The vector representation of Eq. (26) is

$$\mathrm{vec}(\delta X) = \mathrm{vec}(R(\tilde{X})) + \sum_{i=1}^{m} (D_i^\top \tilde{X}^{-1} \otimes A_i) \mathrm{vec}\left((\tilde{X} - \delta X)^{-1} \delta X \right). \tag{27}$$

For the Frobenius norm of the error δX in the calculated approximate solution $\tilde{X}$ it follows from (27)

$$\|\delta X\|_{\mathrm{F}} \leq \|R(\tilde{X})\|_{\mathrm{F}} + \sum_{i=1}^{m} \|D_i^\top \tilde{X}^{-1} \otimes A_i\|_2 \|(\tilde{X} - \delta X)^{-1}\|_2 \|\delta X\|_{\mathrm{F}}. \tag{28}$$

Having in mind that $\|\delta X\|_{\rm F} \le \frac{1}{\|\tilde{X}^{-1}\|_2}$, because of the character of the error δX, for the spectral norm of the inverse of the exact solution X to equation (1), as a function of the calculated approximate solution $\tilde{X}$ and the error δX, we have

$$\|X^{-1}\|_2 = \|(\tilde{X} - \delta X)^{-1}\|_2 \le \frac{\|\tilde{X}^{-1}\|_2}{1 - \|\tilde{X}^{-1}\|_2 \|\delta X\|_{\rm F}}. \tag{29}$$

Then, for the Frobenius norm of the error δX in the approximate solution $\tilde{X}$ of Eq. (1), taking into account (28) and (29), we obtain

$$\|\delta X\|_{\rm F} \le \|R(\tilde{X})\|_{\rm F} + \frac{\sum_{i=1}^m \|D_i^\top \tilde{X}^{-1} \otimes A_i\|_2 \|\tilde{X}^{-1}\|_2}{1 - \|\tilde{X}^{-1}\|_2 \|\delta X\|_{\rm F}} \|\delta X\|_{\rm F}. \tag{30}$$

Denote by $\zeta_i := \|D_i^\top \tilde{X}^{-1} \otimes A_i\|_2 \|\tilde{X}^{-1}\|_2$, $\chi := \|\tilde{X}^{-1}\|_2$ and $r := \|R(\tilde{X})\|_{\rm F}$. We rewrite expression (30) in the form

$$\begin{aligned} \|\delta X\|_{\rm F} &\le r + \frac{\sum_{i=1}^m \zeta_i}{1 - \chi \|\delta X\|_{\rm F}} \|\delta X\|_{\rm F} \le \chi \|\delta X\|_{\rm F}^2 - \left(r\chi - \sum_{i=1}^m \zeta_i \right) \|\delta X\|_{\rm F} + r \\ &\le \chi \|\delta X\|_{\rm F}^2 + q \|\delta X\|_{\rm F} + r, \end{aligned} \tag{31}$$

with $q := \sum_{i=1}^m \zeta_i - r\chi$.

The function $h(\rho) = r + q\rho + \chi\rho^2$ is a Lyapunov majorant for equation (31). The majorant equation for determining a non-local bound ρ for $\|\delta X\|_{\rm F}$ is

$$\chi\rho^2 - (1 - q)\rho + r = 0. \tag{32}$$

Suppose that $r \in \Omega$, where

$$\Omega = \left\{ 0; \sum_{i=1}^m \zeta_i - r\chi + 2\sqrt{r\chi} \le 1 \right\}. \tag{33}$$

Then Eq. (32) has non-negative roots $\rho_1 \le \rho_2$ with $\rho_1 = \theta(r)$,

$$\theta(r) := \frac{2r}{1 - \sum_{i=1}^m \zeta_i + r\chi + \sqrt{(1 - \sum_{i=1}^m \zeta_i + r\chi)^2 - 4r\chi}}.$$

Let for Q_i, W_i, $V, T \in \mathbb{R}^{n \times n}$ — arbitrary given matrices, define the linear operator $\Gamma : \mathscr{R}^n \to \mathscr{R}^{nn}$ as $\Theta(H) = T + \sum_{i=1}^m W_i (V - H)^{-1} H V^{-1} Q_i$. Equation (26) may be rewritten in operator form

$$\delta X = \Theta(\delta X). \tag{34}$$

For $\rho > 0$ denote by $\mathscr{B}(\rho) \subset \mathbb{R}^{n\times n}$ the set of all matrices $M \in \mathbb{R}^{n\times n}$ satisfying $\|M\|_{\mathrm{F}} \le \rho$. For $H \in \mathscr{B}(\rho)$, having in mind (31), we have $\|\Theta(H)\|_{\mathrm{F}} \le r + q\rho + \chi\rho^2$.

The operator Γ maps the closed central ball $\mathscr{B}_\Gamma = \left\{\mathrm{vec}(\delta X) \in \mathbb{R}^{n^2} : \|\mathrm{vec}(\delta X)\|_{\mathrm{F}} \le \theta(r)\right\}$ of radius $\theta(r)$ into itself, where $\theta(r)$ is continuous and $\theta(0) = 0$. Then, according to the Schauder fixed-point principle, there exists a solution $\delta X \in \mathscr{B}_\theta(r)$ of Eq. (34). In what follows, we deduced the following statement.

Theorem 4 *Consider Eq. (1) for which the solution $\tilde{X} = X + \delta X$, obtained by some numerically stable iterative algorithm with residual $R(\tilde{X})$ (8) is an approximation to the exact solution X.*

Let $r := \|R(\tilde{X})\|_{\mathrm{F}}$, $\zeta_i := \|D_i^\top \tilde{X}^{-1} \otimes A_i\|_2 \|\tilde{X}^{-1}\|_2$ and $\chi := \|\tilde{X}^{-1}\|_2$.

For $r \in \Omega$, given in (33), the following bounds for the error δX in the approximate solution $\tilde{X}$ are valid:

- *non-local absolute residual bound*

$$\|\delta X\|_{\mathrm{F}} \le \theta(r), \quad \theta(r) := \frac{2r}{1 - \sum_{i=1}^m \zeta_i + r\chi + \sqrt{(1 - \sum_{i=1}^m \zeta_i + r\chi)^2 - 4r\chi}}$$

- *non-local relative residual bound in terms of the exact solution X*

$$\frac{\|\delta X\|_{\mathrm{F}}}{\|X\|_2} \le \frac{\theta(r)}{\|X\|_2}$$

- *non-local relative residual bound in terms of the computed approximate solution $\tilde{X}$*

$$\frac{\|\delta X\|_{\mathrm{F}}}{\|X\|_2} \le \mathrm{est}_r; \quad \mathrm{est}_r := \frac{\theta(r)/\|\tilde{X}\|_2}{1 - \theta(r)/\|\tilde{X}\|_2}. \tag{35}$$

5 Numerical Example

To illustrate the effectiveness of the residual bounds, defined in Theorem 4 from Sect. 4, we use the experimental setup given in [2, Example 9.2.2 [1]]. A system similar to the M/M/1 queue in a random environment is considered with constant service rate and an arrival rate depending on the state of Markovian environmental process. A label of each arrival indicates the direction taken in the tree. The values of the parameters are chosen as follows: number of children $k = 2$; size of the blocks n; service rate α; stationary arrival rate $\rho = 0.1$ ($\rho < \alpha$ in order to assure stability of the system); fraction of arrivals occurring in phase 1 $f = 0.8$ ($f \in (0, 1)$); $D_1 = \alpha I$; $D_2 = D_1$; $A_1 = diag(a, b, \dots, b)$; $A_2 = bI$; $C = T - D_1 - A_1 - A_2$, with $a = \rho f n$, $b = \rho(1-f)n/(2n-1)$, $T = [T(i,j)]$, $i, j = \overline{1, n}$, $T(i,i) = -1$ for $i = \overline{1, n}$, $T(i, i+1) = 1$ for $i = \overline{1, n-1}$, $T(n, 1) = 1$ and $T(i, j) = 0$ elsewhere.

Table 1 Ratio $\frac{\text{est}_r}{\|\delta X\|_F/\|X\|_2}$ of the residual bound est_r (35) to the estimated value $\|\delta X\|_F/\|X\|_2$

$n=10,\ \alpha=2$	k	1	2	3	4	
	$\frac{\text{est}_r}{\|\tilde{X}-X\|_F/\|X\|_2}$	1.7039	1.6838	1.6838	0	
$n=10,\ \alpha=3$	k	1	2	3	4	
	$\frac{\text{est}_r}{\|\tilde{X}-X\|_F/\|X\|_2}$	1.3765	1.3707	1.3707	0	
$n=20,\ \alpha=3$	k	1	2	3	4	
	$\frac{\text{est}_r}{\|\tilde{X}-X\|_F/\|X\|_2}$	2.3492	2.1648	2.1641	0	
$n=40,\ \alpha=8$	k	1	2	3	4	
	$\frac{\text{est}_r}{\|\tilde{X}-X\|_F/\|X\|_2}$	1.7520	1.6714	1.6709	0	
$n=50,\ \alpha=8$	k	1	2	3	4	5
	$\frac{\text{est}_r}{\|\tilde{X}-X\|_F/\|X\|_2}$	2.3063	2.0111	2.0075	2.0076	0

For different values of block size n and service rate α, the relative residual bound est (35) is compared to the relative error $\|\delta X\|_F/\|X\|_2$ in the approximate solution $\tilde{X}$ to Eq. (1). The approximate solution $\tilde{X}_k$ at iteration step k is obtained by the iterative Newton's method, proposed in [2]. $Y_k + \sum_{i=1}^{2} A_i\tilde{S}_k^{-1}Y_k(-\tilde{S}_k)^{-1}D_i = L_k$, $L_k = \tilde{S}_k - C + \sum_{i=1}^{2} A_i\tilde{S}_k^{-1}D_i$, $\tilde{S}_{k+1} = \tilde{S}_k - Y_k$, $\tilde{S}_0 = C$.

The ratios $\frac{\text{est}_r}{\|\delta X\|_F/\|X\|_2}$ of the relative residual bound est_r (35) to the estimated value $\|\delta X\|_F/\|X\|_2$ from the first step to the step of finding the solution, are listed in Table 1. The above results verify the effectiveness of the relative residual bound proposed in Sect. 4.

6 Conclusion

In this paper, norm-wise, mixed and component-wise condition numbers and based on them easy computable local bounds, as well as norm-wise non-local residual bounds are proposed for the solution to the nonlinear matrix equation (1) connected to a tree-like stochastic processes. The bounds are obtained using the methods of the perturbation analysis. The condition numbers allow to estimate the level of uncertainty in the solution due to errors in the data when applying a numerically stable algorithm to solve the equation. Effective, easy to compute and based on the approximate solution, the residual bounds can be used as a stopping criterion for the iterative algorithm.

References

1. Latouche, G., Ramaswami, V.: Introduction to Matrix Analytic Methods in Stochastic Modeling. ASA-SIAM Series on Statistics and Applied Probability 5. SIAM, Philadelphia (1999)
2. Bini, A.D., Latouche, G., Meini, B.: Solving nonlinear matrix equations arising in Tree-Like stochastic processes. Linear Algebra Appl. **366**, 39–64 (2003)
3. Yeung, R.W., Alfa, A.S.: The Quasi-Birth-Death Type Markov Chain with a Tree Structure. Technical report, University of Manitoba, Winnipeg (1997)
4. Bini, D., Meini, B.: On the solution of a nonlinear matrix equation arising in queueing problems. SIAM J. Matrix Anal. Appl. **17**, 906–926 (1996)
5. Bini, D., Meini, B.: Improved cyclic reduction for solving queueing problems. Numer. Alg. **15**, 5774 (1997)
6. Latouche, G.: Newtons iteration for nonlinear equations in Markov chains. IMA J. Numer. Anal. **14**, 583–598 (1994)
7. Duan, X., Li, C., Lia, A.: Solutions and perturbation analysis for the nonlinear matrix equation $X + \sum_{i=1}^{m} A_i^* X^{-1} A_i = I$. Appl. Math. Comput. **218**, 44584466 (2011)
8. Hasanov, V.I., Hakkaev, S.: Newton's method for a nonlinear matrix equation. Compt. Rend. bulg. Sci. **68**(8), 973–982 (2015)
9. Huang, B.-H.: Ma, C.-F.: Some iterative methods for the largest positive definite solution to a class of nonlinear matrix equation. Numer. Algor. (2017). https://doi.org/10.1007/s11075-017-0432-8
10. Huang, N., Ma, C.: The inversion-free iterative methods for solving the nonlinear matrix equation $X + A^H X^{-1} A + B^H X^{-1} B = I$. Abstract Appl. Anal. (2013). https://doi.org/10.1155/2013/843785
11. Long, J.H., Hu, X.Y., Zhang, L.: On the Hermitian positive definite solution of the nonlinear matrix equation $X + A^* X^{-1} A + B^* X^{-1} B = I$. Bull. Braz. Math. Soc. New Ser. **39**(3), 371–386 (2008)
12. Popchev, I., Konstantinov, M., Petkov, P., Angelova, V.: Condition numbers for the matrix equation $X + A^H X^{-1} A + B^H X^{-1} B = I$. C. R. Acad. Bulgare Sci. **64**(12), 1679–1688 (2011)
13. Popchev, I.P., Petkov, P.H., Konstantinov, M.M., Angelova, V.A.: Perturbation bounds for the nonlinear matrix equation $X + A^H X^{-1} A + B^H x^{-1} B = IA$
14. Vaezzadeh S., Vaezpour, S., Saadati, R., Park, C.: The iterative methods for solving nonlinear matrix equation $X + A^* X^{-1} A + B^* X^{-1} B = Q$. Adv. Differ. Equ. **229** (2013) https://doi.org/10.1186/1687-1847-2013-229.
15. Zhang, Y.: A note on positive definite solution of matrix equation $X + M^* X^{-1} M - N^* X^{-1} N = I$. Linear and Multilinear Algebra (2015). https://doi.org/10.1080/03081087.2015.1068267
16. He, Y., Long, J.H.: On the Hermitian positive definite solution of the nonlinear matrix equation $X + \sum_{i=1}^{m} A_i = I$
17. Engwerda, J.C., Ran, A.C.M., Rijkeboer, A.L.: Necessary and sufficient conditions for the existence of a positive definite solution of the matrix equation $X + A^* X^- 1A = Q$. Linear Algebra Appl. **186**, 255–275 (1993)
18. Ferrante, A., Lev, B.-C.: Hermitian solutions of the equation $X = Q + N X^{-1} N^*$. **247** (1), 359–373, (1996)
19. Hasanov, V.I., Ivanov, I.G.: On two perturbation estimates of the extreme solutions to the equations $X \pm A^* X^{-1} A = Q$. Linear Algebra Appl. **413**(1), 81–92 (2006)
20. Ivanov, I.G., Hasanov, V., Uhlig, F.: Improved methods and starting values to solve the matrix equations $X \pm A^* X^{-1} A = I$ iteratively. Math. Comp. **74**, 263–278 (2004)
21. Konstantinov, M.M., Angelova, V.A., Petkov, P.H., Popchev, P.I.: Comparison of perturbation bounds for the matrix equation $X = A_1 + A_2^{\mathrm{h}} X^{-1} A_2$. Ann. Inst. Arch. Genie Civil Geod. **41** (200-2001), fasc. II Math 75–82 (2003)
22. Meini, B.: Efficient computation of the extreme solutions of $X + A^* X^{-1} A = Q$ and $X - A^* X^{-1} A = Q$. Math. Comp. **71**, 1189–1204 (2001)

23. Popchev, I., Angelova, V.: On the sensitivity of the matrix equations $X \pm A^{*}X^{-1}A = Q$
24. Xu, S.: Perturbation analysis of the maximal solution of the matrix equation $X + A^{*}X^{-}1A = P$. Acta Sci. Nat. Univ. Pekinensis **36**, 29–38 (2000)
25. Konstantinov, M.M., Petkov, P.H., Pelova, G., Angelova, V.A.: Perturbation analysis of differential and difference matrix quadratic equations: a survey. In: Karandzulov, L., Andreev, A. (eds.) Proceedings of Bulgarian-Turkish-Ukrainian Scientific Conference Mathematical analysis, Sunny Beach, Sept. 15–20, 2010, Academic Publishing House Prof. Marin Drinov, 101–110 (2011)
26. Konstantinov, M.M., Petkov, P.H.: Perturbation methods in linear algebra control (survey). Appl. Comput. Math. **7**(2), 141–161 (2008)
27. Yonchev, A.S., Konstantinov, M.M., Petkov, P.H.: Linear perturbation bounds of the continuous-time LMI based H^{∞} quadratic stability problem. Int. J. Appl. Comput. Math. **12**, 133–139 (2013)
28. Cucker, F., Diao, H., Wei, Y.: On mixed and componentwise condition numbers for Moore-Penrose inverse and linear least squares problems. Math. Comput. **76**, 947–963 (2007)
29. Gohberg, I., Koltracht, I.: Mixed, componentwise, and structured condition numbers. SIAM J. Matrix Anal. Appl. **14**, 688–704 (1993)
30. Konstantinov, M.M., Gu, D.W., Mehrmann, V., Petkov, P.H.: Perturbation Theory for Matrix Equations. North-Holland, Amsterdam (2003). [ISBN 0-444-51315-9]

On Two-Way Generalized Nets

Krassimir Atanassov

Abstract Short remarks on Generalized net theory are given. A new extension of the GNs is introduced, called a two-way generalized net. It is proved that the new generalized nets are conservative extensions of the standard generalized nets. Some possible applications of the two-way generalized nets are discussed.

Keywords Data mining · Generalized net · Modelling

1 Introduction

The paper is a part of the authors's plenary talk before BGSIAM'2017, devoted to the origin, the current state of research and the applications in the area of Data Mining (DM) of the *Generalized Nets* (*GNs*, see [1, 3, 5, 7, 11, 14]).

As it is noted in [7], GNs are defined as extension of the ordinary Petri Nets (PNs, [10]) and their modifications, but in a way that is different in principle from the ways of defining the other types of PNs. The additional components in the GN-definition give more and better modelling capabilities and determine the place of GNs among the separate types of PNs, similar to the place of the Turing machine among the finite automata.

In the present paper, a new extension of the GNs called *two-way GNs*, is introduced. A quarter of a century ago, the author published [4], where the same idea was discussed, but the GN-extension was in essentially weaker form that the present one. Here, keeping the older name, we introduce in practice a new GN-extension.

K. Atanassov (✉)
Department of Bioinformatics and Mathematical Modelling,
Institute of Biophysics and Biomedical Engineering - Bulgarian Academy of Sciences,
Acad. G. Bonchev Str., Bl. 105, 1113Sofia, Bulgaria
e-mail: krat@bas.bg

K. Atanassov
Intelligent Systems Laboratory, Prof. Asen Zlatarov University, 8000 Bourgas, Bulgaria

K. Georgiev et al. (eds.), *Advanced Computing in Industrial Mathematics*,
Studies in Computational Intelligence 793,
https://doi.org/10.1007/978-3-319-97277-0_5

2 Definition of the Concept of a Two-Way Generalized Net

The concept of a GN is described in details in [3, 5]. Here, we introduce the concept of a *two-way GN* (*TWGN*) exdending the definition of the GN.

The first basic difference between GNs, especially the TWGNs, and the ordinary PNs is the “place – transition” relation [13]. Here, as in the case of the standard GNs, the transitions are objects of more complex nature. In the GNs, transitions may contain m input places and n output places, where the integers $m, n \geq 1$. In the TWGNs, these places are called *left* and *right* instead of input and output places.

Formally, every TWGN-transition is described by a seven-tuple (Fig. 1):

$$Z = \langle L', L'', t_1, t_2, r, M, \square \rangle,$$

where:

(**a**) L' and L'' are finite, non-empty sets of places (the transition’s left and right places, respectively); for the transition in Fig. 1 these are

$$L' = \{l_1, l_2, \ldots, l_p\}$$

and

$$L'' = \{l_{p+1}, l_{p+2}, \ldots, l_{p+q}\};$$

(**b**) t_1 is the current time-moment of the transition’s firing (activation);

(**c**) t_2 is the current value of the duration of its active state;

(**d**) r is the TWGN-transition’s *condition* determining which tokens will pass (or *transfer*) from the transition’s left and right to its left and right places; it has the form of an Index Matrix (IM; see [2, 6]):

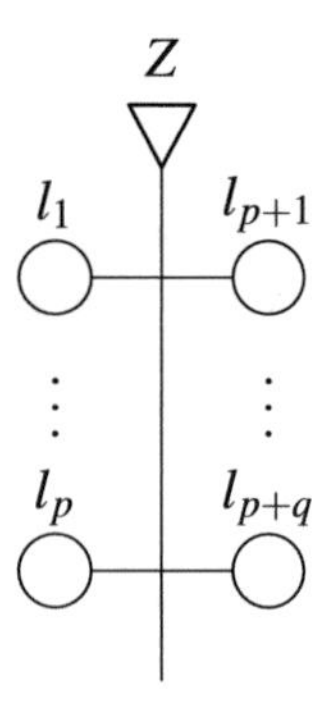

Fig. 1 The form of a TWGN-transition

$$r = \begin{array}{c|c} & l_1 \ldots l_j \ldots l_{p+q} \\ \hline l_1 & \\ \vdots & r_{i,j} \\ l_i & (r_{i,j} \ - \text{predicate}) \\ \vdots & (1 \le i \le p+q, 1 \le j \le p+q) \\ l_{p+q} & \end{array};$$

where $r_{i,j}$ is the predicate that corresponds to the i-th and j-th TWGN-transition places. When its truth value is "*true*", a token from the i-th place transfers to the j-th place; otherwise, this is not possible;

(e) M is an IM of the capacities of transition's arcs:

$$M = \begin{array}{c|c} & l_1 \ldots l_j \ldots l_{p+q} \\ \hline l_1 & \\ \vdots & m_{i,j} \\ l_i & (m_{i,j} \ge 0 - \ \text{natural number}) \\ \vdots & (1 \le i \le p+q, 1 \le j \le p+q) \\ l_{p+q} & \end{array};$$

(f) $\square$ is an object of a form similar to a Boolean expression. It contains as variables the symbols that serve as labels for a TWGN-transition's places. Its semantics is defined as follows:

$\wedge(l_{i_1}, l_{i_2}, \ldots, l_{i_u})$ – every place $l_{i_1}, l_{i_2}, \ldots, l_{i_u}$ must contain at least one token,

$\vee(l_{i_1}, l_{i_2}, \ldots, l_{i_u})$ – there must be at least one token in all places $l_{i_1}, l_{i_2}, \ldots, l_{i_u}$, where $\{l_{i_1}, l_{i_2}, \ldots, l_{i_u}\} \subset L' \cup L''$.

When the value of a type (calculated as a Boolean expression) is "*true*", the transition can become active, otherwise it cannot.

The ordered four-tuple

$$E = \langle\langle A, \pi_A, \pi_L, c, f, \theta_1, \theta_2\rangle, \langle K, \pi_K, \theta_K\rangle, \langle T, t^o, t^*\rangle, \langle X, \Phi, b\rangle\rangle$$

is called a TWGN if:

(a) A is a set of the TWGN-transitions;

(b) π_A is a function giving the priorities of the transitions, i.e., $\pi_A : A \to N$, where $N = \{0, 1, 2, \ldots\} \cup \{\infty\}$;

(c) π_L is a function giving the priorities of the places, i.e., $\pi_L : L \to N$, where $L = pr_1 A \cup pr_2 A$, and $pr_i X$ is the i-th projection of the n-dimensional set, where $n \in N, n \ge 1$ and $1 \le k \le n$ (here, L is the set of all TWGN-places);

(d) c is a function giving the capacities of the places, i.e., $c : L \to N$;

(**e**) f is a function that calculates the truth values of the predicates of the TWGN-transition's conditions (for the TWGN described here, let the function f have the value "*false*" or "*true*", that is, a value from the set $\{0, 1\}$;

(**f**) θ_1 is a function which indicates the next time-moment when a certain transition Z can be activated, that is, $\theta_1(t) = t'$, where $pr_3 Z = t, t' \in [T, T + t^*]$ and $t \leq t'$. The value of this function is calculated at the moment when the transition ceases to function;

(**g**) θ_2 is a function which gives the duration of the active state of a certain transition Z, i.e., $\theta_2(t) = t'$, where $pr_4 Z = t \in [T, T + t^*]$ and $t' \geq 0$. The value of this function is calculated at the moment when the transition starts to function;

(**h**) K is the set of the TWGN's tokens.

(**i**) π_K is a function which gives the priorities of the tokens, that is, $\pi_K : K \rightarrow N$;

(**j**) θ_K is a function which gives the time-moment when a given token can enter the net, that is, $\theta_K(\alpha) = t$, where $\alpha \in K$ and $t \in [T, T + t^*]$;

(**k**) T is the time-moment when the TWGN starts to function. This moment is determined with respect to a fixed (global) time-scale;

(**l**) t^o is an elementary time-step, related to the fixed (global) time-scale;

(**m**) t^* is the duration of the functioning of the TWGN;

(**n**) X is the set of all initial characteristics which the tokens can obtain on entering the TWGN;

(**o**) Φ is the characteristic function that assigns new characteristics to every token when it makes the transfer from one place of a given TWGN-transition to another;

(**p**) b is a function which gives the maximum number of characteristics a given token can obtain, that is, $b : K \rightarrow N$.

For example, if $b(\alpha) = 1$ for any token α, then this token enters the TWGN with some initial characteristic (marked as its zero-characteristic) and subsequently it keeps only its current characteristic. When $b(\alpha) = \infty$, token α keeps all its characteristics. When $b(\alpha) = k < \infty$, except its zero-characteristic, token α keeps its last k characteristics (characteristics older than the last k-th one be "forgotten"). Hence, in general, every token α has $b(\alpha) + 1$ characteristics on leaving the net.

We see that the definitions of the TWGN and of the standard GN are distinguished in three points: (d)–(f) from the definition of a transition. For the standards GN they are, respectively:

(**d**) r is the transition's *condition* determining which tokens will transfer from the transition's inputs to its outputs. Parameter r has the form of an IM:

$$r = \begin{array}{c|c} & l''_1 \ \ldots \ l''_j \ \ldots \ l''_n \\ \hline l'_1 & \\ \vdots & r_{i,j} \\ l'_i & (r_{i,j} \ \ - \text{predicate}) \\ \vdots & (1 \leq i \leq m, 1 \leq j \leq n) \\ l'_m & \end{array} ;$$

where $r_{i,j}$ is the predicate which expresses the condition for transfer from the i-th input place to the j-th output place. When $r_{i,j}$ has truth-value "*true*", then a token from the i-th input place can be transferred to the j-th output place; otherwise, this is impossible;

(e) M is an IM of the capacities of transition's arcs:

$$M = \begin{array}{c|c} & l''_1 \ldots l''_j \ldots l''_n \\ \hline l'_1 & \\ \vdots & m_{i,j} \\ l'_i & (m_{i,j} \geq 0 - \text{ natural number or } \infty) \\ \vdots & (1 \leq i \leq m, 1 \leq j \leq n) \\ l'_m & \end{array};$$

(f) $\square$ is called transition type and it is an object having a form similar to a Boolean expression with variables – the symbols that serve as labels for transition's (only!) input places.

Finally, the arcs in TWGN are not oriented, while in the GNs they are oriented.

3 An Example

In [9], a GN-model of a part of the train-system in Bulgaria, is described. A reduced part of this GN-model is shown on Fig. 2. On Fig. 3 the same model is shown, but by a TWGN. The transitions on both figures represent trains stations. Places l_3, l_6, l_9, l_{12} and l_{15} correspond to the railway platforms of the stations. The trains are represented by GN-tokens, that go from input (or left, in Fig. 3) to output (right, on Fig. 3) places of the GN-transitions.

Now, there are not cycles, as in Fig. 2 and by this reason, a part of the places are omitted. We specially kept the identifiers of the places that continue to exist in the second model to point out the difference between the two models. Obviously, the second one (on Fig. 3) is essentially simpler.

The example demonstrates that the new type of nets give us possibility to construct simpler models.

4 Basic Theorem

Obviously, each standard GN can be perceived as a TWGN in which all tokens are transformed only from input (left) to output (right) places. Therefore, the standard GNs are particular cases of the TWGNs. On the other hand, we prove the following

Theorem 1 *The functioning and the results of the work of each TWGN can be represented by a standard GN.*

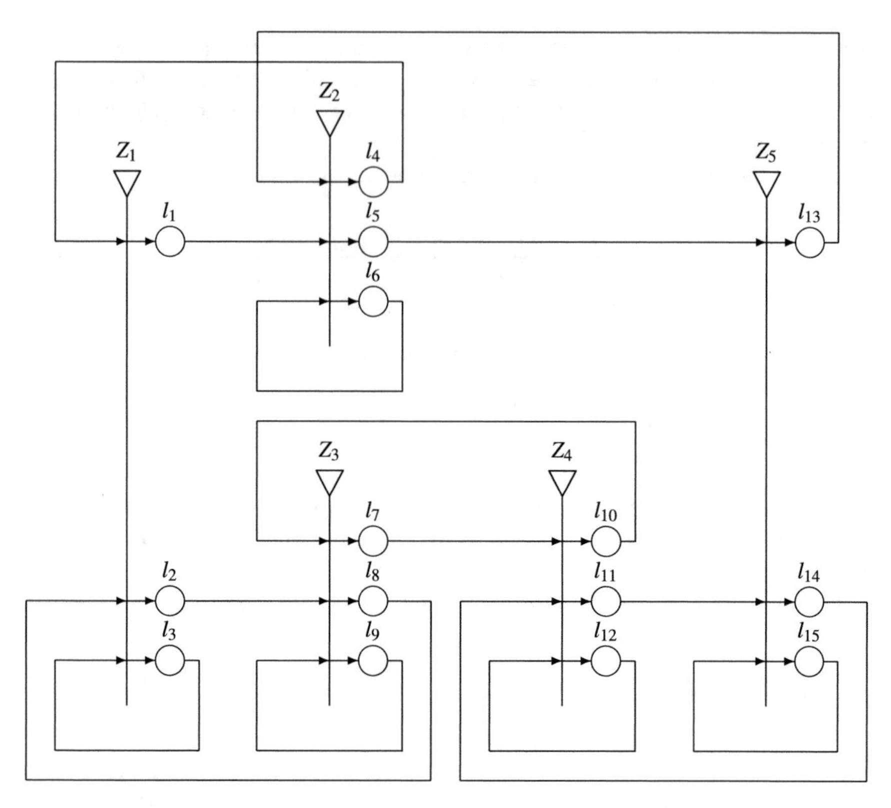

Fig. 2 A GN-model

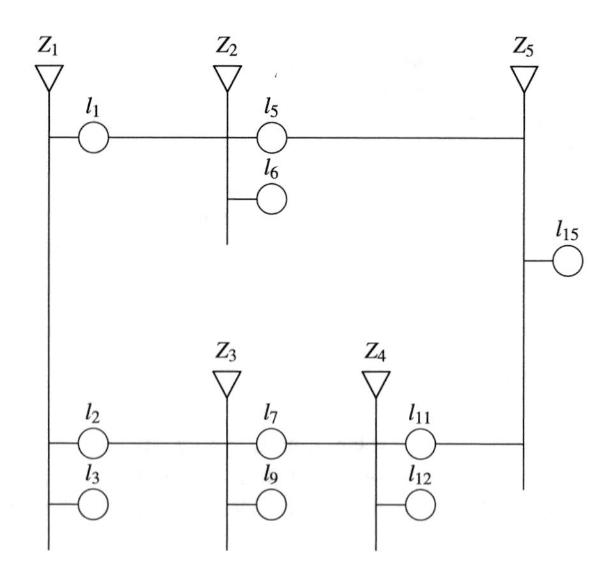

Fig. 3 A TWGN-model

Proof Theorem 5.3.1 in [3], pages 205–210, guarantees that the GNs (and their extensions) are constructed by transitions and if we like to compare the functioning and the results of the work of two different nets - an extended GN and a standard GN, we must show that each transition from the first net functions and gives as results of its work the same results as the part of the second GN, that represents this transition.

Let Z_{TWGN} be an arbitrary transition of the TWGN E_{TWGN}. It is shown on Fig. 1. To it, we juxtapose the subnet, shown on Fig. 4, of a standard GN E. It contains $p+q+2$ transitions $Z_{1,1}$, ..., $Z_{1,p}$, Z, $Z_{2,1}$, ..., $Z_{2,q}$, Z_3. Let us assume that all components of the transition Z_{TWGN} and of the TWGN E_{TWGN} coincide with these from the definitions of the TWGN-transition and of whole TWGN from Sect. 2. We will use them in the description of the E-components.

In places l_1, ..., l_p of Z_{TWGN} can enter tokens from places outside transition Z_{TWGN} and from its places l_{p+1}, ..., l_{p+q} tokens can pass to other transitions. These places are denoted on Fig. 4 by l_1^*, ..., l_{p+q}^*, respectively. So, a pair of places (l_i^*, l_i), $1 \leq i \leq p+q$ in GN E corresponds to place l_i of TWGN E_{TWGN}. We will denote the capacity of the arc between places l_i^* and l_i of E by symbol "∞, because the number of tokens that will transfer from l_i^* to l_i is not known in advance.

For i $(1 \leq i \leq p)$, transition $Z_{1,i}$ has the form:

$$Z_{1,i} = \langle \{l_i^*, m_i, m_{p+q+i}\}, \{l_i\}, \tau, \frac{1}{3}t^0, r_{1,i}, M_{1,i}, \vee(m_i, m_{p+q+i})\rangle,$$

where

$$r_{1,i} = \begin{array}{c|c} & l_i \\ \hline l_i^* & true \\ m_i & true \\ m_{p+q+i} & true \end{array}$$

and

$$M_{1,i} = \begin{array}{c|c} & l_i \\ \hline l_i^* & \infty \\ m_i & 1 \\ m_{p+q+i} & p+q \end{array}.$$

The time-moment τ for firing of transition has two values:

$$\tau = t_1 - \frac{1}{3}t^0$$

when tokens from place l_i^* enter place l_i and

$$\tau = t_1 + \frac{2}{3}t^0$$

when tokens from places m_i or m_{p+q+i} enter place l_i.

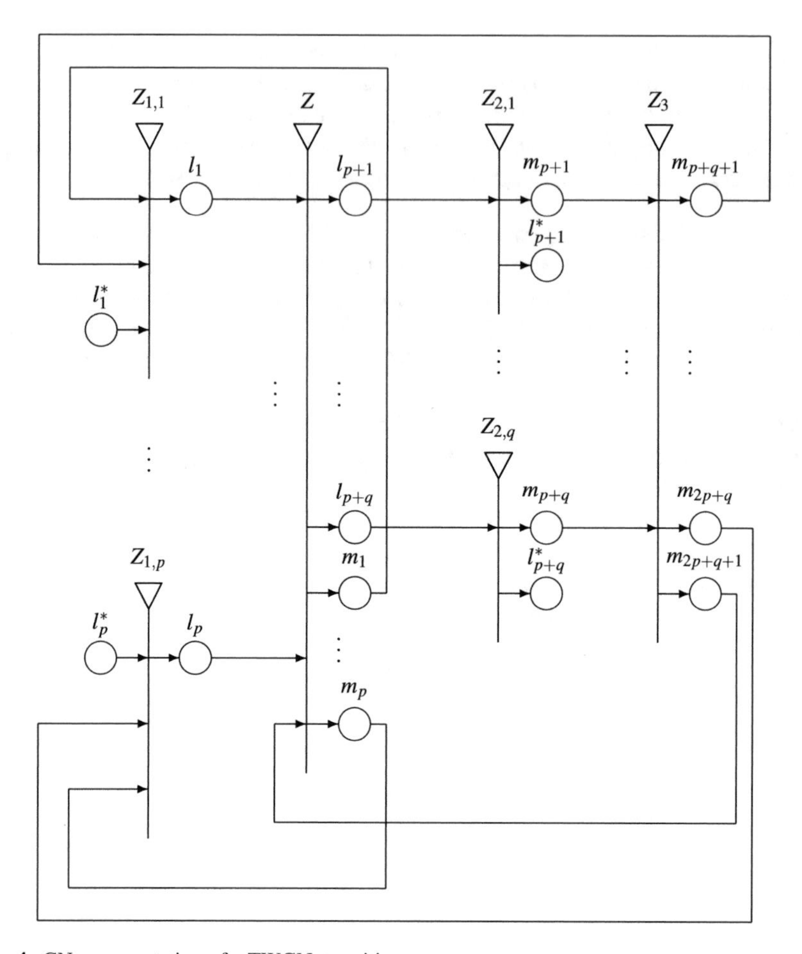

Fig. 4 GN-representation of a TWGN-transition

In the latter case, the tokens enter E's place l_i with the same characteristic as they would obtain in the respective E_{TWGN}'s place.

Let places l_1, ..., l_{p+q} of transition Z have the same characteristic functions, capacities and priorities as the respective places of Z_{TWGN}. We use equal notations for better visualization of the correspondences. Transition Z has the form

$$Z = \langle \{l_1, ..., l_p\}, \{l_{p+1}, ..., l_{p+q}, m_1, ..., m_p\}, t_1, \frac{1}{3}t^0, r, M, \overline{\square_{TWGN}} \rangle,$$

where

$$r = \begin{array}{c|cccccc} & l_{p+1} & \dots & l_{p+q} & m_1 & \dots & m_p \\ \hline l_1 & r_{1,p+1} & \dots & r_{1,p+q} & r_{1,1} & \dots & r_{1,p} \\ \vdots & \vdots & \ddots & \vdots & \vdots & \ddots & \vdots \\ l_p & r_{p,p+1} & \dots & r_{p,p+q} & r_{p,1} & \dots & r_{p,p} \end{array},$$

$$M = \begin{array}{c|cccccc} & l_{p+1} & \dots & l_{p+q} & m_1 & \dots & m_p \\ \hline l_1 & r_{1,p+1} & \dots & r_{1,p+q} & r_{1,1} & \dots & r_{1,p} \\ \vdots & \vdots & \ddots & \vdots & \vdots & \ddots & \vdots \\ l_p & r_{p,p+1} & \dots & r_{p,p+q} & r_{p,1} & \dots & r_{p,p} \end{array},$$

---

and by $\square_{TWGN}$ we denote that from the Boolean expression we have omitted the variables $l_{p+1}, ..., l_{p+q}$.

Tokens from places $l_1, ..., l_p$ enter places $l_{p+1}, ..., l_{p+q}$ with the same characteristic as tokens of TWGN will obtain the same place of E_{TWGN}. On the other hand, if these tokens enter some of places $m_1, ..., m_p$, they will not obtain any characteristic.

For i $(1 \le i \le q)$, transition $Z_{1,p+i}$ has the form:

$$Z_{2,i} = \langle \{l_{p+i}\}, \{m_{p+i}, l^*_{p+i}\}, t_1 + \frac{2}{3}t^0, \frac{1}{3}t^0, r_{2,i}, M_{2,i}, \vee(l_{p+i}) \rangle,$$

where

$$r_{2,i} = \begin{array}{c|cc} & m_{p+i} & l^*_{p+i} \\ \hline l_{p+i} & r_{p+i,1} \vee ... \vee r_{p+i,p+q} & \neg(r_{p+i,1} \vee ... \vee r_{p+i,p+q}) \end{array}$$

and

$$M_{2,i} = \begin{array}{c|cc} & m_{p+i} & l^*_{p+i} \\ \hline l_{p+i} & 1 & \infty \end{array}.$$

Tokens from places l_{p+i} enter places m_{p+i} or l^*_{p+i} without now characteristics.
Finally, transition

$$Z_3 = \langle \{m_{p+1}, ..., m_{p+q}\}, \{m_{p+q+1}, ..., m_{2p+q+1}\}, t_1 + \frac{1}{3}t^0, \frac{1}{3}t^0, r_3, M_3,$$

$$\vee(m_{p+1}, ..., m_{p+q}) \rangle,$$

where

$$r_3 = \begin{array}{c|cccc} & m_{p+q+1} & \dots & m_{2p+q} & m_{2p+q++i} \\ \hline m_{p+1} & r_{p+1,1} & \dots & r_{p+1,p} & r_{p+1,p+1} \vee ... \vee r_{p+1,p+q} \\ \vdots & \vdots & \ddots & \vdots & \vdots \\ m_{p+q} & r_{p+q,1} & \dots & r_{p+q,p} & r_{p+q,p+1} \vee ... \vee r_{p+q,p+q} \end{array}$$

and

$$M_3 = \begin{array}{c|cccc} & m_{p+q+1} & \dots & m_{2p+q} & m_{2p+q++i} \\ \hline m_{p+1} & 1 & \dots & 1 & q \\ \vdots & \vdots & \ddots & \vdots & \vdots \\ m_{p+q} & 1 & \dots & 1 & q \end{array} .$$

Tokens enter the m-places without any characteristic.

The priorities of places l_i^*, m_i, m_{p+i} coincide with that of place l_i for $1 \leq i \leq p+q$.

Now, we will discuss the way of functioning of the new net.

Let us have a token α that in time-moment t_1 will enter place l_i of Z_{TWGN}, $(1 \leq i \leq p)$. In the GN from Fig. 4, the corresponding token α will be in place l_i^*. In moment $t_1 - \frac{1}{3}t^0$, when transition $Z_{1,i}$ will be fired, this token will enter place l_i, obtaining a characteristic that coincides with the characteristic of the corresponding token α in E_{TWGN}.

Token α (in E_{TWGN}) can transfer to each of places l_1, ..., l_{p+q} with respect to the truth-value of predicate $r_{i,j}$ for $1 \leq j \leq p+q$, where it will obtain a next characteristic. Now, in the GN E, the process will flow following one of the next scenaria.

First scenatio: token α from place l_i of E_{TWGN} $(1 \leq i \leq p)$ enters place l_j for $1 \leq j \leq p$. This situation includes the case $j = i$. Now, token α from place l_i enters place m_j in time-moment $t_1 + \frac{1}{3}t^0$, because $r_{i,j} = true$. In a result of this, transition $Z_{1,j}$ will be activated in time-moment $t_1 + \frac{2}{3}t^0$ and token α will enter place l_j obtaining there the same characteristic, as token α will obtain in E_{TWGN}.

Second scenatio: token α from place l_i of E_{TWGN} $(1 \leq i \leq p)$ enters place l_j for $p+1 \leq j \leq p+q$. Now, token α from place l_i of E enters place l_j obtaining there the same characteristic, as token α will obtain in E_{TWGN}.

Third scenatio: token α from place l_i of E_{TWGN} $(p+1 \leq i \leq p+q)$ enters place l_{p+j} for $1 \leq j \leq q$. This situation includes the case $p+j = i$. Now, token α from the corresponding place l_i of E enters place m_{p+j} at time-moment $t_1 + \frac{1}{3}t^0$, because $r_{i,j} = true$ and after this, at time-moment $t_1 + \frac{2}{3}t^0$, it enters place m_{2p+q+1} for the same reason. Finally, at time-moment $t_1 + t^0$ it will enter place l_j with the same characteristic as token α in E_{TWGN}.

Fourth scenatio: token α from place l_i of E_{TWGN} $(p+1 \leq i \leq p+q)$ enters place l_j for $1 \leq j \leq p$. Now, token α from the corresponding place l_i of E enters place m_{p+j} at time-moment $t_1 + \frac{1}{3}t^0$, because $r_{i,j} = true$ and after this, at time-moment $t_1 + \frac{2}{3}t^0$, it enters place m_{p+q+j}. Finally, at time-moment $t_1 + t^0$ it will enter place l_j with the same characteristic as token α in E_{TWGN}.

Therefore, each situation that can arise with token α in E_{TWGN}, will arise with the corresponding token in E and both tokens will obtain equal characteristics. So, Theorem 1 is proved.

Following [3, 5], let us denote by Σ the class of all standard GNs. The fact that all GNs generate a class, but not a set from set-theoretical point of view, is discussed in the cited books.

Let Σ_{TWGN} be the class of all TWGNs.The fact that all TWGNs generate a class, but not a set is proved analogously.

As a corollary of Theorem 1, the validity of the following assertion holds.

Theorem 2 *The class Σ_{TWGN} is a conservative extension of the class Σ.*

5 Conclusion

As we saw from the example in Sect. 3, the TWGN has simpler graphical form than the standard GN. So, the TWGN can be used in complex models with feedbacks. In the area of Data Mining (DM) there are a lot of such processes, for example, the GN-models of the neural networks with back propagation (cf., [12]). In future, it will be interesting, some of these and other models to be re-written in TWGN-form. They can be a basis of new GN-applications, related to modelling, simulation, optimization and control of real processes. In future, we will construct TWGN-models of discrete event systems in theoretical and applied aspects. In near future, we plan to show that the TWGNs are also tools for modelling and simulation of Knowledge Discovery tools and processes.

Acknowledgements This paper has been partially supported by the Bulgarian National Science Fund under the Grant Ref. No. Ref. No. DN 02/10 "New Instruments for Knowledge Discovery from Data and their Modelling".

References

1. Alexieva, J., Choy, E., Koycheva, E.:. Review and bibliography on generalized nets theory and applications. In: Choy, E., Krawczak, M., Shannon, A., Szmidt, E. (eds.) A Survey of Generalized Nets. Raffles KvB Monograph No. 10, 2007, 207–301
2. Atanassov, K.: Generalized index matrices. Compt. Rend. de l'Academie Bulgare des Sci. **40**(11), 15–18 (1987)
3. Atanassov, K.: Generalized Nets. World Scientific, Singapore, New Jersey, London (1991)
4. Atanassov, K.: Two-way generalized nets. Advances in Modelling & Analysis, AMSE Press **19**(2), 29–39 (1994)
5. Atanassov, K.: On Generalized Nets Theory. Prof. M. Drinov Academic Publishing House, Sofia (2007)
6. Atanassov, K.: Index Matrices: Towards an Augmented Matrix Calculus. Springer, Cham (2014)
7. Atanassov, K.: Generalized nets as a tool for the modelling of data mining processes. In: Sgurev, V., Yager, R., Kacprzyk, J., Jotov, V. (eds.) Innovative Issues in Intelligent Systems, pp. 161–215. Springer, Cham (2016)
8. Atanassov, K., Sotirova, E.: Generalized Nets. Drinov Academic Publishing House, Sofia, Prof. M (2017). (in Bulgarian)
9. Fidanova, S., Atanassov, K., Dimov, I.: Generalized Nets as a Tool for Modelling of Railway Networks. In: Georgiev, K., Todorov, M., Georgiev, I. (eds.) Advanced Computing in Industrial Mathematics, pp. 23–35. Springer, Cham (2017)
10. Petri C.-A.: Kommunication mit Automaten. Ph.D. diss., Univ. of Bonn, 1962; Schriften des Inst. fur Instrument. Math., No. 2, Bonn (1962)

11. Radeva, V., Krawczak, M., Choy, E.: Review and bibliography on generalized nets theory and applications. Adv. Stud. Contemp. Math. **4**(2), 173–199 (2002)
12. Sotirov S., Atanassov, K.: *Generalized Nets in Artificial Intelligence, Vol. 6: Generalized Nets and Supervised Neural Networks*. Prof. M. Drinov Academic Publishing House, Sofia (2012)
13. Starke, P.: Petri-Netze. VEB Deutscher Verlag der Wissenschaften, Berlin (1980)
14. Zoteva, D., Krawczak, M.: Generalized nets as a tool for the modelling of data mining processes. A survey. Issues in Intuitionistic Fuzzy Sets and Generalized Nets, vol. 13, pp. 1–60 (2017)

Variational Methods for Stable Time Discretization of First-Order Differential Equations

Simon Becher, Gunar Matthies and Dennis Wenzel

Abstract Starting from the well-known discontinuous Galerkin (dG) and continuous Galerkin-Petrov (cGP) time discretization schemes we derive a general class of variational time discretization methods providing the possibility for higher regularity of the numerical solutions. We show that the constructed methods have the same stability properties as dG or cGP, respectively, making them well-suited for the discretization of stiff systems of differential equations. Additionally, we empirically investigate the order of convergence and performance depending on the chosen method.

1 Introduction

Large systems of stiff ordinary differential equations occur in various fields of application, e.g. after the semi-discretization in space of a parabolic partial differential equation using the finite element method. This gives rise to the need for stable time discretization methods of high order of convergence with low computational costs. In addition, certain applications can demand for regularity of numerical solutions that goes beyond continuity which usually cannot be achieved with classical one-step methods.

Our aim is to construct a new class of methods $\mathrm{G}_r(k)$, $0 \leq r \leq k$, as generalization of the well-known discontinuous Galerkin dG(k) and continuous Galerkin-Petrov cGP(k) methods with the key features *regularity, stability, high order of conver-*

S. Becher · G. Matthies · D. Wenzel (✉)
Technische Universität Dresden, Dresden, Germany
e-mail: dennis.wenzel@tu-dresden.de

S. Becher
e-mail: simon.becher@tu-dresden.de

G. Matthies
e-mail: gunar.matthies@tu-dresden.de

K. Georgiev et al. (eds.), *Advanced Computing in Industrial Mathematics*,
Studies in Computational Intelligence 793,
https://doi.org/10.1007/978-3-319-97277-0_6

gence and *performance*. Here k is the polynomial ansatz order and the parameter r determines the global smoothness of the approximation. We will formulate the methods and verify their properties analytically or empirically when applied to a parabolic model problem.

2 Preliminaries

The underlying problem to be discretized is the, in general only semilinear, initial value problem

$$\begin{aligned} Mu'(t) &= F\big(t, u(t)\big) \qquad \forall t \in (0, T), \\ u(0) &= u_0, \end{aligned} \tag{1}$$

for a right-hand side $F \in C([0, T] \times \mathbb{R}^d, \mathbb{R}^d)$ and $M \in \mathbb{R}^{d\times d}$ regular. Thus, we formally demand that $u \in C^1((0, T), \mathbb{R}^d) \cap C([0, T], \mathbb{R}^d)$. The formulation especially takes into account problems arising from the semi-discretization in spatial variables of a parabolic PDE where a time-constant mass-matrix M can occur as coefficient of the time-derivative u'. Although we could formally multiply by M^{-1} and modify the right-hand side to reach the standard form of a semilinear ODE, we avoid this numerically ill-posed approach and set M to the identity for problems that contain no mass-matrix.

It is well-known that the existence and uniqueness of solutions to (1) generally cannot be guaranteed without further prerequisites. However, we will skip this question and instead restrict ourselves for the numerical experiments to problems that are indeed uniquely solvable, e.g. linear problems with constant coefficients resulting from a semi-discretization of the heat equation with time-independent coefficient functions.

Numerical methods for (1) in general give globally discontinuous solutions that are only locally differentiable. In the presented variational discretization schemes $\mathrm{G}_r(k)$ the regularity of the numerical solution is controlled by the chosen parameter $r \in \mathbb{N}_0$. Of course, in order to achieve higher regularity of the (numerical) solution and for the well-posedness of the method, the right-hand side F has to be sufficiently smooth. In the following we assume a division

$$0 =: t_0 < t_1 < t_2 < \cdots < t_N := T$$

of the time interval and use the notation $I_n := (t_{n-1}, t_n]$ as well as $\tau_n := t_n - t_{n-1}$ for $n = 1, \ldots, N$. The possibly discontinuous numerical solutions will be defined locally on the half-open intervals I_n.

For any piecewise continuous function v given on the interval $(0, T)$ the one-sided limits $v(s^+) := \lim_{t\to s^+} v(t)$ and $v(s^-) := \lim_{t\to s^-} v(t)$ are well-defined for all $s \in (0, T)$. Using this notation we define the jump at s by $[v](s) := v(s^+) - v(s^-)$.

3 Variational Time Discretization Methods

The concept of variational time discretization is based on approaches well-known from the finite element discretization of (partial) differential equations. The numerical solution is locally, i.e. on a certain interval I_n, chosen to be a polynomial of arbitrary but fixed degree $k \in \mathbb{N}_0$. To determine the solution, (local) variational conditions are imposed. Partly also point conditions or collocation conditions are used to couple the intervals. This approach then allows to ensure higher regularity of the solution.

3.1 Discontinuous Galerkin method dG(k)

The discontinuous Galerkin time discretization method dG(k) is well-known and well-studied for a long time. So we omit all details on its derivation and present directly its global formulation:

Find $u_\tau \in L^2([0,T],\mathbb{R}^d)$ with $u_\tau|_{I_n} \in P_k(I_n,\mathbb{R}^d)$, $n = 1,\ldots,N$, such that

$$\sum_{n=1}^{N}\int_{I_n} M u_\tau' v_\tau \,\mathrm{d}t + \sum_{n=1}^{N} M[u_\tau](t_{n-1}) v_\tau(t_{n-1}^+) = \sum_{n=1}^{N}\int_{I_n} F(t,u_\tau) v_\tau \,\mathrm{d}t$$

for all test functions $v_\tau \in L^2([0,T],\mathbb{R})$ with $v_\tau|_{I_n} \in P_k(I_n,\mathbb{R})$, $n = 1,\ldots,N$, where $u_\tau(0^-) = u_0$.

Note that the initial condition $u(0) = u_0$ of (1) is enforced only weakly via the jump $[u_\tau](0)$. Since the test functions v_τ are allowed to be discontinuous at t_n, we can choose v_τ satisfying $v_\tau|_{I_n} \in P_k(I_n,\mathbb{R})$ and $v_\tau|_{[0,T]\setminus I_n} \equiv 0$ for some $n \in \{1,\ldots,N\}$. Hence, the problem localizes and the solution can be computed by a time-marching process where successively only local problems on I_n, $n = 1,\ldots,N$, have to be solved. This leads to the **local problem of dG(k)**:

For given $u_\tau(t_{n-1}^-)$ find $u_\tau|_{I_n} \in P_k(I_n,\mathbb{R}^d)$ such that

$$\int_{I_n} M u_\tau' v_\tau \,\mathrm{d}t + M[u_\tau](t_{n-1}) v_\tau(t_{n-1}^+) = \int_{I_n} F(t,u_\tau) v_\tau \,\mathrm{d}t$$

for all test functions $v_\tau \in P_k(I_n,\mathbb{R})$, where $u_\tau(0^-) = u_0$.

In general it would be difficult to calculate the apparent local integrals exactly, especially if F is nonlinear. Therefore, it is appropriate to use numerical integration. For the dG(k) method the $(k+1)$-point right-sided Gauss-Radau formula, which is well-known to be exact for polynomials up to order $2k$, seems to be suitable. Note that if the right-hand side F of Eq. 1 is affine-linear with a time-independent coefficient, i.e. $F(t, v) = f(t) + Av$, all terms including u_τ are integrated exactly.

For practical reasons the integration points of the $(k + 1)$-point Gauss-Radau formula are reused to serve as degrees of freedom. They define a unique Lagrangian nodal basis of the polynomial (ansatz and test) space $\mathbb{P}_k(I_n, \mathbb{R})$. In view of the local problem it comes in handy that the function value at the right end of each interval is also a quadrature point/degree of freedom since always $u_\tau(t_{n-1}^-)$ is transferred as prior information to the subsequent interval.

3.2 *Continuous Galerkin-Petrov method* cGP(k)

As indicated by its name the continuous Galerkin-Petrov method provides a continuous solution. In order to ensure this, the function value at the beginning of each interval I_n is prescribed by the function value at the end of I_{n-1}. So locally the numerical solution has only k free coefficients and, thus, in contrast to the dG method ansatz and test space are of different order. The **local problem of cGP(k)** is then given by:

For given $u_\tau(t_{n-1}^-)$ find $u_\tau|_{I_n} \in P_k(I_n, \mathbb{R}^d)$ such that $u_\tau(t_{n-1}^+) = u_\tau(t_{n-1}^-)$ and

$$\int_{I_n} M u_\tau' v_\tau \, \mathrm{d}t = \int_{I_n} F(t, u_\tau) v_\tau \, \mathrm{d}t$$

for all test functions $v_\tau \in P_{k-1}(I_n, \mathbb{R})$, where $u_\tau(0^-) = u_0$.

As for the dG method usually numerical integration is applied. Here we choose the $(k + 1)$-point Gauss-Lobatto formula which is exact up to order $2k - 1$ and so again for affine-linear F with time-constant coefficients all solution-dependent terms are integrated exactly. The integration points, i.e. initial and end point of the interval as well as $k - 1$ inner points, again serve as degrees of freedom and determine a nodal basis. Since the function values at the interval boundaries are degrees of freedom, the condition $u_\tau(t_{n-1}^+) = u_\tau(t_{n-1}^-)$ can be expressed easily.

3.3 *General Galerkin method* G$_r(k)$

Earlier we made a wish for a method providing higher regularity of the numerical solution. So we face the need for derivative information at the interval boundaries. For this we use the well-known concept of *collocation equations*.

Definition 1 **(Collocation equation)** Let $s \in [0, T]$ and $U \ni s$ open. Furthermore, let $v \in C^p(U, \mathbb{R}^d)$, $F \in C^{p-1}(U \times \mathbb{R}^d, \mathbb{R}^d)$ for some $p \in \mathbb{N}$ and $M \in \mathbb{R}^{d \times d}$. We say that v fulfills the *collocation equation* of order q for problem (1) at s if

$$Mv^{(q)}(s) = D^{(q-1)}F\big(s, v(s)\big) \tag{2}$$

holds for some $1 \le q \le p$. Here and in the following the notation $D^{(q-1)}F\big(s, v(s)\big)$ is used as abbreviation of $\big(\mathrm{d}^{q-1}/\mathrm{d}t^{q-1}F\big(t, v(t)\big)\big)\,|_{t=s}$.

Extending the approach of the dG and the cGP method, we now combine point and collocation conditions at the boundaries of the intervals with local variational conditions. In order to describe the general Galerkin method $\mathrm{G}_r(k)$ with $r \in \mathbb{N}_0$ we define the quantities

$$D_0 := \left\lfloor \frac{r-1}{2} \right\rfloor \text{ and } D_1 := \left\lfloor \frac{r}{2} \right\rfloor,$$

where $\lfloor z \rfloor$ denotes the largest integer less than or equal to z. If non-negative, D_0 and D_1 represent the number of collocation conditions used at the initial and end point of the underlying interval, respectively. A simple computation shows that $D_0 + D_1 = r - 1$ always holds. Furthermore, when $D_0 \ge 0$ a point condition at the left boundary shall be applied. So, in either case r conditions are stated at the interval boundaries and it remains to impose $k - r + 1$ further conditions, where k is the local ansatz order of the method. Since we are looking for a variational method, thus want to have at least one variational condition, the restriction $r \le k$ is obvious.

For $0 \le r \le k$ the **local problem of $\mathbf{G}_r(k)$** is given as follows:

Find $u_\tau|_{I_n} \in P_k(I_n, \mathbb{R}^d)$ for given $u_\tau(t_{n-1}^-)$ such that

$$\begin{aligned}
u_\tau(t_{n-1}^+) &= u_\tau(t_{n-1}^-), && \text{if } r \ge 1 \Leftrightarrow D_0 \ge 0,\\
Mu_\tau^{(q)}(t_n^-) &= D^{(q-1)}F\big(t_n^-, u_\tau(t_n^-)\big), && \text{for } 1 \le q \le \lfloor \tfrac{r}{2} \rfloor = D_1,\\
Mu_\tau^{(q)}(t_{n-1}^+) &= D^{(q-1)}F\big(t_{n-1}^+, u_\tau(t_{n-1}^+)\big), && \text{for } 1 \le q \le \lfloor \tfrac{r-1}{2} \rfloor = D_0,
\end{aligned}$$

and

$$\int_{I_n} Mu_\tau' v_\tau \,\mathrm{d}t + M[u_\tau](t_{n-1})v_\tau(t_{n-1}^+) = \int_{I_n} F(t, u_\tau)v_\tau \,\mathrm{d}t$$

for all test functions $v_\tau \in P_{k-r}(I_n, \mathbb{R})$, where $u_\tau(0^-) = u_0$.

Note that for $r \ge 1$ the solution is continuous by the point condition. Then the jump term vanishes and could be dropped in the formulation. The above formulation coincides with the method dG(k) for $r = 0$ and cGP(k) for $r = 1$.

Instead of the point and collocation conditions at t_{n-1}^+ we also could demand

$$u_\tau^{(q)}(t_{n-1}^+) = u_\tau^{(q)}(t_{n-1}^-), \qquad \text{for } 0 \le q \le D_0, \tag{3}$$

i.e. explicitly transfer more information from the previous interval. Indeed, if u_τ is continuous and fulfils $Mu_\tau'(s) = F(s, u_\tau(s))$ for $s = t_{n-1}^-$ and t_{n-1}^+, we easily conclude $u_\tau'(t_{n-1}^+) - u_\tau'(t_{n-1}^-) = M^{-1}\left[F\big(t_{n-1}^+, u_\tau'(t_{n-1}^+)\big) - F\big(t_{n-1}^-, u_\tau'(t_{n-1}^-)\big)\right] = 0$. Since $D_1 \ge D_0$ we iteratively get (3). Similarly also the additionally needed derivative information $u_\tau^{(q)}(0)$, $0 \le q \le D_0$, is generated using the collocation condition and u_0. For example $u_\tau'(0)$ can be computed solving the linear system $Mu_\tau'(0) = F(0, u_0)$. From this representation we easily conclude:

Observation 1 ***(Regularity)*** *Let $k, r \in \mathbb{N}_0$ with $r \le k$. Then the solution u_τ of $G_r(k)$ is discontinuous for $r = 0$ and $u_\tau \in C^{D_0}([0, T], \mathbb{R}^d)$ for $r \ge 1$.*

Although the regularity is a straightforward consequence of the construction, we explicitly emphasize the importance of this observation since it is still a key feature of $\mathrm{G}_r(k)$. Most of the classical time discretization methods offer no choice of regularity. In constract the methods $\mathrm{G}_r(k)$ can assure any needed smoothness, given sufficiently smooth data and sufficiently high order k.

Now we want to discuss the issue of numerical integration. Comparing the methods dG(k) and cGP(k) it turns out to be beneficial to use quadrature points and degrees of freedom, respectively, that allow a simple description of the conditions at the interval boundaries.

For our method the values D_0 and D_1 represent the highest order of derivatives used at the initial and end point of an interval, respectively, where the value -1 means that no information is used. Altogether this already gives $D_0 + D_1 + 2 = r + 1$ quadrature conditions at the interval boundary, precisely 0 and 1 in the reference interval $\hat{I} = [0, 1]$, and $k - r$ conditions are left to be chosen. It has been shown in [1] that the quadrature rule is exact of highest possible order, precisely $2k - r$, when the remaining points $\hat{t}_1, \ldots, \hat{t}_{k-r}$ are the roots of the $(k - r)$-th Jacobi polynomial w.r.t. the weight function $(1 + x)^{D_0+1}(1 - x)^{D_1+1}$, transformed to $(0, 1)$.

The combination of derivatives up to order D_0 at 0, function values at $\hat{t}_1, \ldots, \hat{t}_{k-r}$ and derivatives up to order D_1 at 1 then forms a full set of $k + 1$ degrees of freedom on $\hat{I}$. The associated Hermite basis polynomials form a unique nodal basis $\{\hat{\varphi}_0, \ldots, \hat{\varphi}_k\}$ of $\mathbb{P}_k(\hat{I}, \mathbb{R})$ w.r.t. these degrees of freedom.

Defining the quadrature weights $\hat{w}_0, \ldots, \hat{w}_k$ as usual by

$$\hat{w}_j = \int_0^1 \hat{\varphi}_j \,\mathrm{d}t,$$

we obtain an interpolatory quadrature formula $Q_{k,r}$

$$\int_0^1 f \, \mathrm{d}t \approx Q_{k,r}[f] := \sum_{j=0}^{D_0} \hat{w}_j f^{(j)}(0^+) + \sum_{j=1}^{k-r} \hat{w}_{D_0+j} f(\hat{t}_j) + \sum_{j=0}^{D_1} \hat{w}_{D_0+k-r+j+1} f^{(j)}(1^-)$$

which is well-defined for all $f \in C^{D_1}(\hat{I}, \mathbb{R}^d)$. One can easily verify that the previously used methods Gauss-Radau and Gauss-Lobatto are special cases of $Q_{k,r}$ for $r = 0$ and $r = 1$, respectively.

Since the exactness of the quadrature rule $Q_{k,r}$ is $2k - r$, again all term in the variational condition including u_τ are integrated exactly, when F is affine-linear with a time-constant coefficient.

The above findings shall be summarized in an alternative representation of the method. Having defined the basis functions $\{\hat{\varphi}_0, \ldots, \hat{\varphi}_k\}$ of $\mathbb{P}_k(\hat{I}, \mathbb{R})$, we can use the affine transformation $T_n : [0, 1] \to \overline{I}_n$, $\hat{t} \mapsto \tau_n \hat{t} + t_{n-1}$ and if necessary a suitable rescaling to define a local nodal basis $\varphi_0^n, \ldots, \varphi_k^n$ on each interval I_n. Then $u_\tau|_{I_n} \in P_k(I_n, \mathbb{R}^d)$ can locally be expressed as linear combination

$$u_\tau \Big|_{I_n} = \sum_{j=0}^{k} U_j^n \varphi_j^n$$

with coefficients $U_j^n \in \mathbb{R}^d$.

Recalling (3), the first $D_0 + 1$ coefficients in each interval, namely $U_j^n = u_\tau^{(j)}(t_{n-1}^+)$ with $j = 0, \ldots, D_0$, can be transferred from the previous interval I_{n-1} given the fact that $D_1 \geq D_0$ holds for all r. In order to enable this in the first interval, we formally initialize $U_{k-D_1+j}^0 := u_\tau^{(j)}(0)$ for $j = 0, \ldots, D_0$, where the derivative information $u_\tau^{(j)}(0)$ is iteratively computed as described above using collocation conditions and the initial value u_0.

In addition, we used the concept of collocation equations to couple the function value $u_\tau(t_n^-)$ expressed in the last variational equation with the derivatives $u_\tau^{(1)}(t_n^-), \ldots, u_\tau^{(D_1)}(t_n^-)$ at the end point of I_n. Practically this means that we add the collocation equations of order up to D_1 at t_n to our local system on I_n. Due to the choice of the degrees of freedom we have that $U_{k-D_1+q}^n = u_\tau^{(q)}(t_n^-)$ for $q = 0 \ldots, D_1$ and, thus, the collocation equations have a special form.

As we have seen, transferring $D_0 + 1$ coefficients from the previous interval and using D_1 collocation equations at the interval end point, the appropriate number of variational equations is $k + 1 - (D_0 + 1 + D_1) = k - r + 1$. So we choose test functions $\psi_0^n, \ldots, \psi_{k-r}^n$ such that they locally form a basis of $\mathbb{P}_{k-r}(I_n, \mathbb{R})$.
Given this, the **local problem of $\mathbf{G}_r(k)$** is given as follows:

Set $U_j^n := U_{k-D_1+j}^{n-1}$ for $j = 0, \dots, D_0$ and find U_j^n, $j = D_0+1, \dots, k$, such that

$$MU_{k-D_1+q}^n = D^{(q-1)} F\big(t_n^-, \sum\nolimits_{j=0}^k U_j^n \varphi_j^n(t_n^-)\big)$$

for $q = 1, \dots, D_1$ and

$$\sum_{j=0}^k \int_{I_n} MU_j^n (\partial_t \varphi_j^n) \psi_i^n \,\mathrm{d}t + \mathrm{J}_i^n = \int_{I_n} F\big(t, \sum\nolimits_{j=0}^k U_j^n \varphi_j^n\big) \psi_i^n \,\mathrm{d}t$$

for $i = 0, \dots, k-r$, where $\mathrm{J}_i^n := M\big[\sum_{j=0}^k U_j^n \varphi_j^n(t_{n-1}^+) - U_{k-D_1}^{n-1}\big]\psi_i^n(t_{n-1}^+)$.

Note that for $r = 0$ ansatz and test space coincide and we could choose $\psi_i^n = \varphi_i^n$. Furthermore, for $r \geq 1$ all the jump terms J_i^n vanish and so can be dropped. Of course, instead of integrating exactly we could also use appropriate quadrature formulae, e.g. $Q_{k,r}$.

3.4 Stability of $\mathrm{G}_r(k)$

It is well-known that numerical stability of time discretization schemes is crucial for the behaviour when applied to stiff problems, resulting for example from the semi-discretization in space of a parabolic differential equation. In the following part we will investigate this issue in order to prove that the class $\mathrm{G}_r(k)$ in general has suitable stability properties.

Theorem 1 *Let $k, r \in \mathbb{N}_0$ with $r \leq k$.*
Then the methods $G_r(k)$ and $G_{r+2}(k+1)$ have the same stability properties.

Proof ([4])
We observe that the stability properties of the different Galerkin methods are strongly related. For every $k, r \in \mathbb{N}_0$ with $r \leq k$ let $\tilde{k} = k - \lfloor \frac{r}{2} \rfloor$. Then the above theorem implies that $\mathrm{G}_r(k)$ has the same stability properties as $\mathrm{G}_0(\tilde{k})$=dG($\tilde{k}$) if r is even or $\mathrm{G}_1(\tilde{k}) =$ cGP($\tilde{k}$) if r is odd. On the other hand it is well-known that both dG(k) and cGP(k) are A-stable for all $k \in \mathbb{N}$, see e.g. [3].

Corollary 1 ***(Stability)*** *The method $G_r(k)$ is A-stable for all $k, r \in \mathbb{N}_0$ with $r \leq k$.*

Given the stability for the general variational time discretization method $\mathrm{G}_r(k)$ we can conclude that the methods are well-suited for stiff differential equations. Recalling that $\mathrm{G}_r(k)$ provides a C^{D_0} solution, where $D_0 = \lfloor \frac{r-1}{2} \rfloor$, in the following we will use the notation dG-C$D_0(k)$ or cGP-C$D_0(k)$ when the method inherits the stability properties of dG or cGP, respectively. One can easily verify that the notation is well-defined, i.e. for every notation of this type there exists a unique method $\mathrm{G}_r(k)$.

4 Implementation and numerical experiments

The only issue left to discuss is the order of convergence which we shall address empirically by applying our methods to a certain model problem. Firstly, we want to give a few remarks concerning the implementational aspect.

All basis functions φ_j^n and ψ_i^n occuring in the local systems on some interval I_n are defined by transforming functions $\hat{\varphi}_j$ and $\hat{\psi}_i$ from the reference configuration $\hat{I}$ to I_n and rescaling them where required. The quadrature rules $Q_{k,r}$ are also defined on $\hat{I}$ and so we compute the information necessary for the variational equations in a preprocessing step, thus significantly increase the performance for the assembling of the local systems.

Let w be a piecewise continuous function and define

$$||w||_0 := \int_0^T ||w(t)||_{L^2}\,\mathrm{d}t,$$
$$||w||_{\max} := \max_{n=0,\dots,l} ||w(t_n^-)||_{L^2}.$$

The norm $||\cdot||_0$ can be used as a mixed L^1-L^2 error for time-dependent problems while $||\cdot||_{\max}$ represents the maximum error in the discrete time steps.

We have implemented the methods $\mathrm{G}_r(k)$ as time discretization plugin for the finite element software **SOFE** for MATLAB/Octave, see [2]. All computations were done with the following configuration:

- CPU: Intel Core i7-6500U, 4 × 2.50 GHz
- RAM: 8 GB
- Graphics: Intel HD Graphics 520 (Skylake GT2)
- OS: Ubuntu 16.10
- IDE: MATLAB R2016a

As model problem we consider the parabolic equation $u_t(t,x) - u_{xx}(t,x) = f(t,x)$ in the domain $(0,1)\times(0,1)$. We assign homogeneous Neumann boundary conditions and the initial condition $u(0,x) = x^2(1-x)^2$. Setting

$$f(t,x) = -20x^2(1-x)^2\sin(20t) - (2 - 12x + 12x^2)\cos(20t),$$

one can easily verify that the exact solution of this problems is given by

$$u^*(t,x) = x^2(1-x)^2\cos(20t).$$

For the discretization in space we locally use polynomials of order 4, i.e. the term $x^2(1-x)^2$ will be reproduced exactly. Therefore, influences of the space discretization error on the order of convergence are eliminated.

We start by investigating the influence of the polynomial degree on the order of convergence. For this we have done simulations with different methods and for each method we have measured the convergence for different values of k.

We can observe in Fig. 1 that the error w.r.t. $||\cdot||_0$ shows a behaviour that is similar to standard L^2 error estimates for finite element approximation, i.e. convergence of order $k+1$ without any dependence on the given method. For the error w.r.t. $||\cdot||_{\max}$ the results in Fig. 2 show super-convergence but there is a dependence on the chosen method that is not clear from the given examples. To clarify this we will do additional tests for $||\cdot||_{\max}$ with constant k and varying r.

In review of Fig. 3 we can conclude that the super-convergence effects reduce with increasing smoothness of the numerical solution. The first 3 tests show examples where we have altered the parameter r by 1 while the last example shows the two methods with $r=3$ and $r=5$ and thus a difference of 2. This indicates a super-convergence behaviour for $||\cdot||_{\max}$ of order $2k+1-r$.

Observation 2 ***(Convergence)*** *Let $k, r \in \mathbb{N}_0$ with $r \leq k$. Let u^* be the exact solution and u_τ be the solution of $G_r(k)$ on a time grid with maximal step-size τ.*
Then the numerical experiments suggest

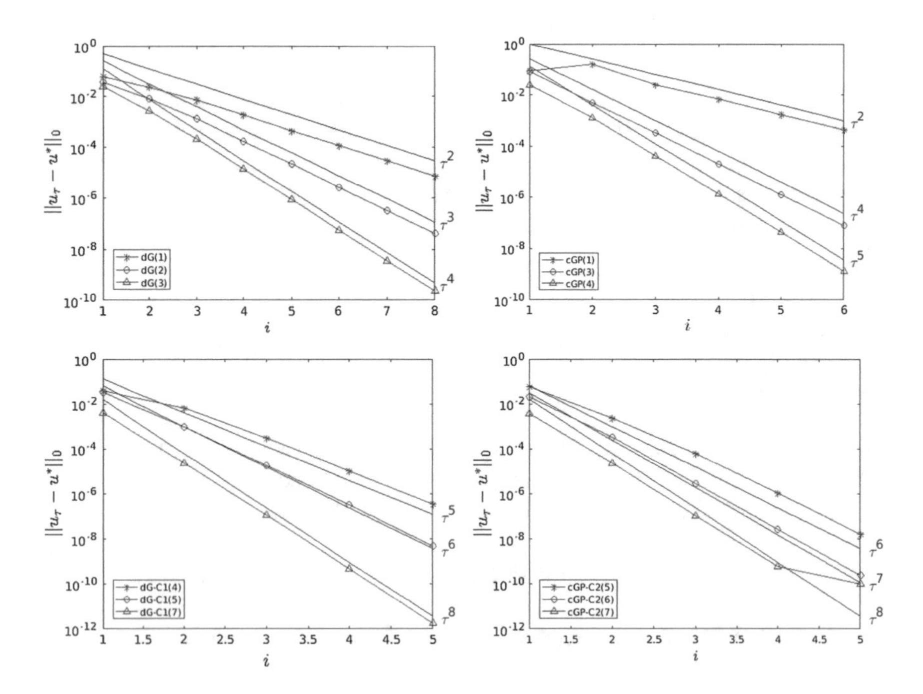

Fig. 1 Convergence w.r.t. $||\cdot||_0$ depending on order k

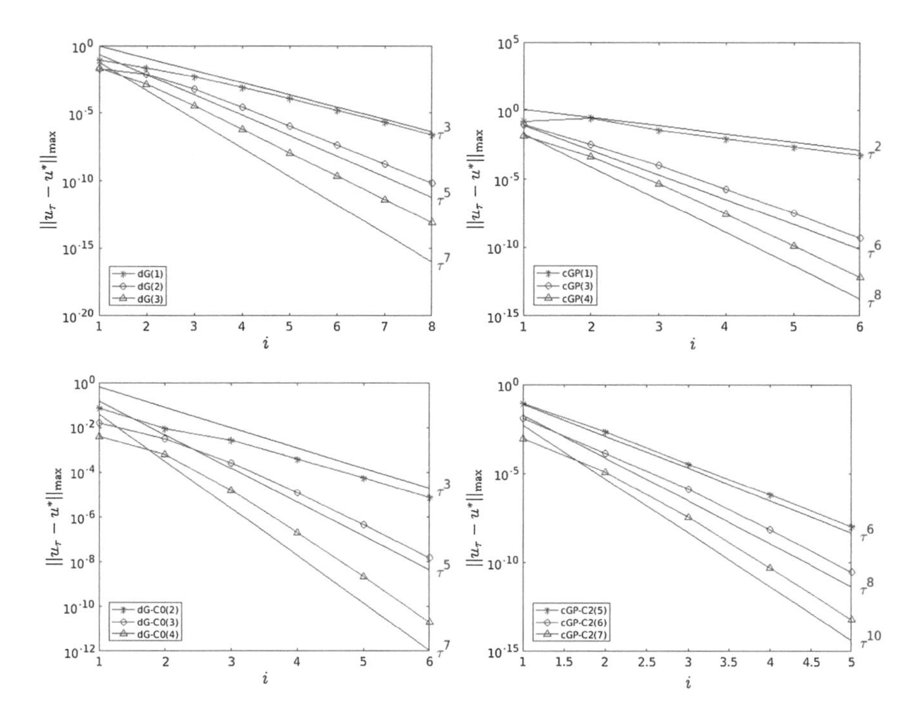

Fig. 2 Convergence w.r.t. $||\cdot||_{\max}$ depending order k

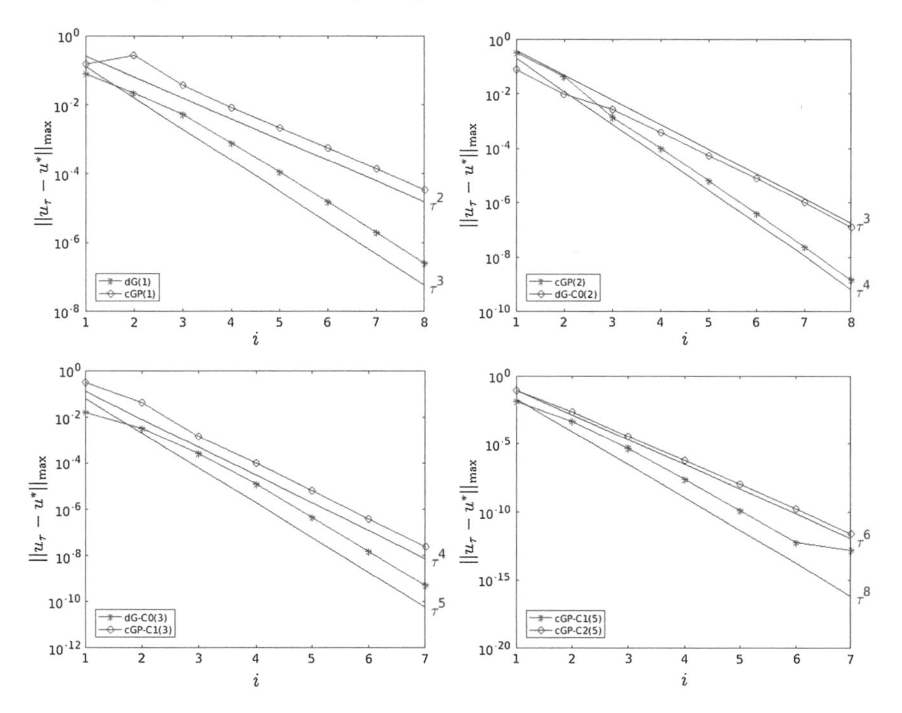

Fig. 3 Super-convergence for $||\cdot||_{\max}$ depending on the method

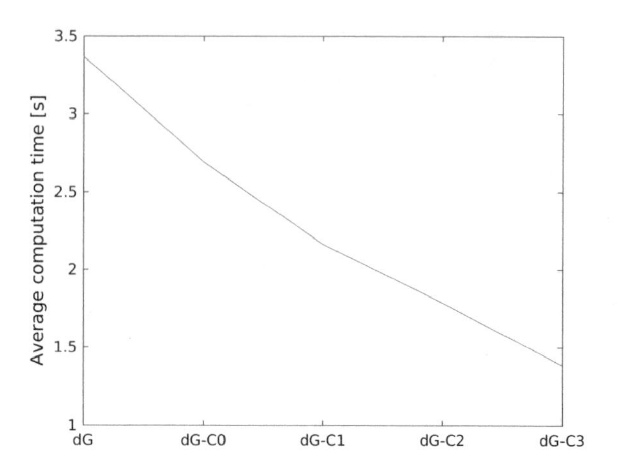

Fig. 4 Average computation time for different dG-type methods of order $k = 10$

$$||u^* - u_\tau||_0 \leq C_0 \tau^{k+1},$$
$$||u^* - u_\tau||_{\max} \leq C_{\max} \tau^{2k+1-r},$$

for constants C_0, $C_{\max} > 0$ *independent on* τ.

The last aspect we want to investigate is the influence of the chosen method on the performance. From the construction of $\mathrm{G}_r(k)$ it follows that for increasing r the number of free coefficients in each interval I_n reduces, indicating reduced computational effort. For this test we fix $k = 10$ and investigate different dG-type methods. We also fix the number of time steps to 32 and artificially increase the general computational effort by using space discretization size $h = 10^{-2}$ to get statistically more significant results. Every computation is performed 5 times and the average computation time is shown in Fig. 4 indicating that the performance indeed increases depending on the smoothness parameter r.

5 Conclusions and Outlook

In this paper we have presented the class $\mathrm{G}_r(k)$ of variational time discretization methods derived as a generalization of the well-known methods dG(k) and cGP(k). We have proven A-stability for all methods and certain regularity of the numerical solution depending on the parameter r. Furthermore, we have empirically investigated the order of convergence for the class $\mathrm{G}_r(k)$ with respect to both a L^1-L^2 mixed error and the maximum error in the discrete time points. The results have shown that the first one depends only on the polynomial degree k while the latter one shows certain super-convergence depending also on the method parameter r. We could observe that upon increased regularity the performance of the methods increases, too.
Future work will include a theoretical investigation of the convergence properties of $\mathrm{G}_r(k)$ with the aim to analytically prove the empirical results.

References

1. Joulak, H., Beckermann, B.: On Gautschi's conjecture for generalized Gauss-Radau and Gauss-Lobatto formulae. J. Comput. Appl. Math. **233**(3), 768–774 (2009)
2. Lars Ludwig. **SOFE**. http://www.math.tu-dresden.de/~ludwigl/, 2017. (Online; Accessed 30 Jan 2018)
3. Matthies, G., Schieweck, F.: Higher order variational time discretizations for nonlinear systems of ordinary differential equations. Preprint 23/2011, Fakultät für Mathematik, Otto-von-Guericke-Universität Magdeburg (2011)
4. Wenzel, D.: Variational Methods for Time Discretization of Parabolic Differential Equations (in german). Master's thesis, Technische Universität Dresden (2017)

Numerical Solutions of Ordinary Fractional Differential Equations with Singularities

Yuri Dimitrov, Ivan Dimov and Venelin Todorov

Abstract The solutions of fractional differential equations (FDEs) have a natural singularity at the initial point. The accuracy of their numerical solutions is lower than the accuracy of the numerical solutions of FDEs whose solutions are differentiable functions. In the present paper we propose a method for improving the accuracy of the numerical solutions of ordinary linear FDEs with constant coefficients which uses the fractional Taylor polynomials of the solutions. The numerical solutions of the two-term and three-term FDEs are studied in the paper.

1 Introduction

In recent years there is a growing interest in applying FDEs for modeling diffusion processes in biology, engineering and finance [1, 2]. The two main approaches to fractional differentiation are the Caputo and Riemann-Liouville fractional derivatives. The Caputo derivative of order α, where $0 < \alpha < 1$ is defined as

Y. Dimitrov (✉)
Department of Mathematics and Physics, University of Forestry, 1756 Sofia, Bulgaria
e-mail: yuri.dimitrov@ltu.bg

I. Dimov · V. Todorov
Institute of Information and Communication Technologies,
Bulgarian Academy of Sciences, Sofia, Bulgaria
e-mail: ivdimov@bas.bg

I. Dimov · V. Todorov
Department of Parallel Algorithms, Acad. Georgi Bonchev Str.,
Block 25 A, 1113 Sofia, Bulgaria

V. Todorov
Institute of Mathematics and Informatics, Bulgarian Academy of Sciences, Department of Information Modeling, Acad. Georgi Bonchev Str., Block 8, 1113 Sofia, Bulgaria
e-mail: vtodorov@math.bas.bg; venelin@parallel.bas.bg

K. Georgiev et al. (eds.), *Advanced Computing in Industrial Mathematics*,
Studies in Computational Intelligence 793,
https://doi.org/10.1007/978-3-319-97277-0_7

$$D_C^{\alpha} y(t) = y^{(\alpha)}(t) = \frac{d^{\alpha}}{dt^{\alpha}} y(t) = \frac{1}{\Gamma(1-\alpha)} \int_0^t \frac{y'(\xi)}{(t-\xi)^{\alpha}} d\xi.$$

The Caputo derivative is a standard choice for a fractional derivative in the models using FDEs. The finite difference schemes for numerical solution of FDEs involve approximations of the fractional derivative. Let $t_n = nh$, where h is a small positive number and $y_n = y(t_n) = y(nh)$. The L1 approximation is an important and commonly used approximation of the Caputo derivative.

$$y_n^{(\alpha)} = \frac{1}{h^{\alpha}\Gamma(2-\alpha)} \sum_{k=0}^{n} \sigma_k^{(\alpha)} y_{n-k} + O\left(h^{2-\alpha}\right), \tag{1}$$

where $\sigma_0^{(\alpha)} = 1,\ \sigma_n^{(\alpha)} = (n-1)^{1-\alpha} - n^{1-\alpha}$ and

$$\sigma_k^{(\alpha)} = (k+1)^{1-\alpha} - 2k^{1-\alpha} + (k-1)^{1-\alpha}, \quad (k = 1, \cdots, n-1).$$

In [3] we obtain the second-order approximation the Caputo derivative

$$y_n^{(\alpha)} = \frac{1}{\Gamma(2-\alpha)h^{\alpha}} \sum_{k=0}^{n} \delta_k^{(\alpha)} y_{n-k} + O\left(h^2\right), \tag{2}$$

where $\delta_k^{(\alpha)} = \sigma_k^{(\alpha)}$ for $2 \leq k \leq n$ and

$$\delta_0^{(\alpha)} = \sigma_0^{(\alpha)} - \zeta(\alpha-1), \delta_1^{(\alpha)} = \sigma_1^{(\alpha)} + 2\zeta(\alpha-1), \delta_2^{(\alpha)} = \sigma_2^{(\alpha)} - \zeta(\alpha-1).$$

When $0 < \alpha < 1$ and the function $y \in C^2[0, t_n]$, the L1 approximation has an accuracy $O\left(h^{2-\alpha}\right)$ [4] and approximation (2) has an accuracy $O\left(h^2\right)$ [3]. The zeta function satisfies $\zeta(0) = -1/2$. From (2) with $\alpha = 1$ we obtain the second-order approximation for the first derivative

$$y'_n = \frac{1}{h}\left(\frac{3}{2}y_n - 2y_{n-1} + y_{n-2}\right) + O\left(h^2\right). \tag{3}$$

The Caputo derivative of order α, where $1 < \alpha < 2$ is defined as

$$D_C^{\alpha} y(t) = y^{(\alpha)}(t) = \frac{d^{\alpha}}{dt^{\alpha}} y(t) = \frac{1}{\Gamma(2-\alpha)} \int_0^t \frac{y''(\xi)}{(t-\xi)^{\alpha-1}} d\xi.$$

In [5] we obtain the expansion formula of order $4-\alpha$ of the L1 approximation of the Caputo derivative. When $0 < \alpha < 2$, approximation (2) has an asymptotic expansion

$$\frac{1}{\Gamma(2-\alpha)h^{\alpha}}\sum_{k=0}^{n}\delta_k^{(\alpha)}y_{n-k}=y_n^{(\alpha)}+\left(D^2y_n^{(\alpha)}-\frac{y_0'}{\Gamma(-\alpha)t^{1+\alpha}}\right)\frac{h^2}{12}-$$
$$\frac{\zeta(\alpha-1)+\zeta(\alpha-2)}{\Gamma(2-\alpha)}y_n'''h^{3-\alpha}+O\left(h^{\min\{3,4-\alpha\}}\right).$$

When $1<\alpha<2$ approximation (2) has an accuracy $O\left(h^{3-\alpha}\right)$. In [6] we obtain the asymptotic expansion formula

$$\frac{1}{\Gamma(-\alpha)h^{\alpha}}\left(\sum_{k=1}^{n-1}\frac{y_{n-k}}{k^{1+\alpha}}-\zeta(1+\alpha)y_n\right)=y_n^{(\alpha)}-\frac{\zeta(\alpha)}{\Gamma(-\alpha)}y_n'h^{1-\alpha}$$
$$+\frac{\zeta(\alpha-1)}{2\Gamma(-\alpha)}y_n''h^{2-\alpha}+O\left(h^{3-\alpha}\right),$$

and an approximation of the Caputo derivative

$$\frac{1}{\Gamma(-\alpha)h^{\alpha}}\sum_{k=0}^{n}\gamma_k^{(\alpha)}z_{n-k}=z_n^{(\alpha)}+O\left(h^{3-\alpha}\right),\tag{4}$$

where $\gamma_k^{(\alpha)}=1/k^{1+\alpha}$, $(k\geq 3)$ and $\gamma_0^{(\alpha)}=-\zeta(1+\alpha)+\frac{3}{2}\zeta(\alpha)-\frac{1}{2}\zeta(\alpha-1)$,

$$\gamma_1^{(\alpha)}=1-2\zeta(\alpha)+\zeta(\alpha-1),\ \gamma_2^{(\alpha)}=\frac{1}{2^{1+\alpha}}+\frac{1}{2}\zeta(\alpha)-\frac{1}{2}\zeta(\alpha-1).$$

Approximation (4) has an accuracy $O\left(h^{3-\alpha}\right)$ when the function $z\in C^3[0,t_n]$ and satisfies $z(0)=z'(0)=z''(0)=0$.

The Miller-Ross sequential derivative for the Caputo derivative of order $n\alpha$ is defined as

$$y^{[n\alpha]}(t)=D^{n\alpha}y(t)=D_C^{\alpha}D_C^{\alpha}\cdots D_C^{\alpha}y(t).$$

The Caputo and Miller-Ross derivatives of order 2α of the function $y(t)=1+t^{\alpha}$ satisfy $y^{[2\alpha]}(t)=0$ and $y^{(2\alpha)}(t)=\Gamma(1+\alpha)t^{-\alpha}/\Gamma(1-\alpha)$. The fractional Taylor polynomials of the function $y(t)$ are defined as

$$T_m^{(\alpha)}(t)=\sum_{k=0}^{m}\frac{y^{[k\alpha]}(0)t^{\alpha k}}{\Gamma(\alpha k+1)}.$$

The fractional Taylor polynomials $T_m^{(\alpha)}(t)$ (polyfractonomials) are defined at the initial point of fractioanal differentiation $t=0$ and are polynomials with respect to t^{α}. An important class of special functions in fractional calculus are the one-parameter and two-parameter Mittag-Leffler functions

$$E_{\alpha}(t)=\sum_{n=0}^{\infty}\frac{t^n}{\Gamma(\alpha n+1)},\quad E_{\alpha,\beta}(t)=\sum_{n=0}^{\infty}\frac{t^n}{\Gamma(\alpha n+\beta)}.$$

The Mittag-Leffler functions generalize the exponential function and appear in the analytical solutions of fractional and integer-order differential equations. The Miller-Ross derivatives of the function $E_{\alpha}\left(t^{\alpha}\right)$ satisfy

$$D^{n\alpha}E_{\alpha}\left(t^{\alpha}\right)=E_{\alpha}\left(t^{\alpha}\right),\quad D^{n\alpha}E_{\alpha}\left(t^{\alpha}\right)\Big|_{t=0}=E_{\alpha}\left(0\right)=1.$$

The two-term equation is called fractional relaxation equation, when $0<\alpha<1$.

$$y^{(\alpha)}(t)+By(t)=F(t),\ y(0)=y_0. \tag{5}$$

The exact solution of Eq. (5) is expressed with the Mittag-Leffler functions as

$$y(t)=y_0E_{\alpha}(-Bt^{\alpha})+\int_0^t\xi^{\alpha-1}E_{\alpha,\alpha}\left(-B\xi^{\alpha}\right)F(t-\xi)d\xi. \tag{6}$$

When the solution of the two-term equation $y\in C^3[0,T]$, the numerical solutions which use approximations (1), (2) and (4) of the Caputo derivative have accuracy $O\left(h^{2-\alpha}\right)$, $O\left(h^2\right)$ and $O\left(h^{3-\alpha}\right)$. The numerical solutions of FDEs which have smooth solutions have been studied extensively in the last three decades. The finite difference schemes involve approximations of the Caputo derivative related to the constructions of the L1 approximation and the Grünwald-Letnikov difference approximation [7–11]. When $F(t)=0$ the two-term Eq. (5) has the solution $y(t)=E_{\alpha}(-Bt^{\alpha})$. When $0<\alpha<1$, the function $E_{\alpha}(-Bt^{\alpha})$ has a singularity at the initial point, because its derivatives are unbounded at $t=0$. This property holds for most ordinary and partial FDEs. The singularity of the solutions of FDEs adds a significant difficulty to the construction of high-order numerical solutions [12–15]. In the present paper we propose a method for transforming linear FDEs with constant coeeficients into FDEs whose solutions are smooth functions. In Sects. 2 and 3 the method is applied for computing the numerical solutions of the two-term and the three-term FDEs.

2 Two-Term FDE

The numerical and analytical solutions of the two-term ordinary FDE are studied in [5, 16–21]. Let N be a positive integer and $h=T/N$. Suppose that

$$\frac{1}{h^{\alpha}}\sum_{k=0}^{n}\lambda_k^{(\alpha)}y_{n-k}=y_n^{(\alpha)}+O\left(h^{\beta(\alpha)}\right) \tag{$*$}$$

is an approximatiom of the Caputo derivative of order $\beta(\alpha)$. Now we drive the numerical solution of two-term Eq. (5), which uses approximation (*) of the Caputo derivative. By approximating the Caputo derivative of Eq. (5) at the point $t_n = nh$ we obtain

$$\frac{1}{h^\alpha}\sum_{k=0}^{n-1}\lambda_k^{(\alpha)} y_{n-k} + By_n = F_n + O\left(h^{\beta(\alpha)}\right).$$

The numerical solution $\{u_n\}_{n=0}^N$ of Eq. (5) is computed with $u_0 = 1$ and

$$\frac{1}{h^\alpha}\sum_{k=0}^{n-1}\lambda_k^{(\alpha)} u_{n-k} + Bu_n = F_n,$$

$$u_n = \frac{1}{\lambda_0^{(\alpha)} + Bh^\alpha}\left(h^\alpha F_n - \sum_{k=1}^{n}\lambda_k^{(\alpha)} u_{n-k}\right). \qquad \text{NS1}(*)$$

When the solution of the two-term equation $y \in C^3[0, T]$, numerical solutions NS1(1), NS1(2), NS1(4) have accuracy $O\left(h^{2-\alpha}\right), O\left(h^2\right), O\left(h^{3-\alpha}\right)$. When the solution of Eq. (5) satisfies $y(0) = y'(0) = y''(0)$ we can choose the initial values of the numerical solution $u_0 = u_1 = u_2 = 0$. The two-term equation

$$y^{(\alpha)}(t) + By(t) = 0, \quad y(0) = 1 \qquad (7)$$

has the solution $y(t) = E_\alpha(-Bt^\alpha)$. When $0 < \alpha < 1$ the first derivative of the solution is undefined at $t = 0$. Numerical solutions NS1(1) and NS1(2) of Eq. (7) have accuracy $O\left(h^\alpha\right)$ [19]. The numerical results for the error and order of numerical solution NS1(1) with $\alpha = 0.3, B = 1$ and $\alpha = 0.5, B = 2$ and NS1(2) with $\alpha = 0.7, B = 3$ on the interval [0, 1] are presented in Table 1. The errors of the numerical methods in Table 1 and the rest of the tables in the paper are computed with respect to the natural (maximum) l_∞ norm. Now we transform Eq. (7) into a two-term equation whose solution has a continuous second derivative. From (7) with $t = 0$ we obtain $y^{(\alpha)}(0) = -By(0) = -B$. By applying fractional differentiation of order α we obtain

$$y^{[2\alpha]}(t) + By^{(\alpha)}(t) = 0, \qquad y^{[2\alpha]}(0) = -By^{(\alpha)}(0) = B^2,$$
$$y^{[3\alpha]}(t) + By^{[2\alpha]}(t) = 0, \qquad y^{[3\alpha]}(0) = -By^{[2\alpha]}(0) = -B^3.$$

By induction we obtain the Miller-Ross derivatives of the solution of Eq. (7) at the initial point $t = 0$:

$$y^{[n\alpha]}(0) = -By^{[(n-1)\alpha]}(0) = (-B)^n.$$

Table 1 Maximum error and order of numerical solutions NS1(1) of Eq. (7) with $\alpha = 0.3, \alpha = 0.5$ and NS1(2) with $\alpha = 0.7$ of order α

h	$\alpha = 0.3, B = 1$		$\alpha = 0.5, B = 2$		$\alpha = 0.7, B = 3$	
	Error	Order	Error	Order	Error	Order
0.00625	0.3344×10^{-1}	0.2089	0.9063×10^{-1}	0.3987	0.6385×10^{-1}	0.6332
0.003125	0.2863×10^{-1}	0.2242	0.6743×10^{-1}	0.4265	0.4046×10^{-1}	0.6583
0.0015625	0.2429×10^{-1}	0.2372	0.4946×10^{-1}	0.4471	0.2536×10^{-1}	0.6741
0.00078125	0.2045×10^{-1}	0.2482	0.3591×10^{-1}	0.4621	0.1578×10^{-1}	0.6840

Substitute

$$z(t) = y(t) - T_m^{(\alpha)}(t) = y(t) - \sum_{n=0}^{m} \frac{(-Bt^\alpha)^n}{\Gamma(\alpha n + 1)}.$$

The function $z(t)$ has a Caputo derivative of order α

$$z^{(\alpha)}(t) = y^{(\alpha)}(t) + B \sum_{n=0}^{m-1} \frac{(-Bt^\alpha)^n}{\Gamma(\alpha n + 1)},$$

and satisfies the two-term equation

$$z^{(\alpha)}(t) + Bz(t) = \frac{(-B)^{m+1} t^{\alpha m}}{\Gamma(\alpha m + 1)}, \quad z(0) = 0. \tag{8}$$

Now we use the uniform limit theorem to show that when $m\alpha > 2$ the solution of Eq. (8) is a twice continuously differentiable function.

Lemma 1 *Let* $\epsilon > 0, T > 0$ *and*

$$M > \max \left\{ \frac{1}{\alpha} \left(2B^{1/\alpha} Te + 2 - \alpha \right), \frac{2}{\alpha} - 1 + \log_2 \left(\frac{2B^{2/\alpha}}{\epsilon} \right) \right\}.$$

Then

$$\left(\frac{B^{1/\alpha} Te}{\alpha M + \alpha - 2} \right)^\alpha < \frac{1}{2} \quad \textit{and} \quad 2B^{2/\alpha} \left(\frac{B^{1/\alpha} Te}{\alpha M + \alpha - 2} \right)^{\alpha M + \alpha - 2} < \epsilon.$$

Proof

$$\left(\frac{B^{1/\alpha} Te}{\alpha M + \alpha - 2} \right)^\alpha < \frac{1}{2} \Leftrightarrow \frac{B^{1/\alpha} Te}{\alpha M + \alpha - 2} < \frac{1}{2^{1/\alpha}} \Leftrightarrow \alpha M + \alpha - 2 > (2B)^{1/\alpha} Te.$$

The first inequality is satisfied when

$$M > \frac{1}{\alpha}\left(2B^{1/a}Te + 2 - \alpha\right).$$

We have that

$$\left(\frac{B^{1/\alpha}Te}{\alpha M + \alpha - 2}\right)^{\alpha M + \alpha - 2} = \left(\frac{B^{1/\alpha}Te}{\alpha M + \alpha - 2}\right)^{\alpha(M+1-2/\alpha)} < \frac{1}{2^{M+1-2/\alpha}} < \frac{\epsilon}{2B^{2/\alpha}},$$

$$2^{M+1-2/\alpha} > \frac{2B^{2/\alpha}}{\epsilon}.$$

The second inequality is satisfied for

$$M > \frac{2}{\alpha} - 1 + \log_2\left(\frac{2B^{2/\alpha}}{\epsilon}\right).$$

Theorem 2 *Let $m\alpha > 2$ and z be the solution of Eq.* (8). *Then $z \in C^2[0, T]$.*

Proof The solution of Eq. (8) satisfies

$$z(t) = y(t) - T_m^{(\alpha)}(t) = \sum_{n=m+1}^{\infty} \frac{(-Bt^\alpha)^n}{\Gamma(\alpha n + 1)},$$

$$z''(t) = \sum_{n=m+1}^{\infty} \frac{(-B)^n t^{\alpha n - 2}}{\Gamma(\alpha n - 1)}.$$

Denote

$$Z_j(t) = \sum_{n=m+1}^{j} \frac{(-B)^n t^{\alpha n - 2}}{\Gamma(\alpha n - 1)},$$

where $j > m$. When $m\alpha > 2$ the functions $Z_j(t)$ are continuous on the interval $[0, T]$. Let $\epsilon > 0$ and

$$\widetilde{M} > M > \max\left\{\frac{1}{\alpha}\left(2B^{1/\alpha}Te + 2 - \alpha\right), \frac{2}{\alpha} - 1 + \log_2\left(\frac{2B^{2/\alpha}}{\epsilon}\right)\right\}.$$

$$Z_{\widetilde{M}}(t) - Z_M(t) = \sum_{n=M+1}^{\widetilde{M}} \frac{(-B)^n t^{\alpha n - 2}}{\Gamma(\alpha n - 1)}.$$

From the triangle inequality

$$|Z_{\widetilde{M}}(t) - Z_M(t)| < \sum_{n=M+1}^{\widetilde{M}} \frac{B^n t^{\alpha n-2}}{\Gamma(\alpha n - 1)} = B^{2/\alpha} \sum_{n=M+1}^{\widetilde{M}} \frac{\left(B^{1/\alpha} T\right)^{\alpha n-2}}{\Gamma(\alpha n - 1)}.$$

The gamma function satisfies [22]

$$\Gamma(1+x)^{2/x} \geq \frac{1}{e^2}(x+1)(x+2) > \left(\frac{x}{e}\right)^2,$$

$$\Gamma(1+x) > \left(\frac{x}{e}\right)^x.$$

Then

$$|Z_{\widetilde{M}}(t) - Z_M(t)| < B^{2/\alpha} \sum_{n=M+1}^{\widetilde{M}} \frac{\left(B^{1/\alpha} Te\right)^{\alpha n-2}}{(\alpha n - 2)^{\alpha n-2}} < B^{2/\alpha} \sum_{n=M+1}^{\widetilde{M}} \frac{\left(B^{1/\alpha} Te\right)^{\alpha n-2}}{(\alpha M + \alpha - 2)^{\alpha n-2}},$$

$$|Z_{\widetilde{M}}(t) - Z_M(t)| < B^{2/\alpha} \left(\frac{B^{1/\alpha} Te}{\alpha M + \alpha - 2}\right)^{\alpha M+\alpha-2} \sum_{n=0}^{\widetilde{M}} \left(\frac{B^{1/\alpha} Te}{\alpha M + \alpha - 2}\right)^{\alpha n}.$$

From Lemma 1:

$$|Z_{\widetilde{M}}(t) - Z_M(t)| < B^{2/\alpha} \left(\frac{B^{1/\alpha} Te}{\alpha M + \alpha - 2}\right)^{\alpha M+\alpha-2} \sum_{n=0}^{\infty} \frac{1}{2^n},$$

$$|Z_{\widetilde{M}}(t) - Z_M(t)| < 2B^{2/\alpha} \left(\frac{B^{1/\alpha} Te}{\alpha M + \alpha - 2}\right)^{\alpha M+\alpha-2} < \epsilon.$$

From the uniform limit theorem, the sequence of functions $Z_j(t)$ converges uniformly to the second derivative of the solution of Eq. (8), which is a continuous function on the interval $[0, T]$.

We can show that when $m\alpha > 3$ then $z \in C^3[0, T]$. The proof is similar to the proof of Theorem 2. In this case numerical solutions NS1(1),NS1(2) and NS1(4) of Eq. (8) have accuracy $O\left(h^{2-\alpha}\right)$, $O\left(h^2\right)$ and $O\left(h^{3-\alpha}\right)$. The numerical results for the maximum error and order of numerical solutions NS1(1),NS1(2) and NS1(4) of Eq. (8) on the inteval $[0, 1]$ are presented in Tables 2, 3 and 4.

Table 2 Maximum error and order of numerical solution NS1(1) of Eq. (8) of order $2-\alpha$

h	$\alpha=0.3, B=1, m=7$		$\alpha=0.5, B=2, m=4$		$\alpha=0.7, B=3, m=3$	
	Error	Order	Error	Order	Error	Order
0.00625	0.7428×10^{-7}	1.6596	0.4811×10^{-3}	1.4858	0.2821×10^{-2}	1.2943
0.003125	0.2338×10^{-7}	1.6681	0.1713×10^{-3}	1.4901	0.1148×10^{-2}	1.2966
0.0015625	0.7322×10^{-8}	1.6746	0.6085×10^{-4}	1.4931	0.4671×10^{-3}	1.2980
0.00078125	0.2286×10^{-8}	1.6797	0.2159×10^{-4}	1.4951	0.1899×10^{-3}	1.2988

Table 3 Maximum error and order of second-order numerical solution NS1(2) of Eq. (8)

h	$\alpha=0.3, B=1, m=8$		$\alpha=0.5, B=2, m=5$		$\alpha=0.7, B=3, m=2$	
	Error	Order	Error	Order	Error	Order
0.00625	0.9068×10^{-6}	1.9692	0.2333×10^{-4}	1.8861	0.2789×10^{-3}	2.0586
0.003125	0.2297×10^{-6}	1.9809	0.6148×10^{-5}	1.9240	0.6715×10^{-4}	2.0545
0.0015625	0.5790×10^{-7}	1.9882	0.1593×10^{-5}	1.9484	0.1597×10^{-4}	2.0718
0.00078125	0.1455×10^{-7}	1.9927	0.4081×10^{-6}	1.9645	0.3771×10^{-5}	2.0826

Table 4 Maximum error and order of numerical solution NS1(4) of Eq. (8) of order $3-\alpha$

h	$\alpha=0.3, B=1, m=8$		$\alpha=0.5, B=2, m=6$		$\alpha=0.7, B=3, m=5$	
	Error	Order	Error	Order	Error	Order
0.00625	0.1468×10^{-5}	2.6281	0.5705×10^{-5}	2.4902	0.8051×10^{-4}	2.2933
0.003125	0.2329×10^{-6}	2.6565	0.1011×10^{-5}	2.4963	0.1638×10^{-4}	2.2968
0.0015625	0.3675×10^{-7}	2.6637	0.1789×10^{-6}	2.4985	0.3331×10^{-5}	2.2984
0.00078125	0.5776×10^{-8}	2.6699	0.3164×10^{-7}	2.4995	0.6767×10^{-6}	2.2992

3 Three-Term FDE

In this section we study the numerical solutions of the three-term equation

$$y^{[2\alpha]}(t)+3y^{(\alpha)}(t)+2y(t)=0,\quad y(0)=3, y^{(\alpha)}(0)=-4, \tag{9}$$

where $0<\alpha<1$. Substitute $w(t)=y^{(\alpha)}(t)+y(t)$. The function $w(t)$ satisfies the two-term equation

$$w^{(\alpha)}(t)+2w(t)=0,\quad w(0)=-1.$$

Then $w(t)=-E_\alpha\left(-2t^\alpha\right)$. The function $y(t)$ satisfies the two-term equation

$$y^{(\alpha)}(t)+y(t)=-E_\alpha\left(-2t^\alpha\right),\quad y(0)=3.$$

From (6)

$$y(t) = 3E_\alpha\left(-t^\alpha\right) - \int_0^t \xi^{\alpha-1} E_{\alpha,\alpha}\left(-\xi^\alpha\right) E_\alpha\left(-2(t-\xi)^\alpha\right) d\xi.$$

From (1.107) in [21] with $\gamma = \alpha, \beta = 1, y = -1, z = -2$ we obtain

$$\int_0^t \xi^{\alpha-1} E_{\alpha,\alpha}\left(-\xi^\alpha\right) E_\alpha\left(-2(t-\xi)^\alpha\right) d\xi = \left(2E_{\alpha,\alpha+1}\left(-2t^\alpha\right) - E_{\alpha,\alpha+1}\left(-t^\alpha\right)\right) t^\alpha$$

$$= t^\alpha \left(2\sum_{n=0}^{\infty} \frac{(-2t^\alpha)^n}{\Gamma(\alpha n+\alpha+1)} - \sum_{n=0}^{\infty} \frac{(-t^\alpha)^n}{\Gamma(\alpha n+\alpha+1)}\right) \qquad [m = n+1]$$

$$= -\sum_{m=1}^{\infty} \frac{(-2t^\alpha)^m}{\Gamma(\alpha m+1)} + \sum_{m=1}^{\infty} \frac{(-t^\alpha)^m}{\Gamma(\alpha m+1)} = E_\alpha\left(-t^\alpha\right) - E_\alpha\left(-2t^\alpha\right).$$

The three-term Eq. (9) has the solution $y(t) = 2E_\alpha\left(-t^\alpha\right) + E_\alpha\left(-2t^\alpha\right)$. When $0 < \alpha < 1$ the first derivative of the solution $y(t)$ of Eq. (9) is undefined at the initial point $t = 0$. The Riemann-Liouville derivative of order α, where $n - 1 \leq \alpha < n$ and n is a nonnegative integer is defined as

$$D_{RL}^\alpha y(t) = \frac{1}{\Gamma(n-\alpha)} \frac{d^n}{dt^n} \int_0^t \frac{y(\xi)}{(t-\xi)^{1+\alpha-n}} d\xi.$$

The Caputo, Miller-Ross and Rieman-Lioville derivatives satisfy [16, 21]:

$$D^\alpha y(t) = D_C^\alpha y(t) = D_{RL}^\alpha y(t) - \frac{y(0)}{\Gamma(1-\alpha)t^\alpha},$$

$$D_{RL}^{2\alpha} y(t) = D_C^{2\alpha} y(t) + \frac{y(0)}{\Gamma(1-2\alpha)t^{2\alpha}} = D^{2\alpha} y(t) + \frac{y(0)}{\Gamma(1-2\alpha)t^{2\alpha}} + \frac{y^{(\alpha)}(0)}{\Gamma(1-\alpha)t^\alpha},$$

where the above identity for the Caputo derivative requires that $0 < \alpha < 0.5$. Three-term Eq. (9) is reformulated with the Riemann-Liouville and Caputo fractional derivatives as

$$D_{RL}^{2\alpha} y(t) + 3D_{RL}^\alpha y(t) + 2y(t) = \frac{3}{\Gamma(1-2\alpha)t^{2\alpha}} + \frac{5}{\Gamma(1-\alpha)t^\alpha}, \tag{10}$$

$$D_C^{2\alpha} y(t) + 3D_C^\alpha y(t) + 2y(t) = -\frac{4}{\Gamma(1-\alpha)t^\alpha}, y(0) = 3,\ 0 < \alpha \leq 0.5. \tag{11}$$

A similar three-term equation which involves the Caputo derivative is studied in Example 5.1 in [14]. The formulation (11) of the three-term equation which uses the Caputo derivative has only one initial condition specified and the values of α, where $0.5 < \alpha < 1$ are excluded. The initial conditions of Eqs. (10) and (11) are inferred from the equations. Formulation (9) of the three-term equation studied in this section has the advantages, to the formulations (10) and (11) which use the Caputo and Riemann-Lioville derivatives, that the initial conditions of Eq. (9) are specified and the analytical solution is obtained with the method described in this section. Now we determine the Miller-Ross derivatives of the solution of Eq. (9). By applying fractional differentiation of order α we obtain

$$y^{[(n+1)\alpha]}(t) + 3y^{[n\alpha]}(t) + 2y^{[(n-1)\alpha]}(t) = 0.$$

Denote $a_n = y^{[n\alpha]}(0)$. The numbers a_n are computed recursively with

$$a_{n+1} + 3a_n + 2a_{n-1} = 0, \quad a_0 = 3, a_1 = -4.$$

Substitute

$$z(t) = y(t) - T_m^{(\alpha)}(t) = y(t) - \sum_{n=0}^{m} \frac{a_n t^{n\alpha}}{\Gamma(\alpha n + 1)}.$$

The function $z(t)$ has fractional derivatives

$$z^{(\alpha)}(t) = y^{(\alpha)}(t) - \sum_{n=1}^{m} \frac{a_n t^{(n-1)\alpha}}{\Gamma(\alpha(n-1)+1)} = y^{(\alpha)}(t) - \sum_{n=0}^{m-1} \frac{a_{n+1} t^{n\alpha}}{\Gamma(\alpha n + 1)},$$

$$z^{[2\alpha]}(t) = y^{[2\alpha]}(t) - \sum_{n=1}^{m-1} \frac{a_{n+1} t^{(n-1)\alpha}}{\Gamma(\alpha(n-1)+1)} = y^{[2\alpha]}(t) - \sum_{n=0}^{m-2} \frac{a_{n+2} t^{n\alpha}}{\Gamma(\alpha n + 1)}.$$

The function $z(t)$ satisfies the condition $z(0) = z^{(\alpha)}(0) = 0$ and its Miller-Ross and Caputo derivatives of order 2α are equal $z^{[2\alpha]}(t) = z^{(2\alpha)}(t)$. The function $z(t)$ satisfies the three-term equation

$$z^{(2\alpha)}(t) + 3z^{(\alpha)}(t) + 2z(t) = F(t), \quad z(0) = 0, z^{(\alpha)}(0) = 0, \tag{12}$$

where

$$F(t) = -\left(2a_{m-1} + 3a_m\right) \frac{t^{(m-1)\alpha}}{\Gamma(\alpha(m-1)+1)} - \frac{2a_m t^{m\alpha}}{\Gamma(\alpha m + 1)}.$$

Now we obtain the numerical solution of three-term Eq. (12) which uses approximation (*) of the Caputo derivative. By approximating the Caputo derivatives of Eq. (12) at the point $t_n = nh$ we obtain

$$\frac{1}{h^{2\alpha}} \sum_{k=0}^{n-1} \lambda_k^{(2\alpha)} z_{n-k} + \frac{3}{h^{\alpha}} \sum_{k=0}^{n-1} \lambda_k^{(\alpha)} z_{n-k} + 2z_n = F_n + O\left(h^{\min\{\beta(\alpha),\beta(2\alpha)\}}\right). \tag{13}$$

Denote by NS2(*) the numerical solution $\{u_n\}_{n=0}^{N}$ of Eq. (12) which uses approximation (*) of the Caputo derivative. From (13)

$$\lambda_0^{(2\alpha)} u_n + 3h^\alpha \lambda_0^{(\alpha)} u_n + 2h^{2\alpha} u_n + \sum_{k=1}^{n-1} \lambda_k^{(2\alpha)} u_{n-k} + 3h^\alpha \sum_{k=1}^{n-1} \lambda_k^{(\alpha)} u_{n-k} = h^{2\alpha} F_n.$$

The numbers u_n are computed with $u_0 = u_1 = 0$ and

$$u_n = \frac{1}{\lambda_0^{(2\alpha)} + 3h^\alpha \lambda_0^{(\alpha)} + 2h^{2\alpha}} \left(h^{2\alpha} F_n - \sum_{k=1}^{n-1} \left(\lambda_k^{(2\alpha)} - 3h^\alpha \lambda_k^{(\alpha)} \right) u_{n-k} \right). \quad \text{NS2}(*)$$

When $y \in C^3[0,T]$ numerical solution NS2(2), which uses approximation (2) of the Caputo derivative has an order $\min\{2, 3-2\alpha\}$. The accuracy of numerical solution NS2(2) is $O\left(h^2\right)$ when $0 < \alpha \leq 0.5$ and $O\left(h^{3-2\alpha}\right)$ for $0.5 < \alpha < 1$. The numerical results for the error and the order of numerical solution NS2(2) of three-term Eq. (12) on the interval [0, 1] with are presented in Table 5. Numerical solution NS2(4) which uses approximation (4) of the Caputo derivative has an accuracy $O\left(h^{3-2\alpha}\right)$ for all values of $\alpha \in (0, 0.5) \cup (0.5, 1)$. Numerical solution NS2(4) is undefined for $\alpha = 0.5$, because the value of $\Gamma(-2\alpha) = \Gamma(-1)$ is undefined. The numerical results for the error and the order of numerical solution NS2(4) of Eq. (12) on the interval [0, 1] are presented in Table 6. Now we obtain the numerical solution NS3(*) of Eq. (12) with $\alpha = 0.5$ which uses approximation (*) for the Caputo derivative.

$$z'(t) + 3z^{(0.5)}(t) + 2z(t) = F(t). \tag{14}$$

By approximating the first derivative with (3) we obtain

$$\frac{1}{h}\left(\frac{3}{2} z_n - 2z_{n-1} + z_{n-2}\right) + \frac{3}{h^{0.5}} \sum_{k=0}^{n} \lambda_k^{(0.5)} z_{n-k} + 2z_n = F_n + O\left(h^{\min\{2, \beta(0.5)\}}\right).$$

The numerical solution $\{u_n\}_{n=0}^{N}$ of Eq. (14) is computed with $u_0 = u_1 = 0$ and

$$\frac{1}{h}\left(\frac{3}{2} u_n - 2u_{n-1} + u_{n-2}\right) + \frac{3}{h^{0.5}} \sum_{k=0}^{n} \lambda_k^{(0.5)} u_{n-k} + 2u_n = F_n,$$

$$u_n = \frac{1}{3 + 6\lambda_0^{(0.5)} h^{0.5} + 4h} \left(2hF_n + 4y_{n-1} - y_{n-2} - 6h^{0.5} \sum_{k=1}^{n} \lambda_k^{(0.5)} u_{n-k} \right) \quad NS3(*)$$

The numerical results for the error and the order of numerical solutions NS3(1), NS3(2) and NS3(4) of Eq. (14) are presented in Table 7.

Table 5 Maximum error and order of numerical solution NS2(2) of Eq. (12) of order $\min\{2, 3-2\alpha\}$

h	$\alpha = 0.3, m = 9$		$\alpha = 0.4, m = 6$		$\alpha = 0.7, m = 5$	
	Error	Order	Error	Order	Error	Order
0.00625	0.7315×10^{-3}	1.9149	0.7261×10^{-4}	1.8455	0.5906×10^{-3}	1.5957
0.003125	0.1902×10^{-3}	1.9429	0.1968×10^{-4}	1.8838	0.1956×10^{-3}	1.5945
0.0015625	0.4886×10^{-4}	1.9610	0.5231×10^{-5}	1.9110	0.6478×10^{-4}	1.5941
0.00078125	0.1245×10^{-4}	1.9728	0.1372×10^{-5}	1.9306	0.2145×10^{-4}	1.5943

Table 6 Maximum error and order of numerical solution NS2(4) of Eq. (12) of order $3 - 2\alpha$

h	$\alpha = 0.3, m = 14$		$\alpha = 0.4, m = 10$		$\alpha = 0.7, m = 6$	
	Error	Order	Error	Order	Error	Order
0.00625	0.1202×10^{-2}	2.4521	0.3205×10^{-3}	2.2431	0.5709×10^{-3}	1.6080
0.003125	0.2202×10^{-3}	2.4481	0.6799×10^{-4}	2.2369	0.1874×10^{-3}	1.6069
0.0015625	0.4052×10^{-4}	2.4423	0.1449×10^{-4}	2.2303	0.6161×10^{-4}	1.6052
0.00078125	0.7486×10^{-5}	2.4363	0.3101×10^{-5}	2.2242	0.2027×10^{-4}	1.6037

Table 7 Maximum error and order of numerical solution NS3(1) of Eq. (14) of order 1.5 and second-order numerical solutions NS3(2) ana NS3(4)

h	$NS3(1), m = 4$		$NS3(2), m = 3$		$NS3(4), m = 5$	
	Error	Order	Error	Order	Error	Order
0.00625	0.8076×10^{-3}	1.4835	0.4197×10^{-3}	1.8868	0.6085×10^{-4}	2.0257
0.003125	0.2872×10^{-3}	1.4915	0.1109×10^{-3}	1.9195	0.1498×10^{-4}	2.0224
0.0015625	0.1019×10^{-3}	1.4955	0.2886×10^{-4}	1.9429	0.3698×10^{-5}	2.0179
0.00078125	0.3608×10^{-4}	1.4975	0.7473×10^{-5}	1.9491	0.9157×10^{-6}	2.0138

4 Conclusion

In the present paper we propose a method for improving the numerical solutions of ordinary fractional differential equations. The method is based on the computation of the fractional Taylor polynomials of the solution at the initial point and transforming the equations into FDEs which have smooth solutions. The method is used for computing the numerical solutions of Eqs. (7) and (9) and it can be applied to other linear fractional differential equations with constant coefficients. The fractional Taylor polynomials have a potential for construction of numerical solutions of linear and nonlinear FDEs which have singular solutions. In future work we are going to generalize the method discussed in the paper to other linear FDEs with constant coefficients and analyze the convergence and the stability of the numerical methods.

Acknowledgements This work was supported by the Bulgarian Academy of Sciences through the Program for Career Development of Young Scientists, Grant DFNP-17-88/2017, Project "Efficient Numerical Methods with an Improved Rate of Convergence for Applied Computational Problems", by the Bulgarian National Fund of Science under Project DN 12/4-2017 "Advanced Analytical and Numerical Methods for Nonlinear Differential Equations with Applications in Finance and Environmental Pollution" and by the Bulgarian National Fund of Science under Project DN 12/5-2017 "Efficient Stochastic Methods and Algorithms for Large-Scale Computational Problems".

References

1. del Cartea, A., Castillo-Negrete, D.: Fractional diffusion models of option prices in markets with jumps. Phys. A **374**(2), 749–763 (2007)
2. Srivastava, V.K., Kumar, S., Awasthi, M.K., Singh, B.K.: Two-dimensional time fractional-order biological population model and its analytical solution. Egypt. J. Basic Appl. Sci. **1**(1), 71–76 (2014)
3. Dimitrov, Y.: A second order approximation for the Caputo fractional derivative. J. Fractional Calc. Appl. **7**(2), 175–195 (2016)
4. Lin, Y., Xu, C.: Finite difference/spectral approximations for the time-fractional diffusion equation. J. Comput. Phys. **225**(2), 1533–1552 (2007)
5. Dimitrov, Y.: Three-point approximation for the Caputo fractional derivative. Commun. Appl. Math. Comput. **31**(4), 413–442 (2017)
6. Dimitrov, Y.: Approximations for the Caputo derivative (I). J. Fractional Calc. Appl. **9**(1), 35–63 (2018)
7. Alikhanov, A.A.: A new difference scheme for the time fractional diffusion equation. J. Comput. Phys. **280**, 424–438 (2015)
8. Chen, M., Deng, W.: Fourth order accurate scheme for the space fractional diffusion equations. SIAM J. Numer. Anal. **52**(3), 1418–1438 (2014)
9. Ding, H., Li, C.: High-order numerical algorithms for Riesz derivatives via constructing new generating functions. J. Sci. Comput. **71**(2), 759–784 (2017)
10. Gao, G.H., Sun, Z.Z., Zhang, H.W.: A new fractional numerical differentiation formula to approximate the Caputo fractional derivative and its applications. J. Comput. Phys. **259**, 33–50 (2014)
11. Ren, L., Wang, Y.-M.: A fourth-order extrapolated compact difference method for time-fractional convection-reaction-diffusion equations with spatially variable coefficients. Appl. Math. Comput. **312**, 1–22 (2017)
12. Jin, B., Zhou, Z.: A finite element method with singularity reconstruction for fractional boundary value problems. ESIM: M2AN **49**, 1261–1283 (2015)
13. Quintana-Murillo, J., Yuste, S.B.: A finite difference method with non-uniform timesteps for fractional diffusion and diffusion-wave equations. Eur. Phys. J. Special Top. **222**(8), 1987–1998 (2013)
14. Zeng, F., Zhang, Z., Karniadakis, G.E.: Second-order numerical methods for multi-term fractional differential equations: smooth and non-smooth solutions. Comput. Methods Appl. Mech. Eng. **327**, 478–502 (2017)
15. Zhang, Y.-N., Sun, Z.-Z., Liao, H.-L.: Finite difference methods for the time fractional diffusion equation on non-uniform meshes. J. Comput. Phys. **265**, 195–210 (2014)
16. Diethelm, K.: The Analysis of Fractional Differential Equations: An Application-Oriented Exposition Using Differential Operators of Caputo Type. Springer (2010)
17. Diethelm, K., Siegmund, S., Tuan, H.T.: Asymptotic behavior of solutions of linear multi-order fractional differential systems. Fractional Calc. Appl. Anal. **20**(5), 1165–1195 (2017)
18. Dimitrov, Y.: Numerical approximations for fractional differential equations. J. Fractional Calc. Appl. **5**(3S), 1–45 (2014)

19. Jin, B., Lazarov, R., Zhou, Z.: An analysis of the L1 scheme for the subdiffusion equation with nonsmooth data. IMA J. Numer. Anal. **36**(1), 197–221 (2016)
20. Li, C., Chen, A., Ye, J.: Numerical approaches to fractional calculus and fractional ordinary differential equation. J. Comput. Phys. **230**(9), 3352–3368 (2011)
21. Podlubny, I.: Fractional Differential Equations. Academic Press, San Diego (1999)
22. Bastero, J., Galve, F., Peña, A., Romano, M.: Inequalities for the gamma function and estimates for the volume of sections of Bnp. Proc. A.M.S., **130**(1), 183–192 (2001)

Testing Performance and Scalability of the Pure MPI Model Versus Hybrid MPI-2/OpenMP Model on the Heterogeneous Supercomputer Avitohol

Vyara Koleva-Efremova

Abstract Over the last years, high performance computers (called clusters) have been used to manage with really complex tasks as to make weather predictions, to run physics simulations and to perform medical researches. Following the Moore's law, the number of cores in the clusters increased significantly until now, where the biggest cluster platforms contain enormous number of cores and have the capability of solving complex problems in a few minutes. The main usage of such massively parallel processing machines is to achieve a performance as high as possible, which depends not only on the hardware architecture model but the used software has also an impact on the performance. In this paper, the performance and scalability of the unified MPI-2 model are compared with the hybrid model of MPI-2 with OpenMP implementation using the high-performance computing system Avitohol [1]. Intel benchmark tests are used: ping-pong operations and one-sided communication [2]. The range of the patterns is from light to heavy communication traffic: point-to-point message passing and shared memory programming. This research discovers the potentials in developing high performance parallel codes on the high-performance computing system "Avitohol".

1 Introduction

1.1 The Unified and Hybrid Models

In this article we compare the performance ratio between pure distributed memory model (MPI) and distributed memory between the nodes plus shared memory model (OpenMP) in each node (Fig. 1), called hybrid model. With such hybrid model we can eliminate domain decomposition in the nodes, can synchronize on the memory instead of barriers and can achieve lower latency and data movement between nodes. The hybrid model reduces the memory traffic and ensures load balancing, but the pure

V. Koleva-Efremova (✉)
Institute of Information and Communication Technologies - BAS, Sofia, Bulgaria
e-mail: viarakoleva@yahoo.com

K. Georgiev et al. (eds.), *Advanced Computing in Industrial Mathematics*,
Studies in Computational Intelligence 793,
https://doi.org/10.1007/978-3-319-97277-0_8

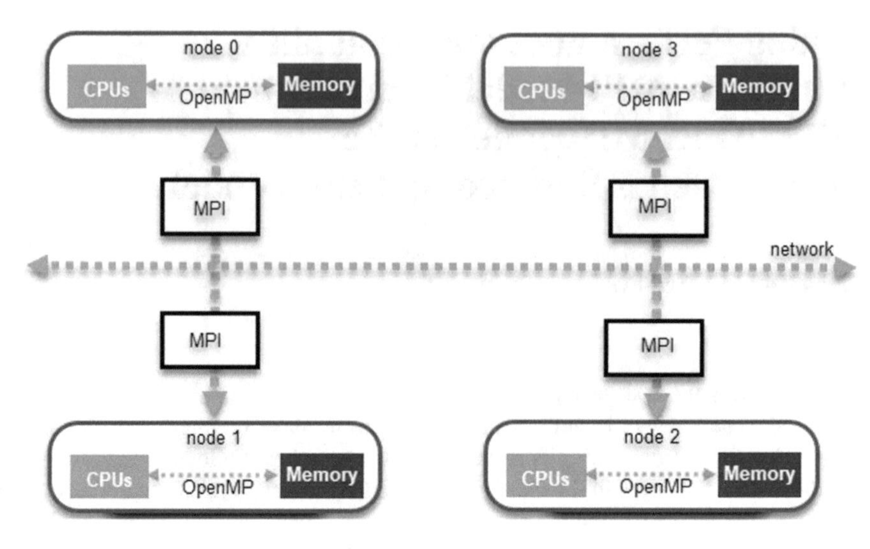

Fig. 1 MPI with OpenMP hybrid model

MPI model is faster than the hybrid programming model [3–6]. This limit should be overcomed in standard MPI-3, which uses another approach for hybrid programming via the new MPI shared memory model. The MPI SHM model, supported by Intel MPI Library Version 5.0.22 enables changes to existing MPI codes incrementally in order to accelerate communication between processes on the shared-memory nodes [7]. The OMP includes implementation of the parallel loops.

1.2 The HPC Avitohol

In our study, we use MPI-2, version 5.0.1.035 with Intel compiler 2.181. The Avitohol consists of 150 computational servers HP SL250s Gen8, equipped with two Intel Xeon E5-2650v2 CPUs and two Intel Xeon Phi 7120P co-processors, 64 GB RAM, two 500 GB hard drives, interconnected with non-blocking FDR InfiniBand running at 56 Gbp/s line speed. The total number of cores is 20,700 and the total RAM is 9600 GB, respectively [8, 9]. High-performance computing system Avitohol was in the TOP 500 list (388th place, November 2015). On each core in the node is assigned two MPI tasks, one MPI task per socket. We used fence and lock-unlock mechanism for MPI one-sided implementations since the passive target communication paradigm is closest to the OpenMP shared memory model. Moreover, the lock-unlock synchronization mechanism would give less overhead than post-start-complete-wait synchronizations since the target process is not involved in the synchronization when using the former [10–12]. Throughout the paper, "*my_rank*" is the rank of the executing process, "*msgSize*" is the message size, "*win*" is the memory window created on the remote process, and "*pedisp*" represents the displacement from the beginning of window "*win*".

2 Benchmarking

Each test has several implementations including *MPI_get*, *MPI_put*, blocking and non-blocking *MPI_send* and *MPI_receive*. For each MPI test we use OpenMP parallelization of the loop nests in the computation part of the MPI code. This approach is also called OpenMP fine-grain or loop level parallelization [13, 14]. Depending on how the data needs to be processed, the MPI has several send modes where the data could be (or not) buffered and could be (or not) synchronized.

The benchmark tests used communication patterns ranging from light to heavy communication traffic: point-to-point message passing and shared memory programming. The synchronization mechanisms are necessary when using one-side communication and overhead an implementation. Each implementation was run using 16 bytes, 32 Kbytes and 1 Mbyte messages. Tests were run with 16, 32, 64, 128, 256 and 320 MPI processes using two MPI processes per node (one MPI process per socket). The tests were also run on Intel Xeon Phi co-processors and the results were compared.

2.1 MPI - One to One Communications

The two basic communication primitives are "*send*" and "*receive*". The first test presents the blocking communication for accessing distant messages of 16 B, 32 KB and 1 MB size. The MPI send/receive "*ping-pong*" operation (the total time is divided by 2 to get the time to "*send*" a message one way) between process 0 and process i is implemented as follows:

```
if (myrank == 0) /* code for process zero */
{
   MPI_Send(message, message_size, MPI_CHAR, i, tag,
                                    MPI_COMM_WORLD);
   MPI_Recv(message, message_size, MPI_CHAR, i, tag,
                                    MPI_COMM_WORLD, &status);
}
else /* code for process i */
{
   MPI_Recv(message, message_size, MPI_CHAR, 0, tag,
                                    MPI_COMM_WORLD, &status);
   MPI_Send(message, message_size, MPI_CHAR, 0, tag,
                                    MPI_COMM_WORLD);
}
```

The average value for all ranks is shown in Table 1 in milliseconds. Here process "*0*" sends message data to process "*i*", while process "*i*" sends and receives message from process "*0*". The total time for this ping-pong operations is measured and divided by the total number of the ranks. The result will present the time for "*accessing distant messages*" for one process. With increasing the total number of the ranks the execution time for this test increases with over 0.29%. With increasing the value of the message size transferred between the two processes the execution time for the same number of ranks increases with over 1–2 times.

In the showed example each "*i*" process waits for a message before being able to send its message. Only process "*0*" firstly sends its message and after that it waits to receive a message. Without process "*0*" we'll be in situation of a deadlock. The blocking communication is very useful in cases of global communication like broadcast for example where the order of the message arrival needs to be correct.

A solution for such deadlock states is to replace the blocking with non-blocking (asynchronous) communication by using "*Isend*" and "*Ireceive*" MPI routines. *MPI_Isend* and *MPI_Irecv* are non-blocking, meaning that the function call returns before the communication is completed. Deadlock then becomes impossible with non-blocking communication, but other precautions must be taken when using them. In particular we want to be sure at a certain point, that our data has effectively arrived. Then we place a *MPI_Wait* call for each send and/or receive we want to be completed before advancing in the program. It is clear that in using non-blocking call in this example, all the exchanges between the tasks occur at the same time. Before the communication, task 0 gathers the sum of all the vectors to sent from each tasks, and prints them out. Similarly after the communication, task 0 gathers all the sum of the vectors received by each task and prints them out along with the communication times. The second benchmark test represents the non-blocking communication for accessing distant messages of 16 B, 32 KB and 1 MB. The implementations using the MPI two-sided routines, *MPI_sendrecv, MPI_isend* and *MPI_irecv* are as follows:

Table 1 Average times in ms for accessing distant messages tests

Average over all ranks of the median times in milliseconds (ms)for the 'accessing distant messages' test						
Number of nodes: 1; Number of processes per node: 16						
Message sizes	MPI_GET (fence)	MPI_GET (locks)	MPI_PUT (fence)	MPI_PUT (locks)	MPI_SEND_RECV	MPI_SENDRECV[a]
16 Bytes	0.029	0.002	0.007	0.002	0.011	0.002
32 Kbyte	0.197	0.201	0.014	0.205	0.038	0.016
1 Mbyte	4.768	1.922	0.122	2.434	0.137	0.160

[a]*MPI_SENDRECV* contains the average times for *MPI_ISEND&MPI_IRECV* combined with *MPI_SENDRECV*

```
if (myrank == 0) /* code for process zero */
{
   MPI_Isend(message, message_size, MPI_CHAR, i, tag,
                                   MPI_COMM_WORLD, &request);
}
if (myrank == i) /* code for process i */
{
   MPI_Irecv(message, message_size, MPI_CHAR, 0, tag,
                                   MPI_COMM_WORLD, &request);
}
MPI_Wait(&request, &status);
```

Another solution to prevent the code from deadlocks is *MPI_SENDRECV* routine. However, even it cannot indeed, guarantee full data security. If there is a cycle of dependencies within a single set of *MPI_SENDRECV* invocations across all involved processes, then send-receive is, indeed, guaranteed to prevent the deadlock. Unfortunately, it is possible to construct a scenario where *MPI_SENDRECV* may deadlock. MPI-2.1 standard says: "*The semantics of a send-receive operation is what would be obtained if the caller forked two concurrent threads, one to execute the send, and one to execute the receive, followed by a join of these two threads*". Here is the implementation for *MPI_Sendrecv*:

```
MPI_Sendrecv( sendBuf, SIZE, MPI_CHAR, next_rank, 0, recvbuf, SIZE,
   MPI_CHAR, prev_rank, MPI_ANY_TAG, MPI_COMM_WORLD, &status );
```

This example shows that *MPI_Send* and *MPI_Recv* are not the most efficient functions to perform this work. Since they work in blocking mode (i.e. wait for the transfer to finish before continuing its execution). In order to receive their vector, each task must post an *MPI_Recv* corresponding to the *MPI_Send* of the sender, and so wait for the receiving to complete before sending sendbuff to the next task. In order to avoid a deadlock, one of the tasks must initiate the communication and post its *MPI_Recv* after the *MPI_Send*. In this example, the last task initiates this cascading process where each consecutive task is waiting for the complete reception from the preceding task before starting to send "*sendbuff*" to the next. The whole process completes when the last task have finished its reception. Before the communication, task 0 gathers the sum of all the vectors to sent from each tasks, and prints them out. Similarly after the communication, task 0 gathers all the sum of the vectors received by each task and prints them out along with the communication times.

2.2 MPI - Remote Memory Access

Fence synchronization is the simplest RMA synchronization pattern and most closely resembles the *MPI_Barrier* call (though advanced users can alter this behavior somewhat by using the assert parameter). Both the start and end of an RMA epoch is defined by all processes calling the collective *MPI_Win_fence* call. RMA communication calls cannot begin until the target process has called *MPI_Win_fence*. When *MPI_Win_fence* is called to terminate an epoch, the call will block until all RMA operations originating at that process complete. The *MPI_Put* implementation using the fence synchronization method is as follows:

```
MPI_Put( recvBuf, SIZE, MPI_CHAR, target, 0, SIZE, MPI_CHAR, win);
MPI_Win_fence(0, win);
```

When using the lock-unlock synchronization method in MPI (also called passive communication), the target is not involved in the communication process. We can initiate multiple passive target epochs to different processes, but the concurrent epochs to the same process are not allowed (affects threads). The *MPI_Put* implementation using the lock-unlock synchronization method is as follows:

```
MPI_Win_lock(MPI_LOCK_EXCLUSIVE, target, MPI_MODE_NOCHECK, win);
MPI_Put(msg, msgSize, MPI_CHAR,target,0,msgSize,MPI_CHAR, win);
MPI_Win_unlock(target, win);
```

MPI_Get allows the calling process to retrieve data from another process, as long as the desired data are contained within the target window and the copied data fits in the origin (calling side) buffer. In essence, we're making a call to *MPI_Irecv* (non-blocking receive) without needing to wait for the other process to call *MPI_Send*. The *MPI_Get* implementation using the fence synchronization method is as follows:

```
MPI_Get(recvBuf,msgSize,MPI_CHAR,target,0,msize,MPI_CHAR,win);
MPI_Win_fence(0, win);
```

The *MPI_Get* implementation using the lock-unlock synchronization method is as follows:

```
MPI_Win_lock(MPI_LOCK_EXCLUSIVE, target, MPI_MODE_NOCHECK, win);
MPI_Get(recvBuf, msgSize, MPI_CHAR, target, 0, msgSize, MPI_CHAR, win);
MPI_Win_unlock(target, win);
```

2.3 OpenMP Implementation

The OpenMP implementation controls the number of threads per test. For this study, the number of threads *threads* per each test is set to 16 as default value.

```
#pragma omp parallel for num_threads(threads)
```

3 Results

In Table 1 are shown the average times in milliseconds for all types of tests, run with different message sizes. In summary, the benchmark applications written using MPI shared programming have an execution time that is smaller than MPI two-sided implementation. The non-blocking (asynchronous) two-sided communication has significant advantage over the blocking MPI two-sided communication. Notice the poor performance of *MPI_Get* (fence) by increasing the message size. Regarding the basic *MPI_Send&MPI_Recv* "*ping-pong*" operations, the hybrid model takes advantage in cases of using Hyperthreading [15]. Basically, the *MPI_Put* routine with fence synchronization has the best performance in the current tests results. Even though, the pure MPI model has smaller execution time than the hybrid model. With passing from light to heavy traffic, the hybrid model shows an improved execution time, smaller than the pure MPI. However the difference is insignificant. We can say, that with message size up to 32 KB, the performance is quite stable and stay almost unchanged.

In Table 2 are shown the *MPI_PUT* with lock/unlock synchronization results. The hybrid model MPI+OMP takes small advantage over the pure model with increasing the message size. However, the performance achieved with small message size(16 KB) is better than this one with the big message size (1 MB).

Table 2 MPI_PUT with lock/unlock synchronization results, comparing pure MPI versus hybrid MPI+OMP model

Average values of	Message 16 B		Message 32 KB		Message 1 MB	
mpi_put_lock()	Time (s)	Time (ms)	Time (s)	Time (ms)	Time (s)	Time (ms)
Nodes=1:16ppn (-np16)	0.0000020	0.0020	0.0002050	0.2050	0.0024340	2.4340
mpi_put_lock() + OMP (16 threads)	Time (s)	Time (ms)	Time (s)	Time (ms)	Time (s)	Time (ms)
Nodes = 1:16ppn (-np16)	0.0000020	0.0020	0.0001850	0.1850	0.0023640	2.3640

On Figs. 2 and 3 are shown the results from the first benchmark test: blocking communication for accessing distant messages of 16 B, 32 KB and 1 MB size. The best time is for the pure *MPI_SEND&MPI_RECV* model. The hybrid model takes double time compared with the pure model, but with hyperthreading included, the time for the hybrid model is over 50% less.

As is seen from Fig. 3, the difference in the achieved performance is almost negligible up to 32 KB message size. The performance is significantly slowing down when increasing the message size to 1 MB.

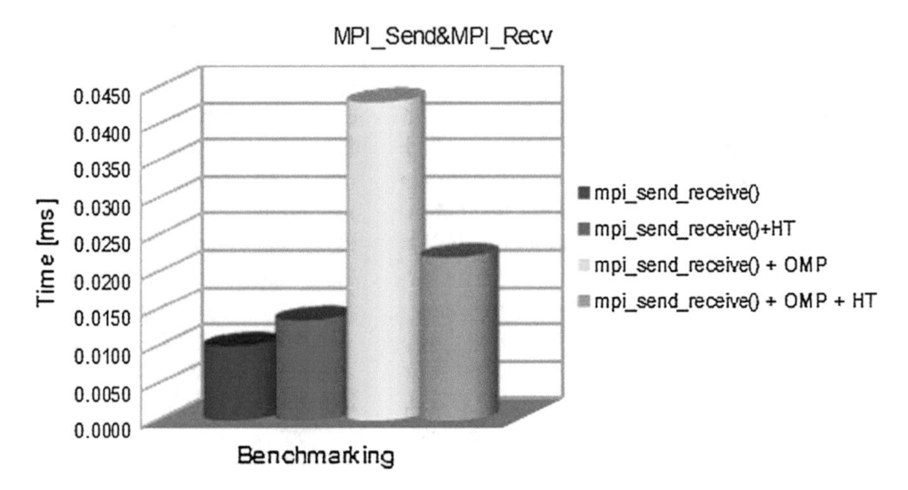

Fig. 2 MPI_Send&MPI_Recv, 16 processes, message size 16, pure MPI model versus Hybrid MPI+OMP model

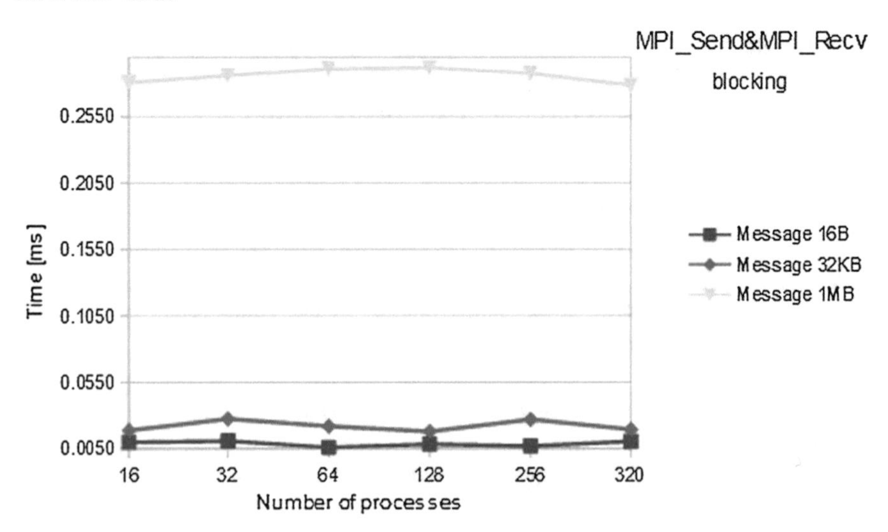

Fig. 3 MPI_Send&MPI_Recv pure model, compare performance for three message sizes

Figure 4 presents the *MPI_Get* with fence synchronisation used. The influence of the number of processes on small amounts of information is insensible and almost constant whereas in cases with a large amount of information such as 1 MB the influence of the number of processes is exponential.

From Fig. 5, we can assume that the *MPI_Get* with hyperthreading used is not efficient for any message size.

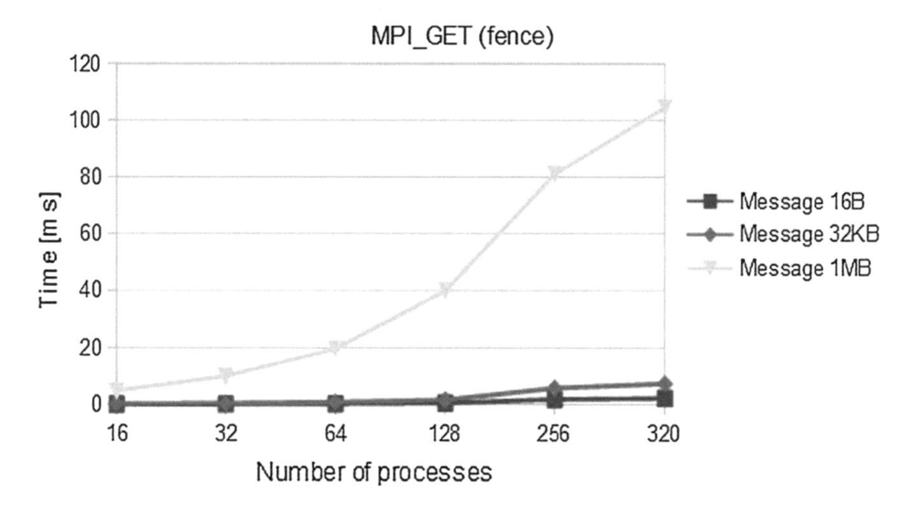

Fig. 4 MPI_Get with fence synchronisation, three different message sizes, 16 processes

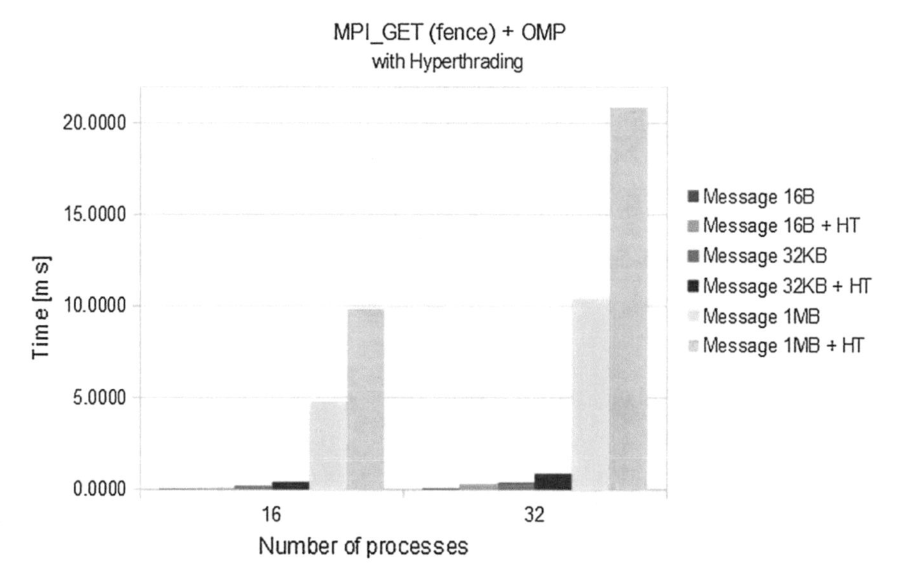

Fig. 5 MPI_Get plus OMP hybrid model compared with Hyperthreading included

From Figs. 6, 7 and 8 we see, that with the hybrid model we achieve better performance, than the pure model for *MPI_Put* using lock/unlock synchronization for big message sizes (1 MB). This effect is improved with increasing the number of processes from 16 to 32.

On Fig. 9 is shown the *MPI_Get* using lock/unlock synchronization results for different message sizes. Here, as like for *MPI_Put* using lock/unlock synchronization, the significant changes on the performance are visible for messages bigger than 32 KB.

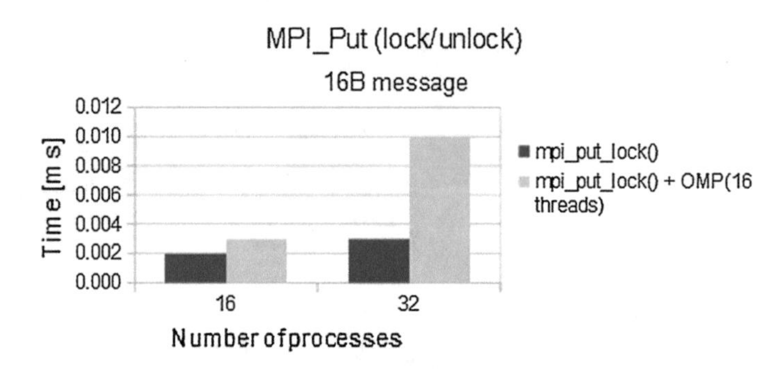

Fig. 6 MPI_Put using lock/unlock synchronization, 16 B message size, pure versus mybrid model

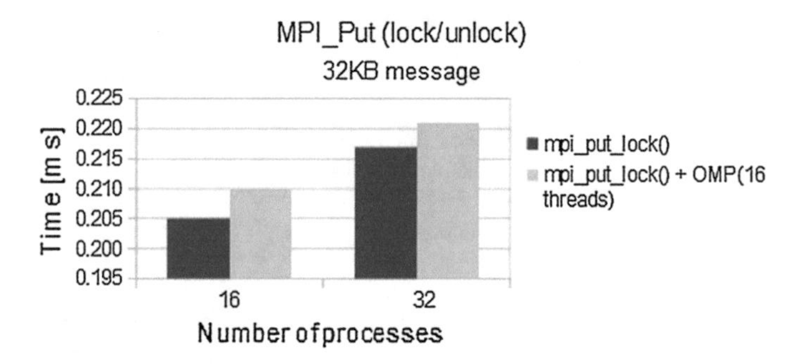

Fig. 7 MPI_Put using lock/unlock synchronization 32 KB message size, pure versus hybrid model

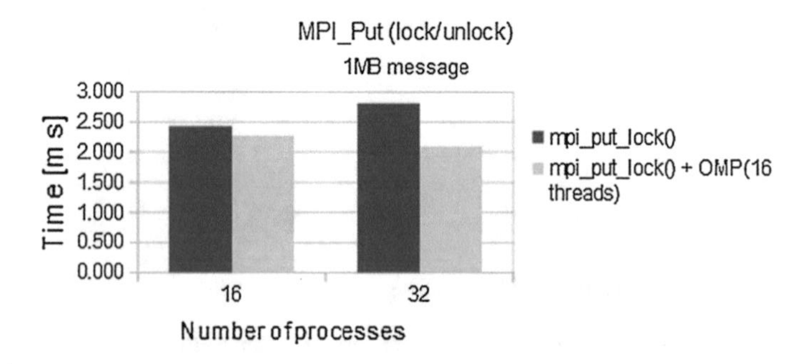

Fig. 8 MPI_Put using lock/unlock synchronization, 1 MB message size, pure versus hybrid model

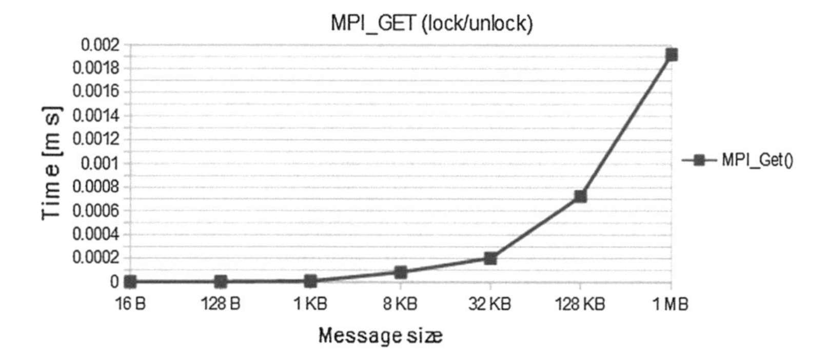

Fig. 9 MPI_Get using lock/unlock synchronization different message sizes

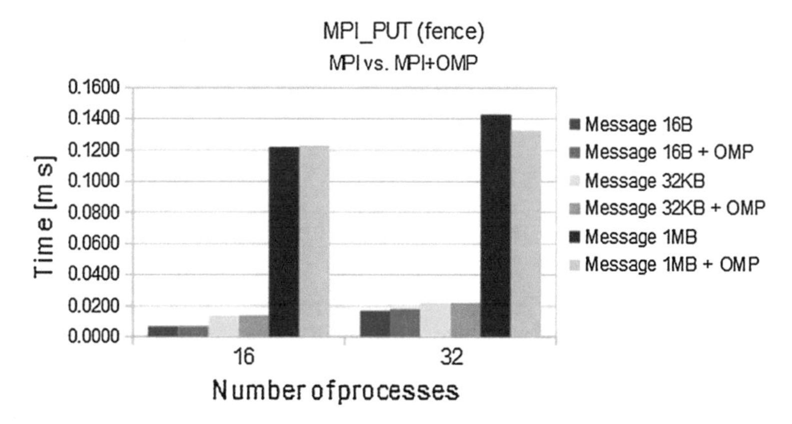

Fig. 10 MPI_Put, using fence synchronization, three different message sizes, pure model versus hybrid model

On Fig. 10 is shown the *MPI_Put*, using fence synchronization results for the three different message sizes. We conclude here, that the hybrid model has advantages over the pure model, only for message size 1 MB and is better to use it when having big number of processes.

4 Conclusions

In this paper it was compared the performance and scalability of the pure MPI model versus the MPI plus OpenMP (called hybrid) model of parallelisation. Several benchmark tests was run on the cluster platform Avitohol. The tests used communication patterns ranging from light to heavy data traffic. The synchronization mechanisms are necessary when using one-side communication and overhead an implementation. Each implementation are run using different message sizes. The results of the tests that we are running on the cluster platform represent the performance achieved with the uniform MPI (shared and two-sided) implementation and shows the difference versus OpenMP and MPI-2 implementations. This information can be very helpful in cases we need to implement specific applications on a Linux cluster like the Human Brain Project (HBP) or for climate simulations [16] and the best implementation method needs to be defined. MPI shared programming has a memory consumption that is considerably less than MPI two-sided programming and comparable to the OpenMP memory consumption. MPI-2 has limitations considering the shared memory routines overcomed in version 3.0. The hybrid OpenMP+MPI model provides better memory consumption compared with the uniform model. The two levels of parallelism provide better load balancing but create conditions for many topology and threads-safety problems. The obtained results are helpful in cases we need to run specific applications on Avitohol and to choose the best approach for parallel implementation.

Acknowledgements This work was supported by the National Science Fund of Bulgaria under Grant DFNI-I02/8.

References

1. Advanced Computing and Data Centre at IICT-BAS. http://www.hpc.acad.bg/system-1/
2. Intel MPI Benchmarks, Document number: 320714-012EN, Revision 4.0 https://www.lrz.de/services/compute/courses/x_lecturenotes/mic_workshop/IMB_Users_Guide.pdf
3. Capello, F., Etiemble, D.: MPI versus MPI+OpenMP on the IBM SP for the NAS Benchmarks, SC '00: Proceedings of the ACM/IEEE Conference on Supercomputing (2000). https://doi.org/10.1109/SC.2000.10001. http://ieeexplore.ieee.org/document/1592725/
4. Grey, A., Henty, D., Smith, L., Hein, J., Booth, S., Bull, J., Reid, F.: A Performance Comparison of HPCx Phase2a to Phase2, EPCC, 20 Feb 2006
5. Hager, G., Jost, G., Rabenseifner, R.: Communication characteristics and hybrid MPI/OpenMP parallel programming on clusters of multi-core SMP nodes (2009). https://cug.org/7-archives/previous_conferences/CUG2009/bestpaper/9B-Rabenseifner/rabenseifner-paper.pdf
6. Negoita, G., Luecke, G., et al.: The performance and scalability of SHMEM and MPI2 one-sided routines on a SGI Origin 2000 and a Cray T3E-600 (2004). https://doi.org/10.1002/cpe.796. https://www.researchgate.net/publication/243134538_The_performance_and_scalability_of_SHMEM_and_MPI2_one-sided_routines_on_a_SGI_Origin_2000_and_a_Cray_T3E-600
7. Dinan, J., Balaji, P., Buntinas, D., et al.: An implementation and evaluation of the MPI 3.0 one-sided communication interface. Concurrency Comput. Pract. Experience (2010). http://www.mcs.anl.gov/uploads/cels/papers/P4014-0113.pdf
8. Atanassov, E., Barth, M., et al.: Best Practice Guide Intel Xeon Phi v2.0 (2017). http://www.prace-ri.eu/IMG/pdf/Best-Practice-Guide-Intel-Xeon-Phi-1.pdf
9. Radenski, A., Gurov, T., et al.: Big data techniques, systems, applications, and platforms: case studies from academia. In: Annals of Computer Science and Information Systems, Vol. 8, Proceedings of the 2016 Federated Conference on Computer Science and Information Systems FedCSIS'16, Sept. 2016, pp. 883–888 (2016). https://doi.org/10.15439/978-83-60810-90-3
10. Adel, M.: Hybrid (MPI+OpenMP) Examples (2012). http://eniac.cyi.ac.cy/pages/viewpage.action?pageId=27951121

11. Gropp, W., Lusk, E., et al.: MPICH2 User's Guide version 1.0.3, Mathematics and Computer Science Division, Argonne National Laboratory (2005)
12. https://www.mpich.org
13. Gropp, W., Lusk, E., Skjellum, A.: Using MPI: Portable Parallel Programming with the Message Passing Interface, 2nd edn. Scientific and Engineering Computation (1999)
14. University of Tennessee: MPI: A Message-Passing Interface Standard Version 2.2. Knoxville, Tennessee (2009)
15. Atanassov, E., Gurov, T., Karaivanova, A., Ivanovska, S., Durchova, M., Dimitrov, D.: On the parallelization approaches for intel MIC architecture. In: AIP Conference Proceedings, vol. 1773, p. 070001 (2016). https://doi.org/10.1063/1.4964983
16. HPC (2018) High Performance Computing—best use examples. https://ec.europa.eu/digital-single-market/en/news/high-performance-computing-best-use-examples

Analyses and Boolean Operation of 2D Polygons

Georgi Evtimov and Stefka Fidanova

Abstract The 2D cutting stock problem (CSP) arises in many industries as paper industry, glass industry, biding construction and so on. In paper and glass industries the cutting shapes are rectangles. In building constructions industry the cutting shapes are postilions, which can have irregular form and can be convex and concave. This increases the difficulty of the problem. In our work we concentrated on 2D cutting stock problem, coming from building industry. Many manufactures companies, which build steel structures, have to cut plates and profiles. The CSP is well known like NP-hard combinatorial optimization problem. The plates are represented like 2D polygons in any CAD environment. The aim of this issue is to apply some mathematical algorithms on each polygon and to prepare them for subsequent processing for solving the main problem. This task is the following basic step (preprocessor) for solving a 2D CSP - arrange all given plates from the project in minimum area. Our preprocessing includes the following main steps: Is polygon clock wise; Remove Wasted Points; Find Intersection Points; Represent the two polygons subtraction by Boolean table.

1 Introduction

The 2D CSP is important industrial problem. In some industries the cutting shapes are rectangler. Rectangular variant of CSP appear in paper and glass industries [1], container loading, Very-large-scale integration (VLSI) design, and on some scheduling problems [2]. When the items have arbitrary shape the problem becomes very complicate. This variant of the problem comes from building constructions in fasteners production, production of clothes, shoes and so on. In [1] is applied hybrid algorithm

G. Evtimov · S. Fidanova (✉)
Institute of Information and Communication Technology,
Bulgarian Academy of Science, Sofia, Bulgaria
e-mail: stefka@parallel.bas.bg

G. Evtimov
e-mail: gevtimov@abv.bg

K. Georgiev et al. (eds.), *Advanced Computing in Industrial Mathematics*,
Studies in Computational Intelligence 793,
https://doi.org/10.1007/978-3-319-97277-0_9

between replacement method and genetic algorithm, the aim is a two-dimensional orthogonal packing problem, where a fixed group of small rectangles must be fitted into a large rectangle so that, most of the material is used, and the unused area of the large rectangle is minimized. In [3] is developed greedy randomized adaptive search procedure with application on two dimensional large-scale cutting problems. In [4] is proposed tabue search algorithm for solving two dimensional non-gillotine cutting problem. In [5] is proposed an exact algorithm based on dynamic programming. This kind of algorithms are appropriate for small problems, because the computational complexity of the problem. Dusberger and Raidl [6, 7] propose two metaheuristic algorithms based on variable neighborhood search for the simplified problem with rectangular items. Parmar et al. [8] made a survey of evolutionary algorithms for cutting stock problem.

In building constructions industry the cutting shapes can be irregular and can be convex and concave. This increases the difficulty of the problem. In our work we concentrated on 2D CSP, coming from building industry. The CSP is well known like NP-hard combinatorial optimization problem. The aim is to arrange all given plates from the project in given minimum area. The plates are represented like 2D polygons in any CAD environment. In this paper we apply some mathematical algorithms on the polygons to prepare them for subsequent processing for solving the 2D CSP. This issue (task) is the first step (preprocessor) for solving a 2D CSP. Our preprocessing includes the following main steps:

- Is polygon clock wise;
- Remove Wasted Points;
- Find Intersections Points;
- Create Boolean tables for the polygons.

The Object of 2D optimization, are the plates from any Steel Structure.

The elements of the steel construction from Fig. 1 have the shape shown on Fig. 2.

Typical shapes of the plates are shown on Fig. 3.

The rest of the paper is organized as follows. In Sect. 2 is formulated the problem we try to solve. In Sect. 3 we propose and compare two methodologies to find the orientation of the polygon. In Sect. 4 is proposed a method for finding intersection between the polygons. In Sect. 5 some conclusions and possibilities for a future work are drown.

2 Problem Formulation

The plates, we will cut, are already drawn in CAD system. Limitation of this analysis is plates with element of arcs, circles, splines and big holes inside it. But for future work we can approximate the arcs, circles and splines into polygons with enough precision, therefore in this work we concentrate on polygons. Each plate is described with N ordered points, this we call a polygon. Each polygon must be non-self crossing. The analyses of polygons start as follows:

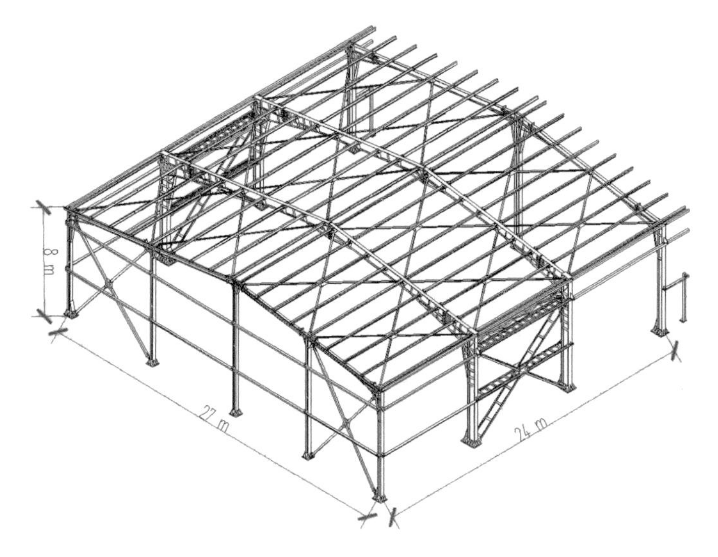

Fig. 1 Steel Structure construction

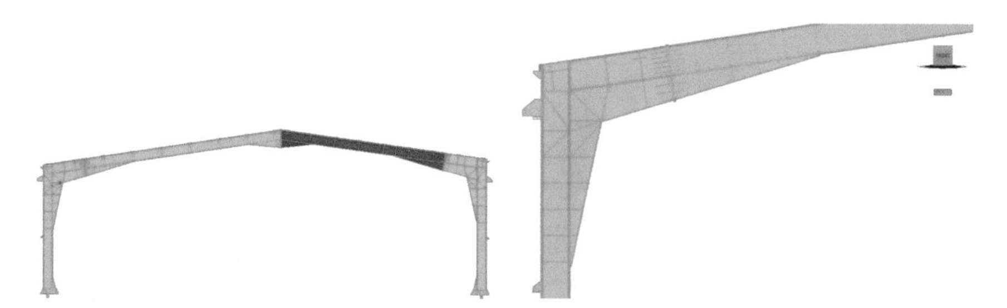

Fig. 2 A views (sections) from Steel Structure

1. **Removing needless points from the given list of points**.
   In the Fig. 4 points V_2 and V_4 are needless, because V_2 is on the line defined by the points V_1, V_3 and V_4 is on the line defined by the points V_3, V_4. How to remove needless points is described in our previous work [9].
2. **Find the orientation of the polygon, ClockWise (CW) or anti-ClockWise (anti-CW)**.
   To find the orientation of polygon we will use two approaches:

   – Classical way: Find oriented area of polygon. If area is greater than zero the polygon is CW, else it is anti-CW.
   – Balanced sum of angles. With one pass through points list balanced sum of angles gives three conclusions:
   a. Is the chosen point is inside the polygon? If the sum of angles is equal to $|2\pi|$ or $|\pi|$ the point is inside, else it is not;

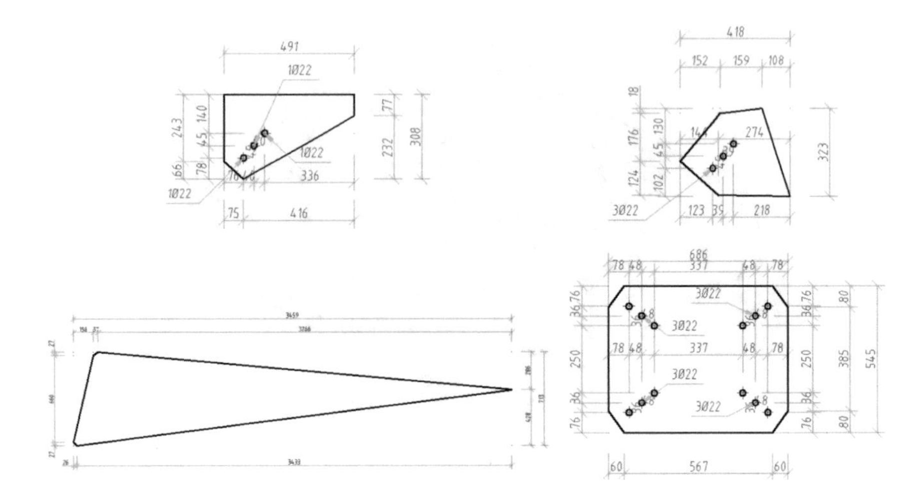

Fig. 3 Considered shapes

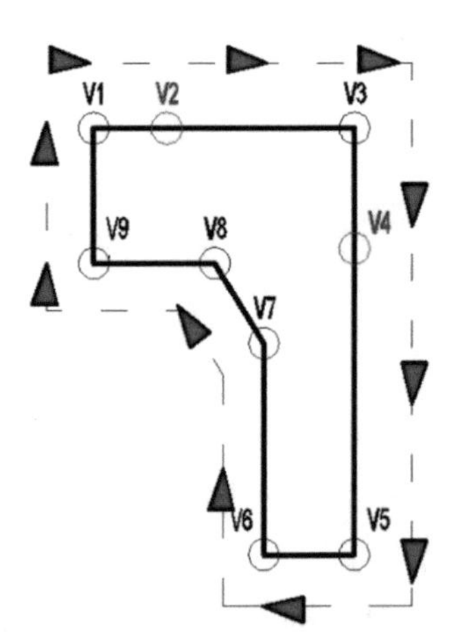

Fig. 4 Polygon with needless nodes

b. Is the polygon is CW or anti-CW? If the sum of all angles is greater than zero the orientation of polygon is anti-CW, else it is CW ;
c. Is the polygon Concave? If one angle from the list of the angles is greater than π, the polygons is concave, else it is convex.

Let is given a polygon P, where: n is the number of vertices of the P, X_i and Y_i are the coordinates of the polygon P, $0 \leq i < n$. Range of coordinate of the points is around x = 50,000, y = 35,000 units. One polygon can have two orientation by CW or anti-CW. Count of vertices is 3373.

3 Orientation of the Poligon

We propose two methods to find the polygon orientation (Fig. 5).

3.1 *Method 1. Oriented Area of Polygon*

The main criteria to find the orientation of the polygon is the following: if area is greater than zero the polygon is CW, it is else anti-CW (Fig. 6).

The problem is how to find oriented area of a polygon. On Fig. 7 is shown example with trapeze. When find the faces of the polygon P, the triangle with hypotenuse P_1, P_m is equal to the triangle with hypotenuse P_m, P_2.

The coordinates of the point P_m are:

$$P_m = \frac{x_i + x_{i+1}}{2}, \frac{y_i + y_{i+1}}{2} \tag{1}$$

The area of ith segment is:

$$A_i = \frac{(x_{i+1} - x_i)(y_{i+1} + y_i)}{2} \tag{2}$$

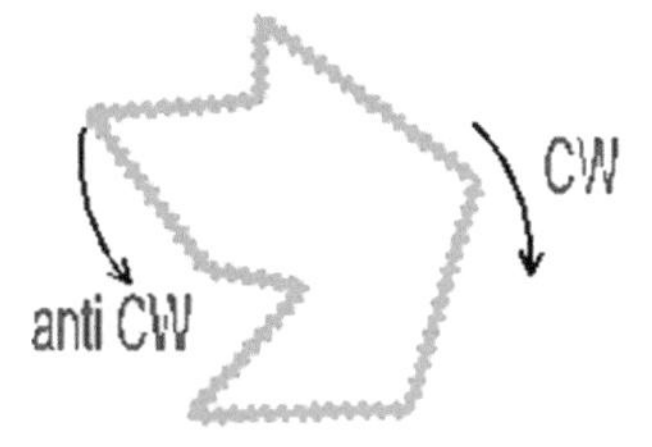

Fig. 5 Polygon orientation

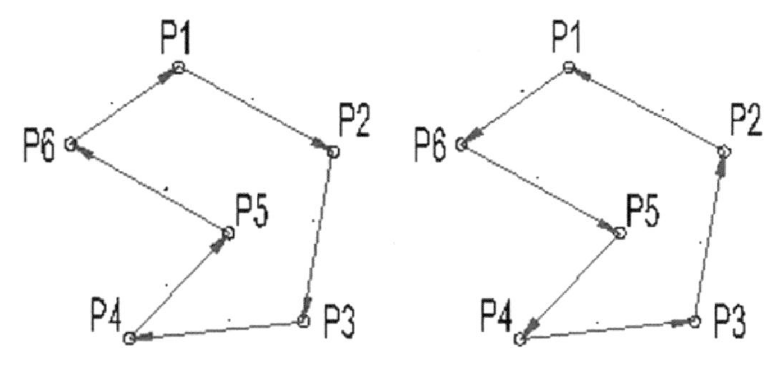

Fig. 6 Polygon orientation

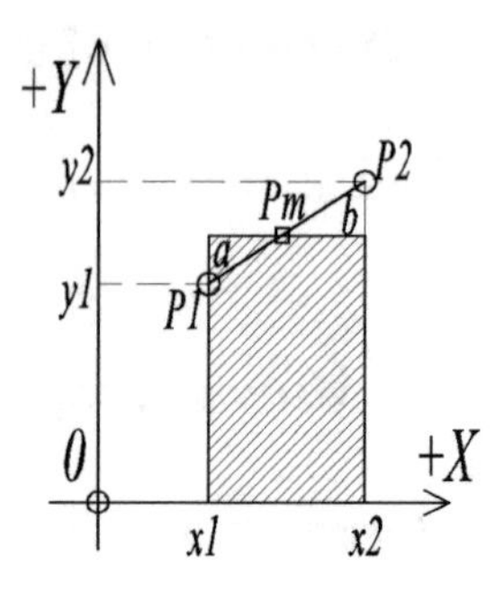

Fig. 7 Oriented area of polygon

Let mention that the sign of A_i depends on the order of the points x_2, x_1 or x_1,x_2. This order of points will give us the oriented area of polygon. The sequence of x coordinates of points of "upper" part from polygon is $x_4 > x_3 > x_2 > x_1 > x_0$. The result is positive sign of the area. An example is shown on Figs. 8, 9 and 10.

The conclusion is: the polygon P is CW.

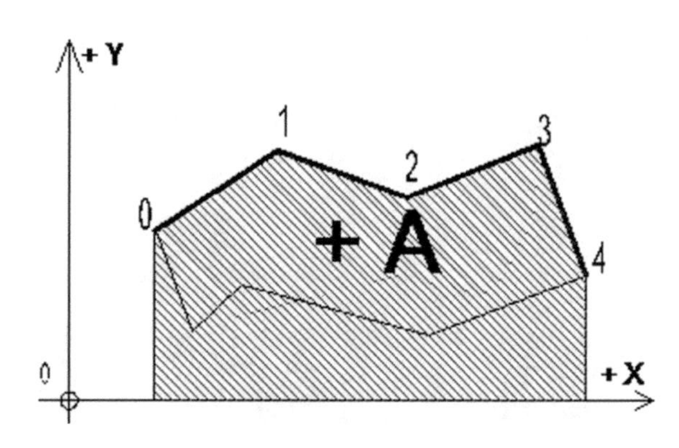

Fig. 8 Positive area

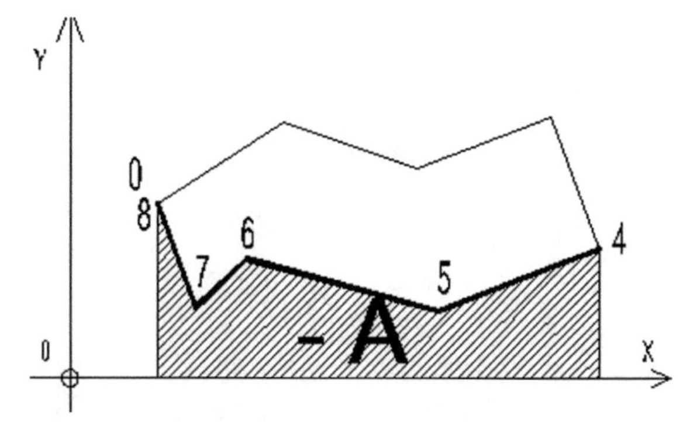

Fig. 9 Negative area

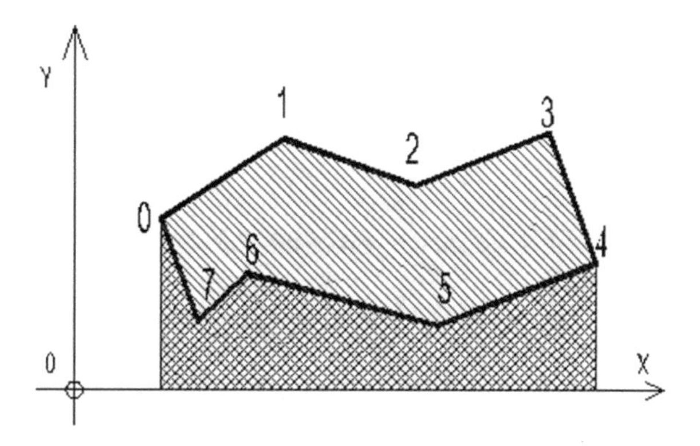

Fig. 10 The added areas

3.2 Method 2. Balanced Sum of Angles

Let's define: α_i is the ith internal angle between any chosen point P_0 and two vertex from the polygon P. The sign of angle α_i depends from order of points from polygon P. So after the calculation of all angles we will receive: $\alpha_1 = -\ldots, \alpha_2 = +\ldots,$ and so on. If we sum all this angles we can make conclusion if the polygon P is CW or anti-CW. The sum of angles is (2π) for CW and $-(2\pi)$ for anti-CW. For triangle the check sum is (π) (Figs. 11 and 12).

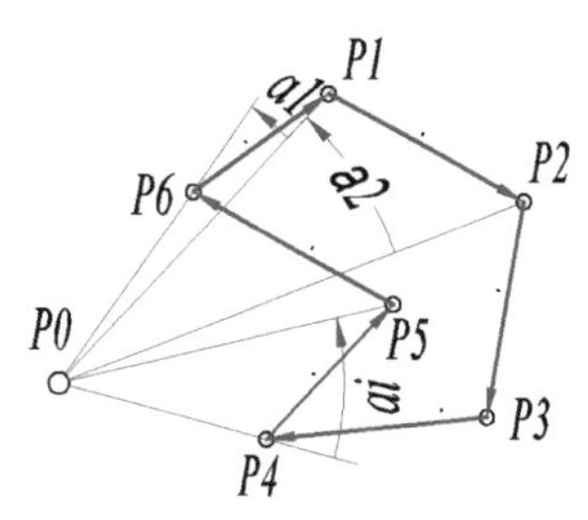

Fig. 11 Point 0 is a basic point

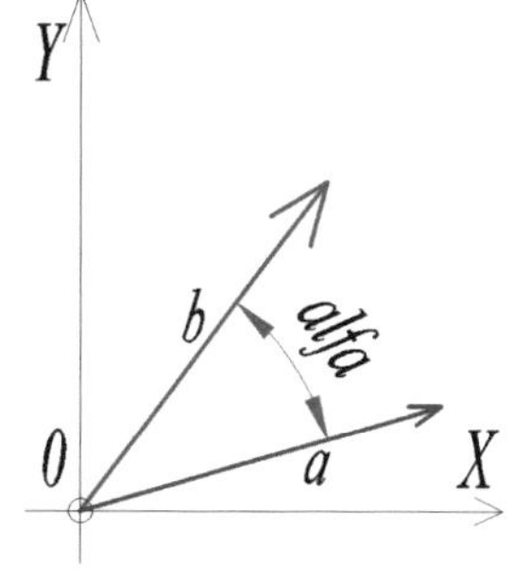

Fig. 12 Calculate inside angle of two vectors

If P_0 is not in position (0, 0) we will translate P_0, P_i, P_{i+1} while point P_0 became with coordinates (0, 0). The inside angle between two vectors is:

$$cos\alpha = \frac{\mathbf{a} \cdot \mathbf{b}}{||a||||b||} \quad (3)$$

In Table 1 are shown part of the calculated angles from the example, based on the point (0, 0).

As we can see that the range of angles is very small. The range is 0.00001. The sum of all angles is 2.83225e−010. Now let's the chosen node is one node from the polygon P. In the Table 2 are the calculated angles based on point from the polygon P.

This range of angles is much more better than above. The range is 1.0. The sum of all angles is −1.51606. According to above definition: balanced sum of angles gives of first base point (list (0, 0)) is positive, but from second chosen point (any vertex) the sum is negative. That means that balanced sum of angles is appropriate method but for small count of vertices.

Let compare Oriented area of polygon and Balanced sum of angles Table 3. The comparison is after testing of 362 polygons with common 374 694 vertices, range of coordinates average $x = 45{,}000$, $y = 30{,}000$ units.

Compose two Boolean tables: for polygon A and for polygon B.

As we can see with the method Balanced sum of angles we can determinate more properties of the given polygon. On the other side, the Oriented area of polygon is faster method if we need only the orientation. According what we need we will chose the method. Thus, when we need only the orientation of the polygon, the Oriented area of polygon is more appropriate method.

Table 1 A part of list of calculated angles, base point (0, 0)

N	Angles (rad)
0	2.91824e−005
1	3.51775e−005
2	−1.09275e−005
3	3.51788e−005
4	5.94318e−005
5	3.51753e−005

Table 2 A part of list of calculated angles, base point one from the polygon P

N	Angles (rad)
0	0.0
1	−0.717395
2	1.24063
3	−0.523233
4	0.0
5	−0.241968

Table 3 Comparison between oriented area and balanced sum

Oriented area of polygon	Balanced sum of angles
Passed for 11 s	Passed for 23 s
Check Is Concave = No	Check Is Concave = Yes
Check Is Point Inside = No	Check Is Point Inside = Yes
Check Is Polygon CW = Yes	Check Is Polygon CW = Yes

4 Find Intersection Points Between Polygons

The next step in solving CSP is to find intersection between polygons. At the beginning the steel sheet from which the polygons are cut of is rectangular, but later its bottom part becomes irregular. Therefore we need an algorithm for finding interaction points between polygons.

On Fig. 13 there are two Crossing points. The coordinates of one point are:

$$X = \frac{\begin{vmatrix} \begin{vmatrix} x_1 & y_1 \\ x_2 & y_2 \end{vmatrix} & \begin{vmatrix} x_1 & 1 \\ x_2 & 1 \end{vmatrix} \\ \begin{vmatrix} x_3 & y_3 \\ x_4 & y_4 \end{vmatrix} & \begin{vmatrix} x_3 & 1 \\ x_4 & 1 \end{vmatrix} \end{vmatrix}}{\begin{vmatrix} \begin{vmatrix} x_1 & 1 \\ x_2 & 1 \end{vmatrix} & \begin{vmatrix} x_1 & 1 \\ x_2 & 1 \end{vmatrix} \\ \begin{vmatrix} x_3 & 1 \\ x_4 & 1 \end{vmatrix} & \begin{vmatrix} x_3 & 1 \\ x_4 & 1 \end{vmatrix} \end{vmatrix}}, Y = \frac{\begin{vmatrix} \begin{vmatrix} x_1 & y_1 \\ x_2 & y_2 \end{vmatrix} & \begin{vmatrix} y_1 & 1 \\ y_2 & 1 \end{vmatrix} \\ \begin{vmatrix} x_3 & y_3 \\ x_4 & y_4 \end{vmatrix} & \begin{vmatrix} y_3 & 1 \\ y_4 & 1 \end{vmatrix} \end{vmatrix}}{\begin{vmatrix} \begin{vmatrix} x_1 & 1 \\ x_2 & 1 \end{vmatrix} & \begin{vmatrix} y_1 & 1 \\ y_2 & 1 \end{vmatrix} \\ \begin{vmatrix} x_3 & 1 \\ x_4 & 1 \end{vmatrix} & \begin{vmatrix} y_3 & 1 \\ y_4 & 1 \end{vmatrix} \end{vmatrix}} \tag{4}$$

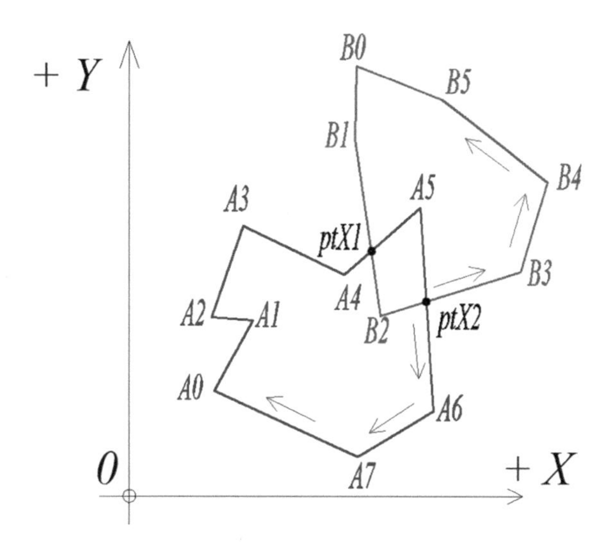

Fig. 13 Interaction between two polygons

We need to know if the point is inside the polygon. We can use two approaches:

1. Balanced sum angles: Chooses point A_0 and check if the sum of angles is $|(2\pi)|$ or $|(\pi)|$ then the point is inside the polygon, else it is outside.
2. Ray crossing method: the criteria is the Number of crossing points between Ray and the given polygon to be odd the point is inside, else it is outside.

Let's compare the both methods, Ray crossing method and Balanced sum angles.

Testing of 120 polygons with common 124 896 vertices, range of coordinates average X = 45,000 ; y = 30,000 units. The test was run as follows:

1. Get first polygon save in var A.
2. Get second polygon save in var B.
3. For each point form polygon A check Is Point Inside Polygon B;
4. For each point form polygon B check Is Point Inside Polygon A.

Balanced sum of angles is powerful method, but it is slow when the polygons consists of many nodes, see Table 4. Therefore we are choosing Ray Crossing method for finding is Point inside the polygon (Fig. 14).

Boolean operation for subtracting two polygons.

Let's compose boolean tables for polygons A and B from Fig. 13:

Table 4 Ray crossing versus balanced sum of angles

Ray crossing method	Balanced sum of angles
Passed for 307 s, about 5 min	More than 50 minutes!

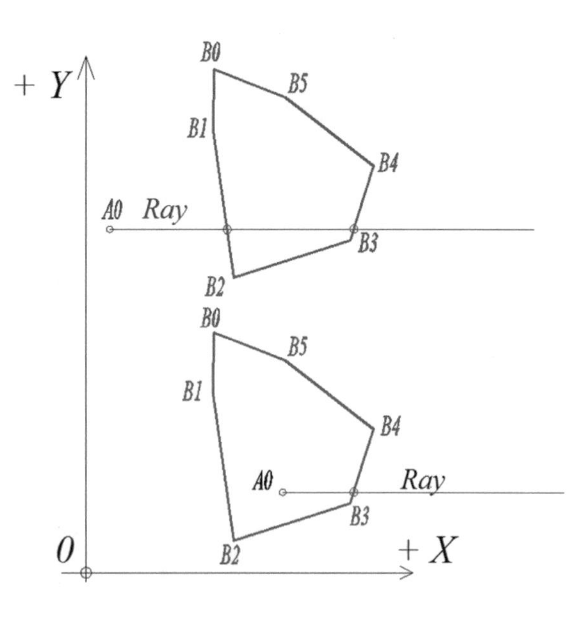

Fig. 14 Ray crossing method

The subtraction of polygon A and B can be find by the following procedure:

1. Get first point from polygon A which is flag Out-True and save in a list Table 5B.
2. Repeat while current point is ptX1.
3. Switch to table B (polygon B) Table 5B Get all point between ptX1 and ptX2.
4. If current point is ptX2 switch to table A.
5. Union of the Table 6.

Table 6 shows the coordinates of the verteces of the polygon, result of the crossing polygon A with polygon B.

Table 5 Boolean tables for polygons A and B

Index	Name	X	Y	Out from B	Index	Name	X	Y	Out from A
0	A0	X	Y	True	0	B0	X	Y	True
1	A1	X	Y	True	1	B1	X	Y	True
2	A2	X	Y	True	2	PtX1	X	Y	True
3	A3	X	Y	True	3	B2	X	Y	False
4	A4	X	Y	True	4	ptX2	X	Y	True
5	ptX1	X	Y	True	5	B3	X	Y	True
6	A5	X	Y	False	6	B4	X	Y	True
7	Ptx2	X	Y	True	7	B5	X	Y	True
8	A6	X	Y	True					
9	A7	X	Y	True					

Table 6 Subtract table for polygons A and B

Index	Name	X	Y	Out from B
0	A0	X	Y	True
1	A1	X	Y	True
2	A2	X	Y	True
3	A3	X	Y	True
4	A4	X	Y	True
Index	Name	X	Y	Out from A
2	PtX1	X	Y	True
3	B2	X	Y	False
4	PtX2	X	Y	True
Index	Name	X	Y	Out from B
8	A6	X	Y	True
9	A7	X	Y	True

5 Conclusion

The cutting stock problem is an important industrial problem. In this paper we focused on preprocessing procedures which helps later for better performance of cutting algorithms. The idea of these procedures comes from the practical experience of the first author. Our preprocessing include the following steps: find the polygon orientation; remove the wasted points; find intersection points; polygon subtraction using boolean tables. These procedures increase the speed of the cutting algorithm preventing unnecessary calculations.

Acknowledgements This work has been funded by the Bulgarian Science Fund under the grants DFNI 02/10 and DFNI 12/5.

References

1. Gonalves, J.F.: A hybrid genetic algorithm-heuristic for a two-dimensional orthogonal packing problem. Eur. J. Oper. Res. **183**(3), 1212–1229 (2007)
2. Lodi, A., Martello, S., Vigo, D.: Recent advances on two-dimensional bin packing problems. Discrete Appl. Math. **123**, 379–396 (2002)
3. Alvarez-Valdes, R., Parajon, A., Tamarit, J.M.: A computational study of heuristic algorithms for two-dimensional cutting stock problems. In: 4th Metaheuristics International Conference (MIC2001), p. 1620 (2001)
4. Alvarez-Valdes, R., Parreno, F., Tamarit, J.M.: A Tabu Search algorithm for two dimensional non-guillotine cutting problems. Eur. J. Oper. Res. **183**(3), 1167–1182 (2007)
5. Cintra, G., Miyazawa, F., Wakabayashi, Y., Xavier, E.: Algorithms for two-dimensional cutting stock and strip packing problems using dynamic programming and column generation. Eur. J. Oper. Res. **191**, 61–85 (2008)
6. Dusberger, F., Raidl, G.R.: A variable neighborhood search using very large neighborhood structures for the 3-staged 2-dimensional cutting stock problem. In: Hybrid Metaheuristics. Lecture Notes in Computer Science, vol. 8457, pp. 85–99. Springer, Berlin (2014)
7. Dusberger, F., Raidl, G.R.: Solving the 3-staged 2-dimensional cutting stock problem by dynamic programming and variable neighborhood search. Electron. Notes Discrete Math. **47**, 133–140 (2015)
8. Parmar, K., Prajapati, H., Dabhi, V.: Cutting stock problem: a survey of evolutionary computing based solution. In: Proceedings of Green Computing Communication and Electrical Engineering (2014). https://doi.org/10.1109/ICGCCEE.2014.6921411
9. Evtimov, G., Fidanova, S.: Heuristic algorithm for 2D cutting stock problem. In: Large Scale Scientific Computing. Lecture Notes in Computer Science, vol. 10665, pp. 350–357 (2018)

Ant Colony Optimization Algorithm for Workforce Planning: Influence of the Algorithm Parameters

Stefka Fidanova, Olympia Roeva and Gabriel Luque

Abstract The workforce planning is a difficult optimization problem. It is important real life problem which helps organizations to determine workforce which they need. A workforce planning problem is very complex and needs special algorithms to be solved using reasonable computational resources. The problem consists to select set of employers from a set of available workers and to assign this staff to the tasks to be performed. The objective is to minimize the costs associated to the human resources needed to fulfil the work requirements. A good workforce planing is important for an organization to accomplish its objectives. The complexity of this problem does not allow the application of exact methods for instances of realistic size. Therefore we will apply Ant Colony Optimization (ACO) method which is a stochastic method for solving combinatorial optimization problems. On this paper we focus on influence of the parameters on ACO algorithm performance.

1 Introduction

Planing the workforce is difficult and important decision making industrial problem. It includes several level of complexity: selection and assignment. First the employers need to be selected from large set of available workers. The second is selected employers to be assigned to the jobs to be performed. The aim is to fulfil the work

S. Fidanova (✉)
Institute of Information and Communication Technology,
Bulgarian Academy of Science, Sofia, Bulgaria
e-mail: stefka@parallel.bas.bg

O. Roeva
Institute of Biophysics and Biomedical Engineering, Bulgarian Academy of Science,
Sofia, Bulgaria
e-mail: olympia@biomed.bas.bg

G. Luque
DLCS University of Mlaga, 29071 Mlaga, Spain
e-mail: gabriel@lcc.uma.es

K. Georgiev et al. (eds.), *Advanced Computing in Industrial Mathematics*,
Studies in Computational Intelligence 793,
https://doi.org/10.1007/978-3-319-97277-0_10

requirements with minimal assignment cost. The human resource management is one of the important parts of the production organization.

It is impossible to apply exact algorithms on realistic instances, because of the complexity of the problem. In [10, 16], deterministic workforce planing problem is studied. In the work [10] workforce planning models that contain non-linear models of human learning are reformulated as mixed integer programs. The authors show that the mixed integer program is much easier to solve than the non-linear program. A model of workforce planning is considered in [16]. The model includes workers differences, as well as the possibility of workers training and upgrading. A variant of the problem with random demands is proposed in [3, 17]. In [3] a two-stage stochastic program for scheduling and allocating cross-trained workers is proposed considering a multi-department service environment with random demands. In to some problems uncertainty has been employed [11, 14, 15, 19, 21]. In this case the corresponding objective function and given constraints is converted into crisp equivalents and then the model is solved by traditional methods [15] or the considered uncertain model is transformed into an equivalent deterministic form as it is shown in [19]. Most of the authors simplifies the problem by omitting some of the constraints. Some conventional methods can be applied on workforce planning problem as mixed linear programming [5], decomposition method [17]. However, for the more complex non-linear workforce planning problems, the convex methods are not applicable. There are some applications of heuristic methods including genetic algorithms [1, 13], memetic algorithms [18], scatter search [1], etc. Metaheuristic algorithms usually are applied when the problem is complex and has a very strong constraints [12, 20]. In this work we propose an Ant Colony Optimization (ACO) algorithm for workforce planning problem [7]. So far the ACO algorithm is proved to be very effecting solving various complex optimization problems [6, 9]. We focus on optimization of the algorithm parameters and finding the minimal number of ants which is need to find the best solution. It is known that when the number of ant is doubles, the computational time and the used memory are doubles. When the number of iteration is doubles, only the computational time is doubles. We look for a minimal product between number of ants and number of iterations that is needed to find the best solution.

We consider the variant of the workforce planning problem proposed in [1]. The rest of the paper is organized as follows. In Sect. 2 the problem is described. In Sect. 3 the ACO algorithm for workforce planing problem is proposed. Section 4 show computational results and analysis of algorithm performance according parameter settings. In Sect. 5 some conclusions and directions for future works are done.

2 Problem Formulation

On this paper we use the description of workforce planing problem given by Glover et al. [8]. There is a set of jobs $J = \{1, \ldots, m\}$, which must be completed during a fixed period. Each job j requires d_j hours to be completed. The set of available workers is $I = \{1, \ldots, n\}$. Every worker must perform every of assigned to him job minimum h_{min} hours, reason of efficiency. The worker i is available s_i hours. The maximal number of assigned jobs to a same worker is j_{max}. The workers have different skills and the set A_i shows the jobs that the worker i is qualified to perform. The maximal number of workers which can be assigned during the planed period is t or at most t workers may be selected from the set I of workers and the selected workers can be capable to complete all the jobs. The aim is to find feasible solution that minimizes the assignment cost.

Every worker i and job j are related with cost c_{ij} of assigning the worker to the job. The workforce planing problem can be represented mathematically as follows:

$$x_{ij} = \begin{cases} 1 \text{ if the worker } i \text{ is assigned to job } j \\ 0 \text{ otherwise} \end{cases}$$

$$y_i = \begin{cases} 1 \text{ if worker } i \text{ is selected} \\ 0 \text{ otherwise} \end{cases}$$

$$z_{ij} = \text{number of hours that worker } i$$

$$\text{is assigned to perform job } j$$

$$Q_j = \text{set of workers qualified to perform job } j$$

$$\text{Minimize} \sum_{i \in I} \sum_{j \in A_i} c_{ij}.x_{ij} \tag{1}$$

Subject to

$$\sum_{j \in A_i} z_{ij} \leq s_i.y_i \quad i \in I \tag{2}$$

$$\sum_{i \in Q_j} z_{ij} \geq d_j \quad j \in J \tag{3}$$

$$\sum_{j \in A_i} x_{ij} \leq j_{max}.y_j \quad i \in I \tag{4}$$

$$h_{min}.x_{ij} \leq z_{ij} \leq s_i.x_{ij} \quad i \in I,\ j \in A_i \tag{5}$$

$$\sum_{i \in I} y_i \leq t \tag{6}$$

$$x_{ij} \in \{0, 1\} \; i \in I,\ j \in A_i$$
$$y_i \in \{0, 1\} \quad i \in I$$
$$z_{ij} \geq 0 \qquad i \in I,\ j \in A_i$$

The assignment cost is the objective function of this problem. The number of hours for each selected worker is limited (inequality 2). The work must be done in full (inequality 3). The number of the jobs, that every worker can perform is limited (inequality 4). There is minimal number of hours that every job must be performed by every assigned worker to can work efficiently (inequality 5). The number of assigned workers is limited (inequality 6).

Different objective functions can be optimized with the same model. In this paper our aim is to minimize the total assignment cost. If $\tilde{c}_{ij}$ is the cost the worker i to performs the job j for one hour, than the objective function can minimize the cost of the hall jobs to be finished (on hour basis).

$$f(x) = \text{Min} \sum_{i \in I} \sum_{j \in A_i} \tilde{c}_{ij}.x_{ij} \tag{7}$$

Some worker can have preference to perform part of the jobs he is qualified and the objective function can be to maximize the satisfaction of the workers preferences or to maximize the minimum preference value for the set of selected workers.

As we mentioned above in this paper the assignment cost is minimized (Eq. 1). The workforce planning problem is difficult to be solved because of very restrictive constraints especially the relation between the parameters h_{min} and d_j. When the problem is structured (d_j is a multiple of h_{min}), it is more easier to find feasible solution, than for unstructured problems (d_j and h_{min} are not related).

3 Ant Colony Optimization Algorithm

The ACO metaheuristic is a method that follows the real ants behaviour when they look for a food. Real ants mark their path with a chemical substance, called pheromone. An isolated ant moves randomly, but when an ant detects a previously laid pheromone it can decide to follow the trail and to reinforce it with additional quantity of pheromone. A trail becomes more attractive if more ants follow it. Thus the ants collectively can find a shorter path between the nest and source of the food. The main idea of the ACO algorithms comes from this natural behaviour.

3.1 Main ACO Algorithm

When a problem is complex in computational point of view it is not practical to use traditional numerical methods. In this case are applied metaheuristic methods. A lot of problems coming from real life and especially from the industry need exponential number of calculations and the only option, when the problem is large, is to be solved by some metaheuristic methods in order to obtain a good solution with a reasonable computational resources [4].

ACO algorithm is proposed by Marco Dorigo [2]. Later some modification are proposed mainly in pheromone updating rules [4]. The ants behaviour is simulated by artificial ants. The problem is represented by graph. The solutions are represented by paths in a graph and we look for shorter path corresponding to given constraints. The requirements of ACO algorithm are as follows:

- Suitable representation of the problem by a graph;
- Suitable pheromone placement on the nodes or on the arcs of the graph;
- Appropriate problem-dependent heuristic function, which manage the ants to improve solutions;
- Pheromone updating rules;
- Transition probability rule, which specifies how to include new nodes in the partial solution.

The structure of the ACO algorithm is shown on Fig. 1.

The transition probability $p_{i,j}$, to choose the node j, when the current node is i, is a product of the heuristic information $\eta_{i,j}$ and the pheromone trail level $\tau_{i,j}$ related with this move, where $i, j = 1, \ldots, n$.

$$p_{i,j} = \frac{\tau_{i,j}^a \eta_{i,j}^b}{\sum\limits_{k \in Unused} \tau_{i,k}^a \eta_{i,k}^b}, \tag{8}$$

where $Unused$ is the set of unused nodes of the graph.

When the value of the heuristic information and the pheromone are higher, the corresponding node of the graph becomes more profitable. At the beginning, the initial pheromone level is the same for all elements of the graph and is set to a small positive constant value τ_0, $0 < \tau_0 < 1$. At the end of every iteration the ants update the pheromone values. Different ACO algorithms adopt different criteria to update the pheromone level [4].

The main pheromone trail update rule is:

Fig. 1 Pseudo-code of ACO algorithm

```
Ant Colony Optimization
Initialize number of ants;
Initialize the ACO parameters;
while not end condition do
        for k = 0 to number of ants
            ant k choses start node;
            while solution is not constructed do
                    ant k selects higher probability node;
            end while
        end for
        Update pheromone trails;
end while
```

$$\tau_{i,j} \leftarrow \rho\tau_{i,j} + \Delta\tau_{i,j}, \tag{9}$$

where ρ decreases the value of the pheromone, like the evaporation in a nature. $\Delta\tau_{i,j}$ is a new added pheromone, which is proportional to the quality of the solution. The quality of the solution is measured by the value of the objective function of the solution constructed by the ant.

An ant starts to construct its solution from a random node of the graph of the problem – this is a diversification of the search. Because the random start, the number of used ants can be small, comparing with other population based methods. The heuristic information represents the prior knowledge of the problem, which we use to better manage the ants. The pheromone is a global experience of the ants to find optimal solution. The pheromone is a tool for concentration of the search around best so far solutions.

3.2 ACO Algorithm for Workforce Planning

Very important part of the ant algorithm is the proper representation of the problem by graph. We propose the graph of the problem to be 3 dimensional and the node (i, j, z) corresponds the worker i to be assigned to the job j for time z. It every iteration an ant starts to construct its solution, from random node of the graph of the problem. Three random numbers are generated for every ant. The first random number is in the interval $[0, \ldots, n]$ and corresponds to the worker we assign. The second random number is in the interval $[0, \ldots, m]$ and corresponds to the job which this worker will perform. The third random number is in the interval $[h_{min}, \ldots, \min\{d_j, s_i\}]$ and corresponds to the number of hours worker i is assigned to performs the job j. The transition probability rule to include next nodes in the partial solution is applied, till the solution is constructed.

We propose the following heuristic information:

$$\eta_{ijl} = \begin{cases} l/c_{ij} & l = z_{ij} \\ 0 & otherwise \end{cases} \tag{10}$$

Thus the most cheapest worker, that can perform the job, is assigned as longer as possible. The ant chooses the node with the highest probability to move. When an ant has several possibilities for next node (several candidates have the same probability to be chosen), then the next node is chosen randomly between them.

When a new node is included we take in to account how many workers are assigned till now, how many time slots every worker is assigned till now and how many time slots are assigned per job till now. When some move of the ant do not meets the problem constraints, then the probability of this move is set to be 0. If it is impossible to include new nodes from the graph of the problem (for all nodes the value of the transition probability is 0), the construction of the solution stops. When the constructed solution is feasible the value of the objective function is the sum of the assignment cost of the assigned workers. If the constructed solution is not feasible, the value of the objective function is set to be equal to -1.

Only the ants, which constructed feasible solution are allowed to add new pheromone to the elements of their solutions. The new added pheromone is equal to the reciprocal value of the objective function.

$$\Delta\tau_{i,j} = \frac{\rho - 1}{f(x)} \tag{11}$$

Thus the nodes of the graph of the problem, which belong to better solutions (with less value of the objective function) receive more pheromone than others and become more desirable in the next iteration.

At the end of every iteration we compare the iteration best solution with the best so far solution. If the best solution from the current iteration is better than the best so far solution (global best solution), we update the global best solution with the current iteration best solution.

The end condition used in our algorithm is the number of iterations.

4 Influence of Algorithm Parameters

In this section we analyse algorithm performance according to the number of ants and the quality of the achieved solutions. The algorithm is written in C language and is run on Pentium desktop computer at 2.8 GHz with 4 GB of memory.

We use the artificially generated problem instances considered in [1].

The ACO algorithm parameters, based on our previous work [7], are shown in Table 1. In [7] we compare obtained by ACO algorithm solutions with the solutions from Genetic algorithm and Scatter search algorithm. The results show that our algorithm achieves better solutions for less execution time compared to the other two metaheuristic algorithms.

The set of test problems consists of ten structured and six unstructured problems. The structured problems are enumerated from $S01$ to $S10$ and unstructured problems

Table 1 Test instances characteristics

Parameters	Value
n	20
m	20
t	10
s_i	[50, 70]
j_{max}	[3, 5]
h_{min}	[10,15]

are enumerated from $U00$ to $U05$. The problem is structured when d_j is proportional to h_{min}.

First we run every one of the tests 30 times with 20 ants and the number of iterations was fixed to be 100.

In this work we focus on algorithm performance according number of ants. If the number of ants increases, the computational time and the used memory increase proportionally. If the number of iterations increases, only the computational time increases. If the computational time is fixed and we vary only the number of ants it means that we vary the number of iteration too, but in opposite direction, or if the time is fixed it is the same to fix the product of number of ants and number of iterations. We apply number of ants from the set {5, 10, 20, 40} and respectively, number of iteration – {400, 200, 100, 50}. Because of stochastic nature of the algorithm we run the algorithm for every test problem with every one of the ants number 30 times. We look for the maximal number of iterations, within the fixed computational time, which we need to find the best solution. We compare the product between the number of ants and number of needed iterations for the best solution with the aim to find best number of ants for this problem. The parameter settings for our ACO algorithm are shown in Table 2.

Tables 3 and 4 show the product between the number of ants and number of iterations needed to find best solution. When the product is the same for different number of ants the best is this with less number of ants, because in this case the used memory is less. The best results are shown in bold.

Lets compare the results reported in Tables 3 and 4. Regarding the structured problems the best execution time is when the number of ants is 5, for 8 test problems. Only for two of the test problems the execution time is better for 40 and 10 ants respectively, but the result is close to this achieved with 5 ants. Regarding unstructured test problems, for four of them the best execution time is when the number of used ants is 5. For two test problems the best results are achieved using 20 ants. For test U05, the result is quite close to the result with 5 ants. Only for test U03 the difference with the results using 5 ants is significant. Thus we can conclude that for this problem the best algorithm performance, using less computational resources, is when the number of used ants is 5.

Table 2 ACO parameter settings

Parameters	Value
Number of iterations	400, 200, 100, 50
ρ	0.5
τ_0	0.5
Number of ants	5, 10, 20, 40
a	1
b	1

Table 3 Computational time for structured problems

Test problem	Product ants × iterations			
	5 ants	10 ants	20 ants	40 ants
S01	195	200	260	**120**
S02	**195**	330	340	640
S03	**475**	490	580	1160
S04	**1540**	1540	1540	1560
S05	415	**320**	420	920
S06	**165**	250	420	520
S07	**570**	720	860	880
S08	**1125**	1130	1140	1120
S09	**855**	860	720	880
S10	**230**	230	520	840

Table 4 Computational time for unstructured problems

Test problem	Product ants × iterations			
	5 ants	10 ants	20 ants	40 ants
U00	**775**	780	780	2060
U01	**330**	340	340	440
U02	**160**	340	340	400
U03	1000	1000	**560**	640
U04	**760**	7 60	820	820
U05	295	295	**280**	280

5 Conclusion

In this paper we solve workforce planning problem applying ACO algorithm. We focus on influence of the algorithm parameters on its performance. We try to find minimal number of ants and minimal execution time and memory, which are needed to find best solution. The results show that for most of the test problems, the minimal computational resources are used when the number of ants is 5. As a future work we will combine our ACO algorithm with appropriate local search procedure for eventual further improvement of the algorithm performance and solutions quality.

Acknowledgements Work presented here is partially supported by the Bulgarian National Scientific Fund under the grants DFNI-DN 12/5 "Efficient Stochastic Methods and Algorithms for Large Scale Problems" and DFNI-DN 02/10 "New Instruments for Data Mining and their Modeling".

References

1. Alba, E., Luque, G., Luna, F.: Parallel metaheuristics for workforce planning. J. Math. Model. Algorithms **6**(3), 509–528 (2007) (Springer)
2. Bonabeau, E., Dorigo, M., Theraulaz, G.: Swarm Intelligence: From Natural to Artificial Systems. Oxford University Press, New York (1999)
3. Campbell, G.: A two-stage stochastic program for scheduling and allocating cross-trained workers. J. Oper. Res. Soc. **62**(6), 1038–1047 (2011)
4. Dorigo, M., Stutzle, T.: Ant Colony Optimization. MIT Press (2004)
5. Easton, F.: Service completion estimates for cross-trained workforce schedules under uncertain attendance and demand. Prod. Oper. Manage. **23**(4), 660–675 (2014)
6. Fidanova, S., Roeva, O., Paprzycki, M., Gepner, P.: InterCriteria analysis of ACO start strategies. In: Proceedings of the 2016 Federated Conference on Computer Science and Information Systems, pp. 547–550 (2016)
7. Fidanova, S., Luquq, G., Roeva, O., Paprzycki, M., Gepner, P.: Ant colony optimization algorithm for workforce planning. In: FedCSIS'2017, IEEE Xplorer, IEEE Catalog Number CFP1585N-ART, pp. 415–419 (2017)
8. Glover, F., Kochenberger, G., Laguna, M., Wubbena, T.: Selection and assignment of a skilled workforce to meet job requirements in a fixed planning period. In: MAEB04, pp. 636–641 (2004)
9. Grzybowska, K., Kovcs, G.: Sustainable supply chain—supporting tools. In: Proceedings of the 2014 Federated Conference on Computer Science and Information Systems, vol. 2, pp. 1321–1329 (2014)
10. Hewitt, M., Chacosky, A., Grasman, S., Thomas, B.: Integer programming techniques for solving non-linear workforce planning models with learning. Eur. J. Oper. Res. **242**(3), 942–950 (2015)
11. Hu, K., Zhang, X., Gen, M., Jo, J.: A new model for single machine scheduling with uncertain processing time. J. Intell. Manuf. **28**(3), 717–725 (2015)
12. Isah, O.R., Usman, A.D., Tekanyi, A.M.S.: A hybrid model of PSO algorithm and artificial neural network for automatic follicle classification. Int. J. Bioautomation **21**(1), 43–58 (2017)
13. Li, G., Jiang, H., He, T.: A genetic algorithm-based decomposition approach to solve an integrated equipment-workforce-service planning problem. Omega **50**, 1–17 (2015)
14. Li, R., Liu, G.: An uncertain goal programming model for machine scheduling problem. J. Intell. Manuf. **28**(3), 689–694 (2014)
15. Ning, Y., Liu, J., Yan, L.: Uncertain aggregate production planning. Soft Comput. **17**(4), 617–624 (2013)
16. Othman, M., Bhuiyan, N., Gouw, G.: Integrating workers' differences into workforce planning. Comput. Industr. Eng. **63**(4), 1096–1106 (2012)
17. Parisio, A., Jones, C.N.: A two-stage stochastic programming approach to employee scheduling in retail outlets with uncertain demand. Omega **53**, 97–103 (2015)
18. Soukour, A., Devendeville, L., Lucet, C., Moukrim, A.: A Memetic algorithm for staff scheduling problem in airport security service. Expert Syst. Appl. **40**(18), 7504–7512 (2013)
19. Yang, G., Tang, W., Zhao, R.: An uncertain workforce planning problem with job satisfaction. Int. J. Mach. Learn. Cybern. (2016) (Springer). https://doi.org/10.1007/s13042-016-0539-6. http://rd.springer.com/article/10.1007/s13042-016-0539-6
20. Zeng, J., Li, Y.: The use of adaptive genetic algorithm for detecting Kiwifruits variant subculture seedling. Int. J. Bioautomation **21**(4), 349–356 (2017)
21. Zhou, C., Tang, W., Zhao, R.: An uncertain search model for recruitment problem with enterprise performance. J Intell. Manuf. **28**(3), 295–704 (2014). https://doi.org/10.1007/s10845-014-0997-1

Convergence of Homotopy Perturbation Method for Solving of Two-Dimensional Fuzzy Volterra Functional Integral Equations

Atanaska Georgieva and Artan Alidema

Abstract In this paper, Homotopy Perturbation Method (HPM) is applied to solve two-dimensional fuzzy Volterra functional integral equations (2D-FVFIE). We use parametric form of fuzzy functions and convert a 2D-FVFIE to a system of Volterra functional integral equations with three variables in crisp case. We use the HPM to find the approximate solution of the converted system, which is the approximate solution for 2D-FVFIE. Also, the existence and uniqueness of the solution and convergence of the proposed methods are proved. The main tool in this discussion is fixed point theorem. The error estimate in this method is also given. Finally, we give some examples to demonstrate the accuracy of the method. The solved problems reveal that the proposed method is effective and simple, and in some cases, it gives the exact solution.

1 Introduction

The topics of fuzzy differential equations and fuzzy integral equations which attracted growing interest for some time, in particular in relation to fuzzy control, have been rapidly developed in recent years. The study of fuzzy integral equations begins with the investigations of Kaleva [18] and Seikkala [23] for the fuzzy Volterra integral equation that is equivalent to the initial value problem for first-order fuzzy differential equations. The solutions of integral equations have a major role in the field of science and engineering. A physical event can be modelled by the differential equation, an integral equation. Since many of these equations cannot be solved explicitly, it is often necessary to resort to numerical techniques which are appropriate combinations of numerical integration and interpolation [3, 19].

A. Georgieva (✉)
FMI, University of Plovdiv Paisii Hilendarski, 24 Tzar Asen, 4000 Plovdiv, Bulgaria
e-mail: atanaska@uni-plovdiv.bg; afi2000@abv.bg

A. Alidema
FMNS, University of Pristina “Hasan Prishtina”, str “Nena Tereze”, 10000 Prishtine, Kosovo
e-mail: artan.alidema@uni-pr.edu

K. Georgiev et al. (eds.), *Advanced Computing in Industrial Mathematics*,
Studies in Computational Intelligence 793,
https://doi.org/10.1007/978-3-319-97277-0_11

The problems posed in the study of fuzzy Volterra integral equations are: the existence, existence and uniqueness, boundendness of the solution and construction numerical methods for approximate solution. The existing numerical methods for fuzzy Volterra integral equations are based on various techniques: successive approximations and iterative methods [4, 5, 7, 10], analytic-numeric methods like Adomian decomposition, homotopy analysis and homotopy perturbation [2, 6, 11], quadrature rules and Nystrum techniques [22].

The HPM proposed by He [14], for solving differential and linear and nonlinear integral equations, has been subject of extensive numerical and analytical studies. He used the HPM to solve Duffing equation [15], Blasius equation [16], and then the idea found its way in sciences and has been used to solve nonlinear quadratic Riccati differential equations [1], wave equations [17], integral equations [9]. Until to now, the HPM have been developed for solving various types of differential and integral equations. The convergence of the proposed method is also proved. For example in [13], Hassan et al. are use the HPM to solve nonlinear differential equations.

In this paper we propose HPM for fuzzy Volterra functional integral equation in two variables

$$u(s,t) = g(s,t) \oplus f(s,t,u(s,t)) \oplus \int_c^t \int_a^s H(s,t,x,y,u(x,y))dxdy, \tag{1}$$

where $g,\ u : A = [a,b] \times [c,d] \to E^1$ are continuous fuzzy-number valued functions, $H : A \times A \times E^1 \to E^1$,$f : A \times E^1 \to E^1$ are continuous functions on E^1. The functions H and f have the following form: $H(s,t,x,y,u(x,y)) = k(s,t,x,y) \odot u^{p_2}(x,y)$, $f(s,t,u(s,t)) = \varphi(s,t) \odot u^{p_1}(s,t)$. For convenience, p_1, p_2 are a positive integers, $k : A \times A \to R_+$ and $\varphi : A \to R_+$. The set E^1 is the set of all fuzzy numbers.

The remainder of this paper is organized as follows: in Sect. 2, we present the basic notations of fuzzy numbers, fuzzy functions and fuzzy integrals. In Sect. 3, the convergence of present method is discussed. We explain the basic idea of the HPM in Sect. 4. In Sect. 5, the parametric from of 2D-FVFIE is introduced and then the method is applied for solving this equation. Some numerical examples are presented in Sect. 6.

2 Preliminaries

In this section, we review some notions and results about fuzzy numbers, fuzzy-number-valued functions and fuzzy integrals.

Definition 1 [12] A fuzzy number is a function $u : R \to [0,1]$ satisfying the following properties: u is upper semi-continuous on R, $u(x) = 0$ outside of some interval $[c,d]$, there are the real numbers a and b with $c \le a \le b \le d$, such that u is increasing on $[c,a]$, decreasing on $[b,d]$ and $u(x) = 1$ for each $x \in [a,b]$ and u is fuzzy convex set i. e. that is $u(\lambda x + (1-\lambda)y) \ge \min\{u(x), u(y)\}$ for all $x, y \in R$ and $\lambda \in [0,1]$.

The set of all fuzzy numbers is denoted by E^1. Any real number $a \in R$ can be interpreted as a fuzzy number $\tilde{a} = \chi_{[a]}$ and therefore $R \subset E^1$. For any $0 < r \leq 1$ we denote the $r-$level set $[u]^r = \{x \in R : u(x) \geq r\}$ that is a closed interval and $[u]^r = [u_-^r, u_+^r]$ for all $r \in [0, 1]$. These lead to the usual parametric representation of a fuzzy number: $u = (u_-^r, u_+^r)$ for all $r \in [0, 1]$, where u_-, u_+ can be considered as functions $u_-, u_+ : [0, 1] \to R$, such that u_- is increasing and u_+ is decreasing.

For $u, v \in E^1$, $k \in R$, the addition and the scalar multiplication are defined by

$$[u \oplus v]^r = [u]^r + [v]^r = [u_-^r + v_-^r, u_+^r + v_+^r] \text{ and}$$
$$[k \odot u]^r = k.[u]^r = \begin{cases} [ku_-^r, ku_+^r], & \text{if } k \geq 0 \\ [ku_+^r, ku_-^r], & \text{if } k < 0 \end{cases} \text{ for all } r \in [0, 1].$$

The neutral element respect to $\oplus$ in E^1, denoted by $\tilde{0} = \chi_{[0]}$. The algebraic properties of addition and scalar multiplication of fuzzy numbers are given in [24].

As a distance between fuzzy numbers we use the Hausdorff metric [24] defined by $D(u, v) = \sup_{r \in [0,1]} \max\{|u_-^r - v_-^r|, |u_+^r - v_+^r|\}$ for any $u, v \in E^1$.

Lemma 1 [24] *The Hausdorff metric has the following properties:*

(i) (E^1, D) *is a complete metric space,*
(ii) $D(u \oplus w, v \oplus w) = D(u, v)$ for all $u, v, w \in E^1$,
(iii) $D(u \oplus v, w \oplus e) \leq D(u, w) + D(v, e)$ for all $u, v, w, e \in E^1$,
(iv) $D(u \oplus v, \tilde{0}) \leq D(u, \tilde{0}) + D(v, \tilde{0})$ for all $u, v \in E^1$,
(v) $D(k \odot u, k \odot v) = |k| D(u, v)$ for all $u, v \in E^1$, for all $k \in R$.
(vi) $D(k_1 \odot u, k_2 \odot u) = |k_1 - k_2| D(u, \tilde{0})$ for all $k_1,\ k_2 \in E^1$, *with* $k_1 k_2 \geq 0$ *and* for all $u \in E^1$.

For all $u, v \in E^1$ with positive supports the multiplication is defined by $[u \odot v]^r = [u_-^r.v_-^r, u_+^r.v_+^r]$. In addition, for all strictly increasing and positive real function ψ, we have $[\psi(u)]^r = [\psi(u_-^r), \psi(u_+^r)]$.

For any fuzzy-number-valued function $f : A \to E^1$ we can define the functions $\underline{f}(.,.,r), \overline{f}(.,.,r) : A \to R$, by $\underline{f}(s, t, r) = (f(s, t))_-^r,\ \overline{f}(s, t, r) = (f(s, t))_+^r$ for each $(s, t) \in A$, for each $r \in [0, 1]$. These functions are called the left and right $r-$level functions of f.

Definition 2 [25] A fuzzy-number-valued function $f : A = [a, b] \times [c, d] \to E^1$ is said to be continuous at $(s_0, t_0) \in A$ if for each $\varepsilon > 0$ there exits $\delta > 0$ such that $D(f(s, t), f(s_0, t_0)) < \varepsilon$ whenever $((s - s_0)^2 + (t - t_0)^2)^{\frac{1}{2}} < \delta$. If f is continuous for each $(s, t) \in A$ then we say that f is continuous on A.

On the set $C(A, E^1) = \{f : A \to E^1;\ f \text{ is continuous}\}$ there is defined the metric $D^*(f, g) = \sup_{(s,t) \in A} D(f(s, t), g(s, t))$, for all $f, g \in C(A, E^1)$. We see that $(C(A, E^1), D^*)$ is a complete metric space.

Definition 3 [25] A fuzzy-number-valued function $f : A \to E^1$ is called bounded iff there is $M \geq 0$, such that $D(f(s, t); \tilde{0}) \leq M$ for all $(s; t) \in A$.

The definition of fuzzy Henstock integral and its properties are given in [21].

Definition 4 [21] Let $f : A \to E^1$, for $\Delta_x^n : a = x_0 < x_1 < \cdots < x_n = b$ and $\Delta_y^m : c = y_0 < y_1 < \cdots < y_m = d$, be two partitions of the intervals $[a, b]$ and $[c, d]$, respectively. Let us consider the intermediate points $\xi_i \in [x_{i-1}, x_i]$ and $\eta_j \in [y_{j-1}, y_j]$, $i = 1, \ldots, n$, $j = 1, \ldots, m$, and the functions $\delta : [a, b] \to R_+$ and $\sigma : [c, d] \to R_+$. The divisions $P_x = ([x_{i-1}, x_i]; \xi_i)$, $i = 1, \ldots, n$, and $P_y = ([y_{j-1}, y_j]; \eta_j)$, $j = 1, \ldots, m$, denoted shortly by $P_x = (\Delta^n, \xi)$ and $P_y = (\Delta^m, \eta)$ are said to be δ-fine and σ-fine, respectively, if $[x_{i-1}, x_i] \subseteq (\xi_i - \delta(\xi_i), \xi_i + \delta(\xi_i))$ and $[y_{j-1}, y_j] \subseteq (\eta_j - \sigma(\eta_j), \eta_j + \sigma(\eta_j))$.

The function f is called two-dimensional Henstock integrable to $I \in E^1$ if for every $\varepsilon > 0$ there are functions $\delta : [a, b] \to R_+$ and $\sigma : [c, d] \to R_+$ such that for any δ-fine and σ-fine divisions we have $D(\sum_{j=1}^{m}\sum_{i=1}^{n}(x_i - x_{i-1})(y_j - y_{j-1}) \odot f(\xi_i, \eta_j), I) < \varepsilon$, where $\sum$ denotes the fuzzy summation. Then, I is called the two-dimensional Henstock integral of f and denoted by $I(f) = (FH)\int_c^d (FH)\int_a^b f(s, t)dsdt$. If the above δ and σ are constant functions, then one recaptures the concept of Riemann integral. In this case, $I \in E^1$ will be called two-dimensional integral of f on A and will be denoted by $(FR)\int_c^d (FR)\int_a^b f(s, t)dsdt$.

Lemma 2 [21] *Let $f : A \to E^1$, then f is (FH)-integrable if and only if $\underline{f}$ and $\overline{f}$ are Henstock integrable for any $r \in [0, 1]$. Moreover*

$$\left[(FH)\int_c^d (FH)\int_a^b f(s, t)dsdt\right]^r =$$
$$= \left[(H)\int_c^d (H)\int_a^b \underline{f}(s, t, r)dsdt, (H)\int_c^d (H)\int_a^b \overline{f}(s, t, r)dsdt\right].$$

Also, if f is continuous then $\underline{f}(., ., r)$ and $\overline{f}(., ., r)$ are continuous for any $r \in [0, 1]$ and consequently, they are Henstock integrable. Using Lemma 2. we deduce that f is (FH)-integrable.

Lemma 3 [21] *If f and g are fuzzy Henstock integrable functions on A and if the function given by $D(f(s, t), g(s, t))$ is Lebesgue integrable, then*

$$D\left((FH)\int_c^d (FH)\int_a^b f(s, t)dsdt, (FH)\int_c^d (FH)\int_a^b g(s, t)dsdt\right) \le$$
$$\le (L)\int_c^d (L)\int_a^b D(f(s, t), g(s, t))dsdt.$$

Lemma 4 [21] *If $f : A \to E^1$ is an integrable bounded function then for any fixed $(u, v) \in A$, the function $\varphi_{u,v} : A \to R_+$ defined by $\varphi_{u,v}(s,t) = D(f(u,v), f(s,t))$ is Lebesgue integrable on A.*

3 Convergence of the HPM

In this section the existence, uniqueness of the solution of equation (1) and convergence of the HPM is proved.

Lemma 5 [20] *Let the functions $g \in C(A, E^1)$ and $h \in C(A, R_+)$. Then the function $h.g : A \to E^1$ given by $(h.g)(s,t) = h(s,t) \odot g(s,t)$, is continuous on A.*

Lemma 6 *Let the functions $k \in C(A \times A, R_+)$, $u \in C(A, E^1)$ and p is positive integer. Then the function $F_u : A \to E^1$ defined by $F_u(s,t) = \int_c^t \int_a^s k(s,t,x,y) \odot u^p(x,y))dxdy$ is continuous on A.*

Proof Analogously of Lemma 8 [8].

Lemma 7 *Let the functions $f, h \in C(A, E^1)$. Then for each positive integer p there exists constant $L > 0$ such that*

$$D(f^p(s,t), h^p(s,t)) \leq pL^{p-1}D(f(s,t), h(s,t)) \quad \text{for all} \quad (s,t) \in A. \tag{2}$$

Proof We prove that $f(s,t)$, $h(s,t)$, satisfy Lipschitz conditions for $p > 0$

$$D(f^p(s,t), h^p(s,t)) = \sup_{r \in [0,1]} \max\{|\underline{f}^p(s,t,r) - \underline{h}^p(s,t,r)|, |\overline{f}^p(s,t,r) - \overline{h}^p(s,t,r)|\} \tag{3}$$

The functions $\underline{f}(s,t,r)$, $\underline{h}(s,t,r)$ are continuous on A therefore are bounded functions. Hence, there exists $\underline{L} > 0$ such that $|\underline{f}(s,t,r)| \leq \underline{L}$ and $|\underline{h}(s,t,r)| \leq \underline{L}$ for each $(s,t) \in A$. Consequently

$$|\underline{f}^p(s,t,r) - \underline{h}^p(s,t,r)| \leq |\underline{f}(s,t,r) - \underline{h}(s,t,r)||\underline{f}^{p-1}(s,t,r) + \underline{f}^{p-2}(s,t,r)\underline{h}(s,t,r)$$
$$+ \cdots + \underline{f}(s,t,r)\underline{h}^{p-2}(s,t,r) + \underline{f}^{p-1}(s,t,r)| \leq p\underline{L}^{p-1}|\underline{f}(s,t,r) - \underline{h}(s,t,r)|.$$

Analogously, the functions $\overline{f}$, $\overline{h}$ are continuous on A therefore are bounded functions. Hence, there exists $\overline{L} > 0$ such that $|\overline{f}(s,t,r)| \leq \overline{L}$ and $|\overline{h}(s,t,r)| \leq \overline{L}$ for each $(s,t) \in A$. Hence $|\overline{f}^p(s,t,r) - \overline{h}^p(s,t,r)| \leq p\overline{L}^{p-1}|\overline{f}(s,t,r) - \overline{h}(s,t,r)|$.
Assume that $L = \max\{\underline{L}, \overline{L}\}$ then from (3) we obtain

$$D(f^p(s,t), h^p(s,t)) \leq pL^{p-1}D(f(s,t), h(s.t))$$

We introduce the following conditions:

(i) $g \in C(A, E^1)$, $\varphi \in C(A, R_+)$ and $k \in C(A \times A, R_+)$;

(ii) $\lambda = M_\varphi p_1 L_{p_1}^{p_1-1} + M_k p_2 L_{p_2}^{p_2-1} \Delta < 1$, where $|k(s,t,x,y)| \leq M_k$ and $|\varphi(s,t)| \leq M_\varphi$ for all $(s,t), (x,y) \in A$, according to the continuity of k, φ, the constants p_1, p_2 are positive integer, $L_{p_1}, L_{p_2} > 0$ are from Lemma 7, respectively for p_1, p_2 and $\Delta = (b-a)(d-c)$;

Theorem 1 *Let the conditions (i)–(ii) are fulfilled. Then the HPM*

$$\begin{aligned} &u_0(s,t) = g(s,t) \\ &u_m(s,t) = g(s,t) \oplus \varphi(s,t) u_{m-1}^{p_1}(s,t) \oplus \\ &\oplus (FR) \int_c^t (FR) \int_a^s k(s,t,x,y) \odot u_{m-1}^{p_2}(x,y) dxdy, \quad m \geq 1 \end{aligned} \tag{4}$$

converges to the unique solution u^* of (1). In addition, the error bound of this method is

$$D^*(u^*, u_m) \leq \frac{\lambda^{m+1}}{1-\lambda} M_0, \tag{5}$$

where $D(g(s,t), \tilde{0}) \leq M_0$ for all $(s,t) \in A$.

Proof Let $\mathscr{F}(A, E^1) = \{f : A \to E^1\}$. We define the operator $A : C(A, E^1) \to \mathscr{F}(A, E^1)$ by $A(u)(s,t) = g(s,t) \oplus \varphi(s,t) \odot u^{p_1}(s,t) \oplus \int_c^t \int_a^s k(s,t,x,y) \odot u^{p_2}(x,y) dxdy$, $(s,t) \in A$, $u \in C(A, E^1)$. Firstly, we prove that $A(C(A, E^1)) \subset C(A, E^1)$. Let $u \in C(A, E^1)$. From Lemma 5 the function $\Phi_u : A \to E^1$ defined by $\Phi_u(s,t) = \varphi(s,t) \odot u^{p_1}(s,t)$ is continuous on A for any $u \in C(A, E^1)$. Applying Lemma 6 it follows that the function $F_u : A \to E^1$ defined by $F_u(s,t) = \int_c^t \int_a^s k(s,t,x,y) \odot u^{p_2}(x,y)) dxdy$ is continuous on A for any $u \in C(A, E^1)$. Since $g \in C(A, E^1)$, we conclude that the operator $A(u)$ is continuous on A for any $u \in C(A, E^1)$.

Now, we prove that $A : C(A, E^1) \to C(A, E^1)$ is a contraction. For this purpose let arbitrary $u, v \in C(A, E^1)$. Applying Lemma 7 it follows that there exist the constants L_{p_1} and L_{p_2} such that $D(u^{p_1}(s,t), v^{p_1}(s,t)) \leq p_1 L_{p_1}^{p_1-1} D(u(s,t), v(s,t))$, $D(u^{p_2}(s,t), v^{p_2}(s,t)) \leq p_2 L_{p_2}^{p_2-1} D(u(s,t), v(s,t))$. Hence from condition (ii) we have

$$D(A(u)(s,t),A(v)(s,t)) \le D(\varphi(s,t)\odot u^{p_1}(s,t),\varphi(s,t)\odot v^{p_1}(s,t))+$$

$$+\int_c^t\int_a^s D\left(k(s,t,x,y)\odot u^{p_2}(x,y),k(s,t,x,y)\odot v^{p_2}(x,y))\right)dxdy \le$$

$$\le |\varphi(s,t)|D(u^{p_1}(s,t),v^{p_1}(s,t)) + \int_c^t\int_a^s |k(s,t,x,y)|D(u^{p_2}(x,y),v^{p_2}(x,y))dxdy \le$$

$$\le M_\varphi p_1 L_{p_1}^{p_1-1}D(u(s,t),v(s,t)) + M_k p_2 L_{p_2}^{p_2-1}\Delta D(u(s,t),v(s,t)) \le \lambda D^*(u,v).$$

Since $\lambda < 1$, the operator A is a contraction map on Banach space $C(A,E^1)$, therefore by using the Banachs fixed point principle theorem infer that equation (1) has a unique solution u^* in $C(A,E^1)$. Also

$$D(u^*(s,t),u_m(s,t)) \le \frac{\lambda^m}{1-\lambda}D(u_0(s,t),u_1(s,t))$$

$$D(u_0(s,t),u_1(s,t)) =$$

$$= D(g(s,t)\oplus\tilde{0}, g(s,t)\oplus\varphi(s,t)\odot u_0^{p-1}(s,t)\oplus\int_c^t\int_a^s k(s,t,x,y)\odot u_0^{p_2}(x,y)dxdy \le$$

$$\le D(\tilde{0},\varphi(s,t)\odot u_0^{p_1}(s,t)) + \int_c^t\int_a^s D(\tilde{0},k(s,t,x,y)\odot u_0^{p_2}(x,y)dxdy) \le$$

$$\le |\varphi(s,t)|D(\tilde{0},u_0^{p_1}(s,t)) + \int_c^t\int_a^s |k(s,t,x,y)|D(\tilde{0},u_0^{p_2}(x,y))dxdy \le$$

$$\le M_\varphi p_1 L_{p_1}^{p_1-1}D(\tilde{0},u_0(s,t)) + M_k p_2 L_{p_2}^{p_2-1}\Delta D(\tilde{0},u_0(s,t)) \le \lambda M_0$$

It is obtained that, $D^*(u^*,u_m)) \le \frac{\lambda^m}{1-\lambda}D(u_0(s,t),u_1(s,t)) \le \frac{\lambda^{m+1}}{1-\lambda}M_0$ which completes the proof.

4 Basic Idea of the HPM

Consider the following differential equation $N[f(\tau)] = 0$, where N is a nonlinear operator, $f(\tau)$ is an unknown function. We construct the zero-order deformation equation

$$(1-q)L[\phi(\tau;q) - f_0(\tau)] = qhH(\tau)N[\phi(\tau;q)], \tag{6}$$

where $q \in [0, 1]$ is the embedding parameter, $h \neq 0$ is an auxiliary parameter, $H(\tau) \neq 0$ an auxiliary function and L is an auxiliary linear operator, $f_0(\tau)$ is an initial quests of $f(\tau)$.

When q = 0 and q = 1, it holds $\phi(\tau; 0) = f_0$, $\phi(\tau; 1) = f(\tau)$, so, as q increase from to 0 to 1, the solution $\phi(\tau; q)$ that is a unknown function varies from the initial quests $f_0(\tau)$ to the solution $f(\tau)$. Due to Taylors theorem

$$\phi(\tau; q) = f_0(\tau) + \sum_{m=1}^{\infty} f_m(\tau) q^m, \tag{7}$$

where,

$$f_m(\tau) = \frac{\partial^m \phi(\tau; q)}{m! \partial q^m}_{|q=0} \tag{8}$$

If the auxiliary linear operator, the initial quests the auxiliary parameter h and $H(\tau)$ be properly chosen the series (7) converges at q = 1, then we have

$$f(\tau) = f_0(\tau) + \sum_{m=1}^{\infty} f_m(\tau) \tag{9}$$

which must be one the solutions of original nonlinear equation. In the HPM $h = -1$ and $H(\tau) = 1$ are chosen, then the Eq. (6) becomes

$$(1 - q)L[\phi(\tau; q) - f_0(\tau)] + qN[\phi(\tau; q)] = 0,$$

where as the solution obtained directly. According to the Eq. (9), the governing equation can be deduced from the zero-order deformation (6). Define the vector $\tilde{f_m} = f_0(\tau), f_1(\tau), \ldots, f_m(\tau)$, by differentiating (6) m times with respect to q and then dividing them by $m!$ and finally setting $q = 0$, we get the following m-th order deformation equation

$$L[f_m(\tau) - \chi_m f_{m-1}(\tau)] = hH(\tau) R_m(\tilde{f}_{m-1}). \tag{10}$$

Applying L^{-1} on both sides of Eq. (10), we get

$$f_m(\tau) = \chi_m f_{m-1}(\tau) + hL^{-1}[H(\tau) R_m(\tilde{f}_{m-1})],$$

where

$$R_m(\tilde{f}_{m-1}) = \frac{1}{(m-1)!} \frac{\partial^{m-1} N[\phi(\tau; q)]}{\partial q^{m-1}}|_{q=0}$$

$$\chi_m = \begin{cases} 0, & m \leq 1 \\ 1, & m > 1 \end{cases}$$

5 HPM for Solving 2D-FVFIE

In this section the parametric form of two-dimensional fuzzy Volterra functional integral equation with respect the definition is introduced. Let $(\underline{u}(s,t,r), \overline{u}(s,t,r))$ and $(\underline{g}(s,t,r), \overline{g}(s,t,r))$, $0 \le r \le 1$ be the parametric form of $u(s,t)$ and $g(s,t)$. So the parametric form of this equation is follows:

$$\begin{aligned}
\overline{u}(s,t,r) &= \overline{g}(s,t,r) + f_2(s,t,\underline{u}(s,t,r),\overline{u}(s,t,r)) \\
&\quad + \int_c^t \int_a^s H_2(s,t,x,y,\underline{u}(x,y,r),\overline{u}(x,y,r))dxdy \\
\underline{u}(s,t,r) &= \underline{g}(s,t,r) + f_1(s,t,\underline{u}(s,t,r),\overline{u}(s,t,r)) \\
&\quad + \int_c^t \int_a^s H_1(s,t,x,y,\underline{u}(x,y,r),\overline{u}(x,y,r))dxdy
\end{aligned}$$

$$H_1(s,t,x,y,\underline{u}(x,y,r),\overline{u}(x,y,r)) = \begin{cases} k(s,t,x,y)\underline{u}^{p_2}(x,y,r), & \text{if } k(s,t,x,y) \ge 0 \\ k(s,t,x,y)\overline{u}^{p_2}(x,y,r), & \text{if } k(s,t,x,y) < 0 \end{cases}$$

$$f_1(s,t,\underline{u}(s,t,r),\overline{u}(s,t,r)) = \begin{cases} \varphi(s,t)\underline{u}^{p_1}(s,t,r), & \text{if } \varphi(s,t) \ge 0 \\ \varphi(s,t)\overline{u}^{p_1}(s,t,r), & \text{if } \varphi(s,t) < 0 \end{cases}$$

$$H_2(s,t,x,y,\underline{u}(x,y,r),\overline{u}(x,y,r)) = \begin{cases} k(s,t,x,y)\overline{u}^{p_2}(x,y,r), & \text{if } k(s,t,x,y) \ge 0 \\ k(s,t,x,y)\underline{u}^{p_2}(x,y,r), & \text{if } k(s,t,x,y) < 0 \end{cases}$$

$$f_2(s,t,\underline{u}(s,t,r),\overline{u}(s,t,r)) = \begin{cases} \varphi(s,t)\overline{u}^{p_1}(s,t,r), & \text{if } \varphi(s,t) \ge 0 \\ \varphi(s,t)\underline{u}^{p_1}(s,t,r), & \text{if } \varphi(s,t) < 0 \end{cases}$$

for $a \le x \le s \le b$, $c \le y \le t \le d$, $0 \le r \le 1$, and p_1, p_2 are positive integers. The HPM method is applied to solve 2D-FVFIE. Let $k(s,t,x,y) \ge 0$ for all $a \le x \le s \le b$, $c \le y \le t \le d$. Then the parametric form of the equation (1) is,

$$\begin{aligned}
\underline{u}(s,t,r) &= \underline{g}(s,t,r) + \varphi(s,t)\underline{u}^{p_1}(s,t,r) + \int_c^t \int_a^s k(s,t,x,y)\underline{u}^{p_2}(x,y,r)dxdy, \\
\overline{u}(s,t,r) &= \overline{g}(s,t,r) + \varphi(s,t)\overline{u}^{p_1}(s,t,r) + \int_c^t \int_a^s k(s,t,x,y)\overline{u}^{p_2}(x,y,r)dxdy.
\end{aligned} \tag{11}$$

To explain the HPM, we consider

$$\begin{aligned}
N(\underline{u}(s,t,r)) &= \underline{u}(s,t,r) - \underline{g}(s,t,r) - \varphi(s,t)\underline{u}^{p_1}(s,t,r) \\
&\quad - \int_c^t \int_a^s k(s,t,x,y)\underline{u}^{p_2}(x,y,r)dxdy,
\end{aligned}$$

According (9) and (10) $\underline{u}(s, t, r) = \underline{u}_0(s, t, r) + \sum_{m=1}^{\infty} \underline{u}_m(s, t, r)$,

$$L[\underline{u}_m(s, t, r) - \chi_m \underline{u}_{m-1}(s, t, r)] = hH(s, t, r)R_m(\tilde{\underline{u}}_{m-1}) \tag{12}$$

where

$$R_m(\tilde{\underline{u}}_{m-1}) = \frac{1}{(m-1)!} \frac{\partial^{m-1} N(\sum_{m=0}^{\infty} \underline{u}_m(s,t,r)q^m)}{\partial^{m-1} q}_{|q=0} =$$

$$= \frac{1}{(m-1)!} \frac{\partial^{m-1}\left(\sum_{m=0}^{\infty} \underline{u}_m(s,t,r)q^m - \underline{g}(s,t,r) - \varphi(s,t)[\sum_{m=0}^{\infty} \underline{u}(s,t,r)q^m]^{p_1} - \int_c^t \int_a^s K(s,t,x,y)[\sum_{m=0}^{\infty} \underline{u}(s,t,r)q^m]^{p_2}\right)}{\partial^{m-1} q}_{|q=0}$$

we take an initial quests $\underline{u}_0(s, t, r) = \underline{g}(s, t, r)$ and auxiliary function $H(s, t, r) = 1$ and the auxiliary linear operator $L[r(\overline{s}, t)] = r(s, t)$ Now the solution of the m-th order deformation equation (12).

$$\underline{u}_m(s, t, r) = \chi_m \underline{u}_{m-1}(s, t, r) + hH(S)R_m(\tilde{\underline{u}}_{m-1}), \quad (m \geq 1), \tag{13}$$

where $h = -1$ and $H(s, t, r) = 1$ with the same procedure it is obtained the approximate solution of (11) as

$$\overline{u}_m(s, t, r) = \chi_m \overline{u}_{m-1}(s, t, r) + hH(S)R_m(\overline{\tilde{u}}_{m-1}), \quad (m \geq 1), \tag{14}$$

where $h = -1$ and $H(s, t, r) = 1$. So the approximate solution $u(s, t) = (\underline{u}(s, t, r), \overline{u}(s, t, r))$

6 Numerical Examples

Example 1 Let $A = [0, 1] \times [0, 1]$. Consider the following 2D-FVFIE.

$$u(s, t) = g(s, t) \oplus \varphi(s, t) \odot u^2(s, t) \oplus (FR) \int_0^t (FR) \int_0^s k(s, t, x, y) \odot u^2(x, y)dxdy,$$

where

$$g(s, t, r) = (st(r - 3) - \frac{1}{3}s^4t^4(r - 3)^2, st(-r - 1) - \frac{1}{3}s^4t^4(-r - 1)^2),$$

$$\varphi(s, t) = \frac{1}{6}s^2t^2 \text{ and } k(s, t, x, y) = tx + sy.$$

The exact solution is $u_{exact}(s, t, r) = (st(r - 3), st(-r - 1))$.

The general form of the equation is

$$\underline{u}(s,t,r) = st(r-3) - \frac{1}{3}s^4t^4(r-3)^2 + \frac{1}{6}\underline{u}^2(s,t,r) + \int_0^t\int_0^s (tx+sy)\underline{u}^2(x,y,r)dxdy,$$

$$\overline{u}(s,t,r) = st(-r-1) - \frac{1}{3}s^4t^4(-r-1)^2 + \frac{1}{6}\overline{u}^2(s,t,r)) + \int_0^t\int_0^s (tx+sy)\overline{u}^2(x,y,r)dxdy.$$

We define a nonlinear operator as,

$$N[\underline{u}(s,t,r)] = \underline{u}(s,t,r) - \underline{g}(s,t,r) - \varphi(s,t)\underline{u}^2(s,t,r) - \int_0^t\int_0^s (tx+sy)\underline{u}^2(x,y,r)dxdy.$$

By using proposed method, we get,

$$\begin{aligned} R(\underline{\tilde{u}}_{m-1}) = {} & \underline{u}_{m-1}(s,t,r) - (1-\chi_m)g^r_-(s,t) - \\ & -\varphi(s,t)\sum_{i=0}^{m-1}\underline{u}_i(s,t,r)\underline{u}_{m-1-i}(s,t,r) \\ & -\int_0^t\int_0^s (tx+sy)\sum_{i=0}^{m-1}\underline{u}_i(x,y,r)\underline{u}_{m-1-i}(x,y,r)dxdy. \end{aligned}$$

The solution of the mth order deformation (12)

$$\underline{u}_m(s,t,r) = \chi_m\underline{u}_{m-1}(s,t,r) + hH(s,t,r)R_m(\underline{\tilde{u}}_{m-1}),\ \ m \geq 1.$$

It start with $\underline{u}_0(s,t,r) = st(r-3) - \frac{1}{3}s^4t^4(r-3)^2$ if $H(s,t,r) = 1$ we can obtain

$$\underline{u}_1(s,t,r) = h(-\frac{1}{6}s^2t^2\underline{u}^2{}_0(s,t,r) - \int_0^t\int_0^s (tx+sy)\underline{u}_0^2(x,y,r)dxdy),$$

$$\underline{u}_m(s,t,r)) = (h+1)\underline{u}_{m-1}(s,t,r)+$$
$$+h(-\frac{1}{6}s^2t^2\sum_{i=0}^{m-1}\underline{u}_i(s,t,r)\underline{u}_{m-1-i}(s,t,r)$$
$$-\int_0^t\int_0^s(tx+sy)\sum_{i=0}^{m-1}\underline{u}_i(x,y,r)\underline{u}_{m-1-i}(x,y,r)dxdy),\ m>1.$$

For $h=-1$ we obtain

$$\underline{u}_1(s,t,r) = \frac{1}{3}s^4t^4(r-3)^2 - \frac{1}{7}s^7t^7(r-3)^3 + \frac{17}{810}s^{10}t^{10}(r-3)^4,$$

$$\underline{u}_2(s,t,r) = \frac{11}{126}s^7t^7(r-3)^3 - \frac{152}{2835}s^{10}t^{10}(r-3)^4+$$
$$+\frac{929}{88452}s^{13}t^{13}(r-3)^5 - \frac{17}{14580}s^{16}t^{16}(r-3)^6,$$

$$\underline{u}_3(s,t,r) = \frac{17}{315}s^{10}t^{10}(r-3)^4 - \frac{739}{15795}s^{13}t^{13}(r-3)^5+$$
$$+(\frac{39811}{2948400}s^{16}t^{16} + \frac{71}{36288}s^{17}t^{17})(r-3)^6$$
$$-(\frac{9421}{629856}s^{19}t^{19} + \frac{17}{20520}s^{20}t^{20})(r-3)^7+$$
$$+(\frac{1343}{10103940}s^{22}t^{22} + \frac{289}{3790800}s^{24}t^{24})(r-3)^8,$$

$$\underline{u}(s,t,r) = (st - \frac{1}{18}s^7t^7(r-3)^2 + \frac{121}{5670}s^{10}t^{10}(r-3)^3 - \frac{15767}{442260}s^{13}t^{13}(r-3)^4+$$
$$+(\frac{327359}{26535600}s^{16}t^{16} + \frac{71}{36288}s^{17}t^{17})(r-3)^5$$
$$-(\frac{9421}{629856}s^{19}t^{19} + \frac{17}{20520}s^{20}t^{20})(r-3)^6+$$
$$+(\frac{1343}{10103940}s^{22}t^{22} + \frac{289}{3790800}s^{24}t^{24})(r-3)^7 + \cdots)(r-3).$$

With the same procedure we can obtain,

$$\begin{aligned}\overline{u}(s,t,r) = (st &- \frac{1}{18}s^7t^7(-r-1)^2 + \frac{121}{5670}s^{10}t^{10}(-r-1)^3 \\ &- \frac{15767}{442260}s^{13}t^{13}(-r-1)^4 + \\ &+ (\frac{327359}{26535600}s^{16}t^{16} + \frac{71}{36288}s^{17}t^{17})(-r-1)^5 \\ &- (\frac{9421}{629856}s^{19}t^{19} + \frac{17}{20520}s^{20}t^{20})(-r-1)^6 + \\ &+ (\frac{1343}{10103940}s^{22}t^{22} + \frac{289}{3790800}s^{24}t^{24})(-r-1)^7 + \cdots)(-r-1).\end{aligned}$$

so the approximate solution $u(s,t,r) = (\underline{u}(s,t,r), \overline{u}(s,t,r))$. The numerical results obtained with the HPM with three iterations are presented in Table 1.

Example 2 Let $A = [0,1] \times [0,1]$. Consider the following 2D-FVFIE.

$$u(s,t) = g(s,t) \oplus \varphi(s,t) \odot u(s,t) \oplus (FR)\int_0^t (FR)\int_0^s k(s,t,x,y) \odot u(x,y)dxdy,$$

where $g(s,t,r) = ((t+s)(1-\frac{1}{3}t^3s^3)(1+r), (t+s)(1-\frac{1}{3}t^3s^3)(3-r))$, $\varphi(s,t) = \frac{1}{6}t^3s^3$ and $k(s,t,x,y) = tsxy$.
The exact solution is $u_{exact}(s,t,r) = ((s+t)(1+r), (s+t)(3-r))$.

The general form of the equation is

$$\underline{u}(s,t,r) = (t+s)(1-\frac{1}{3}t^3s^3)(1+r) + \frac{1}{6}s^3t^3\underline{u}(s,t,r) + \int_0^t\int_0^s tsxy\underline{u}(x,y,r)dxdy,$$

$$\overline{u}(s,t,r) = (t+s)(1-\frac{1}{3}t^3s^3)(3-r) + \frac{1}{6}s^3t^3\overline{u}(s,t,r) + \int_0^t\int_0^s tsxy\overline{u}(x,y,r)dxdy.$$

By using proposed method, we get,

$$\begin{aligned}R(\underline{\tilde{u}}_{m-1}) &= \underline{u}_{m-1}(s,t,r) - (1-\chi_m)\underline{g}(s,t,r) - \\ &- \varphi(s,t)\underline{u}_{m-1}(s,t,r) - \int_0^t\int_0^s tsxy\underline{u}_{m-1}(x,y,r)dxdy.\end{aligned}$$

The solution of the mth order deformation (12)

Table 1 Comparison of approximation solutions with exact solution for $(s_0, t_0) = (0.2, 0.7)$ and $r = 0.5$

m	$\underline{u}_m(0.2, 0.7, 0.5)$	$\overline{u}_m(0.2, 0.7, 0.5)$	$\underline{u}(0.2, 0.7, 0.5)$	$\overline{u}(0.2, 0.7, 0.5)$	$\underline{u}_{exact}(0.2, 0.7, 0.5)$	$\overline{u}_{exact}(0.2, 0.7, 0.5)$
0	−0.350800333	−0.21028812	−0.350800333	−0.21028812	−0.35	−0.21
1	0.000802689	0.000288629	−0.349997645	−0.209999491	−0.35	−0.21
2	−1.444E−06	−3.11379E−07	−0.349999089	−0.209999803	−0.35	−0.21
3	6.13422E−09	7.93108E−10	−0.349999083	−0.209999802	−0.35	−0.21

Table 2 Comparison of approximation solutions with exact solution for $(s_0, t_0) = (0.4, 0.7)$ and $r = 0.5$

m	$\underline{u}_m(0.4, 0.7, 0.5)$	$\overline{u}_m(0.4, 0.7, 0.5)$	$\underline{u}(0.4, 0.7, 0.5)$	$\overline{u}(0.4, 0.7, 0.5)$	$\underline{u}_{exact}(0.4, 0.7, 0.5)$	$\overline{u}_{exact}(0.4, 0.7, 0.5)$
0	1.6379264	2.729877333	1.6379264	2.729877333	1.65	2.75
1	0.012020592	0.02003432	1.649946992	2.749911653	1.65	2.75
2	5.27978E−05	8.79964E−05	1.64999979	2.74999965	1.65	2.75
3	2.09296E−07	3.48827E−07	1.649999999	2.74999965	1.65	2.75
4	6.18072E−11	1.03012E−10	1.649999999	2.749999999	1.65	2.75

$$\underline{u}_m(s,t,r) = \chi_m \underline{u}_{m-1}(s,t,r) + hH(s,t,r)R_m(\tilde{\underline{u}}_{m-1}),\ m \geq 1.$$

It start with $\underline{u}_0(s,t,r) = (1 - \frac{1}{3}s^3t^3)(t+s)(1+r)$ if $H(s,t,r) = 1$ we can obtain

$$\underline{u}_1(s,t,r) = h(-\frac{1}{6}s^3t^3\underline{u}_0(s,t,r) - \int_0^t\int_0^s tsxy\underline{u}_0(x,y,r)dxdy),$$

$$\underline{u}_m(s,t,r) = (1+h)\underline{u}_{m-1}(s,t,r) + \\ + h(-\frac{1}{6}s^3t^3\underline{u}_{m-1}(s,t,r) - \int_0^t\int_0^s tsxy\underline{u}_{m-1}(x,y,r)dxdy),\quad m > 1.$$

For $h = -1$ we obtain

$$\underline{u}_1(s,t,r) = \frac{1}{3}s^3t^3(1 - \frac{1}{5}s^3t^3)(t+s)(1+r),$$
$$\underline{u}_2(s,t,r) = \frac{1}{18}s^6t^6(\frac{6}{5} - \frac{13}{60}s^3t^3)(t+s)(1+r),$$
$$\underline{u}_3(s,t,r) = \frac{13}{1080}s^9t^9(1 - \frac{23}{132}s^3t^3)(t+s)(1+r),$$
$$\underline{u}_4(s,t,r) = \frac{13}{6480}s^{12}t^{12}(\frac{23}{22} - \frac{3}{385}s^3t^3)(t+s)(1+r).$$

Then $\underline{u}(s,t,r) = (s+t)(1+r)$.

With the same procedure $\overline{u}(s,t,r) = (s+t)(3-r)$.

So the approximate solution $u(s,t,r) = (\underline{u}(s,t,r), \overline{u}(s,t,r))$ that is the same exact solution. The numerical results obtained with the HPM with four iterations are presented in Table 2.

7 Conclusion

The HPM has been employed to obtain the approximate solution of the 2D-FVFIE. Also, we have demonstrated the existence of a unique solution for Eq. (1). The convergence and the error estimation of this method are proved.

Acknowledgements Research was partially supported by Fund FP17-FMI-008, Fund Scientific Research, University of Plovdiv Paisii Hilendarski, Bulgaria.

References

1. Abbasbandy, S.: Homotopy perturbation method for quadratic Riccati differential equation and comparison with Adomians decomposition method. Appl. Math. Comput. **172**, 485–490 (2006)
2. Attari, H., Yazdani, A.: A computational method for fuzzy Volterra-Fredholm integral equations. Fuzzy Inf. Eng. **2**, 147156 (2011)
3. Baker, C.T.H.: A perspective on the numerical treatment of Volterra equations. J. Comput. Appl. Math. **125**, 217–249 (2000)
4. Bede, B., Gal, S.G.: Quadrature rules for integrals of fuzzy-number valued functions. Fuzzy Sets Syst. **145**, 359–380 (2004)
5. Behzadi, S.S., Allahviranloo, T., Abbasbandy, S.: Solving fuzzy second-order nonlinear Volterra-Fredholm integro-differential equations by using Picard method. Neural Comput. Appl. **21**(Supp 1), 337–346 (2012)
6. Behzadi, S.S.: Solving fuzzy nonlinear Volterra-Fredholm integral equations by using homotopy analysis and Adomian decomposition methods. J. Fuzzy Set Valued Anal. article ID jfsva-00067 (2011)
7. Bica, A.M., Popescu, C.: Numerical solutions of the nonlinear fuzzy Hammerstein-Volterra delay integral equations. Inf. Sci. **233**, 236255 (2013)
8. Bica, A., Ziari, S.: Iterative numerical method for fuzzy Volterra linear integral equations in two dimensions. Soft Comput. **21**, 1097–1108 (2017)
9. Dong, C., Chen, Z., Jiang, W.: A modified homotopy perturbationmethod for solving the nonlinear mixed Volterra-Fredholm integral equation. J. Comput. Appl. Math. **239**, 359–366 (2013)
10. Georgieva, A., Naydenova, I.: Numerical solution of nonlinear Urisohn-Volterra fuzzy functional integral equations. In: AIP Conference Proceedings, vol. 1910. https://doi.org/10.1063/1.5013992 (2017)
11. Ghanbari, M.: Numerical solution of fuzzy linear Volterra integral equations of the second kind by homotopy analysis method. Int. J. Ind. Math. **2**, 73–87 (2010)
12. Goetschel, R., Voxman, W.: Elementary fuzzy calculus. Fuzzy Sets Syst. **18**, 3143 (1986)
13. Hassan, H.N., El-Tawil, M.A.: A new technique of using homotopy analysis method for second order nonlinear differential equation. Appl. Math. Comput. **219**, 708–728 (2012)
14. He, J.H.: Homotopy perturbation technique. Comput. Methods Appl. Mech. Eng. **178**, 257262 (1999)
15. He, J.H.: Homotopy perturbation method: a new nonlinear analytical technique. Appl. Math. Comput. **135**, 73–79 (2003)
16. He, J.H.: A simple perturbation approach to Blasius equation. Appl. Math. Comput. **140**, 217222 (2003)
17. He, J.H.: Application of homotopy perturbation method to nonlinear wave equations. Chaos Solitons Fractals **26**, 295–700 (2005)
18. Kaleva, O.: Fuzzy differential equations. Fuzzy Sets Syst. **24**, 301–317 (1987)
19. Linz, P.: Analytical and Numerical Methods for Volterra Equations. SIAM, Philadelphia, PA (1985)
20. Sadatrasoul, S., Ezzati, R.: Numerical solution of two-dimensional nonlinear Hammerstein fuzzy integral equations based on optimal fuzzy quadrature formula. J. Comput. Appl. Math. **292**, 430–446 (2016)
21. Sadatrasoul, S., Ezzati, R.: Quadrature rules and iterative method for numerical solution of two-dimensional fuzzy integral equations. Abstr. Appl. Anal. **2014**, 18 (2014)
22. Salehi, P., Nejatiyan, M.: Numerical method for nonlinear fuzzy Volterra integral equations of the second kind. Int. J. Ind. Math. **3**, 169–179 (2011)
23. Seikkala, S.: On the fuzzy initial value problem. Fuzzy Sets Syst. **24**, 319–330 (1987)
24. Wu, C., Gong, Z.: On Henstock integral of fuzzy-number-valued functions (I). Fuzzy Sets Syst. **120**, 523–532 (2001)
25. Wu, C., Wu, C.: The supremum and infimum of these to fuzzy-numbers and its applications. J. Math. Anal. Appl. **210**, 499–511 (1997)

Iterative Method for Numerical Solution of Two-Dimensional Nonlinear Urysohn Fuzzy Integral Equations

Atanaska Georgieva, Albena Pavlova and Svetoslav Enkov

Abstract In this paper, we propose an iterative procedure based on the quadrature formula of Simpson to solve two-dimensional nonlinear Urysohn fuzzy integral equations (2DNUFIE). Moreover, the error estimation of the proposed method in terms of uniform and partial modulus of continuity is given. We extend in the context of using the modulus of continuity, the notion of numerical stability of the solution with respect to the first iteration. Finally, illustrative example is included in order to demonstrate the accuracy and the convergence of the proposed method.

Keywords Urysohn fuzzy integral equations · Iterative method · Modulus of continuity

1 Introduction

The concept of fuzzy integral was initiated by Dubois and Prade [5] and then investigated by Kaleva [10], Goetschel and Voxman [9] and others. Mordeson and Newman [11] started the study of the subject of fuzzy integral equations. The Banach fixed point principle is the powerful tool to investigate of the existence and uniqueness of the solution of fuzzy integral equations. The existence and uniqueness of the solution of fuzzy integral equations can be found in [2, 8, 12]. The iterative techniques are applied to fuzzy integral equation in [3, 13]. Ezzati and Sadatrasoul [7] presented a

A. Georgieva
FMI, University of Plovdiv Paisii Hilendarski, 24 Tzar Asen, 4003 Plovdiv, Bulgaria
e-mail: atanaska@uni-plovdiv.bg

A. Pavlova (✉)
Department of MPC Technical University-Sofia, Plovdiv Branch,
25 Tzanko Djustabanov Str., 4000 Plovdiv, Bulgaria
e-mail: akosseva@gmail.com

S. Enkov
FMI, University of Plovdiv Paisii Hilendarski, 24 Tzar Asen, 4000 Plovdiv, Bulgaria
e-mail: svetoslav.enkov@gmail.com

K. Georgiev et al. (eds.), *Advanced Computing in Industrial Mathematics*,
Studies in Computational Intelligence 793,
https://doi.org/10.1007/978-3-319-97277-0_12

numerical algorithm to solve fuzzy Urysohn integral equations based on successive approximations method.

In the present paper, we investigate the following 2DNUFIE

$$F(s,t) = g(s,t) \oplus (FR)\int_c^d (FR)\int_a^b H(s,t,x,y,F(x,y))dxdy, \quad (1)$$

where $g : [a,b] \times [c,d] \to \mathbf{R}_{\mathscr{F}}$, $H : [a,b] \times [c,d] \times [a,b] \times [c,d] \times \mathbf{R}_{\mathscr{F}} \to \mathbf{R}_{\mathscr{F}}$ are continuous fuzzy-number-valued functions.

Here, we obtain the error estimate in the approximation of the solution of such fuzzy integral equations. The rest of this paper is organized as follows: In Sect. 2, we review some elementary concepts of the fuzzy set theory and modulus of continuity. In Sect. 3, we drive the proposed method to obtain numerical solution of nonlinear fuzzy Fredholm integral equations based on an iterative procedure. The error estimation of the proposed method is obtained in Sect. 4 in terms of uniform and partial modulus of continuity, proving the convergence of the method. In Sect. 5 we prove the numerical stability in the proposed method with respect to the choice of the first iterations. Finally, Sect. 6 includes a numerical example confirming the theoretical results.

2 Preliminaries

First we present some notions and results about fuzzy numbers and fuzzy-number-valued functions, and fuzzy integrals.

Definition 1 [9] A fuzzy number is a function $u : \mathbf{R} \to [0,1]$ satisfying the following properties:

1. u is upper semi-continuous on R,
2. $u(x) = 0$ outside of some interval $[c,d]$,
3. there are the real numbers a and b with $c \leq a \leq b \leq d$, such that u is increasing on $[c,a]$, decreasing on $[b,d]$ and $u(x) = 1$ for each $x \in [a,b]$.

In [5] the fuzzy numbers have the following convexity: u is a convex fuzzy set, i.e. $u(rx + (1-r)y) \geq \min\{u(x), u(y)\}$ for any $x, y \in \mathbf{R}$, $r \in [0,1]$ and possesses compact support $[u]^0 = \overline{\{x \in \mathbf{R} : u(x) > 0\}}$, where $\overline{A}$ denotes the closure of A.

The set of all fuzzy numbers is denoted by $\mathbf{R}_{\mathscr{F}}$ [15]. Any real number $a \in R$ can be interpreted as a fuzzy number $\tilde{a} = \chi(a)$ and therefore $\mathbf{R} \subset \mathbf{R}_{\mathscr{F}}$. The neutral element with respect to $\oplus$ in $\mathbf{R}_{\mathscr{F}}$ is denoted by $\tilde{0} = \chi_{\{0\}}$. For any $0 < r \leq 1$ we denote the r-level set $[u]^r = \{x \in \mathbf{R} : u(x) \geq r\}$ that is a closed interval [15] and $[u]^r = [u_-^r, u_+^r]$ for all $r \in [0,1]$. These lead to the usual LU representation of fuzzy number r-level: $[u]^r = [u_-^r, u_+^r]$ for all $r \in [0,1]$, where u_-, u_+ can be considered as functions $u_-, u_+ : [0,1] \to \mathbf{R}$, such that u_- is increasing and u_+ is decreasing [9].

For $u, v \in \mathbf{R}_{\mathscr{F}}$, $k \in \mathbf{R}$ the addition and the scalar multiplication are defined by

$$[u \oplus v]^r = [u]^r + [v]^r = [u_-^r + v_-^r, u_+^r + v_+^r] \text{ and } [k \odot u]^r = k.[u]^r$$
$$= \begin{cases} [ku_-^r, ku_+^r], & k \geq 0 \\ [ku_+^r, ku_-^r], & k < 0. \end{cases}$$

According to [15], we can summarize the following algebraic properties:

(i) $u \oplus (v \oplus w) = (u \oplus v) \oplus w$ and $u \oplus v = v \oplus u$ for any $u, v, w \in \mathbf{R}_{\mathscr{F}}$,
(ii) $u \oplus \tilde{0} = \tilde{0} \oplus u = u$ for any $u \in \mathbf{R}_{\mathscr{F}}$,
(iii) with respect to $\tilde{0}$, none $u \in \mathbf{R}_{\mathscr{F}} \setminus \mathbf{R}$, $u \neq \tilde{0}$ has opposite in $(\mathbf{R}_{\mathscr{F}}, \oplus)$,
(iv) for any $a, b \in \mathbf{R}$ with $a, b \geq 0$ or $a, b \leq 0$ and any $u \in \mathbf{R}_{\mathscr{F}}$ we have $(a + b) \odot u = a \odot u \oplus b \odot u$,
(v) for any $a \in \mathbf{R}$ and any $u, v \in \mathbf{R}_{\mathscr{F}}$ we have $a \odot (u \oplus v) = a \odot u \oplus a \odot v$,
(vi) for any $a, b \in \mathbf{R}$ and any $u \in \mathbf{R}_{\mathscr{F}}$ we have $a \odot (b \odot u) = (ab) \odot u$ and $1 \odot u = u$.

As a distance between fuzzy numbers we use the Hausdorff metric [15].

Definition 2 [15] For arbitrary fuzzy numbers $u = (u_-^r, u_+^r)$ and $v = (v_-^r, v_+^r)$ the quantity $D(u, v) = \sup_{r \in [0,1]} \max\{|u_-^r - v_-^r|, |u_+^r - v_+^r|\}$ is the distance between u, v.

Theorem 1 [15] *The following properties of the above distance hold:*

1. $(\mathbf{R}_{\mathscr{F}}, D)$ *is a complete metric space,*
2. $D(u \oplus w, v \oplus w) = D(u, v)$, *for all* $u, v, w \in \mathbf{R}_{\mathscr{F}}$,
3. $D(k \odot u, k \odot v) = |k|D(u, v)$, *for all* $u, v \in \mathbf{R}_{\mathscr{F}}$, *for all* $k \in R$,
4. $D(u \oplus v, w \oplus e) = D(u, w) + D(v, e)$, *for all* $u, v, w, e \in \mathbf{R}_{\mathscr{F}}$,
5. $D(u \oplus v, \tilde{0}) \leq D(u, \tilde{0}) + D(v, \tilde{0})$, *for all* $u, v \in \mathbf{R}_{\mathscr{F}}$,
6. $D(k_1 \odot u, k_2 \odot u) = |k_1 - k_2|D(u, \tilde{0})$, *for all* $k_1, k_2 \in R$ *with* $k_1 k_2 \geq 0$ *and* $u \in \mathbf{R}_{\mathscr{F}}$.

For any fuzzy-number-valued function $f : A = [a, b] \times [c, d] \to \mathbf{R}_{\mathscr{F}}$ we can define the functions $\underline{f}(.,.,r)$, $\overline{f}(.,.,r) : A \to \mathbf{R}$, by $\underline{f}(s, t, r) = (f(s, t))_-^r$ and $\overline{f}(s, t, r) = (f(s, t))_+^r$ for all $(s, t) \in A$ and $r \in [0, 1]$. These functions are called the left and right $r-$level functions of f.

Definition 3 [14] A fuzzy-number-valued function $f : A \to \mathbf{R}_{\mathscr{F}}$ is called:

(i) continuous in $(x_0, y_0) \in A$ if for any $\varepsilon > 0$ there exists $\delta > 0$ such that for any $(x, y) \in A$ with $|x - x_0| < \delta$, $|y - y_0| < \delta$ it follows that $D(f(x, y), f(x_0, y_0)) < \varepsilon$. The function f is continuous on A if it is continuous in each $(x, y) \in A$.
(ii) bounded if there exists $M \geq 0$ such that $D(f(x, y), \tilde{0}) \leq M$ for all $(x, y) \in A$.

The set of all continuous functions $f : A \to \mathbf{R}_{\mathscr{F}}$ is denoted by $C(A, \mathbf{R}_{\mathscr{F}})$. On the set $C(A, \mathbf{R}_{\mathscr{F}}) = \{f : A \to \mathbf{R}_{\mathscr{F}} : f \text{ is continuous}\}$, it is defined the metric $D^*(f, g) = \sup_{(s,t) \in A} \{D(f(s, t), g(s, t))\}$, for all $f, g \in C(A, \mathbf{R}_{\mathscr{F}})$. We see that $(C(A, \mathbf{R}_{\mathscr{F}}), D^*)$ is a complete metric space.

Definition 4 [1] Suppose $f : A \to \mathbf{R}_{\mathscr{F}}$, be a bounded mapping, then the functions $\omega^1_{[a,b]}(f, .)$, $\omega^2_{[c,d]}(f, .) : \mathbf{R}_+ \cup 0 \to \mathbf{R}_+$ defined by

$$\omega^1_{[a,b]}(f, \delta) = \sup\{D(f(s_1, t), f(s_2, t)) : |s_1 - s_2| \le \delta;\ t \in [c, d]\},$$
$$\omega^2_{[c,d]}(f, \delta) = \sup\{D(f(s, t_1), f(s, t_2)) : |t_1 - t_2| \le \delta;\ s \in [a, b]\}$$

are called the modulus of oscillation of f on $[a, b]$, $[c, d]$. In addition if $f \in C(A, \mathbf{R}_{\mathscr{F}})$, then $\omega^1_{[a,b]}(f, \delta)$ and $\omega^2_{[c,d]}(f, \delta)$ are called uniform modulus of continuity of f.

Theorem 2 *[3] The following statements, concerning the modulus of oscillation, are true:*

1. $D(f(x, y), f(s, t)) \le \omega^1_{[a,b]}(f, |x - s|) + \omega^2_{[c,d]}(f, |y - t|)$ *for any* $(x, y), (s, t) \in A$,
2. $\omega^1_{[a,b]}(f, \delta)$ *and* $\omega^2_{[c,d]}(f, \delta)$ *are a non-decreasing mapping in* δ,
3. $\omega^1_{[a,b]}(f, 0) = 0$, $\omega^2_{[c,d]}(f, 0) = 0$,
4. $\omega^1_{[a,b]}(f, \delta_1 + \delta_2) \le \omega^1_{[a,b]}(f, \delta_1) + \omega^1_{[a,b]}(f, \delta_2)$ *and* $\omega^2_{[c,d]}(f, \delta_1 + \delta_2) \le \omega^2_{[c,d]}(f, \delta_1) + \omega^2_{[c,d]}(f, \delta_2)$ *for any* $\delta_1, \delta_2 \ge 0$,
5. $\omega^1_{[a,b]}(f, n\delta) \le n\omega^1_{[a,b]}(f, \delta)$ *and* $\omega^2_{[c,d]}(f, n\delta) \le n\omega^2_{[c,d]}(f, \delta)$ *for any* $\delta \ge 0$ *and* $n \in \mathbf{N}$,
6. $\omega^1_{[a,b]}(f, \lambda\delta) \le (\lambda + 1)\omega^1_{[a,b]}(f, \delta)$ *and* $\omega^2_{[c,d]}(f, \lambda\delta) \le (\lambda + 1)\omega^2_{[c,d]}(f, \delta)$ *for any* $\delta, \lambda \ge 0$,
7. *If* $[a_1, b_1] \subseteq [a, b]$ *and* $[c_1, d_1] \subseteq [c, d]$, *then* $\omega^1_{[a_1,b_1]}(f, \delta) \le \omega^1_{[a,b]}(f, \delta)$ *and* $\omega^2_{[c_1,d_1]}(f, \delta) \le \omega^2_{[c,d]}(f, \delta)$ *for all* $\delta \ge 0$.

Definition 5 [14] Let $f : A \to \mathbf{R}_{\mathscr{F}}$. For $\Delta^x_n : a = x_0 < x_1 < \cdots < x_{n-1} < x_n = b$ a partition of the interval $[a, b]$ and $\Delta^y_m : c = y_0 < y_1 < \cdots < y_{m-1} < y_m = d$ a partition of the interval $[c, d]$. Let us consider the intermediate points $\xi_i \in [x_{i-1}, x_i]$, $i = \overline{1, n}$, $\eta_j \in [y_{j-1}, y_j]$, $j = \overline{1, m}$, and the functions $\delta : [a, b] \to \mathbf{R}_+$, $\sigma : [c, d] \to \mathbf{R}_+$. The partitions $P_x = \{([x_{i-1}, x_i]; \xi_i), i = \overline{1, n}\}$ denoted by $P_x = (\Delta^x_n, \xi)$ and $P_y = \{([y_{j-1}, y_j]; \eta_j), j = \overline{1, m}\}$ denoted by $P_y = (\Delta^y_m, \eta)$ are said to be δ-fine if $[x_{i-1}, x_i] \subseteq (\xi_i - \delta(\xi_i), \xi_i + \delta(\xi_i))$, for all $i = \overline{1, n}$ and σ-fine if $[y_{j-1}, y_j] \subseteq (\eta_j - \sigma(\eta_j), \eta_j + \sigma(\eta_j))$, for all $j = \overline{1, m}$ respectively. The function f is said to be fuzzy-Henstock integrable if there exists $I(f) \in \mathbf{R}_{\mathscr{F}}$ with the property that for any $\varepsilon > 0$ there is a function $\delta : [a, b] \to \mathbf{R}_+$ divisions and a function $\sigma : [c, d] \to \mathbf{R}_+$ such that for any δ-fine and σ-fine, we have $D(\sum_{j=1}^m \sum_{i=1}^n (x_i - x_{i-1})(y_j - y_{j-1}) \odot f(\xi_i, \eta_j), I) < \varepsilon$. The fuzzy number I is named the fuzzy-Henstock double integral of f and will be denoted by $I(f) = (FH) \int\limits_c^d (FH) \int\limits_a^b f(s, t) ds dt$.

Remark 1 If the above δ and σ are constant functions then the fuzzy-Riemann double integrability for fuzzy-number-valued functions is obtained. In this case, $I(f) \in \mathbf{R}_{\mathscr{F}}$ is called the fuzzy-Riemann double integral of f on A, being denoted by $I(f) = (FR)\int_c^d (FR)\int_a^b f(s,t)dsdt$, or simply, $\int_c^d (\int_a^b f(s,t)ds)dt$. Consequently, the fuzzy-Riemann double integrability is a particular case of the fuzzy-Henstock double integrability, and therefore the properties of the double integral (FH) will be valid for the double integral (FR), too.

Lemma 1 [14] *If $f : A \to \mathbf{R}_{\mathcal{F}}$ then $(FR)\int_c^d (FR)\int_a^b f(s,t)dsdt$ exists and*

$$\left((FR)\int_c^d (FR)\int_a^b f(s,t)dsdt\right)_-^r = \int_c^d \int_a^b \underline{f}(s,t,r)dsdt,$$
$$\left((FR)\int_c^d (FR)\int_a^b f(s,t)dsdt\right)_+^r = \int_c^d \int_a^b \overline{f}(s,t,r)dsdt.$$

Lemma 2 [14] *If $f : A \to \mathbf{R}_{\mathcal{F}}$ be a Henstock integrable bounded mapping, then for any fixed $u \in [a,b]$ and $v \in [c,d]$ the function $\varphi_{u,v} : A \to \mathbf{R}_+$ defined by $\varphi_{u,v}(s,t) = D(f(u,v), f(s,t))$ is Lebesgue integrable on A.*

Lemma 3 [14] *If f and g are Henstock integrable mappings on A and if $D(f(s,t), g(s,t))$ Lebesgue integrable, then*

$$D\left((FH)\int_c^d (FH)\int_a^b f(s,t)dsdt, (FH)\int_c^d (FH)\int_a^b g(s,t)dsdt\right) \le$$
$$\le (L)\int_c^d (L)\int_a^b D(f(s,t), g(s,t))dsdt.$$

In [15], the authors introduced the concept of the Henstock integral for a fuzzy-number-valued function.

Theorem 3 [6] *Let $f : A \to \mathbf{R}_{\mathcal{F}}$ be Henstock integrable, bounded mappings. Then, for any divisions $a = x_0 < x_1 < \cdots < x_n = b$ and $c = y_0 < y_1 < \cdots < y_{n'} = d$ and any points $\xi_i \in [x_{i-1}, x_i]$, $\eta_j \in [y_{j-1}, y_j]$, one has*

$$D\left((FH)\int_c^d (FH)\int_a^b f(s,t)dsdt, \sum_{j=1}^{n'}\sum_{i=1}^{n}(x_i - x_{i-1})(y_j - y_{j-1}) \odot f(\xi_i, \eta_j)\right) \le$$
$$\le \sum_{j=1}^{n'}\sum_{i=1}^{n}(x_i - x_{i-1})(y_j - y_{j-1})(\omega^1_{[x_{i-1},x_i]}(f, |x_i - x_{i-1}|) + \omega^2_{[y_{j-1},y_j]}(f, |y_j - y_{j-1}|)).$$

Corollary 1 [6] *Let $f : A \to \mathbf{R}_{\mathcal{F}}$ be a a two-dimensional Henstock integrable, bounded mapping. Then*

$$D((FH)\int_c^d (FH)\int_a^b f(s,t)dsdt, \frac{(d-c)(b-a)}{36} \odot (f(a,c) \oplus f(a,d) \oplus f(b,c)$$
$$\oplus f(b,d) \oplus 4 \odot f(a, \frac{c+d}{2}) \oplus 4 \odot f(\frac{a+b}{2}, c) \oplus 4 \odot f(\frac{a+b}{2}, d) \oplus 4$$
$$\odot f(b, \frac{c+d}{2}) \oplus 16 \odot f(\frac{a+b}{2}, \frac{c+d}{2}))) \leq (d-c)(b-a)(\omega^1_{[a,b]}(f, \frac{b-a}{6})$$
$$+ \omega^2_{[c,d]}(f, \frac{d-c}{6})).$$

3 Successive Approximations and the Iterative Algorithm

The following conditions are imposed:

(i) $g \in C(A, \mathbf{R}_{\mathcal{F}}), H \in C(A \times A \times \mathbf{R}_{\mathcal{F}}, \mathbf{R}_{\mathcal{F}})$,

(ii) H is uniformly continuous with respect to $(s,t) \in A$, that is for any $\varepsilon > 0$ there exists $\delta > 0$ such that for any $(s,t), (s',t') \in A$ with $|s-s'| < \delta, |t-t'| < \delta$ it follows that $D(H(s,t,x,y,u), H(s',t',x,y,u)) < \varepsilon$ for all $(x,y) \in A, u \in \mathbf{R}_{\mathcal{F}}$,

(iii) there exists $\alpha \geq 0$, such that $D(H(s,t,x,y,u), H(s,t,x,y,v)) \leq \alpha D(u,v)$ for all $(s,t), (x,y) \in A, u, v \in \mathbf{R}_{\mathcal{F}}$, and $B = \alpha\Delta < 1$, where $\Delta = (b-a)(d-c)$.

Theorem 4 [4] *Under the conditions (i)–(iii) the integral equation (1) has unique solution $F \in C(A, \mathbf{R}_{\mathcal{F}})$ and the sequence of successive approximations $\{F_m\}_{m\in\mathbf{N}} \subset C(A, \mathbf{R}_{\mathcal{F}})$*

$$F_m(s,t) = g(s,t) \oplus (FR)\int_c^d (FR)\int_a^b H(s,t,x,y,F_{m-1}(x,y))dxdy \tag{2}$$

converges to F in $C(A, \mathbf{R}_{\mathcal{F}})$ for any choice of $F_0 \in C(A, \mathbf{R}_{\mathcal{F}})$ and the following error estimates hold:

$$D(F(s,t), F_m(s,t)) \leq \frac{B^m}{1-B} D(F_1(s,t), F_0(s,t)), \text{ for all } (s,t) \in A, m \in \mathbf{N} \tag{3}$$

$$D(F(s,t), F_m(s,t)) \leq \frac{B}{1-B} D(F_m(s,t), F_{m-1}(s,t)), \text{ for all } (s,t) \in A, m \in \mathbf{N}. \tag{4}$$

If $F_0 = g$ then the estimate (3) becomes

$$D(F(s,t), F_m(s,t)) \leq \frac{B^m}{1-B} M_0 \Delta \tag{5}$$

for all $(s,t)\in A$, $m\in\mathbf{N}$, *where* $M_0\geq 0$ *is such that* $D(H(s,t,x,y,g(x,y)),\tilde{0})\leq M_0$ *for all* (s,t), $(x,y)\in A$.

Moreover, the sequence of successive approximations (2) is uniformly bounded and the solution F *is bounded too.*

4 The Error Estimation

We present a numerical method to solve the equation (1) and define uniform partitions of the interval $a=x_0<x_1<x_2<\cdots<x_{2n-1}<x_{2n}=b$, with $x_i=a+i\sigma$, where $\sigma=\frac{b-a}{2n}$ and $c=y_0<y_1<\cdots<y_{2n'-1}<y_{2n'}=d$, with $y_j=b+j\sigma'$, where $\sigma'=\frac{d-c}{2n'}$.

We denoted

$$
\begin{aligned}
&H_A(F)(s,t)=H(s,t,a,c,F(a,c))\oplus H(s,t,b,c,F(b,c))\oplus H(s,t,a,d,F(a,d))\oplus\\
&\oplus H(s,t,b,d,F(b,d))\oplus 4H(s,t,\frac{a+b}{2},c,F(\frac{a+b}{2},c))\oplus 4H(s,t,\frac{a+b}{2},d,F(\frac{a+b}{2},d))\oplus\\
&\oplus 4H(s,t,a,\frac{c+d}{2},F(a,\frac{c+d}{2}))\oplus 4H(s,t,b,\frac{c+d}{2},F(b,\frac{c+d}{2}))\oplus\\
&\oplus 16H(s,t,\frac{a+b}{2},\frac{c+d}{2},F(\frac{a+b}{2},\frac{c+d}{2})).
\end{aligned}
$$

We denoted $A_{ij}=[x_{2i-2},x_{2i}]\times[y_{2j-2},y_{2j}]$, $i=1,2,\ldots,n$ and $j=1,2,\ldots,n'$.

Then the following iterative procedure given the approximate solution of equation (1) in point $(s,t)\in A$, $m=1,2,\ldots,$

$$
\begin{aligned}
&\tilde{F}_0(s,t)=g(s,t),\\
&\tilde{F}_m(s,t)=g(s,t)\oplus\tfrac{\sigma\sigma'}{9}\textstyle\sum_{j=1}^{n'}\sum_{i=1}^{n}H_{A_{ij}}(\tilde{F}_{m-1})(s,t).
\end{aligned}
\tag{6}
$$

We defined

$$
\begin{aligned}
\omega^1_{[a,b]}(H,\sigma)&=\sup\{D(H(s_1,t,x,y,u),H(s_2,t,x,y,u)):|s_1-s_2|\leq\sigma,\\
&\quad(x,y)\in A,t\in[c,d],u\in\mathbf{R}_{\mathcal{F}}\},\\
\omega^2_{[c,d]}(H,\sigma')&=\sup\{D(H(s,t_1,x,y,u),H(s,t_2,x,y,u)):|t_1-t_2|\leq\sigma',\\
&\quad(x,y)\in A,s\in[a,b],u\in\mathbf{R}_{\mathcal{F}}\},\\
\omega^3_{[a,b]}(H,\sigma)&=\sup\{D(H(s,t,\xi_1,y,u),H(s,t,\xi_2,y,u)):|\xi_1-\xi_2|\leq\sigma,\\
&\quad(s,t)\in A,y\in[c,d],u\in\mathbf{R}_{\mathcal{F}}\},\\
\omega^4_{[c,d]}(H,\sigma')&=\sup\{D(H(s,t,x,\eta_1,u),H(s,t,x,\eta_2,u)):|\eta_1-\eta_2|\leq\sigma',\\
&\quad(s,t)\in A,x\in[a,b],u\in\mathbf{R}_{\mathcal{F}}\},
\end{aligned}
$$

$$\omega^{3,1}_{[a,b]}(H_f,\sigma) = \sup\{D(H_f(s,t,\xi_1,y), H_f(s,t,\xi_2,y) : |\xi_1-\xi_2| \le \sigma, \\ (s,t)\in A, y\in[c,d], f\in C(A,\mathbf{R}_{\mathscr{F}})\},$$

$$\omega^{4,2}_{[c,d]}(H_f,\sigma') = \sup\{D(H_f(s,t,x,\eta_1), H_f(s,t,x,\eta_2) : |\eta_1-\eta_2| \le \sigma', \\ (s,t)\in A, x\in[a,b], f\in C(A,\mathbf{R}_{\mathscr{F}})\},$$

where $H_f(s,t,x,y) = H(s,t,x,y,f(x,y))$.

Lemma 4 *Let the conditions (i), (iii) are fulfilled. Then the inequalities are hold*

$$\omega^{3,1}_{[a,b]}(H_{F_m},\sigma) \le \alpha\omega^1_{[a,b]}(g,\sigma) + B\omega^1_{[a,b]}(H,\sigma) + \omega^3_{[a,b]}(H,\sigma), \tag{7}$$

and

$$\omega^{4,2}_{[c,d]}(H_{F_m},\sigma') \le \alpha\omega^2_{[c,d]}(g,\sigma') + B\omega^2_{[c,d]}(H,\sigma') + \omega^4_{[c,d]}(H,\sigma'). \tag{8}$$

Proof First we obtain $\omega^1_{[a,b]}(F_m,\sigma)$. Let $s_1,\ s_2\in[a,b]$, $|s_1-s_2|<\sigma$ and $m\in\mathbf{N}$ then

$$D(F_m(s_1,t),F_m(s_2,t)) \le \\ \le D(g(s_1,t),g(s_2,t)) + \int_c^d\int_a^b D(H(s_1,t,x,y,F_{m-1}(x,y)), \\ H(s_2,t,x,y,F_{m-1}(x,y)))dxdy \le \\ \le \omega^1_{[a,b]}(g,\sigma) + \Delta\omega^1_{[a,b]}(H,\sigma).$$

Hence

$$\omega^1_{[a,b]}(F_m,\sigma) \le \omega^1_{[a,b]}(g,\sigma) + \Delta\omega^1_{[a,b]}(H,\sigma). \tag{9}$$

Let $\xi_1,\ \xi_2\in[a,b]$, $|\xi_1-\xi_2|<\sigma$ and $m\in\mathbf{N}$. We obtain

$$D(H(s,t,\xi_1,y,g(\xi_1,y)), H(s,t,\xi_2,y,g(\xi_2,y))) \le \\ D(H(s,t,\xi_1,y,g(\xi_1,y)), H(s,t,\xi_2,y,g(\xi_1,y))) \\ + D(H(s,t,\xi_2,y,g(\xi_1,y)), H(s,t,\xi_2,y,g(\xi_2,y))) \le \\ \le \omega^3_{[a,b]}(H,\sigma) + \alpha D(g(\xi_1,y),g(\xi_2,y)) \le \omega^3_{[a,b]}(H,\sigma) + \alpha\omega^1_{[a,b]}(g,\sigma).$$

Hence

$$\omega^{3,1}_{[a,b]}(H_g,\sigma) \le \alpha\omega^1_{[a,b]}(g,\sigma) + \omega^3_{[a,b]}(H,\sigma). \tag{10}$$

Then from (9) for $\omega^{3,1}_{[a,b]}(H_{F_m},\sigma)$ we obtain

$$
\begin{aligned}
&D(H(s,t,\xi_1,y,F_m(\xi_1,y)),H(s,t,\xi_2,y,F_m(\xi_2,y)))\leq\\
&D(H(s,t,\xi_1,y,F_m(\xi_1,y)),H(s,t,\xi_2,y,F_m(\xi_1,y)))+\\
&+D(H(s,t,\xi_2,y,F_m(\xi_1,y)),H(s,t,\xi_2,y,F_m(\xi_2,y)))\leq\\
&\leq \omega^3_{[a,b]}(H,\sigma)+\alpha D(F_m(\xi_1,y),F_m(\xi_2,y))\leq\\
&\leq \omega^3_{[a,b]}(H,\sigma)+\alpha\omega^1_{[a,b]}(F_m,\sigma)\leq \alpha\omega^1_{[a,b]}(g,\sigma)+B\omega^1_{[a,b]}(H,\sigma)+\omega^3_{[a,b]}(H,\sigma).
\end{aligned}
$$

Analogous we obtain $\omega^2_{[c,d]}(F_m,\sigma')\leq\omega^2_{[c,d]}(g,\sigma')+\Delta\omega^2_{[c,d]}(H,\sigma')$,

$$
\omega^{4,2}_{[c,d]}(H_g,\sigma')\leq\alpha\omega^2_{[c,d]}(g,\sigma')+\omega^4_{[c,d]}(H,\sigma'), \tag{11}
$$

and (8). □

Theorem 5 *Under the conditions (i)–(iii) the iterative method (6) converges to the unique solution F of (1) and its error estimate is as follows*

$$
\begin{aligned}
&D^*(F,\tilde F_m)\leq \tfrac{B^m}{1-B}M_0\Delta+\tfrac{4\Delta}{3(1-B)}(\alpha(\omega^1_{[a,b]}(g,\sigma)+\omega^2_{[c,d]}(g,\sigma'))+\\
&+B(\omega^1_{[a,b]}(H,\sigma)+\omega^2_{[c,d]}(H,\sigma'))+\omega^3_{[a,b]}(H,\sigma)+\omega^4_{[c,d]}(H,\sigma')).
\end{aligned}\tag{12}
$$

Proof From Lemmas 2 and 3, Corollary 1 and iterative procedure (6), for all $(s,t)\in A$ we have

$$
D(F_1(s,t),\tilde F_1(s,t))=D(g(s,t)\oplus(FR)\int_c^d(FR)\int_a^b H(s,t,x,y,g(x,y))dxdy,
$$

$$
g(s,t)\oplus\frac{\sigma\sigma'}{9}\sum_{j=1}^{n'}\sum_{i=1}^{n}H_{A_{ij}}(g)(s,t))\leq D(g(s,t),g(s,t))
$$

$$
+D((FR)\int_c^d(FR)\int_a^b H(s,t,x,y,g(x,y))dxdy,\frac{\sigma\sigma'}{9}\sum_{j=1}^{n'}\sum_{i=1}^{n}H_{A_{ij}}(g)(s,t))\leq
$$

$$
\leq\sum_{j=1}^{n'}\sum_{i=1}^{n}D((FR)\int_{y_{2j-2}}^{y_{2j}}(FR)\int_{x_{2i-2}}^{x_{2i}}H(s,t,x,y,g(x,y))dxdy,\frac{\sigma\sigma'}{9}H_{A_{ij}}(g)(s,t))\leq
$$

$$
\leq\sum_{j=1}^{n'}\sum_{i=1}^{n}4\sigma\sigma'(\omega^{3,1}_{[x_{2i-2},x_{2i}]}(H_g,\frac{\sigma}{3})+\omega^{4,2}_{[y_{2j-2},y_{2j}]}(H_g,\frac{\sigma'}{3}))\leq
$$

$$
\leq\sum_{j=1}^{n'}\sum_{i=1}^{n}4\sigma\sigma'(\omega^{3,1}_{[a,b]}(H_g,\frac{\sigma}{3})+\omega^{4,2}_{[c,d]}(H_g,\frac{\sigma'}{3}))\leq\frac{4\Delta}{3}(\omega^{3,1}_{[a,b]}(H_g,\sigma)+\omega^{4,2}_{[c,d]}(H_g,\sigma')).
$$

Applying Theorem 2, property 6 of the modulus of continuity and inequality (10) and (11) we obtain

$$D^*(F_1,\tilde{F}_1) \leq \frac{4\Delta}{3}(\alpha(\omega^1_{[a,b]}(g,\sigma)+\omega^2_{[c,d]}(g,\sigma'))+\omega^3_{[a,b]}(H,\sigma)+\omega^4_{[c,d]}(H,\sigma')). \tag{13}$$

Now, for $m=2$, it follows that

$$D(F_2(s,t),\tilde{F}_2(s,t)) = D(g(s,t)\oplus (FR)\int_c^d (FR)\int_a^b H(s,t,x,y,F_1(x,y))dxdy,$$

$$g(s,t)\oplus \frac{\sigma\sigma'}{9}\sum_{j=1}^{n'}\sum_{i=1}^{n} H_{A_{ij}}(\tilde{F}_1)(s,t)) \leq$$

$$\leq D((FR)\int_c^d (FR)\int_a^b H(s,t,x,y,F_1(x,y))dxdy, \frac{\sigma\sigma'}{9}\sum_{j=1}^{n'}\sum_{i=1}^{n} H_{A_{ij}}(\tilde{F}_1)(s,t)) \leq$$

$$\leq \sum_{j=1}^{n'}\sum_{i=1}^{n} D((FR)\int_{y_{2j-2}}^{y_{2j}} (FR)\int_{x_{2i-2}}^{x_{2i}} H(s,t,x,y,F_1(x,y))dxdy, \frac{\sigma\sigma'}{9}H_{A_{ij}}(\tilde{F}_1)(s,t)) \leq$$

$$\leq \sum_{j=1}^{n'}\sum_{i=1}^{n} D((FR)\int_{y_{2j-2}}^{y_{2j}} (FR)\int_{x_{2i-2}}^{x_{2i}} H(s,t,x,y,F_1(x,y))dxdy, \frac{\sigma\sigma'}{9}H_{A_{ij}}(F_1)(s,t))+$$

$$+\sum_{j=1}^{n'}\sum_{i=1}^{n} D(\frac{\sigma\sigma'}{9}H_{A_{ij}}(F_1)(s,t), \frac{\sigma\sigma'}{9}H_{A_{ij}}(\tilde{F}_1))(s,t) \leq$$

$$\leq \sum_{j=1}^{n'}\sum_{i=1}^{n} 4\sigma\sigma'(\omega^{3,1}_{[x_{2i-2},x_{2i}]}(H_{F_1},\frac{\sigma}{3})+\omega^{4,2}_{[y_{2j-2},y_{2j}]}(H_{F_1},\frac{\sigma'}{3}))+\sum_{j=1}^{n'}\sum_{i=1}^{n} 4\sigma\sigma'\alpha D^*(F_1,\tilde{F}_1) \leq$$

$$\leq \Delta(\omega^{3,1}_{[a,b]}(H_{F_1},\frac{\sigma}{3})+\omega^{4,2}_{[c,d]}(H_{F_1},\frac{\sigma'}{3}))+\Delta\alpha D^*(F_1,\tilde{F}_1) \leq$$

$$\leq \frac{4\Delta}{3}(\omega^{3,1}_{[a,b]}(H_{F_1},\sigma)+\omega^{4,2}_{[c,d]}(H_{F_1},\sigma'))+BD^*(F_1,\tilde{F}_1).$$

$$\leq \tfrac{4\Delta}{3}(\omega^{3,1}_{[a,b]}(H_{F_1},\sigma)+\omega^{4,2}_{[c,d]}(H_{F_1},\sigma'))+BD^*(F_1,\tilde{F}_1).$$

From Lemma 4 we obtain

$$D^*(F_2,\tilde{F}_2) \leq \frac{4\Delta}{3}(\alpha(\omega^1_{[a,b]}(g,\sigma)+\omega^2_{[c,d]}(g,\sigma'))+B(\omega^1_{[a,b]}(H,\sigma)+\omega^2_{[c,d]}(H,\sigma'))$$
$$+\omega^3_{[a,b]}(H,\sigma)+\omega^4_{[c,d]}(H,\sigma'))+BD^*(F_1,\tilde{F}_1).$$

We denote $P=\frac{4\Delta}{3}(\alpha(\omega^1_{[a,b]}(g,\sigma)+\omega^2_{[c,d]}(g,\sigma'))+B(\omega^1_{[a,b]}(H,\sigma)+\omega^2_{[c,d]}$ $(H,\sigma'))+\omega^3_{[a,b]}(H,\sigma)+\omega^4_{[c,d]}(H,\sigma'))$.

By induction, for $m\geq 3$, we obtain

$$D^*(F_m,\tilde{F}_m) \leq P + BD^*(F_{m-1},\tilde{F}_{m-1}).$$

So we have

$$D^*(F_m, \tilde{F}_m) \le P + BD^*(F_{m-1}, \tilde{F}_{m-1}),$$
$$D^*(F_{m-1}, \tilde{F}_{m-1}) \le P + BD^*(F_{m-2}, \tilde{F}_{m-2}),$$
$$\dots$$
$$D^*(F_2, \tilde{F}_2) \le P + BD^*(F_1, \tilde{F}_1),$$
$$D^*(F_1, \tilde{F}_1) \le \frac{4\Delta}{3}(\alpha(\omega^1_{[a,b]}(g,\sigma) + \omega^2_{[c,d]}(g,\sigma')) + \omega^3_{[a,b]}(H,\sigma) + \omega^4_{[c,d]}(H,\sigma')).$$

Multiplying these inequalities by $1, B, \dots, B^{m-1}$, respectively, and summing them we have
$D^*(F_m, \tilde{F}_m) \le P(1 + B + \dots + B^{m-2}) + \frac{4\Delta}{3}B^{m-1}(\alpha(\omega^1_{[a,b]}(g,\sigma) + \omega^2_{[c,d]}(g,\sigma'))$
$+ \omega^3_{[a,b]}(H,\sigma) + \omega^4_{[c,d]}(H,\sigma'))$.

Then for $D^*(F_m, \tilde{F}_m)$ we obtained

$$\begin{aligned} &D^*(F_m, \tilde{F}_m) \le \tfrac{4\Delta}{3(1-B)}(\alpha(\omega^1_{[a,b]}(g,\sigma) + \omega^2_{[c,d]}(g,\sigma'))+ \\ &+B(\omega^1_{[a,b]}(H,\sigma) + \omega^2_{[c,d]}(H,\sigma')) + \omega^3_{[a,b]}(H,\sigma) + \omega^4_{[c,d]}(H,\sigma')). \end{aligned} \tag{14}$$

Now $D^*(F, \tilde{F}_m) \le D^*(F, F_m) + D^*(F_m, \tilde{F}_m)$. From (5) of Theorem 4 we obtained (11). □

Remark 2 Since $B < 1$, $\lim_{\sigma\to 0} \omega^1_{[a,b]}(g,\sigma) = 0$, $\lim_{\sigma'\to 0} \omega^2_{[c,d]}(g,\sigma') = 0$, $\lim_{\sigma\to 0} \omega^1_{[a,b]}(H,\sigma) = 0$, $\lim_{\sigma'\to 0} \omega^2_{[c,d]}(H,\sigma') = 0$, $\lim_{\sigma\to 0} \omega^3_{[a,b]}(H,\sigma) = 0$ and $\lim_{\sigma'\to 0} \omega^4_{[c,d]}$ $(H,\sigma') = 0$, is to prove that $\lim_{m\to\infty,\sigma,\sigma'\to 0} D^*(F, \tilde{F}_m) = 0$ that shows the convergence of the method.

5 Numerical Stability Analysis

We study the numerical stability of the iterative algorithm (6) with respect to small changes in the starting approximation. We consider $F_0 = g$ and another starting approximation $G_0 = g^* \in C(A, \mathbf{R}_{\mathcal{F}})$ such that there exists $\varepsilon > 0$ for which $D(F_0(s,t), G_0(s,t)) < \varepsilon$, for all $(s,t) \in A$. The obtained sequence of successive approximations is:

$$G_0(s,t) = g^*(s,t),$$
$$G_m(s,t) = g(s,t) \oplus (FR)\int_c^d (FR)\int_a^b H(s,t,x,y,G_{m-1}(x,y))dxdy,\ m = 1, 2, \dots,$$

and using the iterative method, the terms of produced sequence are:

$$\tilde{G}_0(s,t) = g^*(s,t),$$

$$\tilde{G}_m(s,t) = g(s,t) \oplus \frac{\sigma\sigma'}{9} \sum_{j=1}^{n'} \sum_{i=1}^{n} H_{A_{ij}}(\tilde{G}_{m-1})(s,t),\ m = 1, 2, \ldots$$

Definition 6 The method of successive approximations applied to the integral equation (1) is said to be numerically stable with respect to the choice of the first iteration if for all $(s,t) \in A$ there exist constants $k_1, k_2, k_3, k_4, k_5 > 0$, which are independent by $\sigma = \frac{b-a}{2n}$ and $\sigma' = \frac{d-c}{2n'}$ such that

$$D(\tilde{F}_m(s,t), \tilde{G}_m(s,t)) < k_1\varepsilon + k_2(\omega^1_{[a,b]}(g,\sigma) + \omega^2_{[c,d]}(g,\sigma')) + k_3(\omega^1_{[a,b]}(g^*,\sigma)+$$
$$+ \omega^2_{[c,d]}(g^*,\sigma')) + k_4(\omega^1_{[a,b]}(H,\sigma) + \omega^2_{[c,d]}(H,\sigma')) + k_5(\omega^3_{[a,b]}(H,\sigma) + \omega^4_{[c,d]}(H,\sigma')).$$

Theorem 6 *Under the conditions (i)–(iii) the iterative method is numerically stable with respect to the choice of the first iteration.*

Proof First, we observe that

$$D^*(\tilde{F}_m, \tilde{G}_m) \le D^*(\tilde{F}_m, F_m) + D^*(F_m, G_m) + D^*(G_m, \tilde{G}_m).$$

For $D^*(G_1, \tilde{G}_1) \le \frac{4\Delta}{3}(\alpha(\omega^1_{[a,b]}(g^*,\sigma) + \omega^2_{[c,d]}(g^*,\sigma')) + \omega^3_{[a,b]}(H,\sigma) + \omega^4_{[c,d]}$ $(H,\sigma'))$.

First we obtain $\omega^1_{[a,b]}(G_1,\sigma)$ and $\omega^2_{[c,d]}(G_1,\sigma')$. Let $s_1,\ s_2 \in [a,b]$, $|s_1 - s_2| < \sigma$ then

$$D(G_1(s_1,t), G_1(s_2,t)) \le$$
$$\le D(g(s_1,t), g(s_2,t)) + \int_c^d \int_a^b D(H(s_1,t,x,y,g^*(x,y)),$$
$$H(s_2,t,x,y,g^*(x,y)))dxdy \le$$
$$\le \omega^1_{[a,b]}(g,\sigma) + \Delta\omega^1_{[a,b]}(H,\sigma).$$

Hence $\omega^1_{[a,b]}(G_1,\sigma) \le \omega^1_{[a,b]}(g,\sigma) + \Delta\omega^1_{[a,b]}(H,\sigma)$, and analogous $\omega^2_{[c,d]}(G_1,\sigma')$ $\le \omega^2_{[c,d]}(g,\sigma') + \Delta\omega^2_{[c,d]}(H,\sigma')$.

Now we obtain

$$D(H(s,t,\xi_1,y,G_1(\xi_1,y)), H(s,t,\xi_2,y,G_1(\xi_2,y))) \le$$
$$D(H(s,t,\xi_1,y,G_1(\xi_1,y)), H(s,t,\xi_2,y,G_1(\xi_1,y)))+$$
$$+ D(H(s,t,\xi_2,y,G_1(\xi_1,y)), H(s,t,\xi_2,y,G_1(\xi_2,y))) \le$$
$$\le \omega^3_{[a,b]}(H,\sigma) + \alpha D(G_1(\xi_1,y), G_1(\xi_2,y)) \le$$
$$\le \omega^3_{[a,b]}(H,\sigma) + \alpha\omega^1_{[a,b]}(G_1,\sigma) \le \alpha\omega^1_{[a,b]}(g,\sigma) + B\omega^1_{[a,b]}(H,\sigma) + \omega^3_{[a,b]}(H,\sigma).$$

Hence $\omega^{3,1}_{[a,b]}(H_{G_1},\sigma)\le\alpha\omega^1_{[a,b]}(g,\sigma)+B\omega^1_{[a,b]}(H,\sigma)+\omega^3_{[a,b]}(H,\sigma)$, and analogous we obtain $\omega^{4,2}_{[c,d]}(H_{G_1},\sigma')\le\alpha\omega^2_{[c,d]}(g,\sigma')+B\omega^2_{[c,d]}(H,\sigma')+\omega^4_{[c,d]}(H,\sigma')$.

For $D(G_2(s,t),\tilde{G}_2(s,t))\le$

$$\le D((FR)\int_c^d(FR)\int_a^b H(s,t,x,y,G_1(x,y))dxdy,\frac{\sigma\sigma'}{9}\sum_{j=1}^{n'}\sum_{i=1}^{n}H_{A_{ij}}(\tilde{G}_1)(s,t))\le$$

$$\le\Delta(\omega^{3,1}_{[a,b]}(H_{G_1},\frac{\sigma}{3})+\omega^{4,2}_{[c,d]}(H_{G_1},\frac{\sigma'}{3}))+\Delta\alpha D^*(G_1,\tilde{G}_1)\le$$

$$\le\frac{4\Delta}{3}(\omega^{3,1}_{[a,b]}(H_{G_1},\sigma)+\omega^{4,2}_{[c,d]}(H_{G_1},\sigma'))+BD^*(G_1,\tilde{G}_1).$$

Hence $D^*(G_2,\tilde{G}_2)\le\frac{4\Delta}{3}(\alpha(\omega^1_{[a,b]}(g,\sigma)+\omega^2_{[c,d]}(g,\sigma'))+B(\omega^1_{[a,b]}(H,\sigma)+\omega^2_{[c,d]}(H,\sigma'))+\omega^3_{[a,b]}(H,\sigma)+\omega^4_{[c,d]}(H,\sigma'))+BD^*(G_1,\tilde{G}_1)$.

From inequality (14) and $m=3,\ldots$ we have

$$D^*(G_m,\tilde{G}_m)\le\frac{4\Delta}{3(1-B)}(\alpha(\omega^1_{[a,b]}(g,\sigma)+\omega^2_{[c,d]}(g,\sigma'))+B(\omega^1_{[a,b]}(H,\sigma)$$
$$+\omega^2_{[c,d]}(H,\sigma'))+\omega^3_{[a,b]}(H,\sigma)+\omega^4_{[c,d]}(H,\sigma'))+\frac{4B^m}{3}(\omega^1_{[a,b]}(g^*,\sigma)+\omega^2_{[c,d]}(g^*,\sigma')).$$

By hypothesis, $D(F_0(s,t),G_0(s,t))<\varepsilon$, for all $(s,t)\in A$ and thus

$$D(F_m(s,t),G_m(s,t))\le D(g(s,t),g(s,t))+$$
$$+D((FR)\int_c^d(FR)\int_a^b H(s,t,x,y,F_{m-1}(x,y))dxdy,$$
$$(FR)\int_c^d(FR)\int_a^b H(s,t,x,y,G_{m-1}(x,y))dxdy)\le$$
$$\le\int_c^d\int_a^b D(H(s,t,x,y,F_{m-1}(x,y)),H(s,t,x,y,G_{m-1}(x,y)))dxdy\le$$
$$\le\Delta\alpha D^*(F_{m-1},G_{m-1})=BD^*(F_{m-1},G_{m-1}).$$

Then $D^*(F_m,G_m)\le B^mD^*(F_0,G_0)\le B^m\varepsilon$ for all $(s,t)\in A$, $m\ge1$ and

$$D(\tilde{F}_m(s,t),\tilde{G}_m(s,t))<k_1\varepsilon+k_2(\omega^1_{[a,b]}(g,\sigma)+\omega^2_{[c,d]}(g,\sigma'))+k_3(\omega^1_{[a,b]}(g^*,\sigma)$$
$$+\omega^2_{[c,d]}(g^*,\sigma'))+k_4(\omega^1_{[a,b]}(H,\sigma)+\omega^2_{[c,d]}(H,\sigma'))+k_5(\omega^3_{[a,b]}(H,\sigma))+\omega^4_{[c,d]}(H,\sigma')),$$

where $k_1=1$, $k_2=\frac{8B}{3(1-B)}$, $k_3=\frac{4}{3}$, $k_4=\frac{8\Delta B}{3(1-B)}$, $k_5=\frac{8\Delta}{3(1-B)}$. □

6 Numerical Experiments

In this section, we intent to illustrate the obtained theoretical results on some numerical example testing the convergence of the method and the numerical stability with respect to the choice of the first iteration. The algorithm was implemented using Javascript. The program can be found on the following web address: http://enkov.com/solver2DUErr.

Example Let $A = [0, 1] \times [0, 1]$. For the integral equation

$$F(s,t) = g(s,t) \oplus (FR)\int_c^d (FR)\int_a^b H(s,t,x,y,F(x,y))dxdy,\ (s,t) \in A$$

the exact solution is $F(s,t,r) = ((2+r)s^2t,\ (4-r)s^2t)$. Here

$$g(s,t,r) = ((2+r)s^2t - \frac{2}{15}s^2t\sqrt{2+r},\ (4-r)s^2t - \frac{2}{15}s^2t\sqrt{4-r}),$$
$$H(s,t,x,y,F,r) = (s^2txy\sqrt{\underline{F}(s,t,r)},\ s^2txy\sqrt{\overline{F}(s,t,r)}).$$

Applying the iterative algorithm for various m, n, n', $r_i = ih_r$, $i = 0 \div 5$, $h_r = 0.2$ in the point $(s_0, t_0) = (0.5, 0.5)$. We obtain the computational errors $\underline{E}_m(r_i) = \underline{E}_m(s_0, t_0, r_i) = |\underline{\tilde{F}}_m(s_0, t_0, r_i) - \underline{F}(s_0, t_0, r_i)|$ and $\overline{E}_m(r_i) = \overline{E}_m(s_0, t_0, r_i) = |\overline{\tilde{F}}_m(s_0, t_0, r_i) - \overline{F}(s_0, t_0, r_i)|$. The numerical stability is tested by considering $\varepsilon = 0.1$, and for various m, n, n'. We obtain the computational errors $\underline{D}_m(r_i) = \underline{D}_m(s_0, t_0, r_i) = |\underline{\tilde{F}}_m(s_0, t_0, r_i) - \underline{\tilde{G}}_m(s_0, t_0, r_i)|$ and $\overline{D}_m(r_i) = \overline{D}_m(s_0, t_0, r_i) = |\overline{\tilde{F}}_m(s_0, t_0, r_i) - \overline{\tilde{G}}_m(s_0, t_0, r_i)|$. The results are expressed in the Table 1.

Table 1 Numerical errors in (0.5,0.5)

	m=5 $n = n'$=7				$m = 10$ $n = n' = 10$			
r_i	$\underline{E}_m(r_i)$	$\overline{E}_m(r_i)$	$\underline{D}_m(r_i)$	$\overline{D}_m(r_i)$	$\underline{E}_m(r_i)$	$\overline{E}_m(r_i)$	$\underline{D}_m(r_i)$	$\overline{D}_m(r_i)$
0.0	1.178E−6	1.648E−6	1.225E−8	2.275E−9	4.858E−7	6.772E−7	2.831E−15	1.110E−16
0.2	1.234E−6	1.608E−6	9.729E−9	2.578E−9	5.083E−7	6.606E−7	1.776E−15	1.665E−16
0.4	1.287E−6	1.566E−6	7.880E−9	2.941E−9	5.299E−7	6.436E−7	1.166E−15	1.665E−16
0.6	1.338E−6	1.524E−6	6.490E−9	3.381E−9	5.505E−7	6.262E−7	7.216E−16	2.220E−16
0.8	1.387E−6	1.480E−6	5.422E−9	3.919E−9	5.704E−7	6.082E−7	5.551E−16	2.776E−16
1.0	1.434E−6	1.434E−6	4.585E−9	4.585E−9	5.896E−7	5.896E−7	3.886E−16	3,886E−16

Acknowledgements This research was partially supported by Fund FP17-FMI-008, Fund Scientific Research, University of Plovdiv Paisii Hilendarski.

References

1. Anastassiou, G.A., Gal, S.G.: Approximation Theory: Moduli of Contnuity and Global Smoothness Preservation. Springer Science Business Media, LLC (2000)
2. Balachandran, K., Kanagarajan, K.: Existence of solutions of general nonlinear fuzzy Volterra-Fredholm integral equations. J. Appl. Math. Stoch. Anal. **3**, 333–343 (2005)
3. Bede, B., Gal, S.: Quadrature rules for integrals of fuzzy-number-valued functions. Fuzzy Sets Syst. **145**, 359–380 (2004)
4. Bica A., Popescu C.: Fuzzy trapezoidal cubature rule and application to two-dimensional fuzzy Fredholm integral equations. Soft Comput. **21**, 1229–1243. https://doi.org/10.1007/s00500-015-1856-5 (Springer)
5. Dubois, D., Prade, H.: Towards fuzzy differential calculus. Part 2: Integration of fuzzy intervals. Fuzzy Sets Syst. **8**, 105–116 (1982)
6. Enkov, S., Georgieva, A., Pavlova, A.: Quadrature rules and iterative numerical method for two-dimensional nonlinear Fredholm fuzzy integral equations. Commun. Appl. Anal. **21**, 479–498 (2017)
7. Ezzati, R., Sadatrasoul, S.M.: On numerical solution of two-dimensional nonlinear Urysohn fuzzy integral equations based on fuzzy Haar wavelets. Fuzzy Sets Syst. **309**, 145–164 (2017)
8. Friedman, M., Ma, M., Kendal, A.: Solutions to fuzzy integral equations with arbitrary kernels. Int. J. Approx. Reason **20**, 249–262 (1999)
9. Goetschel, R., Voxman, W.: Elementary fuzzy calculus. Fuzzy Sets Syst. **18**, 31–43 (1986)
10. Kaleva, O.: Fuzzy differential equations. Fuzzy Sets Syst. **24**, 301–317 (1987)
11. Mordeson, J., Newman, W.: Fuzzy integral equations. Inf. Sci. **87**, 215–229 (1995)
12. Park, J.Y., Jeong, J.U.: On the existence and uniqueness of solutions of fuzzy Volttera Fredholm integral equations. Fuzzy Sets Syst. **115**, 425–431 (2000)
13. Park, J.Y., Lee, S.Y., Jeong, J.U.: The approximate solution of fuzzy functional integral equations. Fuzzy Sets Syst. **110**, 79–90 (2000)
14. Sadatrasoul, S., Ezzati, R.: Quadrature rules and iterative method for numerical solution of two-dimensional fuzzy integral equations. In: Abstract and Applied Analysis, vol. 2114, 18 p. (2014)
15. Wu, C., Gong, Z.: On Henstock integral of fuzzy-number-valued functions (1). Fuzzy Sets Syst. **120**, 523–532 (2001)

Comparison Analysis on Two Numerical Solvers for Fractional Laplace Problems

Stanislav Harizanov and Svetozar Margenov

Abstract This study is inspired by the rapidly growing interest in the numerical solution of factional diffusion problems, strongly motivated by their advanced applications. Several different techniques are proposed to localize the nonlocal fractional diffusion operator. First three of them are based on transformation of the original problem to local elliptic or pseudoparabolic problem, or to an integral representation of the solution, thus increasing the dimension of the computational domain. More recently, an alternative method was proposed, based on best uniform rational approximations (BURA) of the function $t^{\beta-\alpha}$ for $0 < t \leq 1$ and natural β. Despite their different origin and approximation properties, BURA and the method based on integral representation have very similar implementation. For test problems with different regularity of the solution, the achieved accuracy is analysed, assuming settings of the two studied methods, leading to equal computational complexity.

1 Introduction

The interest in fractional diffusion models is motivated by numerous applications related to Hamiltonian chaos [32], nonlocal continuum physics [16], discontinuities and long-range forces in elasticity [27], anomalous diffusion in complex systems [5], nonlocal electromagnetic fluid flows [24], fractional Cahn Hilliard equation [4], contaminant transport in porous media [8], materials science [7], fractional diffusion in silicon, [33], modeling of soft tissues [13], image processing [1, 18], to name just a few.

The nonlocal continuum field theories are concerned with materials whose behavior at any interior point depends on the state of all other points - rather than only on

S. Harizanov (✉) · S. Margenov
Institute of Information and Communication Technologies,
Bulgarian Academy of Sciences, Sofia, Bulgaria
e-mail: sharizanov@parallel.bas.bg

S. Margenov
e-mail: margenov@parallel.bas.bg

K. Georgiev et al. (eds.), *Advanced Computing in Industrial Mathematics*,
Studies in Computational Intelligence 793,
https://doi.org/10.1007/978-3-319-97277-0_13

an effective field resulting from these points - in addition to its own state and the state of some calculable external field. Such kind of applications lead to fractional order partial differential equations that involve in general non-symmetric elliptic operators see, e.g. [21]. An important subclass of this topic are the fractional powers of self-adjoint elliptic operators, which are nonlocal but self-adjoint. In particular, the fractional Laplacian [25] describes an unusual diffusion process associated with random excursions. In general, the parabolic equations with fractional derivatives in time are associated with sub-diffusion, while the fractional elliptic operators are related to super-diffusion.

At the beginning, let us comment briefly on the integral definition of fractional Laplacian problem in bounded domain $\Omega \subset \mathbb{R}^d$, i.e., $(-\Delta)^\alpha u = f$, in Ω and $u = 0$ in $\Omega^c = \mathbb{R}^d \setminus \Omega$, given by

$$(-\Delta)^\alpha u(x) = C(d,\alpha) P.V. \int_{\mathbb{R}^d} \frac{u(x) - u(x')}{|x - x'|^{d+2\alpha}} dx' = f(x), \tag{1}$$

P.V. stands for principle value. In [3] error bounds in the energy norm and numerical experiments (in 2D) are presented, demonstrating an accuracy of the order of $h^{\frac{1}{2}} \log h$ for solutions obtained by means of quasi uniform mesh. In contrast to elliptic PDEs, the numerical solution of problems involving such a non-local operator is rather complicated. There are at least two reasons for that: the handling of highly singular kernels and the need to cope with an unbounded region of integration. Even in 2D, this approach could be very expensive.

In what follows we will consider the alternative definition, based on spectral decomposition of the elliptic operator. The weak formulation of the boundary value problem is: find $u \in V$ such that

$$a(u, v) := \int_\Omega (\mathbf{a}(x) \nabla u(x) \cdot \nabla v(x) + q(x))\, dx = \int_\Omega f(x) v(x) dx, \quad \forall v \in V, \tag{2}$$

where $V := \{v \in H^1(\Omega) : \ v(x) = 0 \ \text{ on } \ \Gamma_D\}$, $\Gamma = \partial\Omega$, and $\Gamma = \bar{\Gamma}_D \cup \bar{\Gamma}_N$. We assume that Γ_D has positive measure, $q(x) \geq 0$ in Ω, and $\mathbf{a}(x)$ is an SPD $d \times d$ matrix, uniformly bounded in Ω. Then, $\mathcal{L}^\alpha$, $0 < \alpha < 1$ is introduced through its spectral decomposition, i.e.

$$\mathcal{L}^\alpha u(x) = \sum_{i=1}^{\infty} \lambda_i^\alpha c_i \psi_i(x), \qquad u(x) = \sum_{i=1}^{\infty} c_i \psi_i(x), \tag{3}$$

where $\{\psi_i(x)\}_{i=1}^\infty$ are the eigenfunctions of $\mathcal{L}$, orthonormal in L_2-inner product and $\{\lambda_i\}_{i=1}^\infty$ are the corresponding positive real eigenvalues. There is still ongoing research about the relations of the different definitions and their applications, see, e.g. [6]. Let us assume again that linear elements are used to get the FEM approximation u_h. In the case of full regularity, the best possible convergence rate for $f \in L^2(\Omega)$ is, cf. [11],

$$\|u - u_h\|_{L^2(\Omega)} \le Ch^{2\alpha}|\ln h|\|f\|_{L^2(\Omega)}. \tag{4}$$

This estimate well illustrates the lower FEM accuracy for fractional diffusion problems, depending on $\alpha \in (0, 1)$. Therefore, some additional mesh refinement will be needed to get a targeted accuracy. This means stronger requirements to the solution methods, strengthening the motivations of this study.

As already noticed, the numerical solution of nonlocal problems is rather expensive. The following four approaches (A1–A4) are based on transformation of the original problem $\mathcal{L}^\alpha u = f$ to some auxiliary local problems. The goal is to make possible real life applications of 2D and 3D fractional diffusion models in computational domains with general geometry.

A1 Extension to a mixed boundary value problem in the semi-infinite cylinder.

In [14], the solution of fractional Laplacian problem is obtained by $u(x) = v(x, 0)$ where $v : \Omega \times \mathbb{R}_+ \to \mathbb{R}$ is a solution of the equation

$$-div\left(y^{1-2\alpha}\nabla v(x, y)\right) = 0, \quad (x, y) \in \Omega \times \mathbb{R}_+,$$

where $v(\cdot, y)$ satisfies the boundary conditions of (2) $\forall y \in \mathbb{R}_+$, $\lim_{y\to\infty} v(x, y) = 0$, $x \in \Omega$, as well as $\lim_{y\to 0^+}\left(-y^{1-2\alpha}v_y(x, y)\right) = f(x)$, $x \in \Omega$. The finite element approximation uses the rapid decay of the solution $v(x, y)$ in the y direction, thus enabling truncation of the semi-infinite cylinder to a bounded domain of modest size. The proposed multilevel method is based on the Xu-Zikatanov identity [31].

A2 Transformation to a pseudo-parabolic problem
The problem considered in [30] satisfies the boundary condition $a(x)\frac{\partial u}{\partial n} + \mu(x)u = 0$, $x \in \partial\Omega$, which ensures $\mathcal{L} = \mathcal{L}^* \ge \delta\mathcal{I}$, $\delta > 0$. Then the solution of fractional power diffusion problem u can be found as $u(x) = w(x, 1)$, $w(x, 0) = \delta^{-\alpha} f$, where $w(x, t)$, $0 < t < 1$, is the solution of pseudo-parabolic equation

$$(t\mathcal{D} + \delta\mathcal{I})\frac{dw}{dt} + \alpha\mathcal{D}w = 0,$$

and $\mathcal{D} = \mathcal{L} - \delta\mathcal{I} \ge 0$. Stability conditions are obtained for the fully discrete schemes under consideration.

A3 Integral representation of the solution
The following representation of the solution is used in [11]

$$\mathcal{L}^{-\alpha} = \frac{2\sin(\pi\alpha)}{\pi}\int_0^\infty t^{2\alpha-1}\left(\mathcal{I} + t^2\mathcal{L}\right)^{-1} dt,$$

introducing an exponentially convergent quadrature scheme. Then, the approximate solution of fractional Laplacian only involves evaluations of $(\mathcal{I} + t_i\mathbb{A})^{-1}f$, where $t_i \in (0, \infty)$ is related to the current quadrature node, and where $\mathcal{I}$ and $\mathbb{A}$ stand for the identity and the finite element stiffness matrix corresponding to the

Laplacian. Different types of quadrature rules, such as Gauss-Jacobi, have also been considered in the literature [2, 17, 29]. Here, the fractional power of the Laplace operator is approximated using Padé techniques.

A4 Best uniform rational approximation (BURA)
A class of optimal solvers for the linear system

$$\mathcal{A}^{\alpha}\mathbf{u} = \mathbf{f}, \quad \text{where} \quad 0 < \alpha < 1 \tag{5}$$

is proposed in [19], where $\mathcal{A}$ is a normalized symmetric and positive definite (SPD) matrix generated by a finite element or finite difference approximation of some self-adjoint elliptic problem. Instead of (5), the system $\mathcal{A}^{\alpha-\beta}\mathbf{u} = \mathcal{A}^{-\beta}\mathbf{f} := \mathbf{F}$, $\beta \geq 1$ an integer, is considered. Then $\mathcal{A}^{\beta-\alpha}\mathbf{F}$ is approximated by a set of solutions of systems with $\mathcal{A}$ and $\mathcal{A} - d_j\mathcal{I}$, for $j = 1, \dots, k$, where $k \geq 1$ is the number of partial fractions of BURA $r_{\alpha}^{\beta}(t)$ of $t^{\beta-\alpha}$, $t \in (0, 1]$.

Some further developments of these approaches can be found, e.g., in [9, 10, 12, 20, 23], as well as in [15, 22], where first efficient parallel algorithms are presented.

One can observe the algorithmic similarity of A3 and A4. More precisely, the method from [11] can be viewed as a particular rational approximation of $\mathcal{A}^{-\alpha}$. At the same time, the two methods have different theoretical background, and, as we will see in the comparison analysis, each of them has its specific convergence behavior.

The rest of the paper is organized as follows. In Sect. 2 we provide a brief introduction to the two methods [11, 19] including some new estimates related to the case of finite difference discretization of elliptic PDEs. The major contributions are presented in Sect. 3. The comparison analysis is based on two sets of numerical tests for problems with different regularity. Short concluding remarks are given at the end.

2 Description of the Solvers

Denote by $\mathcal{R}(k, k) := \{P_k/Q_k \ : \ P_k, Q_k \in \mathcal{P}_k\}$ the class of (k, k) rational functions. The (k, k) best uniform approximation $r_{\alpha}(t)$ of $t^{1-\alpha}$, $t \in [0, 1]$ and its approximation error $E_{\alpha,k}$ are defined as follows:

$$r_{\alpha} := \operatorname*{argmin}_{r\in\mathcal{R}(k,k)} \max_{t\in[0,1]} \left|t^{1-\alpha} - r(t)\right|, \qquad E_{\alpha,k} := \max_{t\in[0,1]} \left|t^{1-\alpha} - r_{\alpha}(t)\right|.$$

Let $\mathbb{A}$ be the $N \times N$ finite difference stiffness matrix corresponding to the elliptic operator $\mathcal{L}$ over a quasi-uniform mesh with mesh size h and

$$C := \|\mathbb{A}\|_{\infty} = \max_{1\leq i\leq N} \sum_{j=1}^{N} |a_{ij}|$$

be its matrix sup norm. Following [19], we introduce the normalized matrix $\mathcal{A} := C^{-1}\mathbb{A}$, which is symmetric, positive definite and its spectrum lies inside (0, 1]. Then, the original problem $\mathbb{A}^\alpha \mathbf{u}_h = \mathbf{f}$ can be rewritten as $\mathcal{A}^\alpha \mathbf{u}_h = C^{-\alpha}\mathbf{f}$ and the true solution $\mathbf{u}_h$ can be approximated via

$$\mathbf{u}_r := C^{-\alpha}\mathcal{A}^{-1}r_\alpha(\mathcal{A})\mathbf{f},$$

for which computation no information about the typically dense matrix $\mathcal{A}^{-\alpha}$ is required (see [19]). Note that the eigenvalues of $\mathcal{A}$ are $\{C^{-1}\lambda_i\}_1^N$, where $\lambda_1 \leq \lambda_2 \cdots \leq \lambda_N$ are the eigenvalues of $\mathbb{A}$ and for $\mathbf{f} = \psi_i$ – eigenvector of $\mathbb{A}$ we have

$$C^{-\alpha}\mathcal{A}^{-1}r_\alpha(\mathcal{A})\psi_i = C^{1-\alpha}\frac{r_\alpha\left(C^{-1}\lambda_i\right)}{\lambda_i}\psi_i, \quad \Longrightarrow \quad \frac{\|\mathbf{u}_r - \mathbf{u}_h\|_2}{\|\mathbf{f}\|_2} \leq \frac{C^{1-\alpha}E_{\alpha,k}}{\lambda_1}. \tag{6}$$

In [28] the asymptotic behavior of $E_{\alpha,k}$ as $k \to \infty$ is studied. Their result allows us to complete the error analysis for the BURA solver and derive the estimate

$$\|\mathbf{u}_r - \mathbf{u}_h\|_2 \leq \frac{4^{2-\alpha}C^{1-\alpha}}{\lambda_1}\sin(\pi\alpha)e^{-2\pi\sqrt{(1-\alpha)k}}\|\mathbf{f}\|_2, \qquad k \to \infty. \tag{7}$$

For the numerical computation of $\mathbf{u}_r$ we use the fractional decomposition of $t^{-1}r_\alpha(t)$ and independently solve $k+1$ linear systems based on nonnegative diagonal shifts of $\mathbb{A}$, namely

$$\mathbf{u}_r = \sum_{j=0}^{k} c_j(\mathcal{A} + d_j\mathcal{I})^{-1}\mathbf{f} = C\sum_{j=0}^{k} c_j(\mathbb{A} - Cd_j\mathcal{I})^{-1}\mathbf{f},$$

where $0 = d_0 > d_1 \cdots > d_k$ are the poles of r_α plus the additional pole at zero, and $c_j > 0$ for every j (see [26] for details).

The solver, proposed by Bonito and Pasciak in [11] incorporates an exponentially convergent quadrature scheme for the approximate computation of the integral solution representation A3

$$\mathbf{u}_Q := \frac{2k'\sin(\pi\alpha)}{\pi}\sum_{\ell=-m}^{M} e^{2(\alpha-1)\ell k'}\left(\mathbb{A} + e^{-2\ell k'}I\right)^{-1}\mathbf{f},$$

$m = \lceil \pi^2/(4\alpha k'^2)\rceil$, $M = \lceil \pi^2/(4(1-\alpha)k'^2)\rceil$. The parameter $k' > 0$ controls the accuracy of the approximant $\mathbf{u}_Q$ and the number of linear systems that has to be solved. For example $k' = 1/3$ gives rise to 120 systems for $\alpha = \{0.25, 0.75\}$, respectively 91 systems for $\alpha = 0.5$ and guarantees $\|\mathbf{u}_Q - \mathbf{u}_h\|_2 \leq O(2^{-7})\|\mathbf{f}\|_2$. The error analysis, developed in [11] states

$$\|\mathbf{u}_Q - \mathbf{u}_h\|_2 \leq \frac{2\sin(\pi\alpha)}{\pi}\left(\frac{1}{\alpha} + \frac{1}{(1-\alpha)\lambda_1}\right)e^{-\pi^2/(2k')}\|\mathbf{f}\|_2, \qquad k' \to 0.$$

To further increase the analogy between $\mathbf{u}_r$ and $\mathbf{u}_Q$, we set

$$k_Q := \frac{\pi^2}{4\alpha(1-\alpha)k'^2}$$

and consider only integer values for it. Indeed, $m = \lceil(1-\alpha)k_Q\rceil$, $M = \lceil\alpha k_Q\rceil$ and the number of linear systems that needs to be solved for $\mathbf{u}_Q$ are either $k_Q + 1$ or $k_Q + 2$, provided αk_Q is integer or not. Furthermore, although the number of linear systems is a function of the continuous variable k', for rational α it changes value only when k_Q is an integer. Therefore, both endpoints of every interval for $k_Q(k')$, corresponding to a fixed number $M + m + 1$ of summands for $\mathbf{u}_Q$ are integers, meaning that such an assumption on k_Q is natural and does not restrict the scope of the conducted numerical experiments. On the other hand k_Q plays similar role as k, indicating the computational efficiency of the approach (number of linear systems to be solved).

In terms of k_Q the error estimate becomes

$$\|\mathbf{u}_Q - \mathbf{u}_h\|_2 \leq \frac{2\sin(\pi\alpha)}{\pi}\left(\frac{1}{\alpha} + \frac{1}{(1-\alpha)\lambda_1}\right)e^{-\pi\sqrt{\alpha(1-\alpha)k_Q}}\|\mathbf{f}\|_2, \qquad k_Q \to \infty. \tag{8}$$

Comparing (7) with (8) we observe that both error estimates depend linearly on λ_1^{-1} and exponentially on the number of linear systems to be solved. The exponential order of the BURA estimate is at least twice higher than the one for the quadrature rule, but there is a multiplicative factor $C^{1-\alpha}$ in (7), which depends on the mesh size h and $C \to \infty$ as $h \to 0$. This implies trade-off between numerical accuracy and computational efficiency for the BURA method. The choice for the degree k of r_α should be synchronized with h, while the size of h does not affect the choice of k_Q. Another difference between the two approaches is that the error bound in (7) can be reached only for $\mathbf{f} = \psi_1$ and only if $C^{-1}\lambda_1$ is an extreme point for r_α (see (6)), while the error bound in (8) is more uniform. Hence, the BURA error heavily depends on the right-hand-side $\mathbf{f}$ and possesses a wide range of values, while the quadrature error is relatively independent on $\mathbf{f}$.

In the next section we numerically investigate the advantages and disadvantages of the two solvers on the fractional Laplace problem.

3 Comparison Analysis

We consider the following model problem

$$(-\Delta)^\alpha \mathbf{u} = \mathbf{f}, \qquad \Omega = [0, 1]^d, \quad d = 1, 2, \tag{9}$$

with homogeneous Dirichlet boundary conditions. Experimental data on the behavior of the BURA solver for (9) is available in [20, Sect. 5] for $d = 1$, and in [19, Sect. 5.2] for $d = 2$. The quadrature solver has also been tested on (9) for $d = 2$ in [11, Sect. 4.2]. This motivates our choice for model problem. Moreover, this paper extends the scope of the previously conducted numerical experiments, increases the depth of the performed analysis, and can be seen as a natural continuation of the other three.

On a uniform grid with mesh-size $h = 1/(N+1)$, the discrete 3-point stencil approximation of $-\Delta$ in 1D has the form $\mathbb{A} = h^{-2} tridiag(-1, 2, -1)$, while in 2D we work with the discrete 5-point stencil approximation

$$\mathbb{A} = h^{-2} tridiag\left(-I_N, \mathbb{A}_{1,1}, -I_N\right), \quad \mathbb{A}_{1,1} = tridiag(-1, 4, -1).$$

For $(-\Delta)^\alpha$ we consider the corresponding discrete approximation $\mathbb{A}^\alpha$.

In the one dimensional setting we know the eigenvalues and the eigenvectors of $\mathbb{A}$, namely

$$\lambda_i = \frac{4}{h^2} \sin^2 \frac{i\pi h}{2}, \qquad \psi_i = \{\sqrt{2h} \sin(\pi i j h)\}_{j=1}^N, \qquad i = 1, \ldots, N.$$

Since $\mathbb{A}^{-\alpha}\psi_i = \lambda_i^{-\alpha}\psi_i$ for each eigenvector, we have explicit formula for the exact solution $\mathbf{u}_h$ and we can compute the corresponding approximate solutions $\mathbf{u}_r$ and $\mathbf{u}_Q$ using a direct matrix solver, like the Thomas algorithm. Therefore, we can compare the theoretical error estimates (7)–(8) with the true approximation error, which is the maximal error over the eigenvectors.

Using that $C = \|\mathbb{A}\|_\infty = 4h^{-2}$ and $\lambda_1 \to \pi^2$, as $h \to 0$ we compute the estimates in Table 1 for $h = 10^{-3}$ and $\alpha = \{0.25, 0.50, 0.75\}$. We observe that they are overestimates of the actual ℓ_2 errors. For the BURA setting, this is due to the fact that $C^{-1}\lambda_1$ is not among the extremal points of r_α. For the quadrature setting, this is due to the asymptotic origin of the error bound and the fact that we apply it for relatively small values of k_Q. The quadrature error for ψ_i monotonically decreases as i increases and its overall order range increases with α. This range decreases for higher values

Table 1 Comparing the theoretical error estimates (7)–(8) with the actual ℓ_2 errors in 1D for $h = 10^{-3}$

$(\alpha, k = k_Q)$	Solver	Relative ℓ_2 error		
		Estimate	$\max_i \|\mathbf{u} - \mathbb{A}^{-\alpha}\psi_i\|_2$	$\min_i \|\mathbf{u} - \mathbb{A}^{-\alpha}\psi_i\|_2$
(0.25, 9)	$\mathbf{u} = \mathbf{u}_r$	5.900e−3	9.086e−4	5.074e−11
	$\mathbf{u} = \mathbf{u}_Q$	3.144e−2	9.726e−3	9.153e−3
(0.50, 8)	$\mathbf{u} = \mathbf{u}_r$	5.653e−3	5.677e−4	2.188e−11
	$\mathbf{u} = \mathbf{u}_Q$	1.649e−2	4.324e−3	4.776e−4
(0.75, 7)	$\mathbf{u} = \mathbf{u}_r$	4.451e−3	2.855e−3	2.182e−12
	$\mathbf{u} = \mathbf{u}_Q$	2.140e−2	2.610e−3	9.958e−6

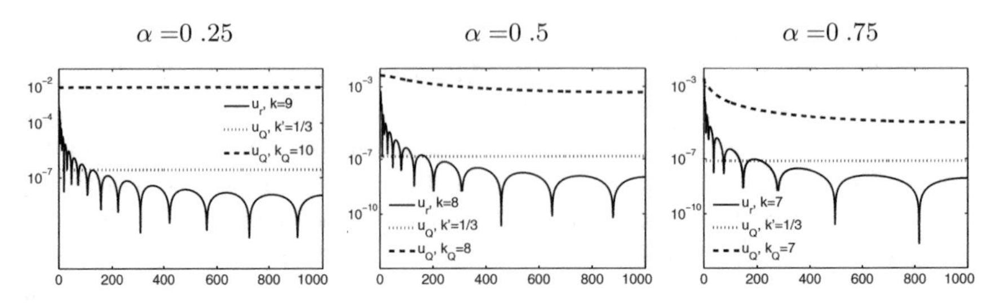

Fig. 1 1D ℓ_2-relative error analysis over the eigenvectors of $\mathbb{A}$ ($h = 10^{-3}$)

of k_Q. For example, when $\alpha = \{0.25, 0.50, 0.75\}$ and we take $k_Q = \{120, 91, 120\}$ (guaranteed accuracy of order 7) the min/max errors $\|\mathbf{u}_Q - \mathbf{u}_h\|_2$ over $\{\psi_i\}$ are 2.793e−7/2.873e−7, 1.417e−7/1.560e−7, and 7.797e−8/1.065e−7, respectively. The maximal BURA error is again attained for the first eigenvectors. The overall error order range, however, is significantly larger, as the error for $\mathbf{f} = \psi_i$ depends on λ_i, not λ_1 (see (6)). The error is affected by the oscillating behavior of r_α and is not monotone with respect to i. The accuracy of the BURA solver exceeds the one of the quadrature rule with the same computational complexity, and except for the low spectrum of $\mathbb{A}$ is comparable to the one of the benchmark u_Q for $k' = 1/3$ (see Fig. 1). Only for ψ_1 and $\alpha = 0.75$ we have $\|\mathbf{u}_r - \mathbf{u}_h\|_2 > \|\mathbf{u}_Q - \mathbf{u}_h\|_2$, but already for ψ_2 we get $\|\mathbf{u}_r - \mathbf{u}_h\|_2 = O(10^{-4})$ and for ψ_8: $\|\mathbf{u}_r - \mathbf{u}_h\|_2 < \min_i \|\mathbf{u}_Q - \mathbb{A}^{-0.75}\psi_i\|_2$.

In 2D the eigenvectors of $\mathbb{A}$ are element-wise products of pairs of $\{\psi_i\}$, while the eigenvalues are sums of the corresponding two 1D eigenvalues. Here, $\lambda_1 = 8h^{-2}\sin^2(\pi h/2) \approx 2\pi^2$ and $C = \|\mathbb{A}\|_\infty = 8h^{-2}$. We consider two different test functions f_1 and f_2 in (1)

$$f_1(x, y) = \begin{cases} 1, & \text{if } (x - 0.5)(y - 0.5) > 0, \\ -1, & \text{otherwise.} \end{cases} \qquad f_2(x, y) = \cos(\pi h x)\cos(\pi h y) \tag{10}$$

where $(x, y) \in (0, 1)^2$, and their corresponding discretizations $\mathbf{f}_1, \mathbf{f}_2$ in (9). The first one is the checkerboard function. It has a jump discontinuity along $x = 0.5$ and $y = 0.5$ and has already been used as a test function in this framework [11, 19]. The second one is a tensor product of cosine functions, thus highly regular. Due to the homogeneous boundary conditions, there are jump discontinuities along $\partial\Omega$ in both examples (Fig. 2). Although theoretically possible, the numerical derivation of the exact solution $\mathbf{u}_h$ is quite expensive. Therefore, we do not do it, but use $\mathbf{u}_Q$ with $k' = 1/3$ as a reference solution, instead.

We conduct two series of numerical experiments for two different choices of norms – the ℓ_2 norm and the ℓ_∞ norm. In the first one, we compare pure approximation error with respect to $\mathbf{u}_r$ and $\mathbf{u}_Q$ on coarser and finer meshes. For this purpose, we set $h = 2^{-n}$, $n = 8, \ldots, 12$ and compute the reference solution on the current mesh. Results are summarized in Table 2. For the second one, we additionally take into account the discretization error (4) and analyze the relative errors with respect to the

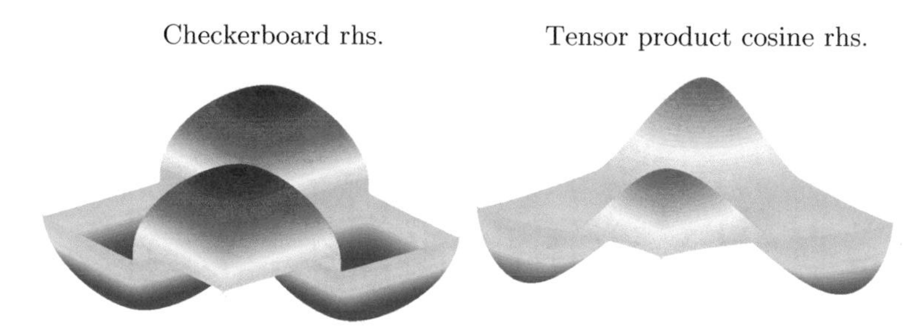

Fig. 2 BURA-approximation of $\mathbb{A}^{-\alpha}\mathbf{f}_1$ and $\mathbb{A}^{-\alpha}\mathbf{f}_2$ for $(\alpha, k) = (0.25, 9)$, $h = 2^{-11}$

corresponding projection onto the coarse mesh of the reference solution, computed for $h = 2^{-12}$. Results are summarized in Table 3.

The collected values in the two tables are the relative errors $\|\mathbf{u}_r - \mathbf{u}_h\|_2/\|\mathbf{f}\|_2$, respectively $\|\mathbf{u}_r - \mathbf{u}_h\|_\infty/\|\mathbf{f}\|_\infty$, where $\mathbf{u}_h$ is the reference solution. The $\mathbf{u}_Q$ error behaves stably with respect to the change of mesh size and norm choice, remaining practically the same. This is due to the independence of λ_1 from h and the small error amplitudes among the eigenvectors of $\mathbb{A}$. On the other hand, the $\mathbf{u}_r$ error varies from mesh to mesh and between the norm choices. Indeed, the spectrum of $\mathbb{A}$ changes with h, thus the relation between eigenvalues and extremal points of r_α, and as $C = 8h^{-2}$ the mesh size explicitly affects the BURA accuracy, according to (7). Furthermore, we observe that the sup-norm is more sensitive to the different order of BURA errors for ψ_1 and the rest, since $\|\mathbf{u}_r - \mathbf{u}_h\|_\infty/\|\mathbf{f}\|_\infty$ is always worse than $\|\mathbf{u}_r - \mathbf{u}_h\|_2/\|\mathbf{f}\|_2$. The effect of the interaction of all these phenomena is clearly visible for $\mathbf{f}_1$, $\alpha = 0.25$, $h = 2^{-11}, 2^{-12}$, where the relative sup-norm error for $\mathbf{u}_r$ between the two meshes increases 20 times and becomes comparable to the $\mathbf{u}_Q$ error. In general, we observe that for both examples the BURA method is more accurate than the other one, when the same computational complexity is considered. The presence of additional regularity in the right-hand-side does not seem to affect the error, as those for $\mathbf{f}_1$ and $\mathbf{f}_2$ are quite similar.

When the discretization error is also taken into account the order of the BURA relative error decreases, while the one of the quadrature solver remains the same. This indicates that for the BURA method and the choice of k the discretization error majorates the approximating one, while for the other method and the choice of k_Q it is vice versa. As h decreases, the BURA approximation error increases, while the discretization error decreases. We observe that for discontinuous f the overall error decreases as h changes from 2^{-8} to 2^{-11}, while for smooth one – increases.

Finally, on Fig. 3 we compare computational complexity for fixed accuracy. More precisely, for different k and α we compute the relative ℓ_2 BURA error and we find the smallest k_Q, for which the corresponding ℓ_2 quadrature error is smaller. We set $h = 2^{-10}$. When $\alpha = 0.25, 0.5$ those k_Q values are almost the same for $\mathbf{f}_1$ and $\mathbf{f}_2$, while for $\alpha = 0.75$ and discontinuous right-hand-side the advantage of the BURA

Table 2 Relative ℓ^2 errors for various discretization levels. We consider $k_Q = k$ and u_Q for $k' = 1/3$ and the current mesh size h as a reference solution

(α, k)	h	Checkerboard rhs				Tensor product cosine rhs			
		BURA (A4)		Quadrature (A3)		BURA (A4)		Quadrature (A3)	
		ℓ_2	ℓ_∞	ℓ_2	ℓ_∞	ℓ_2	ℓ_∞	ℓ_2	ℓ_∞
(0.25, 9)	2^{-8}	4.745e−5	1.238e−4	9.374e−3	9.569e−3	4.235e−5	5.116e−5	6.712e−3	6.784e−3
	2^{-9}	3.566e−5	7.559e−5	9.375e−3	9.568e−3	5.062e−5	7.052e−5	6.713e−3	6.786e−3
	2^{-10}	1.756e−4	2.951e−4	9.375e−3	9.568e−3	2.176e−4	2.870e−4	6.713e−3	6.789e−3
	2^{-11}	2.689e−4	5.551e−4	9.375e−3	9.568e−3	4.879e−4	7.706e−4	6.713e−3	6.790e−3
	2^{-12}	4.883e−3	1.019e−2	9.374e−3	9.568e−3	4.425e−3	5.031e−3	6.713e−3	6.790e−3
(0.50, 8)	2^{-8}	1.016e−4	2.026e−4	4.021e−3	4.094e−3	9.597e−5	1.089e−4	1.912e−3	1.912e−3
	2^{-9}	2.317e−4	5.033e−4	4.014e−3	4.093e−3	2.153e−4	2.435e−4	1.907e−3	1.913e−3
	2^{-10}	3.833e−4	8.679e−4	4.010e−3	4.093e−3	3.617e−4	4.203e−4	1.906e−3	1.913e−3
	2^{-11}	2.411e−4	6.327e−4	4.008e−3	4.093e−3	3.310e−4	4.232e−4	1.905e−3	1.913e−3
	2^{-12}	1.424e−3	2.817e−3	4.007e−3	4.093e−3	1.337e−3	1.613e−3	1.904e−3	1.913e−3
(0.75, 7)	2^{-8}	1.069e−5	4.141e−5	1.505e−3	1.825e−3	4.505e−5	5.613e−5	2.005e−3	1.959e−3
	2^{-9}	1.560e−4	3.240e−4	1.502e−3	1.824e−3	1.477e−4	1.833e−4	1.998e−3	1.959e−3
	2^{-10}	4.180e−4	9.017e−4	1.500e−3	1.823e−3	3.727e−4	4.060e−4	1.994e−3	1.959e−3
	2^{-11}	5.215e−4	1.190e−3	1.500e−3	1.823e−3	4.668e−4	5.147e−4	1.992e−3	1.960e−3
	2^{-12}	6.560e−5	1.420e−4	1.499e−3	1.823e−3	1.810e−4	2.771e−4	1.991e−3	1.960e−3

Table 3 Relative ℓ^2 errors for various discretization levels. We consider $k_Q = k$ and u_Q for $k' = 1/3$ and mesh size $h = 2^{-12}$ as a reference solution

(α, k)	h	Checkerboard rhs				Tensor product cosine rhs			
		BURA (A4)		Integral (A3)		BURA (A4)		Integral (A3)	
		ℓ_2	ℓ_∞	ℓ_2	ℓ_∞	ℓ_2	ℓ_∞	ℓ_2	ℓ_∞
(0.25, 9)	2^{-8}	5.863e−3	5.236e−2	1.080e−2	4.285e−2	2.781e−4	2.600e−3	6.823e−3	9.381e−3
	2^{-9}	2.823e−3	3.234e−2	9.707e−3	2.425e−2	1.441e−4	1.813e−3	6.752e−3	8.586e−3
	2^{-10}	1.253e−3	1.785e−2	9.436e−3	1.870e−2	2.268e−4	1.210e−3	6.726e−3	7.984e−3
	2^{-11}	5.027e−4	7.443e−3	9.381e−3	1.383e−2	4.888e−4	7.707e−4	6.717e−3	7.412e−3
(0.50, 8)	2^{-8}	1.349e−3	6.822e−3	4.214e−3	7.478e−3	9.589e−5	1.809e−4	1.926e−3	1.976e−3
	2^{-9}	6.671e−4	3.495e−3	4.056e−3	5.545e−3	2.161e−4	2.438e−4	1.912e−3	1.927e−3
	2^{-10}	4.683e−4	1.624e−3	4.018e−3	4.662e−3	3.615e−4	4.202e−4	1.907e−3	1.916e−3
	2^{-11}	2.569e−4	6.592e−4	4.009e−3	4.261e−3	3.310e−4	4.232e−4	1.905e−3	1.914e−3
(0.75, 7)	2^{-8}	4.194e−4	1.277e−3	1.558e−3	2.276e−3	4.461e−5	5.485e−5	2.007e−3	1.961e−3
	2^{-9}	2.510e−4	6.038e−4	1.514e−3	1.984e−3	1.479e−4	1.836e−4	1.998e−3	1.956e−3
	2^{-10}	4.264e−4	9.644e−4	1.503e−3	1.887e−3	3.727e−4	4.059e−4	1.994e−3	1.956e−3
	2^{-11}	5.222e−4	1.206e−3	1.500e−3	1.843e−3	4.668e−4	5.147e−4	1.992e−3	1.956e−3

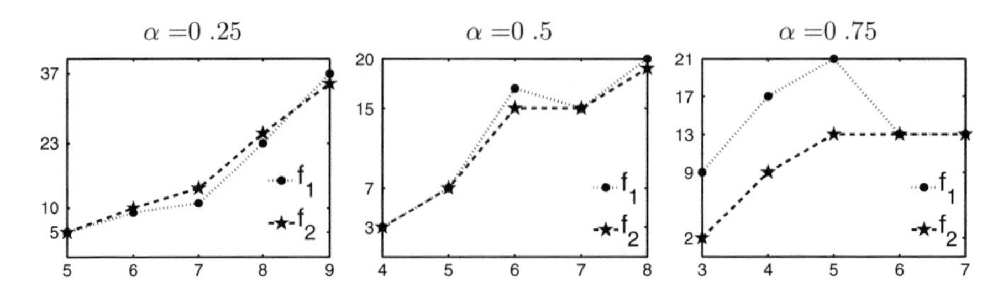

Fig. 3 Minimal k_Q as a function of k which gives rise to smaller ℓ_2 relative error than the corresponding (k, k)- BURA error ($h = 2^{-10}$)

solver is stronger indicated. Due to the presence of $\sqrt{\alpha}$ in the exponential order of the error estimate (8), unlike (7), larger k_Q are needed for matching the BURA error for smaller α. For example, $(\alpha, k) = (0.25, 9)$ gives rise to $k_Q = 37$, meaning that 39 different linear systems need to be computed for the derivation of a more accurate $\mathbf{u}_Q$ than the $\mathbf{u}_r$. The latter incorporates only 10 linear systems. The last remark is devoted to the case $\alpha = 0.75$. It seems that $k = 5$ is a quite good choice for the BURA method when solving fractional Laplace problems, as the relative error is smaller than the corresponding ones with $k = 6, 7$. In 1D this was already observed in [20].

4 Concluding Remarks

In this paper a comparison analysis on both numerical accuracy and computational efficiency was performed for two numerical solvers of fractional Laplace problems with similar complexity. It was shown that the accuracy of both solvers increases exponentially with the number of solved linear systems. The exponential rate of the BURA accuracy is higher than the one of the quadrature accuracy, and the effect increases as α decreases. On the other hand, the BURA accuracy decreases with the mesh size h of the finite difference discretization, while the quadrature accuracy is independent of h. The conducted numerical experiments confirmed the theoretical analysis and indicated that the BURA solver outperforms the quadrature one, provided the number of linear systems to be solved is chosen appropriately with respect to h.

Acknowledgements The work has been partially supported by the Bulgarian National Science Fund under grants No. BNSF-DM02/2 and BNSF-DN12/1.

References

1. Abirami, A., Prakash, P., Thangavel, K.: Fractional diffusion equation-based image denoising model using cngl scheme. Int. J. Comput. Math. 1–18 (2017)
2. Aceto, L., Novati, P.: Rational approximation to the fractional Laplacian operator in reaction-diffusion problems. SIAM J. Sci. Comput. **39**(1), A214–A228 (2017)
3. Acosta, G., Borthagaray, J.: A numerical study of the homogeneous elliptic equation with fractional order boundary conditions. A Fractional Laplace Equation: Regularity of Solutions and Finite Element Approximations **55**(2), 472–495 (2017)
4. Ainsworth, M., Mao, Z.: Analysis and approximation of a fractional cahn-hilliard equation. SIAM J. Numer. Anal. **55**(4), 1689–1718 (2017)
5. Bakunin, O.G.: Turbulence and Diffusion: Scaling Versus Equations. Springer Science & Business Media (2008)
6. Bates, P.W.: On some nonlocal evolution equations arising in materials science. Nonlinear Dyn. Evol. Equ. **48**, 13–52 (2006)
7. Bates, P.: On some nonlocal evolution equations arising in materials science. Nonlinear dynamics and evolution equations. Fluid Inst. Commun. **48**(1–40), 10 (2006)
8. Bear, J., Cheng, A.H.D.: Modeling Groundwater Flow and Contaminant Transport, vol. 23. Springer, Berlin (2010)
9. Bolin, D., Kirchner, K.: The SPDE approach for Gaussian random fields with general smoothness. Submitted, posted as arXiv:1612.04846 (Nov 2017)
10. Bonito, A., Borthagaray, H., Nochetto, R., Otarola, E., Salgado, A.: Numerical methods for fractional diffusion. Submitted, posted as arXiv:1707.01566 (July 2017)
11. Bonito, A., Pasciak, J.: Numerical approximation of fractional powers of elliptic operators. Math. Comput. **84**(295), 2083–2110 (2015)
12. Bonito, A., Pasciak, J.: Numerical approximation of fractional powers of regularly accretive operators. IMA J. Numer. Anal. **37**(3), 1245–1273 (2016)
13. Bueno-Orovio, A., Kay, D., Grau, V., Rodriguez, B., Burrage, K.: Fractional diffusion models of cardiac electrical propagation: role of structural heterogeneity in dispersion of repolarization. J. R. Soc. Interface **11**(97) (2014)
14. Chen, L., Nochetto, R., Enrique, O., Salgado, A.J.: Multilevel methods for nonuniformly elliptic operators and fractional diffusion. Math. Comput. **85**, 2583–2607 (2016)
15. Ciegis, R., Starikovicius, V., Margenov, S., Kriauziene, R.: Parallel solvers for fractional power diffusion problems. Concurrency Comput. Pract. Experience **29**(24) (2017)
16. Eringen, A.C.: Nonlocal Continuum Field Theories. Springer, Berlin (2002)
17. Frommer, A., Güttel, S., Schweitzer, M.: Efficient and stable Arnoldi restarts for matrix functions based on quadratures. SIAM J. Sci. Comput. **35**(2), 661–683 (2014)
18. Gilboa, G., Osher, S.: Nonlocal operators with applications to image processing. Multiscale Model. Simul. **7**(3), 1005–1028 (2008)
19. Harizanov, S., Lazarov, R., Margenov, S., Marinov, P., Vutov, Y.: Optimal solvers for linear systems with fractional powers of sparse SPD matrices. Numer. Linear Algebra Appl. e2167 (2018)
20. Harizanov, S., Margenov, S.: Positive approximations of the inverse of fractional powers of SPD M-matrices. To appear in Springer, Lecture Notes in Economics and Mathematical Systems, posted as arXiv:1706.07620 (June 2017)
21. Kilbas, A., Srivastava, H., Trujillo, J.: Theory and Applications of Fractional Differential Equations. Elsevier, Amsterdam (2006)
22. Kosturski, N., Margenov, S., Vutov, Y.: Performance analysis of MG preconditioning on Intel Xeon Phi: towards scalability for extreme scale problems with fractional laplacians. In: Large-Scale Scientific Computing. LNCS, vol. 10665, pp. 304–312. Springer, Berlin (2017)
23. Lazarov, R., Vabishchevich, P.: A numerical study of the homogeneous elliptic equation with fractional order boundary conditions. Fractional Calc. Appl. Anal. **20**(2), 337–351 (2017)
24. McCay, B., Narasimhan, M.: Theory of nonlocal electromagnetic fluids. Arch. Mech. **33**(3), 365–384 (1981)

25. Pozrikidis, C.: The Fractional Laplacian. Chapman and Hall/CRC (2016)
26. Saff, E.B., Stahl, H.: Asymptotic distribution of poles and zeros of best rational approximants to x^α on [0, 1]. In: "Topics in Complex Analysis", Banach Center Publications. vol. 31. Institute of Mathematics, Polish Academy of Sciences, Warsaw (1995)
27. Silling, S.: Reformulation of elasticity theory for discontinuities and long-range forces. J. Mech. Phys. Solids **48**(1), 175–209 (2000)
28. Stahl, H.: Best uniform rational approximation of x^α on [0, 1]. Bull. Am. Math. Soc. **28**(1), 116–122 (1993)
29. Vabishchevich, P.: Numerical solution of time-dependent problems with fractional power elliptic operator. Comput. Methods Appl. Math. **18**(1), 111–128 (2017)
30. Vabishchevich, P.N.: Numerically solving an equation for fractional powers of elliptic operators. J. Comput. Phys. **282**, 289–302 (2015)
31. Xu, J., Zikatanov, L.: The method of alternating projections and the method of subspace corrections in Hilbert space. J. Am. Math. Soc. **15**(3), 573–597 (2002)
32. Zaslavsky, G.M.: Chaos, fractional kinetics, and anomalous transport. Phys. Rep. **371**(6), 461–580 (2002)
33. Zijlstra, E., Kalitsov, A., Zier, T., Garcia, M.: Fractional diffusion in silicon. Adv. Mater. **25**(39), 5605–5608 (2013)

Initial Calibration of MEMS Accelerometers, Used for Measuring Inclination and Toolface

Tihomir B. Ivanov and Galina S. Lyutskanova-Zhekova

Abstract In the present work, we consider calibrating MEMS accelerometers for the purpose of determining orientation in space. We propose a new objective function whose minimization gives an estimate of the calibration coefficients. The latter takes into account the specifics of measuring toolface and inclination in seek of better accuracy, when the device is used for this purpose. To the best of our knowledge, such an objective function has not been mentioned in the literature. The calibration algorithm is described in detail because, even though, some of the steps are standard from the point of view of a numerical analyst, this could be helpful for an engineer or an applied scientist, looking to make a concrete implementation for applied purposes. On the basis of numerical experiments with sensor data, we compare the accuracy of the proposed algorithm with a classical method. We show that the proposed one has an advantage when sensors are to be used for orientation purposes.

1 Introduction

Microelectromechanical systems (MEMS) sensors have wide range of applications. Amongst their advantages are their miniature size and low-cost. One particular application of accelerometer sensors is determining orientation in space [10]. For instance, in directional drilling, the orientation of the borehole is determined by measuring three angles—toolface, inclination, and azimuth angles [8]. In the present work, we are interested in the problem of measuring the first two of them (see Fig. 1).

Let us denote the acceleration vector, acting on a three-axial accelerometer sensor, with

T. B. Ivanov · G. S. Lyutskanova-Zhekova (✉)
Faculty of Mathematics and Informatics, Institute of Mathematics and Informatics, Bulgarian Academy of Sciences, Sofia University, 5 James Bourchier Blvd., 1164 Sofia, Bulgaria
e-mail: g.zhekova@fmi.uni-sofia.bg

T. B. Ivanov
e-mail: tbivanov@fmi.uni-sofia.bg

K. Georgiev et al. (eds.), *Advanced Computing in Industrial Mathematics*, Studies in Computational Intelligence 793,
https://doi.org/10.1007/978-3-319-97277-0_14

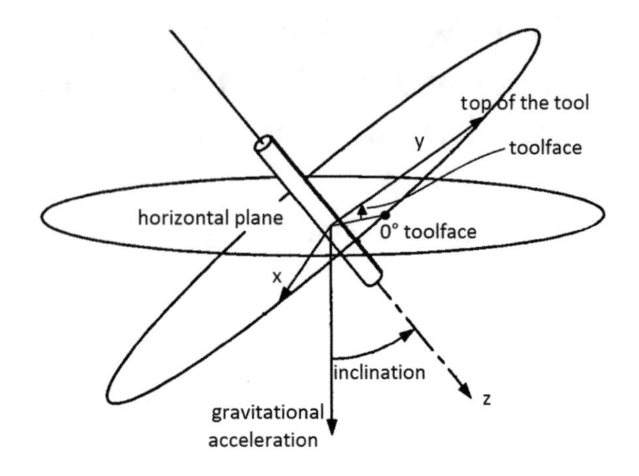

Fig. 1 Toolface and inclination angles

$$\mathbf{a} = (a_x, a_y, a_z)^T,$$

where the corresponding x-, y- and z-directions are defined as in Fig. 1. Let us note that the coordinate system is "attached" to the device.

When the device is in still position, it should be measuring only the gravitational acceleration. Then, using simple geometric considerations, the toolface (tf) and inclination (*incl*) can be easily computed, as follows[1]:

$$tf(\mathbf{a}) = \begin{cases} \frac{360^\circ}{2\pi} atan2(a_y, a_x) & \text{if } atan2(a_y, a_x) \geq 0, \\ 360^\circ + \frac{360^\circ}{2\pi} atan2(a_y, a_x) & \text{otherwise,} \end{cases} \tag{1}$$

where $atan2$ computes the angle which the segment through the point (x, y) and the origin makes with the positive x-semi-axis:

$$atan2(x, y) := \begin{cases} \arctan(\frac{y}{x}) & \text{if } x > 0, \\ \arctan(\frac{y}{x}) + \pi & \text{if } x < 0 \text{ and } y \geq 0, \\ \arctan(\frac{y}{x}) - \pi & \text{if } x < 0 \text{ and } y < 0, \\ +\frac{\pi}{2} & \text{if } x = 0 \text{ and } y > 0, \\ -\frac{\pi}{2} & \text{if } x = 0 \text{ and } y < 0, \\ \text{undefined} & \text{if } x = 0 \text{ and } y = 0. \end{cases} \tag{2}$$

The inclination is given by

[1]Let us remark that when using formula (1) in computer arithmetic, it is wise to substitute the condition for the first case on the right-hand side with $atan2(a_y, a_x) \geq -\varepsilon$ for some $\varepsilon > 0$ in order to avoid large errors for near-zero angles.

$$incl(\mathbf{a}) = \begin{cases} 90^\circ - \dfrac{360^\circ}{2\pi} \arcsin \sqrt{\dfrac{a_x^2 + a_y^2}{a_x^2 + a_y^2 + a_z^2}} & \text{if } 90^\circ - \dfrac{360^\circ}{2\pi} \arccos \dfrac{a_z}{\sqrt{a_x^2 + a_y^2 + a_z^2}} > 60^\circ, \\ 90^\circ - \dfrac{360^\circ}{2\pi} \arccos \dfrac{a_z}{\sqrt{a_x^2 + a_y^2 + a_z^2}} & \text{otherwise.} \end{cases} \quad (3)$$

The above definition is split into two cases that are theoretically equivalent in order to achieve better numerical accuracy [9].

It is well-known, however, that MEMS accelerometers are subject to different sources of error that should be accounted for before they can be used in practice. There are two main types of errors—deterministic and random.

The deterministic error sources include the bias (offset) and the scale factor errors [2]. Another issue lies in the fact that formulae (1) and (3) are only valid if the three axes of the sensor are perfectly orthogonal. This, however, can never be the case in practice. The random errors include bias-drifts or scale factor drifts, and the rate at which these errors change with time. Furthermore, all the errors are sensitive to different environmental factors, especially to temperature variations [1].

In the present work, we are interested in the initial calibration in laboratory conditions of the deterministic sources of error. There are numerous studies, concerning this problem (see e.g. [1–3, 5–7] and the references therein). Most of them are based on a direct comparison between the vector of gravitational acceleration and the raw output from the sensors. This, however, might lead to a certain problem. If the inclination is close to 90°, the x- and y-components of the acceleration are small. Inaccuracies in measuring them, after calibrating the device, will not be crucial if it is used, e.g., in an inertial navigation system. However, they might have a more significant impact on determining the toolface from (1). Thus, we suggest an algorithm for compensating the errors due to non-orthogonalities, shifts and scale factors, taking into account the specifics of this particular application.

Apart from this, random errors should be accounted for by using an appropriate stochastic model. Many authors suggest using Alan variance as a tool for studying those errors [7]. Also, a temperature model should be used to account for the temperature variations [1].

The paper is structured as follows. In Sect. 2, we describe the general methodology that we study. A complete algorithm is developed and then formulated in Sect. 2.3. Numerical experiments compare its applicability with a classical calibration approach in Sect. 3.

2 General Description of the Calibration Methodology

We assume the following classical linear relation between a raw measurement, $\hat{\mathbf{a}}$, and the calibrated one, $\mathbf{a}$ [2]:

$$\mathbf{a} = M.\hat{\mathbf{a}} + \mathbf{b}, \quad (4)$$

where M is a full 3×3 matrix,

$$M = \begin{bmatrix} m_{xx} & m_{xy} & m_{xz} \\ m_{yx} & m_{yy} & m_{yz} \\ m_{zx} & m_{zy} & m_{zz} \end{bmatrix},$$

wherein the diagonal elements account for the scaling errors and the off-diagonal elements—for the errors due to non-orthogonalities. The vector

$$\mathbf{b} = \begin{bmatrix} b_x \\ b_y \\ b_z \end{bmatrix}$$

contains the offsets.

2.1 Formulation of the Parametric Identification Problem

In order to compute the twelve unknown parameters in (4), we use a set of s known measurements $\hat{\mathbf{a}}_i$ with corresponding toolfaces tf_i and inclinations $incl_i$, $i = \overline{1, s}$, and define the following objective function:

$$\varepsilon(M, \mathbf{b}) = \sum_{i=1}^{s} \left(\left(tf(M.\hat{\mathbf{a}}_i + \mathbf{b}) - tf_i \right)^2 + \left(incl(M.\hat{\mathbf{a}}_i + \mathbf{b}) - incl_i \right)^2 \right), \quad (5)$$

where tf and $incl$ are defined with (1) and (3). We obtain the calibration parameters, minimizing the latter with respect to M and $\mathbf{b}$.

Let us denote the vector of unknown parameters as follows:

$$\mathbf{p} = (m_{xx}, m_{xy}, m_{xz}, m_{yx}, m_{yy}, m_{yz}, m_{zx}, m_{zy}, m_{zz}, b_x, b_y, b_z) =: (p_1, p_2, \ldots, p_{12}).$$

Then, we rewrite the minimization problem (5) as

$$\min_{\mathbf{p}} \|\mathbf{r(p)}\|_2^2, \quad (6)$$

where the vector of residuals $\mathbf{r} \in \mathbb{R}^{2n}$ is

$$\begin{aligned} r_{2i-1} &= tf(M.\hat{\mathbf{a}}_i + \mathbf{b}) - tf_i, \\ r_{2i} &= incl(M.\hat{\mathbf{a}}_i + \mathbf{b}) - incl_i, \quad i = \overline{1, s}. \end{aligned} \quad (7)$$

2.2 Optimization Algorithms

We shall use classical numerical methods for solving the nonlinear least squares problem.

2.2.1 Gauss–Newton Method

The Gauss–Newton method for problem (6) is based on a sequence of linear approximations of $\mathbf{r}(\mathbf{p})$ [4]. If $\mathbf{p}_k$ denotes the current approximation, then a correction $\bar{\mathbf{p}}_k$ is computed as a solution of the linear least squares problem

$$\min_{\bar{\mathbf{p}}} \|\mathbf{r}(\mathbf{p}_k) + J(\mathbf{p}_k)\bar{\mathbf{p}}\|_2^2, \tag{8}$$

and the new approximation is

$$\mathbf{p}_{k+1} = \mathbf{p}_k + \alpha_k \bar{\mathbf{p}}_k. \tag{9}$$

In the latter, α_k is chosen, such that it satisfies the Armijo–Goldstein step length principle, i.e. it is the largest number in the sequence $1, 1/2, 1/4, \ldots$ for which the inequality

$$\|\mathbf{r}(\mathbf{p}_k)\|_2^2 - \|\mathbf{r}(\mathbf{p}_k + \alpha_k \bar{\mathbf{p}}_k)\|_2^2 \geq \frac{1}{2}\alpha_k \|J(\mathbf{p}_k)\bar{\mathbf{p}}_k\|_2^2 \tag{10}$$

holds [4]. For our particular problem, the Jacobian matrix $J(\mathbf{p}_k)$ is computed in the Appendix.

2.2.2 Linear Least Squares

We consider the linear least squares problem (8):

$$J\bar{\mathbf{p}} = -\mathbf{r}. \tag{11}$$

Since our numerical experiments show that most of the basic methods for solving the latter problem (see, e.g., [4]) are unstable since they lead to solving a very ill-conditioned linear algebraic system, we first compute the singular value decomposition of J [4, 11].

Let $J \in \mathbb{R}^{m\times n}$ be a matrix of rank r. Then, there exist orthogonal matrices $U \in \mathbb{R}^{m\times m}$ and $V \in \mathbb{R}^{n\times n}$ such that the singular value decomposition is

$$J = U\Sigma V^T, \quad \Sigma = \begin{bmatrix} \Sigma_r & 0 \\ 0 & 0 \end{bmatrix},$$

where $\Sigma_r = \mathrm{diag}(\sigma_1, \ldots, \sigma_r)$ and $\sigma_1 \geq \sigma_2 \geq \cdots \geq \sigma_r > 0$.

The minimum-norm least squares solution of (11) is given with

$$\bar{\mathbf{p}} = -J^{\dagger}\mathbf{b} = V \begin{bmatrix} \Sigma_r^{-1} & 0 \\ 0 & 0 \end{bmatrix} \mathbf{c},$$

where

$$\mathbf{c} = -U^T \mathbf{r}.$$

It is important to note, however, that the rank r of the matrix can be wrongly estimated because of round-off errors. Thus, we keep only those singular values σ_i that satisfy

$$|\sigma_i| < 100 * \varepsilon * |\sigma_1|,$$

where ε is the machine precision.

2.3 Numerical Algorithm

Here, we summarize the algorithm for calibrating a three-axial MEMS accelerometer.

Step 1. Choose an initial guess for the parameter vector, $\mathbf{p}_0$.
Step 2. For $k = 0, 1, 2, \ldots$:

- Compute $\mathbf{r}(\mathbf{p}_k)$, using (7);
- Compute $J(\mathbf{p}_k)$, using the formulae in the Appendix;
- Solve the linear least squares problem (11) (see Sect. 2.2.2) and let its solution be $\bar{\mathbf{p}}_k$;
- Compute $\mathbf{p}_{k+1}$ from (9), using condition (10).

We iterate until either of the following stopping criteria is met:

- (Convergence) $\|\mathbf{p}_{k+1} - \mathbf{p}_k\| < tol$, where tol is a chosen error tolerance;
- (Divergence) $k > max\_iterations$.

3 Numerical Experiments

In this section, we shall present numerical results, obtained by using the method presented in the previous section. In order to verify its applicability, we compare the results to those, obtained by using a classical calibration method, i.e. minimizing the following objective function:

$$\overline{\varepsilon}(M, \mathbf{b}) = \sum_{i=1}^{s} \left(\left(\frac{err_{i,x}}{g_{i,x}} \right)^2 + \left(\frac{err_{i,y}}{g_{i,y}} \right)^2 + \left(\frac{err_{i,z}}{g_{i,z}} \right)^2 \right), \tag{12}$$

where $\mathbf{err}_i(M, \mathbf{b})$ is defined with

$$\mathbf{err}_i = M.\widehat{\mathbf{a}}_i + \mathbf{b} - \mathbf{g}_i.$$

In the latter, $\mathbf{g}_i$ denotes the gravitational vector, corresponding to tf_i and $incl_i$, and it can be easily computed. This goal function is well-known in literature (see [1] and the references therein).

For the numerical experiments, we use three different datasets, which we would refer to as $dataset_1$, $dataset_2$, $dataset_3$. Each of them contains raw measurements from a three-axial FXOS accelerometer.

Out of 10 experiments that we have done with the two algorithms, here, we shall comment on three representative ones—one for each dataset. Let us mention, nevertheless, that from all experiments in only one the algorithm based on minimizing (12) has given better results.

Experiment 1. We have used 19 observations for calibration and 76 as a test set for $dataset_1$. In Fig. 2, the absolute errors for the inclination (on the left) and toolface (on the right) are given for each observation from the test set. The maximum errors for the inclination, using the classical method and the modified one, are 1.05° and 0.73°, respectively. The corresponding mean errors are 0.27° and 0.24°.

For the toolface, the respective maximum errors are 2.80° and 2.22°, and the mean ones are 1.08° and 0.88°.

Experiment 2. For $dataset_2$, we have used 14 observations for calibration and the remaining 70 are used as a test set. The results are given in Fig. 3. The maximum errors for the inclination, using the classical method and the modified one, are 0.80° and 0.43°, respectively. The corresponding mean errors are 0.24° and 0.19°.

For the toolface, the respective maximum errors are 2.33° and 2.12°, and the mean ones are 0.57° and 0.51°.

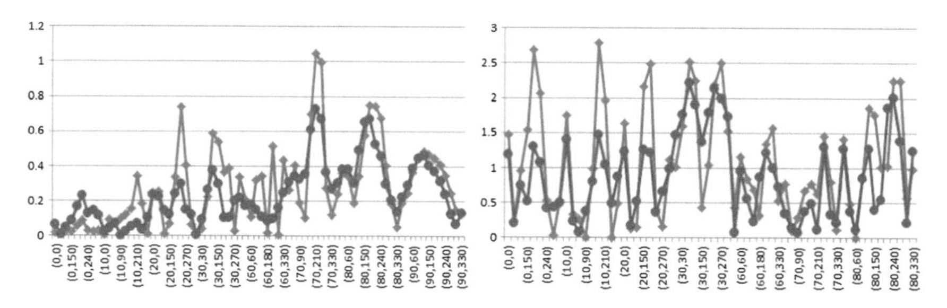

Fig. 2 Absolute errors for inclination (on the left) and toolface (on the right) obtained by minimizing (5) (circles) and (12) (rhombuses) for $dataset_1$. On the abscissa (*inclination*, *toolface*)-pairs are given

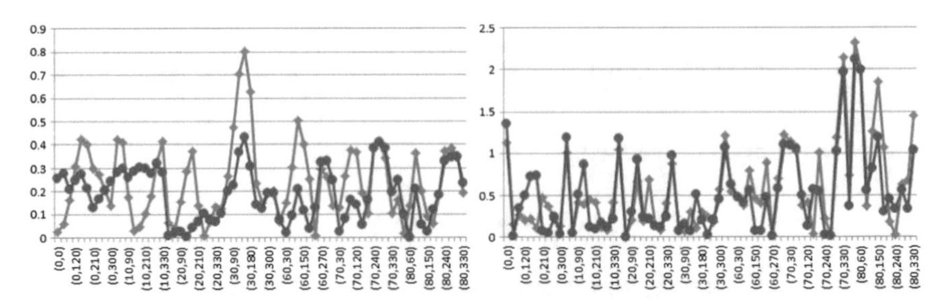

Fig. 3 Absolute errors for inclination (on the left) and toolface (on the right) obtained by minimizing (5) (circles) and (12) (rhombuses) for *dataset*$_2$. On the abscissa (*inclination*, *toolface*)-pairs are given

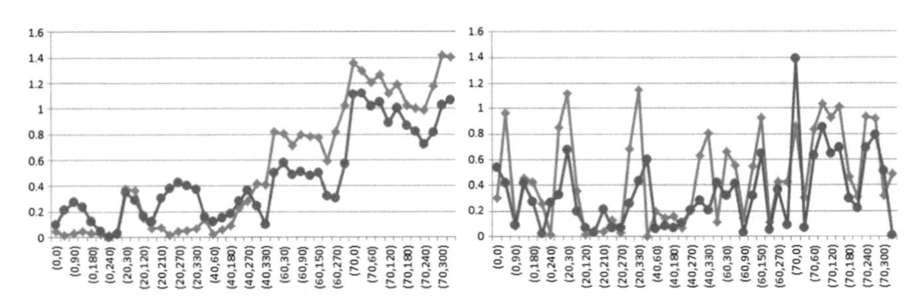

Fig. 4 Absolute errors for inclination (on the left) and toolface (on the right) obtained by minimizing (5) (circles) and (12) (rhombuses) for *dataset*$_3$. On the abscissa (*inclination*, *toolface*)-pairs are given

Experiment 3. For *dataset*$_3$, we have used 14 observations for calibration and the remaining 46 are used as a test set. In Fig. 4, the results are depicted for each observation from the test set. The maximum errors for the inclination, using the classical method and the modified one, are 1.42° and 1.12°, respectively. The corresponding mean errors are 0.54° and 0.46°.

For the toolface, the respective maximum errors are 1.15° and 1.39°, and the mean ones are 0.47° and 0.34°.

Comparison analysis. Based on the conducted experiments, the mean error for the inclination improves with the modified method between 10 and 20%, while the maximum error is improved between 20 and 50%.

The mean error of the toolface is reduced between 10 and 25%. In the first two experiments, the maximum error is reduced between 10 and 20%. In the third experiment, the maximum error with the classical model is less than when the modified one is used. However, we argue that this might be due to a wrong measurement since the problem is only for 70° inclination and 0° toolface (see Fig. 4).

In general, the results, obtained for the inclination, are more accurate than the ones for the toolface. This is due to the fact that when the inclination is close to 90°, there are greater errors for the toolface. This is the reason why in the third experiment (where the inclination is not greater than 70°) the maximum error for the toolface is about 1°.

Even though the newly proposed algorithm is slower than the classical one, it computes the calibration parameters in a reasonable time. In particular, the results, presented in the last section, required about 100 iterations of the Gauss–Newton method. The computational times, recorded on a personal computer, for the classical method and the modified one were about 1 ms and 2 s, respectively. The latter is sufficiently fast for initial calibration purposes.

4 Conclusion

In the present work, we have considered calibrating MEMS accelerometers for the purpose of determining orientation in space. We propose the new objective function (5), whose minimization gives an estimate of the calibration coefficients. The errors in inclination and toolface, obtained using the latter, are less than the ones obtained for a classical goal function (12) in almost all cases. Thus, we believe that it is more appropriate to use this goal function, if the purpose is to compute inclination and toolface angles with higher precision.

In order to further investigate the possible benefits of the proposed algorithm, additional experiments with more accurate accelerometers can be made.

Acknowledgements The work of the authors has been partially supported by the Sofia University "St. Kl. Ohridski" under contract No. 80.10-11/2017.

Appendix

Here, we shall give formulas for computing the Jacobian matrix, associated with the linearization of the minimization problem (6).

We have

$$M.\hat{\mathbf{a}}_i + \mathbf{b} = \begin{bmatrix} m_{xx}.\hat{a}_{i,x} + m_{xy}.\hat{a}_{i,y} + m_{xz}.\hat{a}_{i,z} + b_x \\ m_{yx}.\hat{a}_{i,x} + m_{yy}.\hat{a}_{i,y} + m_{yz}.\hat{a}_{i,z} + b_y \\ m_{zx}.\hat{a}_{i,x} + m_{zy}.\hat{a}_{i,y} + m_{zz}.\hat{a}_{i,z} + b_z \end{bmatrix}.$$

Let us define for $i = \overline{1, n}$ and $j = \overline{1, 12}$

$$
\begin{aligned}
J_{2i-1,j} &= \frac{\partial r_{2i-1}}{\partial p_j} = \frac{360^\circ}{2\pi} \frac{\partial}{\partial p_j} tf(a_i) \\
&= \frac{360^\circ}{2\pi} \frac{\partial}{\partial p_j} \arctan\left(\frac{m_{xx}.\hat{a}_{i,x} + m_{xy}.\hat{a}_{i,y} + m_{xz}.\hat{a}_{i,z} + b_x}{m_{yx}.\hat{a}_{i,x} + m_{yy}.\hat{a}_{i,y} + m_{yz}.\hat{a}_{i,z} + b_y} \right) \\
&= \frac{360^\circ}{2\pi} \frac{\partial}{\partial p_j} \arctan\left(\frac{p_1.\hat{a}_{i,x} + p_2.\hat{a}_{i,y} + p_3.\hat{a}_{i,z} + p_{10}}{p_4.\hat{a}_{i,x} + p_5.\hat{a}_{i,y} + p_6.\hat{a}_{i,z} + p_{11}} \right) \\
&=: \frac{360^\circ}{2\pi} \frac{\partial}{\partial p_j} \arctan\left(\frac{A}{B} \right).
\end{aligned}
\tag{13}
$$

Let us denote for further use:

$$
\begin{aligned}
B_1 &= p_1\hat{a}_{i,x} + p_2\hat{a}_{i,y} + p_3\hat{a}_{i,z} + p_{10}, \\
B_2 &= p_4\hat{a}_{i,x} + p_5\hat{a}_{i,y} + p_6\hat{a}_{i,z} + p_{11},
\end{aligned}
$$

$$
\begin{aligned}
B_3 &= p_7\hat{a}_{i,x} + p_8\hat{a}_{i,y} + p_9\hat{a}_{i,z} + p_{12}, \\
B &= \sqrt{B_1^2 + B_2^2 + B_3^2}, \quad A = B_1^2 + B_2^2.
\end{aligned}
$$

After straightforward (but rather lengthy) computations that we omit due to lack of space, one can obtain:

$$
\begin{aligned}
J_{2i-1,1} &= \frac{360^\circ \hat{a}_{i,x}}{2\pi B \left(1 + \frac{A^2}{B^2}\right)}, \quad J_{2i-1,2} = \frac{360^\circ \hat{a}_{i,y}}{2\pi B \left(1 + \frac{A^2}{B^2}\right)}, \\
J_{2i-1,3} &= \frac{360^\circ \hat{a}_{i,z}}{2\pi B \left(1 + \frac{A^2}{B^2}\right)}, \quad J_{2i-1,10} = \frac{360^\circ}{2\pi B \left(1 + \frac{A^2}{B^2}\right)}, \\
J_{2i-1,4} &= -\frac{360^\circ A\hat{a}_{i,x}}{2\pi B^2 \left(1 + \frac{A^2}{B^2}\right)}, \quad J_{2i-1,5} = -\frac{360^\circ A\hat{a}_{i,y}}{2\pi B^2 \left(1 + \frac{A^2}{B^2}\right)}, \\
J_{2i-1,6} &= -\frac{360^\circ A\hat{a}_{i,z}}{2\pi B^2 \left(1 + \frac{A^2}{B^2}\right)}, \quad J_{2i-1,11} = -\frac{360^\circ A}{2\pi B^2 \left(1 + \frac{A^2}{B^2}\right)}, \\
J_{2i-1,7} &= J_{2i-1,8} = J_{2i-1,9} = J_{2i-1,12} = 0.
\end{aligned}
$$

Further, if

$$
\frac{360^\circ}{2\pi} \left(\arccos\left(\frac{\hat{a}_{i,z}}{\sqrt{\hat{a}_{i,x}^2 + \hat{a}_{i,y}^2 + \hat{a}_{i,z}^2}} \right) \right) > 60^\circ \tag{14}
$$

holds true, we have:

$$J_{2i,1} = -\frac{360^\circ}{2\pi}\frac{-\frac{2\hat{a}_{i,x}B_1A}{B^4} + \frac{2\hat{a}_{i,x}B_1}{B^2}}{2\sqrt{\frac{A}{B^2}}\sqrt{1-\frac{A}{B^2}}},\ J_{2i,2} = -\frac{360^\circ}{2\pi}\frac{-\frac{2\hat{a}_{i,y}B_1A}{B^4} + \frac{2\hat{a}_{i,y}B_1}{B^2}}{2\sqrt{\frac{A}{B^2}}\sqrt{1-\frac{A}{B^2}}},$$

$$J_{2i,3} = -\frac{360^\circ}{2\pi}\frac{-\frac{2\hat{a}_{i,z}B_1A}{B^4} + \frac{2\hat{a}_{i,z}B_1}{B^2}}{2\sqrt{\frac{A}{B^2}}\sqrt{1-\frac{A}{B^2}}},\ J_{2i,10} = -\frac{360^\circ}{2\pi}\frac{-\frac{2B_1A}{B^4} + \frac{2B_1}{B^2}}{2\sqrt{\frac{A}{B^2}}\sqrt{1-\frac{A}{B^2}}},$$

$$J_{2i,4} = -\frac{360^\circ}{2\pi}\frac{-\frac{2\hat{a}_{i,x}B_2A}{B^4} + \frac{2\hat{a}_{i,x}B_2}{B^2}}{2\sqrt{\frac{A}{B^2}}\sqrt{1-\frac{A}{B^2}}},\ J_{2i,5} = -\frac{360^\circ}{2\pi}\frac{-\frac{2\hat{a}_{i,y}B_2A}{B^4} + \frac{2\hat{a}_{i,y}B_2}{B^2}}{2\sqrt{\frac{A}{B^2}}\sqrt{1-\frac{A}{B^2}}},$$

$$J_{2i,6} = -\frac{360^\circ}{2\pi}\frac{-\frac{2\hat{a}_{i,z}B_2A}{B^4} + \frac{2\hat{a}_{i,z}B_2}{B^2}}{2\sqrt{\frac{A}{B^2}}\sqrt{1-\frac{A}{B^2}}},\ J_{2i,11} = -\frac{360^\circ}{2\pi}\frac{-\frac{2B_2A}{B^4} + \frac{2B_2}{B^2}}{2\sqrt{\frac{A}{B^2}}\sqrt{1-\frac{A}{B^2}}},$$

$$J_{2i,7} = \frac{360^\circ}{2\pi}\frac{\hat{a}_{i,x}AB_3}{B^4\sqrt{\frac{A}{B^2}}\sqrt{1-\frac{A}{B^2}}},\ J_{2i,8} = \frac{360^\circ}{2\pi}\frac{\hat{a}_{i,y}AB_3}{B^4\sqrt{\frac{A}{B^2}}\sqrt{1-\frac{A}{B^2}}},$$

$$J_{2i,9} = \frac{360^\circ}{2\pi}\frac{\hat{a}_{i,z}AB_3}{B^4\sqrt{\frac{A}{B^2}}\sqrt{1-\frac{A}{B^2}}},\ J_{2i,12} = \frac{360^\circ}{2\pi}\frac{AB_3}{B^4\sqrt{\frac{A}{B^2}}\sqrt{1-\frac{A}{B^2}}}.$$

Otherwise, if condition (14) is not fulfilled, then:

$$J_{2i,1} = -\frac{360^\circ}{2\pi}\frac{\hat{a}_{i,x}B_1B_3}{B^3\sqrt{1-\frac{B_3^2}{B^2}}},\ J_{2i,2} = -\frac{360^\circ}{2\pi}\frac{\hat{a}_{i,y}B_1B_3}{B^3\sqrt{1-\frac{B_3^2}{B^2}}},$$

$$J_{2i,3} = -\frac{360^\circ}{2\pi}\frac{\hat{a}_{i,z}B_1B_3}{B^3\sqrt{1-\frac{B_3^2}{B^2}}},\ J_{2i,10} = -\frac{360^\circ}{2\pi}\frac{B_1B_3}{B^3\sqrt{1-\frac{B_3^2}{B^2}}},$$

$$J_{2i,4} = -\frac{360^\circ}{2\pi}\frac{\hat{a}_{i,x}B_2B_3}{B^3\sqrt{1-\frac{B_3^2}{B^2}}},\ J_{2i,5} = -\frac{360^\circ}{2\pi}\frac{\hat{a}_{i,y}B_2B_3}{B^3\sqrt{1-\frac{B_3^2}{B^2}}},$$

$$J_{2i,6} = -\frac{360^\circ}{2\pi}\frac{\hat{a}_{i,z}B_2B_3}{B^3\sqrt{1-\frac{B_3^2}{B^2}}},\ J_{2i,11} = -\frac{360^\circ}{2\pi}\frac{B_2B_3}{B^3\sqrt{1-\frac{B_3^2}{B^2}}},$$

$$J_{2i,7} = -\frac{360^\circ}{2\pi}\frac{-\frac{\hat{a}_{i,x}B_3^2}{B^3} + \frac{\hat{a}_{i,x}}{B}}{\sqrt{1-\frac{B_3^2}{B^2}}},\ J_{2i,8} = -\frac{360^\circ}{2\pi}\frac{-\frac{\hat{a}_{i,y}B_3^2}{B^3} + \frac{\hat{a}_{i,y}}{B}}{\sqrt{1-\frac{B_3^2}{B^2}}},$$

$$J_{2i,9} = -\frac{360^\circ}{2\pi}\frac{-\frac{\hat{a}_{i,z}B_3^2}{B^3} + \frac{\hat{a}_{i,z}}{B}}{\sqrt{1-\frac{B_3^2}{B^2}}},\ J_{2i,12} = -\frac{360^\circ}{2\pi}\frac{-\frac{B_3^2}{B^3} + \frac{1}{B}}{\sqrt{1-\frac{B_3^2}{B^2}}}.$$

References

1. Aggarwal, P., Syed, Z., Niu, X., El-Sheimy, N.: Thermal Calibration of Low Cost MEMS Sensors for Integrated Positioning. The Institute of Navigation National Technical Meeting, Navigation Systems, p. 2224 (2007)
2. Aggarwal, P., Syed, Z., Niu, X., El-Sheimy, N.: A standard testing and calibration procedure for low cost MEMS inertial sensors and units. J. Navig. **61**, 323–336 (2008)
3. Aydemir, G.A., Saranli, A.: Characterization and calibration of MEMS inertial sensors for state and parameter estimation applications. Measurement **45**, 12101225 (2012)
4. Björck, Å.: Numerical Methods for Least Squares Problems. SIAM (1996)
5. Forsberg, T., Grip, N., Sabourova, N.: Non-iterative calibration for accelerometers with three non-orthogonal axes and cross-axis interference. Research Report No. 8, Department of Engineering Sciences and Mathematics, Division of Mathematics, Lulea University of Technology (2012)
6. Georgieva, I., Hofreither, C., Ilieva, T., Ivanov, T., Nakov, S.: Laboratory calibration of a MEMS accelerometer sensor. ESGI'95 Problems and Final Reports, pp. 61–86 (2013)
7. Hou, H.: Modeling Inertial Sensors Errors using Allan Variance. Library and Archives Canada (2005)
8. Illfelder, H., Hamlin, K., McElhinney, G.: A gravity-based measurement-while-drilling technique determines borehole azimuth from toolface and inclination measurements. In ADDE 2005 National Technical Conference and Exhibition, Houston, Texas (2005)
9. Kang, J., Wang, B., Hu, Z., Wang, R., Liu, T.: Study of drill measuring system based on MEMS accelerative and magnetoresistive sensor. In: The Ninth International Conference on Electronic Measurement and Instruments (ICEMI2009)
10. Luczak, S., Oleksiuk, W., Bodnicki, M.: Sensing tilt with MEMS accelerometers. IEEE Sens. J. **6**, 1669–1675 (2006)
11. Strang, G.: Introduction to Linear Algebra, 3rd edn. Wellesley-Cambridge Press, Wellesley (2003)

Design and Implementation of Moving Average Calculations with Hardware FPGA Device

Vladimir Ivanov and Todor Stoilov

Abstract The article examines the design and implementation of the "moving average" procedure in the structure of FPGAs. It demonstrates that its implementation in the form of a recursive filter and the use of the DSP48A1 digital signal processing unit, embedded in the Xilinx Spartan 6 FPGA series, guarantee minimal logical resources. An estimation of the speed and dynamic range of the device is provided.

1 Introduction

There are various methods for processing a temporary series used in experimental data for detection of non-random, determinate constituents, and tendencies. Also the methods are used for filtering noise and to convert the data into a relatively smooth curve. The method used in this paper is going to be Moving Average (MA), which in essence consists of replacing the actual values of time series with a calculated mean values.

There are many varieties for the Moving Average, and they all use an uniform concept. This is a simple and apparent way of obtaining an average movement of the observed parameter for a certain time interval. The MA calculations are widely applied for real time assessment of dynamical series of numerical data which take place in forecasting of parameters in control systems, market estimation of security returns, for portfolio optimization. The MA calculations provide simple and apparent way of obtaining an average movement of the observed parameters for a certain time interval. The real time calculations of MA is key point for many control applications. By definition, the MA is a common name for a family of functions whose meanings in each point are equal to the mean value of the function of the previous period.

V. Ivanov (✉) · T. Stoilov
Institute of Information and Communication Technology,
Bulgarian Academy of Science, Acad. G. Bonchev str. Bl 2, 1113 Sofia, Bulgaria
e-mail: ivanov.vladi@gmail.com

T. Stoilov
e-mail: todor@hsi.iccs.bas.bg

K. Georgiev et al. (eds.), *Advanced Computing in Industrial Mathematics*,
Studies in Computational Intelligence 793,
https://doi.org/10.1007/978-3-319-97277-0_15

Typically, they are used to smooth short-term fluctuations and to separate major trends or cycles into different data. With its help it is possible to trace the end of the current and the beginning of a new trend as the strength of the trend on its slope. For this reason, the moving average should be considered as a basic indicator of technical analysis, which determines its great popularity as one of the tools widely used by investment companies, auctions, stock exchanges and more. There are several varieties of moving average, among which the most popular are the following types:

1. Simple;
2. Weighted;
3. Exponential.

These methods use the same principles, have their own pros and cons but differ only in the formulas for their calculation. The simple moving average is the simplest and most primitive indicator of technical analysis. It is calculated according to the formulae [1, 2]:

$$\hat{y}(i) = \frac{1}{k}\sum_{j=0}^{k} x(i-j) \tag{1}$$

The use of the simple type of MA is based on a transition from the initial meanings of time order to average meanings of any preselected interval containing g the member of the process line. In this case, the dispersion of the new line turns out to be less than that of the original order. Thus, smoothing of random and periodic fluctuations is achieved, and a clear picture of the general tendency in the behavior of the order is obtained. The weighted MA is used in cases where the line contains a substantial nonlinear trend. The formulae to which the smoothing takes place are strongly dependent on the length of the g interval and the degree of polynomial in which the smoothing is performed. In case when a second order polynomial and $g = 5$ are used, the formula for a weighted MA has the form:

$$\bar{y}(t) = \frac{1}{35}(-3y(t-2) + 12y(t-1) + 17y(t) + 12y(t+1) - 3y(t-2)) \tag{2}$$

for $t = 3, \ldots n-2$.

A peculiar feature of the simple MA is the uniform values of weight coefficients (usually they are equal to 1). Applying the exponential MA method, the indicated drawback of the other types of the methods can be removed. In this case, the highest weight factor is attributed to the most recent observation, whereby the expected estimate for the $\bar{y}(t+1)$ moment is calculated by the expression:

$$\bar{y}(t+1) = \alpha y(t) + (1-\alpha)\bar{y(t)} \tag{3}$$

for $t = 1, 2 \ldots n-1$ and $\alpha \in (0 \leq 1)$. Thus, the evaluation for $t = 1$ is calculated recurrently based on the value of the previous evaluation and the evaluation of the current value. The calculations under this formula usually begin with $t = 1$, whereby as initial values of the calculations are taken the average of the first few members of the order or other reasonable meaning are taken. From a computational point of view, the representation of the expression describing exponential smoothing is more convenient to be presented in the form of:

$$\bar{y}(t+1) = y(t) + \alpha(y(t) - \bar{y(t)}) \quad (4)$$

Thus, the evaluation of a member $\bar{y}(t+1)$ is obtained as a sum of the previous value and some part of the error $(y(t) - \bar{y(t)})$ of the previous forecast. The magnitude of the error used to correct the forecast is determined by the smoothing constant α . As the value of is approaching 1, much of the difference between the forecast and the actual value is considered to be accurate and used for correction in the calculations. Conversely, as the value of α approaches zero, the greater the difference between the forecast and the actual value is considered to be random, and accordingly, a small part of it is used for correction. In this aspect, exponential smoothing becomes an example of an adaptive model use. In practice, there are no clear formal criteria for choosing the value of α. Most often, its value range from 0.1 to 0.3 and reflects the subjective opinion of the researcher on the relative resistance to change in the indicator. Relatively rarely, in the case of large measurement errors as a smoothing tool, the least squares or the negative exponential smoothing methods are used [3, 4]. The MA method is based on the replacement of the initial values of the time series elements with the mean values calculated over the preselected interval that crawl in order to allow for both casual and periodic fluctuations to be smoothed out. Due to the averages being executed, the so obtained times series turn out to be smoother than the initial one. The results achieved through with the MA method are directly related to the length of the preselected interval for calculating of the average value. Therefore, in order to exclude the occurrence of cyclical variations, the interval length must be an integer multiplied by the length of the loop [5–8]. The starting point for the practical solution for the tasks relating to the implementation of super-large-scale real-time information processing systems using the MA method is the use of Field Programmable Gate Array (FPGA) devices. The application of this class of devices allows parallel computing procedures to be implemented, resulting in a much more efficient use of these devices than specialized signal processing processors. In this respect, the present article becomes an attempt to show a possible realization of MA based devices with the tools of modern FPGA devices. The hardware system solution for evaluation of the MA parameters of a dynamical data is commonly applied for solving technical problems. This research tries to extend the potential usage of such hardware system in business and portfolio optimization. The hardware system can easily and in real time estimate the currencies trends which are needed for the definition of portfolio characteristics. This domain application is not presented in the research below, but provides comments about this new area of potential application for the system of moving average computation in real time.

2 Designing an Moving Average Device with an FPGA Device

The classic description of the "moving average" procedure used to "smooth" data is given by expression (1). This presentation of the MA procedure is equivalent to the description of a non-recursive filter with single weight coefficients. The practical implementation of this procedure allows a conversion within an FPGA device, but requires the storage for a $2K$ data. At large K values, the volume of the necessary logical resources proves to be proportional to the magnitude of the interval, which in these cases makes the realization not profitable. For this reason, the "moving average" procedure is more convenient to be realized as a recursive algorithm described by a differential equation of the type:

$$y(n) = y(n-1) + x(n) - x(n-2k-1) \tag{5}$$

where n and k are respectively the number of the current sample and the number of averages. Practical implementation of this algorithm significantly reduces the amount of logical resources required, but requires the presence of two algebraic adders. Another approach for the practical realization of the MA procedure in an FPGA device, which further minimizes the necessary logical resources, is the use of the built-in specialized DSP blocks and the block memory [9–11]. A key role for the practical realization of this approach is assigned to the digital signal processing unit. In the architecture of popular FPGA devices from the Spartan-6 family, this block is called the DSP48A1. Its structure is shown in Fig. 1. The basic elements that build the structure of this block are a preliminary algebraic adder, two input multiplier, 48-bit ALU, input ports, multiplexers, registers and logic. The preliminary algebraic adder executes D – B or D + B operations on data passing through the single inputs with the same names. The multiplier realized in the DSP48A1 unit works in two compliments code. It accepts two 18-bit operands and offers 36 bit result. The output of the multiplier is connected to a register. The output of this register is feed to the multiplexer that pass the multiplication result to the input of the ALU. The ALU accepts the information to be processed from the X and Z multiplexers, which in the cases when the multiplier is not used enable the performance of multiply-accumulate operation. The main purpose of the input, output and control inputs is to feed the relevant data to the processing elements and to adjust their operating mode. The data to be processed in the internal structure of the DSP48A1 block arrives via A, B, C, and D ports. The data flow control and the modes of the two algebraic addresses are determined by the 8-bit OPMODE signal. When implementing this algorithm in the structure of the DSP48A1, the MA process is encapsulated in its own resources but requires a delay line in which the data $x(n-2k-1)$ of the expression (2) is stored. The structural diagram of the device performing the MA operation in to the resources of the DSP48A1 block is shown in Fig. 2. Its analytical description has been constructed on the basis of Eq. (2) written as:

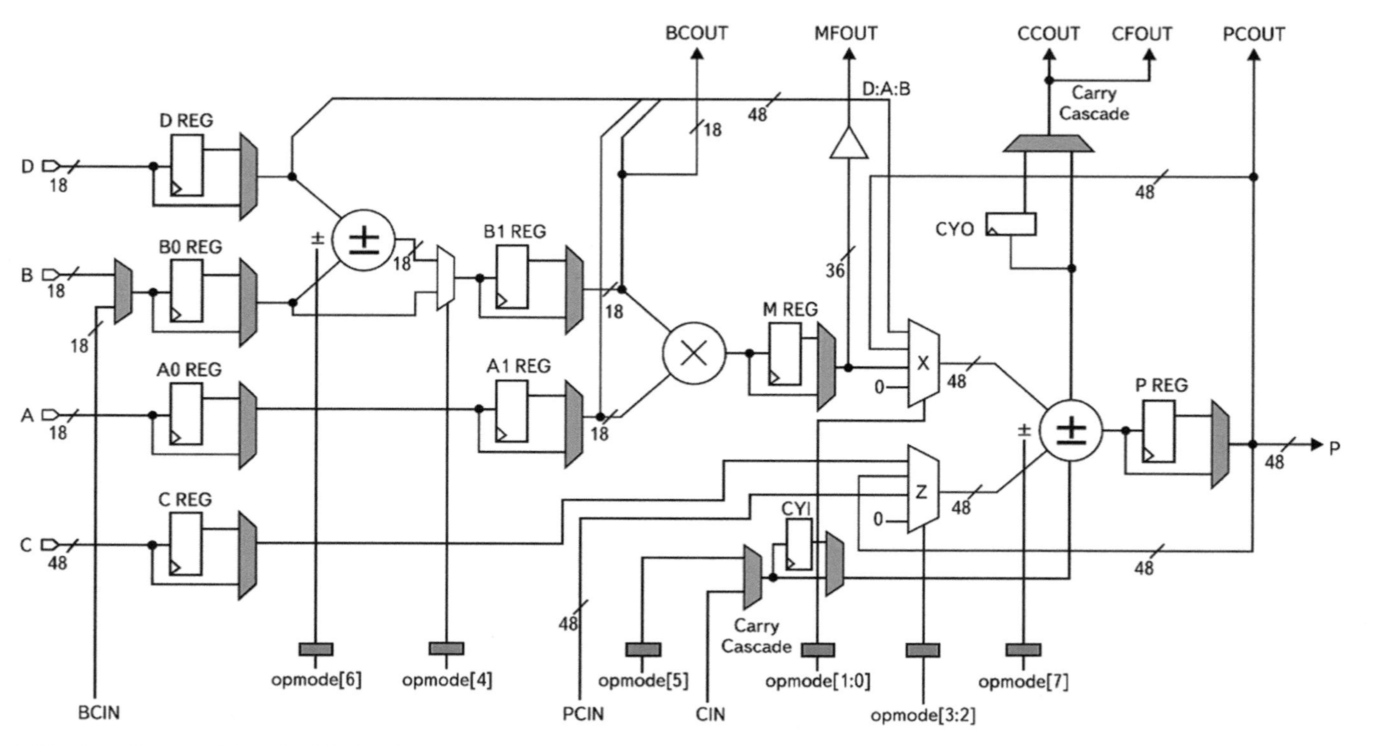

Fig. 1 Structure diagram of the DSP48A1 block

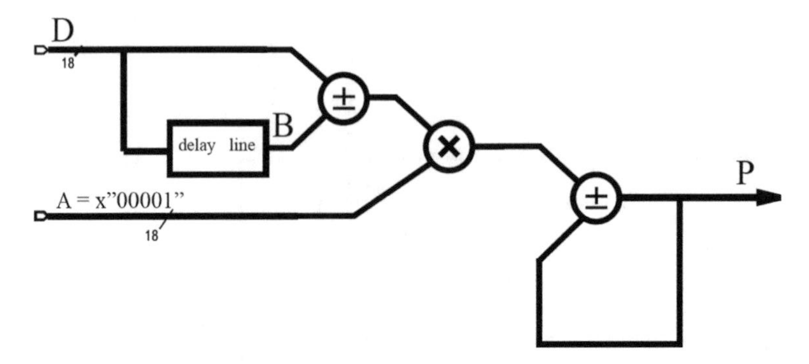

Fig. 2 A block diagram of the device according to the expression (3)

$$y(n) = y(n-1) + [x(n) - x(n-2k-1)] \tag{6}$$

This representation allows the calculation of the expression in square brackets to be made with the resources of the first algebraic adder and the accumulation of data with the second algebraic adder.

Under this block diagram, a practical modeling, design, and simulation of a MA device were performed within the Xilinx web pack environment. The smallest FPGA XC6SLX4 from the Spartan-6 family was used as the built-in object for the designed device. The block diagram of the developed device presented with the resources of the Web Pack environment is shown in Fig. 3. The simulation of the project concerning the average of 8 neighborhood values is shown in Fig. 4. In order to improve the representation of the idea of the whole project work, in the simulation, besides the base signals A, B, D and P the signals of the outputs of the Z and X multiplexes are also displayed.

The location of the project in the resources of the smallest Spartan6 (XC6SLX4) chip is shown in Fig. 5.

3 Results and Analysis

A feature of the developed device is the use of arithmetic in the integer field, allowing the error-free computations [12]. It is imperative to perform a dynamic range estimation of the device being developed, when the processing of incoming information involves making a large number of summaries. The estimation of the dynamic range of the preliminary adder is clear. It is obtained as a difference between the arithmetic device's disintegration and that of the incoming data. Thus, the 18 bit pre-adder can handle 17 bit data without problems. The dynamic range of the second algebraic adder is fixed by its 48 bits, which allow the accumulation of a significant amount of data. As far as the speed of the processed information is concerned, it is determined by the parameters of the used Spartan6 XC6SLX4 FPGA device, its value is limited

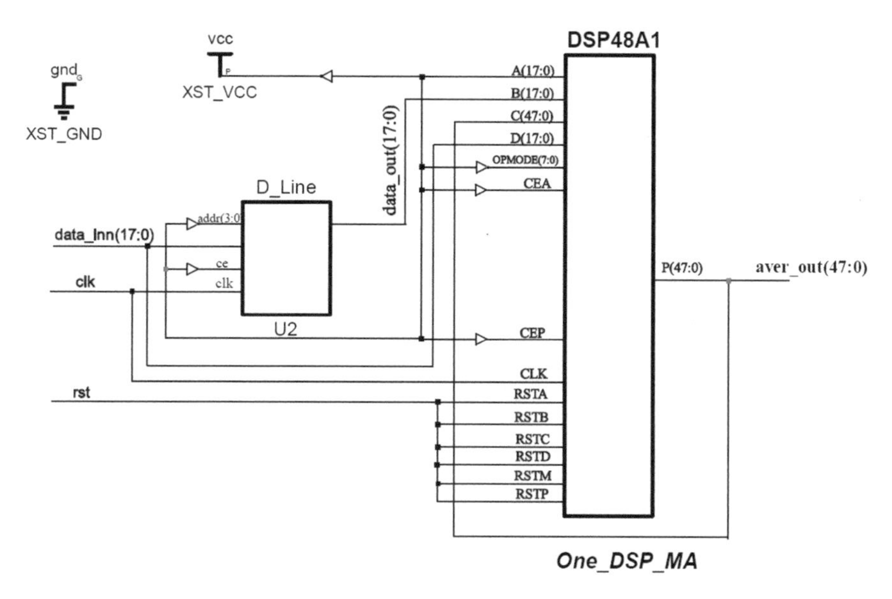

Fig. 3 Block diagram of the developed device presented with the resources of the Web Pack environment

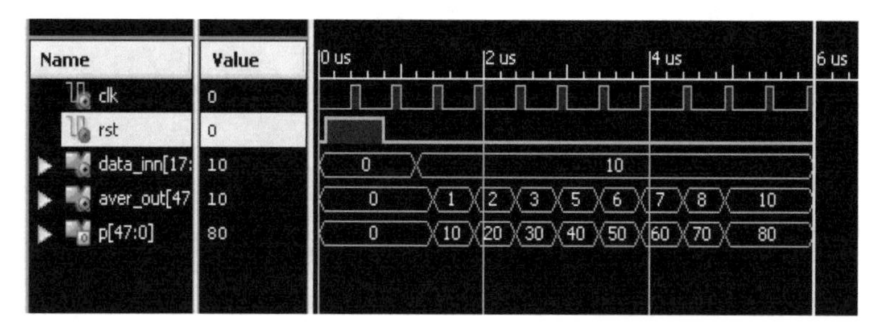

Fig. 4 Simulation of the project

to 300 MHz. Due to the large dynamic range of the ALU 2^{48} built-in embedded DSP blocks, the use of FPGA devices, make no sense the implementation of floating arithmetic in conventional and dedicated processors to build large-scale data processing devices in real-time mode.

4 Conclusion

The paper presents an approach for implementing the MA device using a DSP48A1 digital signal processing unit presented in the FPGA devices XC6SLX4 from the Spartan-6 family. A simulation of the projected device was made in conditions closest

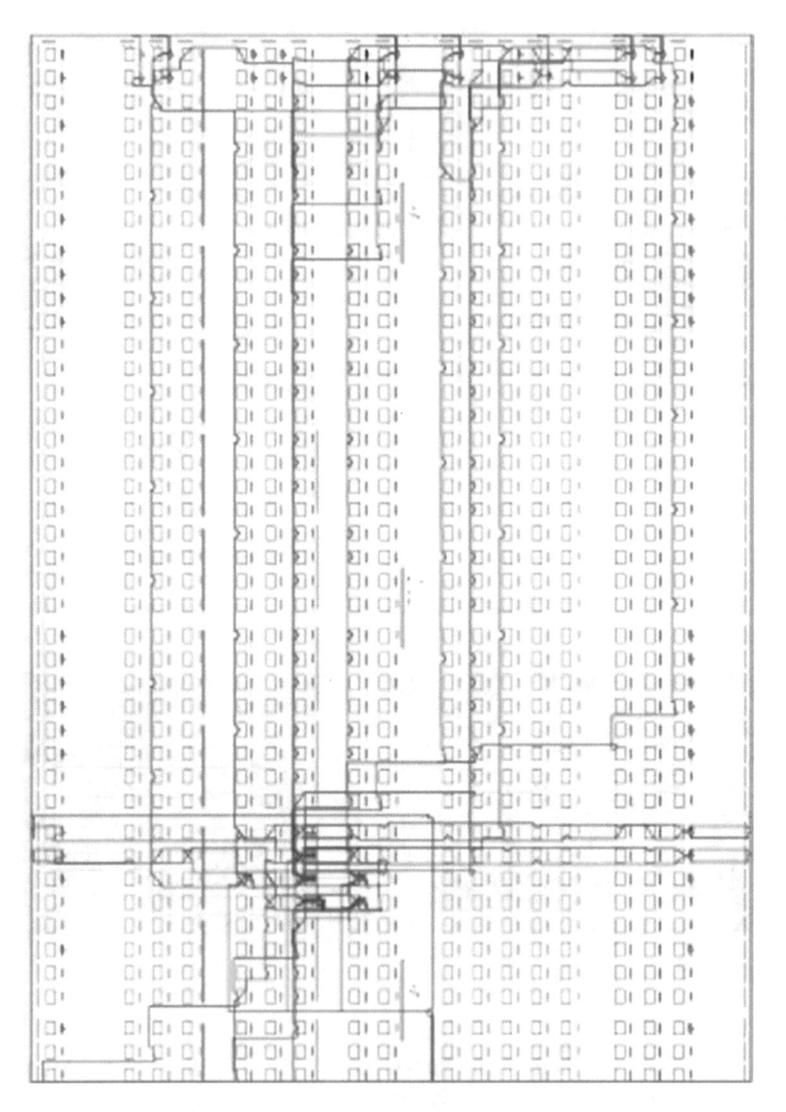

Fig. 5 Project placement in the resources of the XC6SLX4 chip

to the real ones. An evaluation of the dynamic range of the device being developed and its performance has been made, which allows the use of the presented material for the purposes of the training. This research targets the application of a system solution for evaluation of moving average values of dynamical series of data. The data used can concern the trends of currencies, which have places on stock markets. The hardware application of the evaluation of moving average values supports the real time and automatic generation of indicators, which estimate the directions of the currency trends. The currencies commonly participate as assets in portfolio optimization of financial investments. Thus the hardware design of moving average calculator can, with a certain amount of success, be applied, also in non-technical areas but for business and investment management.

Acknowledgements This work has been partly supported by project H12/8, 14.07.2017 of the Bulgarian National Science fund: Integrated bi-level optimization in information service for portfolio optimization contract H12/10, 20.12.2017.

References

1. Rabiner, L., Gold, B.: Theory and Application of Digital Signal Processing. Prentice-Hall, Englewood Cliffs (1975)
2. Arce, G.R.: Nonlinear Signal Processing: A Statistical Approach. Wiley, New Jersey, USA (2005)
3. Kantz, H., Thomas, S., Nonlinear Time Series Analysis. Cambridge University Press, London (2004). ISBN 978-0521529020
4. Madsen, H.: Time Series Analysis. Chapman Hall (2008). ISBN 978-1-4200-5967-0
5. Box, G., Jenkins, G.: Time Series Analysis: Forecasting and Control. McGraw-Hill, Holden-Day, San Francisco (1970)
6. Velleman, P.F., Hoaglin, D.C.: Applications, Basics, and Computing of Exploratory Data Analysis (1981)
7. Shihovtsev, I.V., Jacobov, V.P.: Statistical Radiophysics Novosibirsk (2011) (In Russian)
8. Litjuk, V.I., Litjuk, L.V.: Methods of Digital Multiprocessing of Ensembles of Radio Signals Development. Solon-Press, Moscow (2007) (In Russian)
9. Spartan-6 FPGA DSP48A1 User Guide, UG389, May 2014. www.xilinx.com
10. Zotov, V.: Features architects new generations PLIS FPGA company Xilinx series Spartan-6, Components and technologies, 9 (2009) (In Russian)
11. Shypullin, C.H., Hrapov, V.J.: Characteristics of the projected digital circuit of PLIS—Chip News, 5 (1996) (In Russian)
12. Gregory, R.T., Krishnamurthy, E.V.: Methods and Applications of Error-Free Computations. Springer, New York (1984)

On the Exact Traveling Wave Solutions of a Hyperbolic Reaction-Diffusion Equation

Ivan P. Jordanov and Nikolay K. Vitanov

Abstract We discuss a class of hyperbolic reaction-diffusion equations and apply the modified method of simplest equation in order to obtain an exact solution of an equation of this class (namely the equation that contains polynomial nonlinearity of fourth order). We use the equation of Bernoulli as a simplest equation and obtain traveling wave solution of a kink kind for the studied nonlinear reaction-diffusion equation.

1 Introduction

Differential equations arise in mathematical analysis of many problems from natural and social sciences. The reason for this is that the differential equations relate quantities with their changes and such relationships are frequently encountered in many disciplines such as meteorology, fluid mechanics, solid-state physics, plasma physics, ocean and atmospheric sciences, mathematical biology, chemistry, material science, etc. [9, 17, 33, 34, 37, 39, 52, 53]. The qualitative features and mechanisms of many phenomena and processes in the above-mentioned research areas can be studied by means of exact solutions of the model nonlinear differential equations. Examples are phenomena such as existence and change of different regimes of functioning of complex systems, spatial localization, transfer processes, etc. In addition the exact solutions can be used to test computer programs for numerical simulations. Because of all above exact solutions of nonlinear partial differential equations are studied very intensively [1–3, 5, 6, 11, 13, 19, 20, 27, 28, 36–38,

I. P. Jordanov
Institute of Mechanics, Bulgarian Academy of Sciences,
University of National and World Economy, Sofia, Bulgaria
e-mail: jordanov@email.bg

N. K. Vitanov (✉)
Institute of Mechanics, Bulgarian Academy of Sciences, Sofia, Bulgaria
e-mail: vitanov@imbm.bas.bg

K. Georgiev et al. (eds.), *Advanced Computing in Industrial Mathematics*,
Studies in Computational Intelligence 793,
https://doi.org/10.1007/978-3-319-97277-0_16

40, 41]. The nonlinear PDEs are integrable or nonlitegrable. Well known methods exist for obtaining exact solutions of integrable nonlinear PDEs, e.g., the method of inverse scattering transform or the method of Hirota [2, 3, 13, 18, 35]. Many approaches for obtaining exact special solutions of nonintegrable nonlinear PDEs have been developed in the recent years, e.g., [10, 16, 22, 29, 42, 56, 57]. In this chapter we shall use a version of the method of simplest equation called modified method of simplest equation [22, 24, 25, 42, 45, 51]. The method of simplest equation uses a procedure analogous to the first step of the test for the Painleve property [21, 23, 26]. In the modified method of simplest equation [43, 44, 46–50] instead of this procedure (the procedure requires work in the space of complex numbers) one uses one or several balance equations. The modified method of simplest equation has shown its effectiveness on the basis of numerous applications, such as obtaining exact traveling wave solutions of generalized Swift - Hohenberg equation and generalized Rayleigh equation [45], generalized Degasperis - Procesi equation and b-equation [46], extended Korteweg-de Vries equations [48, 51], generalized Fisher equation, generalized Huxley equation [43], generalized Kuramoto - Sivashinsky equation, reaction - diffusion equation, reaction - telegraph equation [42], etc. [49, 50, 54].

Below we shall discuss hyperbolic reaction-diffusion equation of the kind

$$\tau \frac{\partial^2 Q}{\partial t^2} + \frac{\partial Q}{\partial t} = D \frac{\partial^2 Q}{\partial x^2} + \sum_{i=1}^{n} \alpha_i Q^i, \tag{1}$$

where $Q = Q(x, t)$, n is a natural number and τ, D, and α_i, $i = 1, 2, ..., n$ are parameters. The difference between Eq. (1) and the classic nonlinear reaction-diffusion equation is in the term $\tau \frac{\partial^2 Q}{\partial t^2}$. Equation of class (1) are known also as damped nonlinear Klein-Gordon equations [4, 7, 12, 14, 55]. We note that reaction-diffusion equations have many applications for describing different kinds of processes in physics, chemistry, biology, etc. [8, 15, 58]. Traveling wave solutions of these equations are of special interest as they describe the motion of wave fronts or the motion of boundary between two different states existing in the studied system. Below we apply the modified method of simplest equation (described in Sect. 2) for obtaining exact traveling solutions of nonlinear reaction-diffusion PDE with polynomial nonlinearity of fourth order (Sect. 3). The obtained waves are discussed in Sect. 3 and several concluding remarks are summarized in Sect. 4.

2 The modified method of simplest equation

Below we shall apply the modified method of simplest equation. The current version of the methodology used by our research group is based on the possibility of use of more than one simplest equation [54]. We shall describe this version of the methodology and below we shall use the particular case when the solutions of the studied nonlinear PDE are obtained by use of a single simplest equation. The steps of the methodology are as follows.

1. By means of appropriate ansätze (below we shall use a traveling-wave ansatz but in principle there can be one or several traveling-wave ansätze such as $\xi = \alpha x + \beta t$; $\zeta = \gamma x + \delta t, \ldots$. Other kinds of ansätze may be used too) the solved nonlinear partial differential equation is reduced to a differential equation E, containing derivatives of one or several functions

$$E\left[a(\xi), a_\xi, a_{\xi\xi}, \ldots, b(\zeta), b_\zeta, b_{\zeta\zeta}, \ldots\right] = 0 \tag{2}$$

2. In order to make transition to the solution of the simplest equation we assume that any of the functions $a(\xi)$, $b(\zeta)$, etc., is a function of another function, i.e.

$$a(\xi) = G[f(\xi)]; \quad b(\zeta) = F[g(\zeta)]; \ldots \tag{3}$$

3. We note that the kind of the functions $F, G, \ldots$ is not prescribed. Often one uses a finite-series relationship, e.g.,

$$a(\xi) = \sum_{\mu_1=-\nu_1}^{\nu_2} q_{\mu_1}[f(\xi)]^{\mu_1}; \quad b(\zeta) = \sum_{\mu_2=-\nu_3}^{\nu_4} r_{\mu_2}[g(\zeta)]^{\mu_2}, \ldots \tag{4}$$

   where $q_{\mu_1}, r_{\mu_2}, \ldots$ are coefficients. However other kinds of relationships may be used too. Below we shall work on the basis of relationships of kind (4).
4. The functions $f(\xi)$, $g(\zeta)$ are solutions of simpler ordinary differential equations called simplest equations. For several years the methodology of the modified method of simplest equation was based on use of one simplest equation. The new version of the methodology allows the use of more than one simplest equation. The idea for use of more than one simplest equation can be traced back two decades ago to the articles of Martinov and Vitanov [30–32].
5. Equation (3) is substituted in Eq. (2) and let the result of this substitution be a polynomial containing $f(\xi)$, $g(\zeta)$, ... Next we have to deal with the coefficients of this polynomial.
6. A balance procedure is applied that has to ensure that all of the coefficients of the obtained polynomial of $f(\xi)$ and $g(\zeta)$ contain more than one term. This procedure leads to one or several balance equations for some of the parameters of the solved equation and for some of the parameters of the solution. Especially the coefficients ν_i from Eq. (4) as well as the parameters connected to the order of nonlinearity of the simplest equations are terms in the balance equations. Note that the coefficients of all powers of the polynomials have to be balanced (and not only the coefficient of the largest power). This is why the extended balance may require more than one balance equation.

7. Equation (3) represent a candidate for solution of Eq. (2) if all coefficients of the obtained polynomial of are equal to 0. This condition leads to a system of nonlinear algebraic equations for the coefficients of the solved nonlinear PDE and for the coefficients of the solution. Any nontrivial solution of this algebraic system leads to a solution of the studied nonlinear partial differential equation. Usually the system of algebraic equations contains many equations that have to be solved by means of a computer algebra system.

Below we shall search a solution of the studied equation of the kind

$$Q(\xi) = \sum_{i=0}^{n} a_i [\phi(\xi)]^i, \quad \xi = x - vt \tag{5}$$

where $\phi(\xi)$ is a solution of the Bernoulli differential equation

$$\frac{d\phi}{d\xi} = a\phi(\xi) + b[\phi(\xi)]^k \tag{6}$$

where k is a positive integer. We shall use the following solutions of the Bernoulli equation

$$\phi(\xi) = \sqrt[k-1]{\frac{a e^{a(k-1)(\xi+\xi_0)}}{1 - b e^{a(k-1)(\xi+\xi_0)}}}, \quad \phi(\xi) = \sqrt[k-1]{-\frac{a e^{a(k-1)(\xi+\xi_0)}}{1 + b e^{a(k-1)(\xi+\xi_0)}}} \tag{7}$$

for the cases $b < 0$, $a > 0$ and $b > 0$, $a < 0$ respectively. Above ξ_0 is a constant of integration.

3 Studied hyperbolic reaction-diffusion equation and application of the method

Below we shall solve the equation

$$\tau \frac{\partial^2 Q}{\partial t^2} + \frac{\partial Q}{\partial t} = D \frac{\partial^2 Q}{\partial x^2} + \sum_{i=1}^{4} \alpha_i Q^i. \tag{8}$$

Reaction-diffusion equation of this kind (with polynomial nonlinearity of fourth order) was used to model the propagation of wave fronts in populations systems [40, 41]. We apply the ansatz $Q(\xi) = Q(x - vt)$ and then we substitute $Q(\xi)$ by Eq. (5) where $n = 2$, i.e.,

$$Q[\phi(\xi)] = a_0 + a_1\phi(\xi) + a_2\phi(\xi)^2. \tag{9}$$

The balance procedure leads to a simplest equation of fourth order:

$$\frac{d\phi(\xi)}{d\xi} = b_0 + b_1[\phi(\xi)] + b_2[\phi(\xi)]^2 + b_3[\phi(\xi)]^3 + b_4[\phi(\xi)]^4. \qquad (10)$$

Above quantities a_0, a_1, a_2, b_0, b_1, b_2, b_3 and b_4 are parameters. We note here that we shall use a particular case of this simplest equation where $b_0 = b_2 = b_3 = 0$. The substitution of the traveling wave ansatz and Eqs. (9),(10) in Eq. (8) leads to the following system of 9 algebraic equations:

$$\begin{aligned}
&10\,(D - v^2\tau)\,a_2\,{b_4}^2 + \alpha_4\,{a_2}^4 = 0,\\
&18\,(D - v^2\tau)\,a_2\,b_3\,b_4 = 0,\\
&4\,\alpha_4\,a_0\,{a_2}^3 + \alpha_3\,{a_2}^3 + (D - v^2\tau)\,(8\,a_2\,{b_3}^2 + 16\,a_2\,b_2\,b_4) = 0,\\
&(D - v^2\tau)\,(14\,a_2\,b_1\,b_4 + 14\,a_2\,b_2\,b_3) + 2\,v\,a_2\,b_4 = 0,\\
&(D - v^2\tau)\,(12\,a_2\,b_1\,b_3 + 12\,a_2\,b_0\,b_4 + 6\,a_2\,{b_2}^2) + (3\,\alpha_3\,a_0\, + \alpha_2 +\\
&6\,\alpha_4\,{a_0}^2\,)\,{a_2}^2 + 2\,v\,a_2\,b_3 = 0,\\
&2\,v\,a_2\,b_2 + (D - v^2\tau)\,(10\,a_2\,b_0\,b_3 + 10\,a_2\,b_1\,b_2) = 0,\\
&3\,\alpha_3\,{a_0}^2\,a_2 + 2\,\alpha_2\,a_0\,a_2 + 4\,\alpha_4\,{a_0}^3\,a_2 + (D - v^2\tau)\,(8\,a_2\,b_0\,b2 + 4\,a_2\,{b_1}^2) +\\
&2\,va_2b_1 + \alpha_1\,a_2 = 0,\\
&2\,v\,a_2\,b_0 + 6\,(D - v^2\tau)\,a_2\,b_0\,b_1 = 0,\\
&2\,(D - v^2\tau)\,a_2\,{b_0}^2 + \alpha_4\,{a_0}^4 + \alpha_3\,{a_0}^3 + \alpha_1\,a_0 + \alpha_2\,{a_0}^2 = 0. \qquad (11)
\end{aligned}$$

A nontrivial solution of the system (11) is:

$$\begin{aligned}
b_1 &= \frac{[\alpha_3^3(49\alpha_3^3\tau - 640\alpha_4^2)]^{1/2}}{640\alpha_4^2 D^{1/2}}\\
b_4 &= -\frac{7\alpha_3^2\tau a_2(49\alpha_3^3\tau - 640\alpha_4^2)\left[\frac{\alpha_4^4 D a_2\left(320+\frac{49\alpha_3^3\tau-320\alpha_4^2}{\alpha_4^2}\right)}{\alpha_3^3\tau(49\alpha_3^3\tau-640\alpha_4^2)}\right]^{1/2}}{80\alpha_4^4 D\left(320 + \frac{49\alpha_3^3\tau-320\alpha_4^2}{\alpha_4^2}\right)}\\
b_0 &= b_2 = b_3 = 0,\\
v &= \frac{\alpha_4^2 D^{1/2}\left(320 + \frac{49\alpha_3^3\tau-320\alpha_4^2}{\alpha_4^2}\right)}{7\alpha_3\tau[\alpha_3(49\alpha_3^3\tau - 540\alpha_4^2)]^{1/2}},\\
\alpha_2 &= \frac{3}{8}\frac{\alpha_3^2}{\alpha_4}, \quad \alpha_1 = \frac{3}{64}\frac{\alpha_3^3}{\alpha_4^2}, \quad a_0 = -\frac{1}{4}\frac{\alpha_3}{\alpha_4}, \quad a_1 = 0 \qquad (12)
\end{aligned}$$

Then the solution of the simplest equation becomes ($k = 4$)

$$\phi(\xi) = \Bigg\{ \Bigg[\frac{[\alpha_3^3(49\alpha_3^3\tau - 640\alpha_4^2)]^{1/2}}{640\alpha_4^2 D^{1/2}} \exp\Bigg[3\frac{[\alpha_3^3(49\alpha_3^3\tau - 640\alpha_4^2)]^{1/2}}{640\alpha_4^2 D^{1/2}}(\xi + \xi_0)\Bigg]\Bigg]$$

$$\Bigg/ \Bigg[\Bigg[1 + \frac{7\alpha_3^2\tau a_2(49\alpha_3^3\tau - 640\alpha_4^2)\left[\frac{\alpha_4^4 D a_2\left(320+\frac{49\alpha_3^3\tau-320\alpha_4^2}{\alpha_4^2}\right)}{\alpha_3^3\tau(49\alpha_3^3\tau-640\alpha_4^2)}\right]^{1/2}}{80\alpha_4^4 D\left(320 + \frac{49\alpha_3^3\tau-320\alpha_4^2}{\alpha_4^2}\right)}\Bigg]$$

$$\times \exp\Bigg[3\frac{[\alpha_3^3(49\alpha_3^3\tau - 640\alpha_4^2)]^{1/2}}{640\alpha_4^2 D^{1/2}}(\xi + \xi_0)\Bigg]\Bigg]\Bigg\}^{1/3},$$

$$\phi(\xi) = \Bigg\{ \Bigg[\frac{[\alpha_3^3(49\alpha_3^3\tau - 640\alpha_4^2)]^{1/2}}{640\alpha_4^2 D^{1/2}} \exp\Bigg[3\frac{[\alpha_3^3(49\alpha_3^3\tau - 640\alpha_4^2)]^{1/2}}{640\alpha_4^2 D^{1/2}}(\xi + \xi_0)\Bigg]\Bigg]$$

$$\Bigg/ \Bigg[\Bigg[1 - \frac{7\alpha_3^2\tau a_2(49\alpha_3^3\tau - 640\alpha_4^2)\left[\frac{\alpha_4^4 D a_2\left(320+\frac{49\alpha_3^3\tau-320\alpha_4^2}{\alpha_4^2}\right)}{\alpha_3^3\tau(49\alpha_3^3\tau-640\alpha_4^2)}\right]^{1/2}}{80\alpha_4^4 D\left(320 + \frac{49\alpha_3^3\tau-320\alpha_4^2}{\alpha_4^2}\right)}\Bigg]$$

$$\times \exp\Bigg[3\frac{[\alpha_3^3(49\alpha_3^3\tau - 640\alpha_4^2)]^{1/2}}{640\alpha_4^2 D^{1/2}}(\xi + \xi_0)\Bigg]\Bigg]\Bigg\}^{1/3}, \quad (13)$$

for the cases $b_4 < 0$, $b_1 > 0$ and $b_4 > 0$, $b_1 < 0$ respectively. Thus the solutions of Eq. (8) are

$$Q(\xi) = -\frac{1}{4}\frac{\alpha_3}{\alpha_4} + a_2$$

$$\times \Bigg\{ \Bigg[\frac{[\alpha_3^3(49\alpha_3^3\tau - 640\alpha_4^2)]^{1/2}}{640\alpha_4^2 D^{1/2}} \exp\Bigg[3\frac{[\alpha_3^3(49\alpha_3^3\tau - 640\alpha_4^2)]^{1/2}}{640\alpha_4^2 D^{1/2}}(\xi + \xi_0)\Bigg]\Bigg]$$

$$\Bigg/ \Bigg[\Bigg[1 + \frac{7\alpha_3^2\tau a_2(49\alpha_3^3\tau - 640\alpha_4^2)\left[\frac{\alpha_4^4 D a_2\left(320+\frac{49\alpha_3^3\tau-320\alpha_4^2}{\alpha_4^2}\right)}{\alpha_3^3\tau(49\alpha_3^3\tau-640\alpha_4^2)}\right]^{1/2}}{80\alpha_4^4 D\left(320 + \frac{49\alpha_3^3\tau-320\alpha_4^2}{\alpha_4^2}\right)}\Bigg]$$

$$\times \exp\Bigg[3\frac{[\alpha_3^3(49\alpha_3^3\tau - 640\alpha_4^2)]^{1/2}}{640\alpha_4^2 D^{1/2}}(\xi + \xi_0)\Bigg]\Bigg]\Bigg\}^{2/3},$$

$$
\begin{aligned}
Q(\xi) = & -\frac{1}{4}\frac{\alpha_3}{\alpha_4} + a_2 \\
& \times \Bigg\{ \Bigg[\frac{[\alpha_3^3(49\alpha_3^3\tau - 640\alpha_4^2)]^{1/2}}{640\alpha_4^2 D^{1/2}} \exp\Bigg[3\frac{[\alpha_3^3(49\alpha_3^3\tau - 640\alpha_4^2)]^{1/2}}{640\alpha_4^2 D^{1/2}}(\xi + \xi_0) \Bigg] \Bigg] \\
& \Bigg/ \Bigg[\Bigg[1 - \frac{7\alpha_3^2\tau a_2(49\alpha_3^3\tau - 640\alpha_4^2)\left[\frac{\alpha_4^4 D a_2\left(320 + \frac{49\alpha_3^3\tau - 320\alpha_4^2}{\alpha_4^2}\right)}{\alpha_3^3\tau(49\alpha_3^3\tau - 640\alpha_4^2)}\right]^{1/2}}{80\alpha_4^4 D\left(320 + \frac{49\alpha_3^3\tau - 320\alpha_4^2}{\alpha_4^2}\right)} \Bigg] \\
& \times \exp\Bigg[3\frac{[\alpha_3^3(49\alpha_3^3\tau - 640\alpha_4^2)]^{1/2}}{640\alpha_4^2 D^{1/2}}(\xi + \xi_0) \Bigg] \Bigg] \Bigg\}^{2/3}
\end{aligned} \tag{14}
$$

for the cases $b_4 < 0$, $b_1 > 0$ and $b_4 > 0$, $b_1 < 0$ respectively.
The obtained solutions (14) describe kink waves. Several of the waves are shown in Figs. 1, 2 and 3. The parameters of the solutions are the same except the parameter α_3 that has different values for the three kinks. As one can observe the decrease of the value of the parameter α_3 leads to: (i) change of the values of Q (from negative to positive); (ii) decrease of the width of the ink, and (iii) increase of the amplitude of the kink. Similar effects can be observed also in the case when the values of other parameters of the solution are varied.

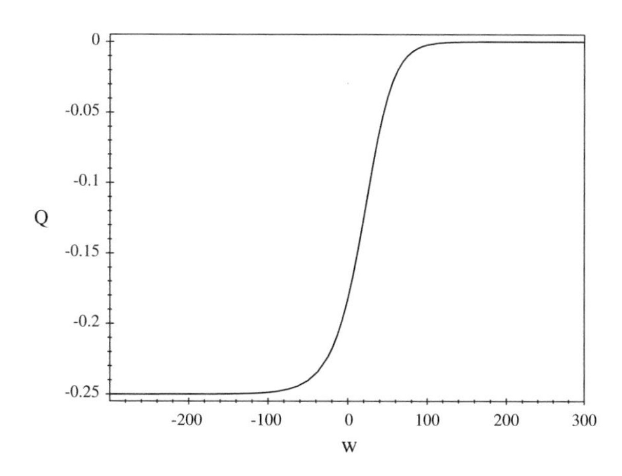

Fig. 1 Solution of equation (14). The values of parameters are: $\tau = 20$; $D = 2$; $a_2 = 1$; $\alpha_3 = 1$; $\alpha_4 = 1$; $\xi_0 = 0$. $w = \xi + \xi_0$

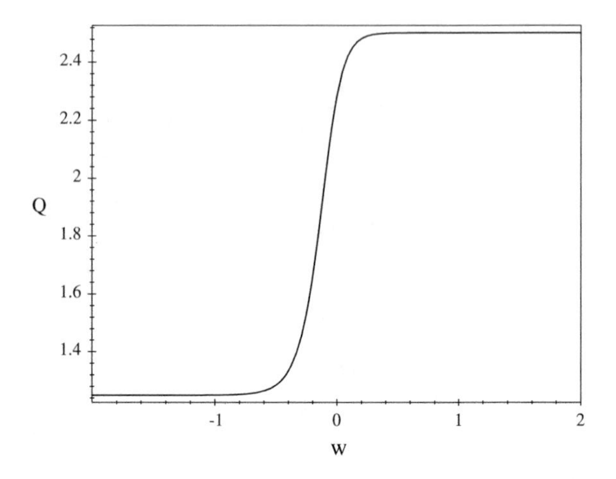

Fig. 2 Solution of equation (14). The values of parameters are: $\tau = 20$; $D = 2$; $a_2 = 1$; $\alpha_3 = -5$; $\alpha_4 = 1$; $\xi_0 = 0$. $w = \xi + \xi_0$

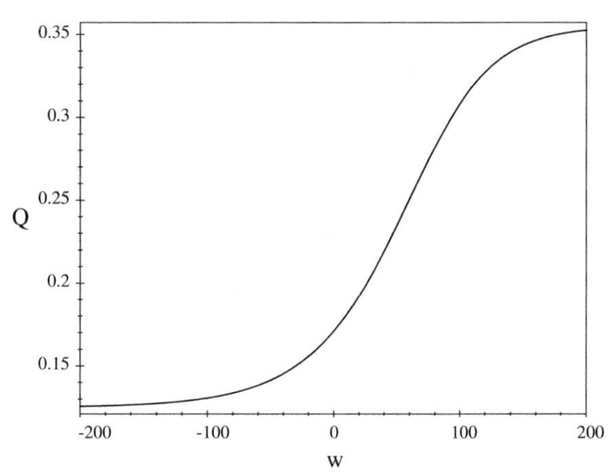

Fig. 3 Solution of equation (14). The values of parameters are: $\tau = 20$; $D = 2$; $a_2 = 1$; $\alpha_3 = -1/2$; $\alpha_4 = 1$; $\xi_0 = 0$. $w = \xi + \xi_0$

4 Concluding remarks

In this chapter we have discussed the nonlinear hyperbolic reaction-diffusion equation (8). It can be related to the nonlinear reaction-diffusion equation that was used to model systems from population dynamics [40, 41]. We note that:

- The obtained solutions of the hyperbolic reaction-diffusion equation do not contain as particular cases the kink solutions of the reaction-diffusion equation discussed in [40, 41]. This is easily seen from the relationship for b_4 in Eq. (12). If we set there $\tau = 0$ then $b_4 = 0$ and we cannot construct a kink solution of the kind $Q(\xi) = b_1 + b_4 \phi(\xi)^2$.
- Figures 4 and 5 show the influence of increasing values of the parameter τ on the obtained kink solutions of the nonlinear hyperbolic reaction-diffusion equation. As it can be seen from the figures the influence of the increasing value of τ on the kink profile is: (i) to decrease the amplitude of the kink, and (ii) to make

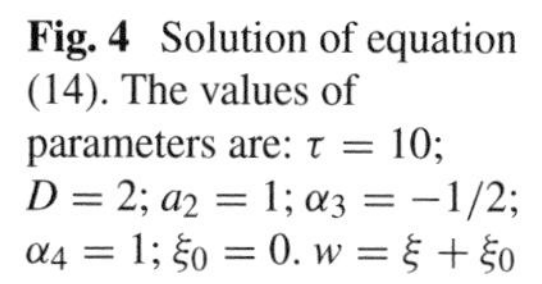

Fig. 4 Solution of equation (14). The values of parameters are: $\tau = 10$; $D = 2$; $a_2 = 1$; $\alpha_3 = -1/2$; $\alpha_4 = 1$; $\xi_0 = 0$. $w = \xi + \xi_0$

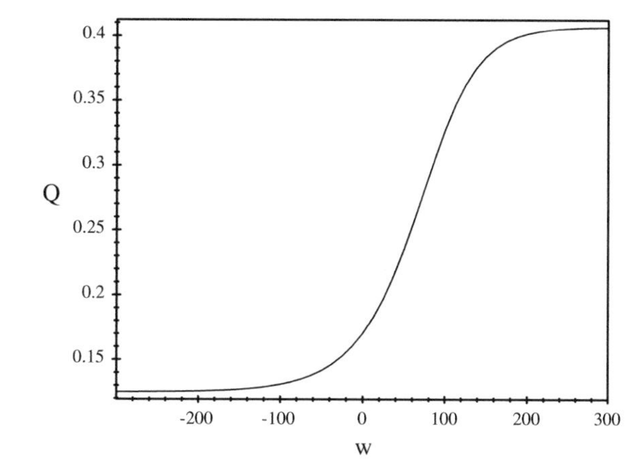

Fig. 5 Solution of equation (14). The values of parameters are: $\tau = 1000$; $D = 2$; $a_2 = 1$; $\alpha_3 = -1/2$; $\alpha_4 = 1$; $\xi_0 = 0$. $w = \xi + \xi_0$

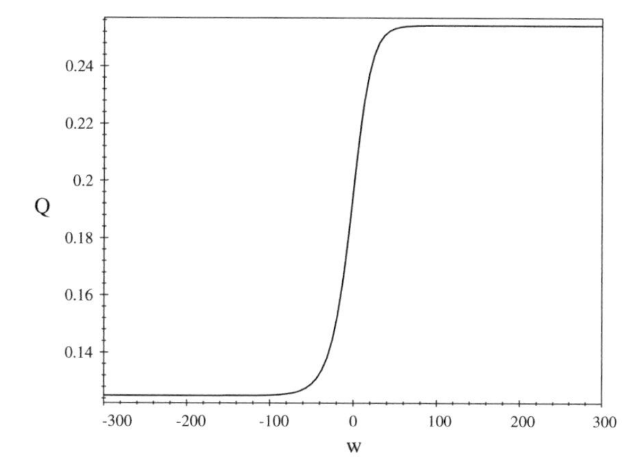

the transition between the areas of lower and higher values of the kink more concentrated (i.e. this transition happens in the smaller interval of values of w).

- The exact solution of the studied nonlinear partial differential equation was obtained by means of the modified method of simplest equation. We have shown that this method is an effective method for obtaining particular exact solutions of nonlinear partial differential equations that do not belong to the class of integrable equations.

Acknowledgements This study contains results, which are supported by the UNWE project for scientific research with grant agreement No. NID NI – 21/2016

References

1. Ablowitz, M.J., Kaup, D.J., Newell, A.C.: Nonlinear evolution equations of physical significance. Phys. Rev. Lett. **31**, 125–127 (1973)
2. Ablowitz, M.J., Kaup, D.J., Newell, A.C., Segur, H.: Inverse scattering transform - fourier analysis for nonlinear problems. Stud. Appl. Math. **53**, 249–315 (1974)
3. Ablowitz, M., Clarkson, P.A.: Solitons, Nonlinear Evolution Equations and Inverse Scattering. Cambridge University Press, Cambridge (1991)
4. Al-Ghoul, M., Eu, B.C.: Hyperbolic reaction-diffusion equations, patterns, and phase speeds for the Brusselator. J. Phys. Chem. **100**, 18900–18910 (1996)
5. Ames, W.F.: Nonlinear Partial Differential Equations in Engineering. Academic Press, New York (1972)
6. Benkirane, A., Gossez, J.-P. (eds.): Nonlinear Patial Differential Equations. Addison Wesley Longman, Essex, UK (1996)
7. Burq, N., Raugel, R., Schlag, W.: Long-time dynamics for damped Klein-Gordon equations. hal-01154421 (2015)
8. Cantrell, R.S., Costner, C.: Spatial Ecology Via Reaction-Diffusion Equations. Wiley, Chichester (2003)
9. Debnath, L.: Nonlinear Partial Differential Equations for Scientists and Engineers. Springer, New York (2012)
10. Fan, E., Hon, Y.C.: A series of travelling wave solutions for two variant Boussinesq equations in shallow water waves. Chaos, Solitons Fractals **15**, 559–566 (2003)
11. Galaktionov, V.A., Svirhchevskii, S.R.: Exact Solutions and Invariant Subspaces of Nonlinear Partial Differential Equations in Mechanics and Physics. Chapman & Hall/CRC, Bora Raton, FL (2007)
12. Gallay, T., Raugel, R.: Scaling variables and stability of hyperbolic fronts. SIAM J. Math. Anal. **32**, 1–29 (2000)
13. Gardner, C.S., Greene, J.M., Kruskal, M.D., Miura, R.R.: Method for solving Korteweg- de Vries equation. Phys. Rev. Lett. **19**, 1095–1097 (1967)
14. Gonzalez, J.A., Oliveira, F.A.: Nucleation theory, the escaping processes, and nonlinear stability. Phys. Rev. B **59**, 6100–6105 (1999)
15. Grzybowski, B.A.: Chemistry in Motion: Reaction-Diffusion Systems for Micro- and Nanotechnology. Wiley, Chichester (2009)
16. He, J.-H., Wu, X.-H.: Exp-function method for nonlinear wave equations. Chaos, Solitons Fractals **30**, 700–708 (2006)
17. Hirsch, M., Devaney, R.L., Smale, S.: Differential Equations, Dynamical Systems, and an Introduction to Chaos. Academic Press, New York (2004)
18. Hirota, R.: Exact solution of Korteweg-de Vries equation for multiple collisions of solitons. Phys. Rev. Lett. **27**, 1192–1194 (1971)
19. Holmes, P., Lumley, J.L., Berkooz, G.: Turbulence, Coherent Structures, Dynamical Systems and Symmetry. Cambridge University Press, Cambridge (1996)
20. Kudryashov, N.A.: Exact solutions of the generalized Kuramoto - Sivashinsky equation. Phys. Lett. A **147**, 287–291 (1990)
21. Kudryashov, N.A.: Exact solitary waves of the Fisher equation. Phys. Lett. **342**, 99–106 (2005)
22. Kudryashov, N.A.: Simplest equation method to look for exact solutions of nonlinear differential equations. Chaos Solitons Fractals **24**, 1217–1231 (2005)
23. Kudryashov, N.A., Demina, M.V.: Polygons of differential equations for finding exact solutions. Chaos Solitons Fractals **33**, 480–496 (2007)
24. Kudryashov, N.A., Loguinova, N.B.: Extended simplest equation method for nonlinear differential equations. Appl. Math. Comput. **205**, 396–402 (2008)
25. Kudryashov, N.A.: Solitary and periodic wave solutions of generalized Kuramoto-Sivashinsky equation. Regul. Chaotic Dyn. **13**, 234–238 (2008)
26. Kudryashov, N.A.: Meromorphic solutions of nonlinear ordinary differential equations. Commun. Nonlinear Sci. Numer. Simul. **15**, 2778–2790 (2010)

27. Logan, J.D.: An Introduction to Nonlinear Partial Differential Equations. Wiley, New York (2008)
28. Leung, A.W.: Systems of Nonlinear Partial Differential Equations. Applications to Biology and Engineering. Kluwer, Dordrecht (1989)
29. Malfliet, W., Hereman, W.: The tanh method: I. exact solutions of nonlinear evolution and wave equations. Phys. Scr. **54**, 563–568 (1996)
30. Martinov, N., Vitanov, N.: On some solutions of the two-dimensional sine-Gordon equation. J. Phys. A Math. Gen. **25**, L419–L425 (1992)
31. Martinov, N., Vitanov, N.: Running wave solutions of the two-dimensional sine-Gordon equation. J. Phys. A: Math. Gen. **25**, 3609–3613 (1992)
32. Martinov, N.K., Vitanov, N.K.: New class of running-wave solutions of the (2+ 1)-dimensional sine-Gordon equation. J. Phys. A: Math. Gen. **27**, 4611–4618 (1994)
33. Murray, J.D.: Lectures on Nonlinear Differential Equation Models in Biology. Oxford University Press, Oxford, UK (1977)
34. Perko, L.: Differential Equations and Dynamical Systems. Springer, New York (1991)
35. Remoissenet, M.: Waves Called Solitons. Springer, Berlin (1993)
36. Scott, A.C.: Nonlinear Science. Emergence and Dynamics of Coherent Structures. Oxford University Press, Oxford, UK (1999)
37. Strauss, W.A.: Partial Differential Equations: An Introduction. Wiley, New York (1992)
38. Tabor, M.: Chaos and Integrability in Dynamical Systems. Wiley, New York (1989)
39. Verhulst, F.: Nonlinear Differential Equations and Dynamical Systems. Springer, Berlin (1990)
40. Vitanov, N.K., Jordanov, I.P., Dimitrova, Z.I.: On nonlinear dynamics of interacting populations: coupled kink waves in a system of two populations. Commun. Nonlinear Sci. Numer. Simulat. **2009**(14), 2379–2388 (2009)
41. Vitanov, N.K., Jordanov, I.P., Dimitrova, Z.I.: On nonlinear population waves. Appl. Math. Comput. **215**, 2950–2964 (2009)
42. Vitanov, N.K., Dimitrova, Z.I., Kantz, H.: Modified method of simplest equation and its application to nonlinear PDEs. Appl. Math. Comput. **216**, 2587–2595 (2010)
43. Vitanov, N.K.: Application of simplest equations of Bernoulli and Riccati kind for obtaining exact traveling wave solutions for a class of PDEs with polynomial nonlinearity. Commun. Nonlinear Sci. Numer. Simulat. **15**, 2050–2060 (2010)
44. Vitanov, N.K., Dimitrova, Z.I.: Application of the method of simplest equation for obtaining exact traveling-wave solutions for two classes of model PDEs from ecology and population dynamics. Commun. Nonlinear Sci. Numer. Simulat. **15**, 2836–2845 (2010)
45. Vitanov, N.K.: Modified method of simplest equation: powerful tool for obtaining exact and approximate traveling-wave solutions of nonlinear PDEs. Commun. Nonlinear Sci. Numer. Simulat. **16**, 1176–1185 (2011)
46. Vitanov, N.K., Dimitrova, Z.I., Vitanov, K.N.: On the class of nonlinear PDEs that can be treated by the modified method of simplest equation. Application to generalized Degasperis - Processi equation and b-equation. Commun. Nonlinear Sci. Numer. Simulat. **16**, 3033 – 3044 (2011)
47. Vitanov, N.K.: On modified method of simplest equation for obtaining exact and approximate solutions of nonlinear PDEs: the role of the simplest equation. Commun. Nonlinear Sci. Numer. Simul. **16**, 4215–4231 (2011)
48. Vitanov, N.K., Dimitrova, Z.I., Kantz, H.: Application of the method of simplest equation for obtaining exact traveling-wave solutions for the extended Korteweg-de Vries equation and generalized Camassa-Holm equation. Appl. Math. Comput. **219**, 7480–7492 (2013)
49. Vitanov, N.K., Dimitrova, Z.I., Vitanov, K.N.: Traveling waves and statistical distributions connected to systems of interacting populations. Comput. Math. Appl. **66**, 1666–1684 (2013)
50. Vitanov, N.K., Dimitrova, Z.I.: Solitary wave solutions for nonlinear partial differential equations that contain monomials of odd and even grades with respect to participating derivatives. Appl. Math. Comput. **247**, 213–217 (2014)
51. Vitanov, N.K., Dimitrova, Z.I., Vitanov, K.N.: Modified method of simplest equation for obtaining exact analytical solutions of nonlinear partial differential equations: further development of the methodology with applications. Appl. Math. Comput. **269**, 363–378 (2015)

52. Vitanov, N.K.: Science Dynamics and Research Production. Indicators, Indexes, Statistical Laws and Mathematical Models. Springer, Cham (2016)
53. Vitanov, N.K., Vitanov, K.N.: Box model of migration channels. Math. Soc. Sci. **80**, 108–114 (2016)
54. Vitanov N.K., Dimitrova Z.I.: Modified method of simplest equation and the nonlinear Schrödinger equation. J. Theor. Appl. Mech. **48**, 58–69 (2018)
55. Wang, Q.F., Cheng, D.Z.: Numerical solution of damped nonlinear Klein-Gordon equations using variational method and finite element approach. Appl. Math. Comput. **162**, 381–401 (2005)
56. Wazwaz, A.-M.: The tanh method for traveling wave solutions of nonlinear equations. Appl. Math. Comput. **154**, 713–723 (2004)
57. Wazwaz, A.-M.: Partial Differential Equations and Solitary Waves Theory. Springer, Dordrecht (2009)
58. Wilhelmson, H., Lazzaro, E.: Reaction-Diffusion Problem in the Physics of Hot Plasmas. IOP Publishing, Bristol (2000)

Generalized Nets: A New Approach to Model a Hashtag Linguistic Network on Twitter

Kristina G. Kapanova and Stefka Fidanova

Abstract In the last few years the micro-blogging platform Twitter has played a significant role in the communication of civil uprisings, political events or natural disasters. One of the reasons is the adoption of the hashtag, which represents a short word or phrase that follows the hash sign (#). These semantic elements captured the topics behind the tweets and allowed the information flow to bypass traditional social network structure. The hashtags provide a way for users to embed metadata in their posts achieving several important communicative functions: they can indicate the specific semantic domain of the post, link the post to an existing topic, or provide a range of complex meanings in social media texts. In this paper, Generalized nets are applied as a tool to model the structural characteristics of a hashtag linguistic network through which possible communities of interests emerge, and to investigate the information propagation patterns resulting from the uncoordinated actions of users in the underlying semantic hashtag space. Generalized nets (GN) are extensions of the Petri nets by providing functional and topological aspects unavailable in Petri nets. The study of hashtag networks from a generalized nets perspective enables us to investigate in a deeper manner each element of the GN, substituting it with another, more detailed network in order to be examined in depth. The result is an improved understanding of topological connections of the data and the ability to dynamically add new details to expand the network and as a result discover underlying structural complexities unable to be discovered through traditional network analysis tool due to the prohibitive computational cost. Analysis is performed on a collection of Tweets and results are presented.

K. G. Kapanova · S. Fidanova (✉)
Institute of Information and Communication Technology,
Bulgarian Academy of Science, Sofia, Bulgaria
e-mail: stefka@parallel.bas.bg

K. G. Kapanova
e-mail: kkapanova@gmail.com

K. Georgiev et al. (eds.), *Advanced Computing in Industrial Mathematics*,
Studies in Computational Intelligence 793,
https://doi.org/10.1007/978-3-319-97277-0_17

1 Introduction

In the past decade social media platforms like Twitter and Facebook have become an effective and popular way for users to generate content and engage in discussions with others on a wide variety of social topics and activities through predominantly textually-mediated and multimodal practices [1]. Such practice in this regard is the use of hashtags, which are social annotations used to indicate messages content. First proposed by Chris Messina [2] in 2007 to improve Twitter conversational flow, the shorthand convention serves as an intuitive and flexible instrument to make an immense collection of posts searchable, easily organized and retrievable by new members of the community. The inherent ability of hashtags to bypass the network's structure limitation during the communication process contribute to the easy dissemination of information beyond a users network [3]. Resultantly, hashtags assume a dual role in the communication process, serving both as a metadata for archival and retrieval purposes, and as an integral part in the generation of ad-hoc thematic discursive spaces, where a subset of interested individuals communicate. In turn the duality of hashtags functions in a semiotic manner for social signaling, defining a shared context for specific topics.

The hashtags, often composed of natural language n-grams or abbreviations (i.e. *#imwithher*), serve as a method to promote posts with the goal of extending the reach to bigger community of readers. Since they are developed by the users themselves, a new social event can lead to the simultaneous emergence of multitude of different tags. Those hashtags in turn can be either accepted by the other users or not, leading some hashtags to propagate the network and others to be confined only to a few messages. Similarly, the terminology of users language can provide lexical innovations [3] - including the creation of new words, abbreviations, specific jargon to refer to important events or artifacts, adaptation of phrases, inside jokes - differing from the accepted natural language usage.

The diffusion of such linguistic innovation occurs through a cascade in which the network members decide to reshare, adopt and propagate a behavior in a given moment [4] and therefore acting as an open-ended information network. In this work we investigate through generalized nets (GN) the diffusion of a hashtag in a linguistic network, the structural characteristics of a hashtag linguistic network through which possible communities of interests emerge and the propagation patterns resulting from the uncoordinated actions of users in the underlying semantic hashtag space. We exploit two advantages of generalized nets to describe the characteristics, models and results of a hashtag linguistic network. Foremost, we utilize the inherent functionality of GNs to model parallel processes. Furthermore, we benefit from the capacity of the model to dynamically add new details to expand the network and as thereupon discover underlying structural complexities incapable of being identified through traditional network analysis tool due to the prohibitive computational cost.

The remainder of the paper is structured as follows. In the next section we briefly introduce key concepts and terms about Generalized nets. Then we follow with a description of the methods of data collection and analysis of the results. The concluding remarks discuss possible future directions and importance of investigating hashtags linguistic networks through generalized nets.

2 Generalized Nets

Recently researchers have extensively examined the communication practices on Twitter. Widely used social networking practices were outlined in three structural layers by Brun and Moe [5], while Boyd [6] considered retweeting practices and represented them as a conversational activity. In [7] the content of the reply messages on the platform and the interactional aspects among users were examined. Others have addressed the information diffusion on the network through "internet memes" [8]. Zappavigna [9] investigated the function of hashtags from a linguistic perspective and have proposed the term "searchable talk" to describe the communicative relationship between users. Other investigations ascertained the social dimensions of hashtags phrases within a tweet [10] and explored the social relevance of hashtags. Studies on data collected on the basis of hashtag searches, related to a specific event have been conducted, for example for the 2016 presidential election [11], Occupy Wall Street protests [12], the Egypt revolution [13].

In this work we propose to model a hashtag linguistic network with Generalized nets. Generalized nets (GN), as an extension of the Petri nets provide many functional and topological advantages, among them the ability to model parallel processes [14–16]. Since 1993 many GN applications have been developed to model processes in areas such as Artificial intelligence, expert systems, data bases, optimization algorithms, pattern recognition, as well as medical, biological and economical problems [17]. Generalized nets have been also utilized for text recognition and translation tasks by modeling the process in an abstract form [18].

One of the biggest advantages of GNs is their ability to describe models of complex systems, consisting of heterogeneous components and with concurrent activities. The reader should consider there are several main components of generalized nets. The static structures can be expressed by objects (also known as transitions), which have input and output places. While two transitions are allowed to share a place, a place can become an input and an output of only one transition. The dynamic structure is described by tokens, which serve as information pipes and through each stage of the generalized net can occupy a single place. A transition arc in the model represents a pass of the token through an input to an output place. The change of the tokens is regulated by predicates, described in a predicate matrix of the transition. According to specific states, the values of the tokens change in time and are described through characteristic functions. The latter can be regarded as a way to assign new characteristics to the incoming tokens. Additional functionality of the tokens is their potentiality to split and merge in the place therefore creating the expandability of generalized nets. By this, one can replace each place on the GN with a new net, consequently improving the understanding of the underlying structural functionalities of the model.

In our model, we describe the hashtags as a complex co-occurrence network with each module of hashtags depicted by transition and each group of hashtags will be described by tokens with different characteristics, representing the possibility of hashtags being part of multiple small subgroups, or the inherent semantic connection between groups of hashtags.

3 Generalized Net Models of Hashtag Linguistic Network

The ability to describe the hashtag co-occurrence network as a graph empowers us to utilize a generalized net model. We begin with no tokens inside the GN-model. At the moment of activation only one token, representing the collected data from Twitter, arrives in the net in place l_1 with characteristic "user actions" (post content with hashtag, favorite, retweet, etc). The token is split then into two other tokens which enter places - l_2 and l_3. The token entry l_2 with characteristics "post attributes". For instance, consider the collected Twitter data to have several attributes, including a dimension for source, target, timestamp, location, user mentions, hashtags, type, content, as well as further information about the users. The token entry l_3 with characteristics "graph parameters", in which we include degree centrality, betweenness centrality, page rank, etc. The constructed GN consists of the subsequent set of transactions: $T = \{Z_1, Z_2, Z_3, Z_4, Z_5\}$, where those transactions describe the chosen action of the user.

The form of the first transition of the GN model - Z_1 describes the action by the user as follows:

$$Z_1 = \langle \{l_1\}, \{l_2, l_3\}, \frac{\begin{array}{c|cc} & l_2 & l_3 \end{array}}{\begin{array}{c|cc} l_1 & W_{1,2} & W_{1,3} \end{array}} \rangle,$$

where
$W_{1,2} =$ "user posts with hashtag",
$W_{1,3} =$ "user retweets/favorites a tweet with hashtag".

The token in place l_2 splits to three tokens that enter places l_4, l_5 and l_6, with characteristics "development of hashtag network", "frequency distribution" and "bipartite network parameters", respectively.

$$Z_2 = \langle \{l_2\}, \{l_4, l_5, l_6\}, \frac{\begin{array}{c|ccc} & l_4 & l_5 & l_6 \end{array}}{\begin{array}{c|ccc} l_2 & W_{2,4} & W_{2,5} & W_{2,6} \end{array}} \rangle,$$

where
$W_{2,4} =$ "separate hashtags and create hashtags network",
$W_{2,5} =$ "calculate the hashtag frequency distribution",
$W_{2,6} =$ "create a bipartite network between hashtags and users".

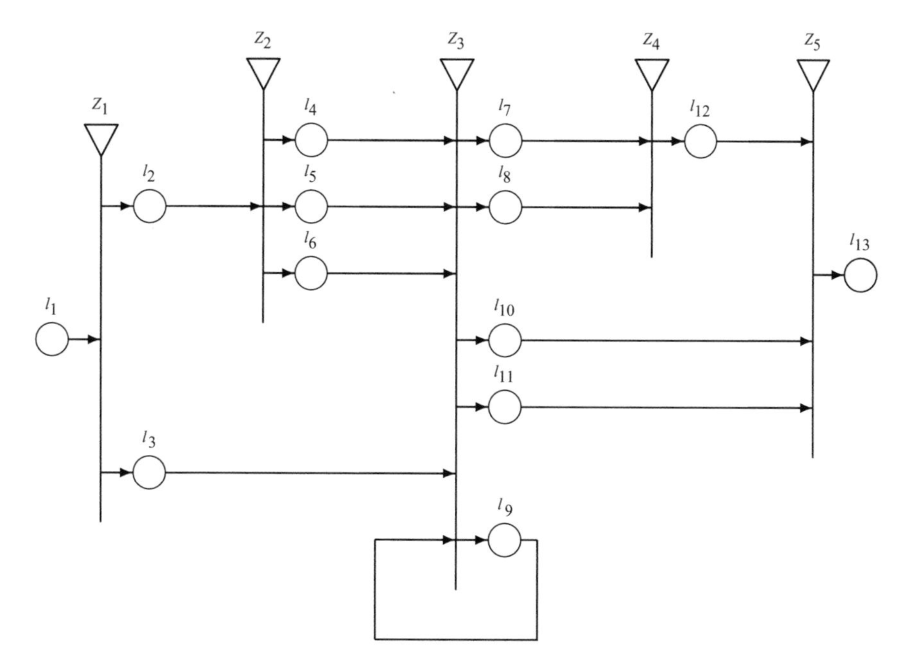

Fig. 1 Graphical representation of the GN-model

Moreover, the tokens from places l_3, l_4, l_5 and l_6 enter Z_3. The results from transition Z_3 go to l_7, l_8, l_9, $l_1 0$ and $l_1 1$. Token from place l_9 creates a loop which is used to calculated the modularity of multiple different settings (see Fig. 1). They have the respective characteristics: "shortest path calculation"; "network diameter calculation"; "modularity calculation"; "connected components determination" and "clustering coefficient estimation". The second transition of the model is described below:

$$Z_3 = \langle \{l_3, l_4, l_5, l_6, l_9\}, \{l_7, l_8, l_9, l_{10}, l_{11}\}, \begin{array}{c|ccccc} & l_7 & l_8 & l_9 & l_{10} & l_{11} \\ \hline l_3 & W_{3,7} & W_{3,8} & W_{3,9} & W_{3,10} & W_{3,11} \\ l_4 & W_{4,7} & W_{4,8} & W_{4,9} & W_{4,10} & W_{4,11} \\ l_5 & \mathit{false} & \mathit{false} & \mathit{false} & \mathit{false} & W_{5,11} \\ l_6 & W_{6,7} & W_{6,8} & W_{6,9} & W_{6,10} & W_{6,11} \\ l_9 & \mathit{false} & \mathit{false} & W_{9,9} & \mathit{false} & \mathit{false} \end{array} \rangle,$$

where
$W_{3,7} = W_{4,7} = W_{6,7} =$ "find the shortest path from a source node to all other nodes contained within the graph",
$W_{3,8} = W_{4,8} = W_{6,8} =$ "identify the network diameter",
$W_{3,9} = W_{4,9} = W_{6,9} =$ "calculate modularity of the network",
$W_{3,10} = W_{4,10} = W_{6,10} =$ "connected components",
$W_{3,11} = W_{4,11} = W_{5,11} = W_{6,11} =$ "average clustering coefficient".

Tokens from places l_7 and l_8 are combined in place l_{12} with characteristic "calculate average weight degree".

$$Z_4 = \langle \{l_7, l_8\}, \{l_{12}\}, \begin{array}{c|c} & l_{12} \\ \hline l_7 & true \\ l_8 & true \end{array} \rangle.$$

Tokens from places l_{10}, l_{11}, and l_{12}, enter the final transition Z_5, merging into place l_{13} which has the characteristic "visualize the output".

$$Z_5 = \langle \{l_{10}, l_{11}, l_{12}\}, \{l_{13}\}, \begin{array}{c|c} & l_{13} \\ \hline l_{10} & true \\ l_{11} & true \\ l_{12} & true \end{array} \rangle.$$

4 Data and Results

The incoming data to the GN-model was obtained with a custom build script through the Twitter API. In this particular work, the script has been developed to collect posts, which have in their content the hashtag *clexa*.[1] The total dataset of random sample of publicly available Twitter messages consists of 1823 nodes (representing hashtags) and 10134 edges defining their co-occurrence in a single tweet.

Figure 2 depicts the hashtags distribution data, which is part of the place l_4 in the GN model. This particular place provides information of which are the most widely used hashtags in the data sample in relation to the observable hashtag *clexa*. Expectedly the most frequently used hashtag correspond closely to the community of people. Noticeably, the collected hashtags observe a close distribution with respect to the Zipf's law [19].

With the goal being the investigation of the emergent semantic properties of the data sample by focusing on the relations of co-occurrence among hashtags we need to use a place on the GN-model to create the co-occurrence network, which in this case is labeled by l_4. The importance of such graphs stems from the fact that tagging is an inclusive process, meaning all users can develop and share hashtags publicly, and a large overlap among resources is often possible, thus the co-occurrence relations among hashtags expose certain semantic aspects, including hierarchical relations. In this respect, place l_4 is responsible for the formation of the co-occurrence network depicted as an undirected graph $G = (V, E)$ with V number of nodes and E number of links developed from a set of V hashtags. Each node $\vartheta \in V$ represents a hashtag from V. The edges define the different semantic associations (adjacency relations) of the hashtag formed in the post's tagging space, i.e. $e \equiv (v_i, v_j) \in E$ being the

[1] The particular hashtag *clexa* represents a specific sub-community of people, who are fans of the tv-series "The 100".

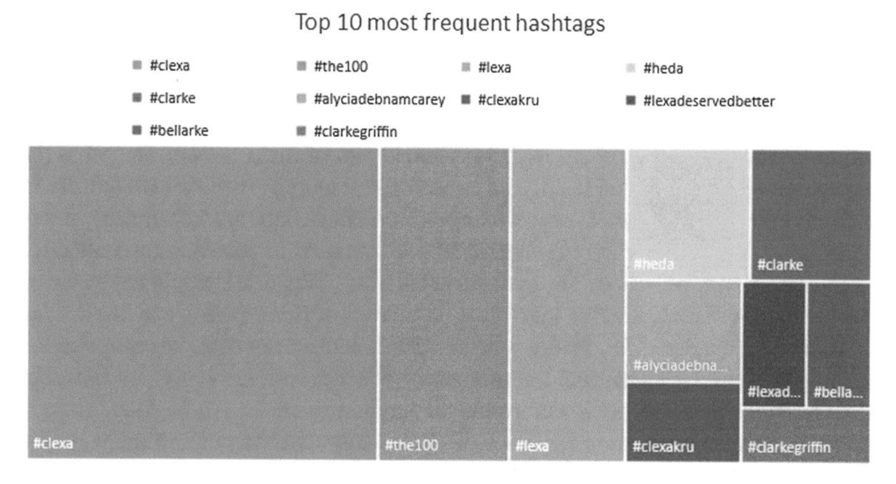

Fig. 2 The figure represents the 10 most used hashtags from the collected sample. They are all related to the topic-the community of fans representing *clexa*

association between the hashtags v_i and v_j. The network is weighted and the edge value depicts how often any two hashtags co-occur in a single post. In this case the length, or weight of the a path P will represent the sum of the weights of edges of P.

The next transition and place on the GN-model present the basic network metrics. In studying real networks it was discovered that they display relatively short path length values, known as the *small world* phenomenon [20]. In calculating the average path length, one needs to find the shortest path from a source node to all other nodes contained within the graph, where the average shortest path [21] is defined as:

$$a = \sum_{s,t \in V} \frac{d(s,t)}{n(n-1)} \tag{1}$$

with $d(s,t)$ representing the shortest path from s to t, and n being the number of nodes in the data set. The place l_7 is activated after the construction of the graph from the previous transition and the average path length is computed, which in our networks corresponds to 2.2. The result in this sense is consistent with the small world phenomenon and therefore indicating the hashtags are very interconnected.

The place l_{11} being responsible for the computation of the clustering coefficient measure is necessary part of the GN-model, since the coefficient provides an insight into how and whether the nodes tend to form cliques. The function examines the proportion of a node's neighbors tend to be also neighbors of each other. The output result shows, in this particular implementation, a relatively high clustering coefficient, similar to those in real networks, at 0.69, showcasing the existence of triad formation process.

The reader should note that tagging behavior varies significantly between communities consisting of people sharing a similar interest. The reasons include on the one side the particular design features of the technological system, and on the other the patterns of behavior emerging in a response to specific events. A user tagging a resource with a specific tag situates the resource in the context of the already similarly tagged shared resources. For example, certain hashtags can be considered "conversational", meaning the hashtag itself is valuable part of the message. Therefore the tag can serve as a indexical label, as a for additional communication possibilities in order to carry out an asynchronous many-person conversation. Furthermore, it can be utilized as a community organizational tool in terms of focusing the collective attention to a specific action. The distinctive language categories, employed by users, can function as a recognizable identification of level of participation and knowledge of other communities. To explore whether this is indeed the case, we examine the hashtag network by measuring the propensity of a network to be divided into separate relationship clusters. To determine the modularity class of a hashtag, place l_9 is responsible for computing the heuristic community detection algorithm [22]. The algorithm first assigns each node to a different community. Then, for every node v, v is moved to its neighbors u of v, where the observable gain is maximum. The processes is iterated until no further improvement can be achieved. Next, the nodes become the communities obtained from the previous step. The weights of the edges represent the sum of the weights of the links between nodes in the corresponding two communities. The algorithm iterates between the two phases until no improvement is possible, which is then considered the maximum modularity attained. This inherent iteration of the algorithm is assigned with a self-loop on the GN-model. While a high modularity indicate a dense connection between nodes within a module and sparse connection of nodes from different modules, our modularity results indicate a lower modularity (0.319, see Fig. 3).

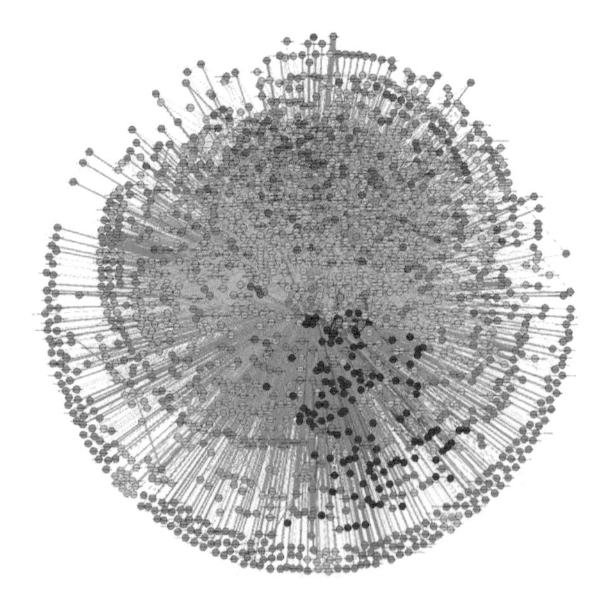

Fig. 3 The figure represents the separation of hashtags into different clusters. Each color depicts new cluster

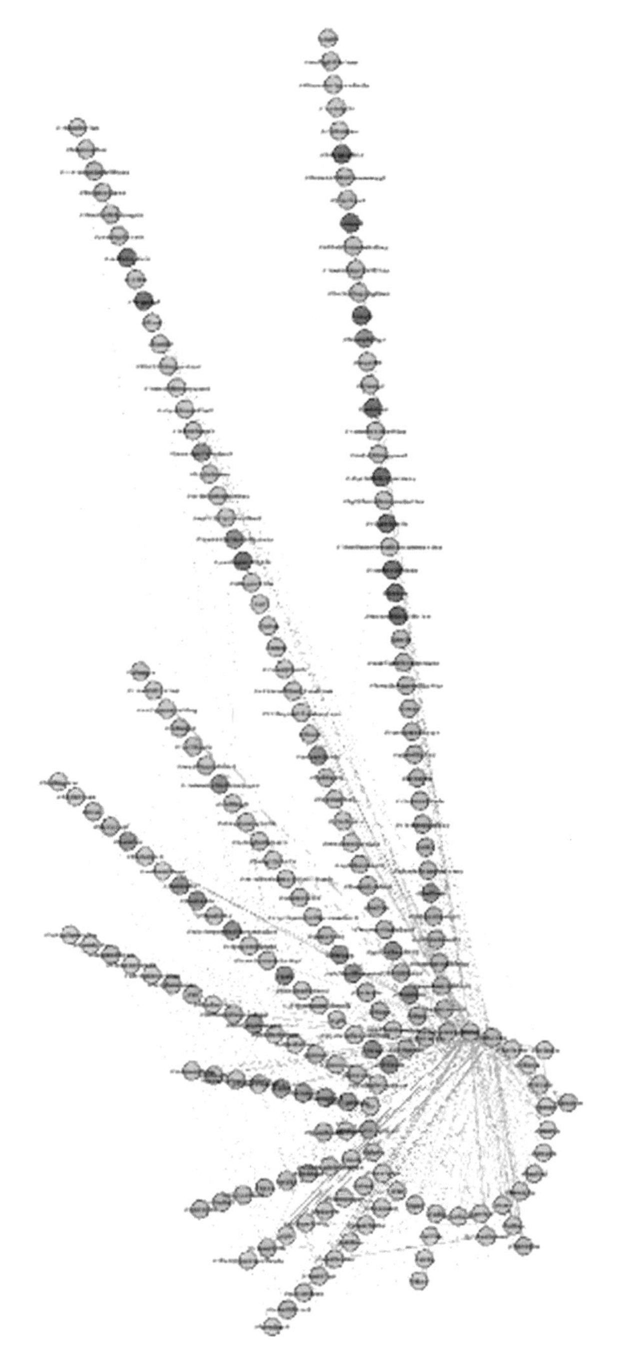

Fig. 4 The figure represents the most popular hashtags (from the data with hashtags appearing more than 10 times) and their modularity visualized on the basis of betweenness centrality

The developed GN can be further extended by replacing some of the places with another GN or transitions. For instance, the place l_9 for the modularity parameter can be replaced with another GN representing the modularity not only of the entire network, but with specific subnetworks and can provide new places with specific computational characteristics. In this case to better understand the connections between the hashtags in the different communities, we remove all nodes that have been used less than 10 times and recalculate the network parameters. We then create 4 main communities of hashtags with slightly lower modularity coefficient of 0.254 - depicted on Fig. 4. Then we can include further details by replacing places and transitions with GN, providing the ability to further drill down the data and make it more detailed. The same possibility is applied to the place l_{13} responsible for the visualization of the data. Substituting with a new GN transition can provide a better visualization capabilities by utilizing several different data visualization techniques. For instance, Figs. 3 and 4 implement two visualization methods - force-directed layout and axis based algorithms respectively. Further implementations, especially for the time-evolution analysis can include a place with transition for Fruchterman-Reingold or random graph layouts which can appended to the new GN model.

5 Conclusion

The proposed GN model can be incorporated in many different data mining scenarios for social networking data. While we focused only on the hashtag co-occurrence analysis an interesting endeavor will be to incorporate into the model the relationship between users, hashtags and locational information. This first step nevertheless provides the tools to extend the study and the GN to incorporate the process of hashtag evolution in a time-series environment.

Acknowledgements This work has been funded by the Bulgarian Science Fund under grants DFNI 02/10 and DFNI 12/5.

References

1. Lee, C.K., Barton, D.: Constructing glocal identities through multilingual writing practices on Flickr.com. Int. Multilingual Res. J. **5**(1), 39–59 (2011)
2. Hurlock, J., Wilson, M.L.: Searching Twitter: separating the Tweet from the Chaff. InICWSM, pp. 161–168 (2011 Jul 11)
3. Chang, H.C.: A new perspective on Twitter hashtag use: diffusion of innovation theory. Proc. Assoc. Inf. Sci. Technol. **47**(1), 1–4 (2010)
4. Romero, D.M., Meeder, B., Kleinberg, J.: Differences in the mechanics of information diffusion across topics: idioms, political hashtags, and complex contagion on twitter. In: Proceedings of the 20th International Conference on World Wide Web, pp. 695–704. ACM. (2011)

5. Brun, A., Moe, H.: Stuctural Layers of Communication on Twitter. In: Weller, K., Bruns, A., Burgess, A.J., Mahrt, M., Puschmann, C. (eds.) Twitter and Society. Peter Lang, New York, pp. 1528 (2014)
6. Boyd, D., Golder, S., Lotan, G.: Tweet, tweet, retweet: conversational aspects of retweeting on twitter. In: Proceedings of the 43rd Hawaii International Conference on System Sciences, Washington, DC: IEEE, pp. 110 (2010)
7. Naaman, M., Boase, J., Lai, C.H.: Is It really about me? message content in social awareness streams. In: Proceedings of the 2010 ACM conference on Computer Supported Cooperative Work, pp. 189–192. ACM, NY (2010)
8. Naaman, M., Becker, H., Gravano, L.: Hip and trendy: characterizing emerging trends on Twitter. J. Assoc. Inf. Sci. Technol. **62**(5), 902–18 (2011)
9. Zappavigna, M.: Searchable talk: the linguistic functions of hashtags. Soc. Semiotics **25**(3), 274–291 (2015)
10. Rightler-McDaniels, J.L., Hendrickson, E.M.: Hoes and hashtags: constructions of gender and race in trending topics. Soc. Semiot. **24**(2), 175–190 (2014)
11. Schmidbauer, H., Roesch, A., Stieler, F.: The 2016 US presidential election and media on instagram: who was in the lead? Comput. Human Behavior **30**(81), 148–60 (2018)
12. Tremayne, M.: Anatomy of protest in the digital era: a network analysis of Twitter and Occupy wall street. Social Movement Studies **13**(1), 110–26 (2014)
13. Starbird, K., Palen, L.: (How) will the revolution be retweeted?: information diffusion and the 2011 Egyptian uprising. In: Proceedings of the Acm 2012 Conference on Computer Supported Cooperative Work, pp. 7–16. ACM (2012 Feb 11)
14. Alexieva, J., Choy. E., Koycheva E.: Review and bibloigraphy on generalized nets theory and applications. In: Choy, E., Krawczak, M., Shannon A., Szmidt, E. (eds.) A Survey of Generalized Nets. Raffles KvB Monograph, no. 10, pp. 207–301 (2007)
15. Atanassov, K.: Generalized Nets. World Scientific, Singapore, London (1991)
16. Atanassov, K.: On Generalized Nets Theory. Prof. M. Drinov Academic Publ. House, Sofia (2007)
17. Atanassov, K.: On Generalized Nets Theory. Prof. M. Drinov Academic Publishing House, Sofia (2007)
18. Atanassov, K.: Applications of Generalized Nets. World Scientific, Singapore, London (1993)
19. Montemurro, M.A.: Beyond the ZipfMandelbrot law in quantitative linguistics. Phys. A: Stat. Mech. Appl. **300**(3–4), 567–78 (2001)
20. Watts, D.J.: Networks, dynamics, and the small-world phenomenon. Am. J. Soc. **105**(2), 493–527 (1999)
21. Poschko J.: Exploring Twitter Hashtags. arXiv preprint arXiv:1111.6553 (2011 Nov 28)
22. Blondel, V.D., Guillaume, J.L., Lambiotte, R., Lefebvre, E.: Fast unfolding of communities in large networks. J. Stat. Mech.: Theory Exp. **2008**(10), P10008 (2008)

On the Group Analysis of Differential Equations on the Group $SL(2, R)$

Georgi Kostadinov and Hristo Melemov

Abstract We get a method for constructing explicitly the exact solution of the initial value problem defined by the linear differential equation $\dot{g}(t) = A(t)g(t),\ \ g(0) = E$, where $g(t) \in SL(2, R)$ and $A(t)$ is a polynomial matrix belonging to the Lie algebra $sl(2, R)$ obeying some constraints. The analysis is carried out by using the nilpotent elements of $sl(2, R)$ and Lie algebraic techniques. The solution is presented in the form of finite product of exponentials.

1 Introduction

The aim of this paper is to analyse the existence of polynomial solutions of the initial value problem defined by the linear differential equation

$$\dot{g}(t) = A(t)g(t), \quad g(0) = E \tag{1}$$

where $g(t) \in G = SL(2, R)$, the special linear group and $A(t)$ is a polynomial matrix belonging to the Lie algebra $sl(2, R)$, (the tangent space at the unity E). Obviously (1) determines a curve in $sl(2, R)$

$$A(t) = dR^{-1}_{g(t)}\dot{g}(t),$$

where $R_{g(t)}$ denotes the right translation in G. We present a method for constructing the solution as a product of polynomial exponentials. Many investigations are devoted

G. Kostadinov · H. Melemov (✉)
University of Plovdiv Paisii Hilendarski, 24 Tzar Asen, 4000 Plovdiv, Bulgaria
e-mail: hristomelemov@yahoo.com

G. Kostadinov
e-mail: geokostbg@yahoo.com

K. Georgiev et al. (eds.), *Advanced Computing in Industrial Mathematics*,
Studies in Computational Intelligence 793,
https://doi.org/10.1007/978-3-319-97277-0_18

to different aspects of the problem (1), [2–4]. In [2] the solution is represented in an infinite product of exponentials:

$$g(t) = e^{P(t)}e^{P_1(t)}\dots e^{P_n(t)},$$

where

$$P(t) = \int_0^t A(s)\,ds, \dots, P_n(t) = \int_0^t A_n(s)\,ds$$

and

$$A_n = e^{-P_{n-1}(t)}A_{n-1}(t)e^{P_{n-1}(t)} - \int_{-1}^{0} e^{sP_{n-1}(t)}A_{n-1}(t)e^{-sP_{n-1}(t)}\,ds$$

In essence these formulae have a theoretical meaning. An alternative method presents the Magnus expansion:

$$g(t) = e^{\Omega(t)},$$

where the operator $\Omega(t)$ is expressed as an infinite series $\Omega(t) = \sum_{k=1}^{\infty} \Omega_k(t)$. These methods are widely applied for obtaining numerical algorithms in [1].

The investigations of Riccati equation and Schrodinger equation lead to the similar problem [3, 4, 6].

2 Relations in $sl(2, R)$ and $sl(2, R)$

Choose the following basis in $sl(2, R)$:

$$H = \begin{pmatrix} 1 & 0 \\ 0 & -1 \end{pmatrix},\ X = \begin{pmatrix} 0 & 1 \\ 0 & 0 \end{pmatrix},\ Y = \begin{pmatrix} 0 & 0 \\ 1 & 0 \end{pmatrix}$$

with commutators $[X, Y] = H,\ [H, X] = 2X,\ [H, Y] = -2Y.$

To the nilpotent elements X and Y correspond 1-parameter subgroups $\exp tX = \begin{pmatrix} 1 & t \\ 0 & 1 \end{pmatrix}$ and $\exp tY = \begin{pmatrix} 1 & 0 \\ t & 1 \end{pmatrix},\quad t \in R.$

The adjoint representation of the Lie group G in the corresponding Lie algebra is defined as a homomorphism

$$Ad : G \to Aut(g),\ \ Ad : W \to gWg^{-1},\ \ g \in G,\ \ W \in sl(2, R).$$

In this paper, we suggest a method for finding polynomial solutions as a product of polynomial exponentials $g(t) = g_1(t)\dots g_k(t)$ in the case when $A(t)$ is a polynomial matrix obeying some constraints.

Lemma 1 *The adjoint representation of $G = SL(2, R)$ in $sl(2, R)$ has the form*

$$Ad(\exp tX) = \begin{pmatrix} 1 & 0 & t \\ -2t & 1 & -t^2 \\ 0 & 0 & 1 \end{pmatrix} \quad \text{and} \quad Ad(\exp tY) = \begin{pmatrix} 1 & -t & 0 \\ 0 & 1 & 0 \\ 2t & -t^2 & 1 \end{pmatrix}, \quad t \in R.$$

Proof In order to establish these formulae we write, [5]

$$\begin{pmatrix} 1 & t \\ 0 & 1 \end{pmatrix} \begin{pmatrix} e^\tau & 0 \\ 0 & e^{-\tau} \end{pmatrix} \begin{pmatrix} 1 & -t \\ 0 & 1 \end{pmatrix} = \begin{pmatrix} e^\tau & -te^\tau + te^{-\tau} \\ 0 & e^{-\tau} \end{pmatrix},$$

$$\begin{pmatrix} 1 & t \\ 0 & 1 \end{pmatrix} \begin{pmatrix} 1 & \tau \\ 0 & 1 \end{pmatrix} \begin{pmatrix} 1 & -t \\ 0 & 1 \end{pmatrix} = \begin{pmatrix} 1 & \tau \\ 0 & 1 \end{pmatrix},$$

$$\begin{pmatrix} 1 & t \\ 0 & 1 \end{pmatrix} \begin{pmatrix} 1 & 0 \\ \tau & 1 \end{pmatrix} \begin{pmatrix} 1 & -t \\ 0 & 1 \end{pmatrix} = \begin{pmatrix} 1 + t\tau & -t^2\tau \\ \tau & 1 - t\tau \end{pmatrix}.$$

Differentiating with respect to τ, at $\tau = 0$ we get respectively

$$Ad(\exp tX)H = \begin{pmatrix} 1 & -2t \\ 0 & -1 \end{pmatrix} = \begin{pmatrix} 1 & 0 \\ 0 & -1 \end{pmatrix} - 2t \begin{pmatrix} 0 & 1 \\ 0 & 0 \end{pmatrix} = H - 2tX,$$

$$Ad(\exp tX)X = X,$$

$$Ad(\exp tX)Y = \begin{pmatrix} t & -t^2 \\ 1 & -t \end{pmatrix} = t \begin{pmatrix} 1 & 0 \\ 0 & -1 \end{pmatrix} - t^2 \begin{pmatrix} 0 & 1 \\ 0 & 0 \end{pmatrix} + \begin{pmatrix} 0 & 0 \\ 1 & 0 \end{pmatrix} = tH - t^2X + Y.$$

Thus the first formula is proved. For the second formula we proceed by the same way. It is important to emphasize that in these calculations the argument t is supposed to be constant and $\begin{pmatrix} 1 & t \\ 0 & 1 \end{pmatrix}$ is just an element of $SL(2, R)$. In the following we will consider as well elements of the form $\begin{pmatrix} 1 & t^k \\ 0 & 1 \end{pmatrix}$, k-positive integer.

Lemma 2 *For the adjoint representation of $G = SL(2, R)$ in $sl(2, R)$ we have properties:*

(i) $Ad(\exp tX)$ *and* $Ad(\exp tY)$ *are 1-parametric subgroups of transformations of* $sl(2, R)$,

(ii) $Ad(\exp tX)^{-1} = Ad(\exp -tX) = \begin{pmatrix} 1 & 0 & -t \\ 2t & 1 & -t^2 \\ 0 & 0 & 1 \end{pmatrix}.$

Proof By direct calculations we get:

$$Ad(\exp tX).Ad(\exp sX) = Ad(\exp(t+s)X).$$

Our assertions follow immediately.

3 Differential Equations on $G = SL(2, R)$

As is well known the curves $g_1(t) = exptX$ and $g_2(t) = exptY$, $(X, Y \in sl(2, R))$ satisfy respectively the equations:

$$\dot{g}(t) = \begin{pmatrix} 0 & 1 \\ 0 & 0 \end{pmatrix} g(t) \ \ and \ \ \dot{g}(t) = \begin{pmatrix} 0 & 0 \\ 1 & 0 \end{pmatrix} g(t).$$

Proposition 1 *The curve $g(t) = \exp tX \exp tY$ is a solution of the equation:*

$$\dot{g}(t) = B(t)g(t)$$

with $B(t) = X + Ad(\exp tX)Y$.

Proof We have

$$\dot{g}(t) = \dot{g}_1(t)g_2(t) + g_1\dot{g}_2(t)$$

$$\dot{g}(t)g^{-1}(t) = \dot{g}_1(t)g_1^{-1}(t) + g_1(t)\dot{g}_2(t)g_2^{-1}(t)g_1^{-1}(t) = X + Ad(\exp tX)Y.$$

From this $B(t) = X + Ad(\exp tX)Y$.

More generally the curve

$$g(t) = \exp tX_1 \exp tX_2 \cdots \exp tX_n \tag{2}$$

is a solution of the equation $\dot{g}(t) = A(t)g(t)$ with

$$A(t) = X_1 + Ad(\exp tX_1)X_2 + \cdots + Ad(\exp tX_1 \exp tX_2 \cdots \exp tX_{n-1})X_n. \tag{3}$$

Here X_i, $i = 1, 2 \cdots, n$ stand for the nilpotent elements X and Y of $sl(2, R)$.

Let us have $A(t) = \begin{pmatrix} a(t) & b(t) \\ c(t) & -a(t) \end{pmatrix}$,
where $a(t) = a_0 + a_1 t + \cdots + a_k t^k$, $b(t) = b_0 + b_1 t + \cdots + b_l t^l$ and $c(t) = c_0 + c_1 t + \cdots + c_m t^m$.
We identify this polynomial matrix with the curve in $sl(2, R)$,

$$\alpha(t) = a(t)H + b(t)X + c(t)Y = (a(t), b(t), c(t)). \tag{4}$$

By using the adjoint representation of the group $G = SL(2, R)$ in the algebra $sl(2, R)$ we may state the following necessary conditions for the existence of polynomial solutions of the form (2).

Theorem 1 *If the system (1) admits a polynomial solution (2), the following conditions are fulfilled:*

(i) The leading terms of $A(t)$ *satisfy the equality:* $\frac{a_k t^k}{c_m t^m} = \frac{b_l t^l}{-a_k t^k}$*,*
(ii) The initial value of the curve $\alpha(t)$ *is* $\alpha(0) = bX + cY$*, where* $b, c \in R$*,*
(iii) The derivatives of the vector function $\alpha : R \to sl(2, R)$ *satisfy at* $t = 0$ $\alpha^{(j)} = (*, 0, 0)$ *or* $\alpha^{(j)} = (0, *, *)$*, where* $j = 1, \cdots, n$*. The stars stand for real numbers.*

Proof Since $Ad(\exp tX)$ and $Ad(\exp tY)$ have the form in Lemma 2, from this (i) and (ii) follows immediately.

We proceed to prove (iii) by induction with respect to the number of factors in (3). We have

$$Ad(\exp tX)Y = \begin{pmatrix} 1 & 0 & t \\ -2t & 1 & -t^2 \\ 0 & 0 & 1 \end{pmatrix} \begin{pmatrix} 0 \\ 0 \\ 1 \end{pmatrix} = \begin{pmatrix} t \\ -t^2 \\ 1 \end{pmatrix}$$

and according to (4), we have $A_1(t) = \begin{pmatrix} t \\ -t^2 + 1 \\ 1 \end{pmatrix} = tH + (-t^2 + 1)X + Y.$

The powers by H are odd, respectively by X and Y are even. Next

$$Ad(\exp tY)Ad(\exp tX)Y = \begin{pmatrix} 1 & -t & 0 \\ 0 & 1 & 0 \\ 2t & -t^2 & 1 \end{pmatrix} \begin{pmatrix} t \\ -t^2 + 1 \\ 1 \end{pmatrix} =$$

$$= t \begin{pmatrix} 1 \\ 0 \\ 2t \end{pmatrix} + (-t^2 + 1) \begin{pmatrix} -t \\ 1 \\ -t^2 \end{pmatrix} + \begin{pmatrix} 0 \\ 0 \\ 1 \end{pmatrix},$$

$$A_2(t) = \begin{pmatrix} t^3 \\ -t^2 + 1 \\ t^4 + t^2 + 2 \end{pmatrix}.$$

Suppose $A_n(t) = \begin{pmatrix} a(t) & b(t) \\ c(t) & -a(t) \end{pmatrix}$, with $a(t) = \sum_{k=1}^{n} a_{2k-1} t^{2k-1}$ and $b(t) = \sum_{k=0}^{n} b_{2k} t^{2k}$, $c(t) = \sum_{k=1}^{n} c_{2k-2} t^{2k-2}$.

In order to find $A_{n+1}(t)$ we calculate

$$Ad(\exp tY)(A_n(t)) = \begin{pmatrix} 1 & -t & 0 \\ 0 & 1 & 0 \\ 2t & -t^2 & 1 \end{pmatrix} \cdot \begin{pmatrix} \cdots + a_{2k-1}t^{2k-1} \\ \cdots + b_{2k}t^{2k} \\ \cdots + c_{2k-2}t^{2k-2} \end{pmatrix} = \begin{pmatrix} \cdots - b_{2k}t^{2k+1} \\ \cdots + b_{2k}t^{2k} \\ \cdots - b_{2k}t^{2k+2} \end{pmatrix}.$$

The differentiation at $t = 0$ finishes the proof.

Corollary 1 *If the system (1) admits a polynomial solution of the form:*

$$g(t) = \exp t\beta_1 X \exp t\gamma_1 Y \cdots \exp t\beta_k X \exp t\gamma_k Y$$

then the curve $\alpha(t)$ satisfies $\alpha(0) = bX + cY$, where $b = \sum_{i=1}^{k} \beta_i$, $c = \sum_{i=1}^{k} \gamma_i$.

Example 1 Let us have the problem (1) with

$$A(t) = \begin{pmatrix} -9t^3 & 9t^4 + 3t^2 + 2 \\ -9t^2 + 3 & 9t^3 \end{pmatrix}$$

Then $\alpha(t) = (-9t^3, 9t^4 + 3t^2 + 2, -9t^2 + 3)$, $\alpha(0) = (0, 2, 3)$,

$\alpha'(0) = (0, 0, 0)$, $\alpha''(0) = (0, 6, -18)$, $\alpha'''(0) = (-54, 0, 0)$, $\alpha'^{v}(0) = (0, 216, 0)$.

We have $\frac{9t^3}{9t^2} = t$,

$$A_1(t) = Ad(\exp tX)^{-1}(\alpha(t) - (0, 1, 0)) = (-3t, 1, -9t^2 + 3).$$

We denote $\alpha_1(t) = (-3t, 1, -9t^2 + 3)$. Next $\frac{-3t}{-9t^2} = \frac{1}{3t}$,

$$A_2(t) = Ad(\exp 3tY)^{-1}(\alpha_1(t) - (0, 0, 3)) = (0, 1, 0).$$

Thus

$$g(t) = \begin{pmatrix} 1 & t \\ 0 & 1 \end{pmatrix} \begin{pmatrix} 1 & 0 \\ 3t & 1 \end{pmatrix} \begin{pmatrix} 1 & t \\ 0 & 1 \end{pmatrix} = \begin{pmatrix} 1 + 3t^2 & 3t^3 + 2t \\ 3t & 1 + 3t^2 \end{pmatrix}.$$

Thus we get a recurrent way for constructing explicitly polynomial solutions. One may consider as well k-th order "tangent" vectors of $sl(2, R)$: $\tilde{X}(t) = \begin{pmatrix} 0 & t^k \\ 0 & 0 \end{pmatrix}$ and $\tilde{Y}(t) = \begin{pmatrix} 0 & 0 \\ t^k & 0 \end{pmatrix}$.

In this case we get polynomial curves of the form $\tilde{g}(t) = \exp t^{k_1} X_1 \exp t^{k_2} X_2 \cdots \exp t^{k_n} X_n$, where k_i, $i = 1, 2 \cdots, n$ are positive integers, but the Theorem is not true for the corresponding system. This situation requires an additional analysis.

We may formulate the following Proposition.

Proposition 2 *Let* $\tilde{X}(t) = \begin{pmatrix} 0 & t^k \\ 0 & 0 \end{pmatrix}$ *and* $\tilde{Y}(t) = \begin{pmatrix} 0 & 0 \\ t^l & 0 \end{pmatrix}$. *Let furthermore*

$$\tilde{g}_1(t) = \begin{pmatrix} 1 & t^{k+1} \\ 0 & 1 \end{pmatrix} \text{ and } \tilde{g}_2(t) = \begin{pmatrix} 1 & 0 \\ t^{l+1} & 1 \end{pmatrix}.$$

The curve $\tilde{g}(t) = \tilde{g}_1(t)\tilde{g}_2(t)$ *is a solution of the differential equation* $\dot{\tilde{g}}(t) = \tilde{B}(t)\tilde{g}(t)$ *with* $\tilde{B}(t) = (k+1)\tilde{X}(t) + Ad(\tilde{g}_1(t))(l+1)\tilde{Y}(t)$.

Proof First we note that

$$\tilde{g}_1(t) = \exp\{(k+1)\int \begin{pmatrix} 0 & t^k \\ 0 & 0 \end{pmatrix} dt\}$$

satisfies the equation $\dot{\tilde{g}}(t) = (k+1)\begin{pmatrix} 0 & t^k \\ 0 & 0 \end{pmatrix}\tilde{g}(t)$ and $\dot{\tilde{g}}_2(t)\tilde{g}_2^{-1}(t) = (l+1)\begin{pmatrix} 0 & 0 \\ t^l & 0 \end{pmatrix}$ $= (l+1)\tilde{Y}(t)$.

Consider the curve $\tilde{g}(t) = \tilde{g}_1(t)\tilde{g}_2(t)$. We have

$$\dot{\tilde{g}}(t)\tilde{g}^{-1}(t) = \dot{\tilde{g}}_1(t)\tilde{g}_1^{-1}(t) + Ad(\tilde{g}_1(t))\dot{\tilde{g}}_2(t)\tilde{g}_2^{-1}(t) = (k+1)\tilde{X}(t) + Ad(\tilde{g}_1(t))(l+1)\tilde{Y}(t).$$

This finishes the proof.

Example 2 Let us have the problem (1) with

$$A(t) = \begin{pmatrix} t^2 & -t^4 + 2t \\ 1 & -t^2 \end{pmatrix}.$$

From $\frac{t^2}{1} = \frac{-t^4}{-t^2} = t^2$ we conclude that the first matrix is $\tilde{g}_1(t) = \begin{pmatrix} 1 & t^2 \\ 0 & 1 \end{pmatrix}$.
From Lemma 1. by replacing with t^2 we get

$$Ad(\tilde{g}_1(t)) = \begin{pmatrix} 1 & 0 & t^2 \\ 2t^2 & 1 & -t^4 \\ 0 & 0 & 1 \end{pmatrix}.$$

Then using Lemma 2. we have

$$Ad(\tilde{g}_1(t))^{-1}(t^2, -t^4, 1) = \begin{pmatrix} 1 & 0 & -t^2 \\ -2t^2 & 1 & -t^4 \\ 0 & 0 & 1 \end{pmatrix}\begin{pmatrix} t^2 \\ -t^4 \\ 1 \end{pmatrix} = \begin{pmatrix} 0 \\ 0 \\ 1 \end{pmatrix}.$$

The curve $\alpha_2(t) = (0, 0, 1)$ corresponds to the 1-parametric group $\begin{pmatrix} 1 & 0 \\ t & 1 \end{pmatrix}$. From this we get the solution in the form

$$\tilde{g}(t) = \tilde{g_1}(t)\tilde{g_2}(t) = \begin{pmatrix} t^3 + 1 & t^2 \\ t & 1 \end{pmatrix}.$$

Comment. In the theory of homogeneous spaces $M = G/H$, H is the isotropic subgroup. The vector fields on M corresponding to the action of H vanish at the initial point. Here, we have $\tilde{X}(0)$, $\tilde{Y}(0)$ vanish and by that there arise changes in the behaviour of the corresponding curve $\tilde{\alpha}(t)$. Our approach has a certain advantage because it proposes the calculation of the exponentials only for the basic nilpotent elements of the Lie algebra.

Acknowledgements Research was partially supported by Fund FP17-FS-011, Fund Scientific Research, University of Plovdiv Paisii Hilendarski, Bulgaria.

References

1. Blanes, S., Casas, F.: Optimization of Lie group methods for differential equations. Future Generation Comput. Syst. **19**, 331–339 (2003)
2. Fer, F.: Resolution de lequation matritiel $\dot{u} = pu$ par produit dexponentielles. Bull Classe Sci. Acad. R. Belg. **44**, 818–829 (1958)
3. Turbiner, A.: Quasi-Exactlly-Sovable Problems and $sl(2)$ Algebra. Commun. Math. Phys. **118**, 467–474 (1988)
4. Turbiner A.: Lie-algebraic approach to the theory of polynomial solutions. Commun. Math. Phys., CPT-92/P.2679
5. Vilenkin N.: Special Function and Representation Group Theory. Moskow Nauka (1965)
6. Winternitz P.: Nonlinear action of Lie group and superposition principles for nonlinear differntial equations. Centre de recherche de mathematiques appliqees. Universite de Montreal, Montreal, Quebec. H3C307, CRMA-1044 (1981)

On Digital Watermarking for Audio Signals

Hristo Kostadinov and Nikolai L. Manev

Abstract We investigate the possibility of embedding watermarks robust against compression in musical audio files. The process of embedding and retrieving a watermark can be regarded as a binary communication channel. We investigate its statistic to give recommendation how to choose the embedding parameters and what error correcting codes to be used. The investigation covers the case of AAC and MP3 compression. The described method of embedding is based on a combination of key-dependable dither modulation and Haar wavelet transform. We analyze the whole process for its embedding capacitate and robustness. A practical method for choosing the embedding parameters is proposed.

1 Introduction

The goal of data hiding techniques is to provide a way to embed extra information within the original digital contents, without serious degradation of its quality. Data hiding techniques are formally divided into two classes: watermarking and steganography.

Digital watermarking is a data hiding technique of imperceptibly altering a digital object to embed a message about that object. The goal of watermarking is to prevent piracy or to prove the ownership, of course, with a trade-off between the embedding capacity, robustness and quality.

In many practical situations we do not need to provide strong security against removing or modification of the hidden message but it is very important to conceal its existence. *Steganography* is a technique of altering the object that assures

H. Kostadinov (✉)
Institute of Mathematics and Informatics, Bulgarian Academy of Sciences (IMI-BAS),
Acad. G. Bonchev Str., Bl. 8, 1113 Sofia, Bulgaria
e-mail: hristo@math.bas.bg

N. L. Manev
IMI-BAS, Acad. G. Bonchev Str., Bl. 8, 1113 Sofia, Bulgaria
e-mail: nlmanev@math.bas.bg

K. Georgiev et al. (eds.), *Advanced Computing in Industrial Mathematics*,
Studies in Computational Intelligence 793,
https://doi.org/10.1007/978-3-319-97277-0_19

this desired undetectable communication between partners, that is, no one but the intended recipient to be able to detect this altering.

Many audio file formats are used recently but compressed forms dominate in the way of distributing and storing audio content. **MP3** was the first commonly accepted format of compressed digital audio but modern ISO-MPEG codecs such as the **Advanced Audio Coding (AAC)** family began to dominate in recent years. Both are considered to be common audio signal manipulations. That is the reason to address both formats in our investigation.

The general structure of watermark embedding in audio files is given in Fig. 1. One can recognize three nested communication channels:

- outer binary channel - includes error control codes
- medium binary channel - dither modulation and wavelet transformation
- inner real number channel (in Fig. 2).

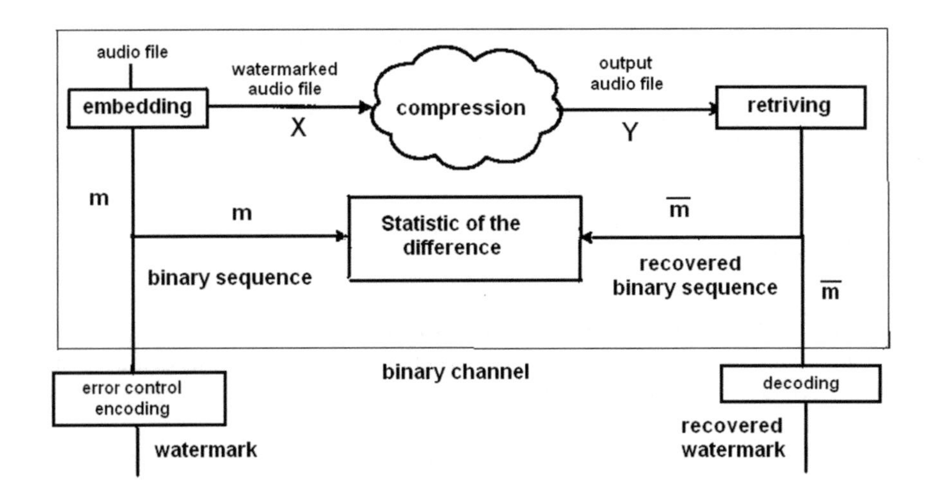

Fig. 1 The general structure of watermark embedding

Fig. 2 The inner (real number) channel

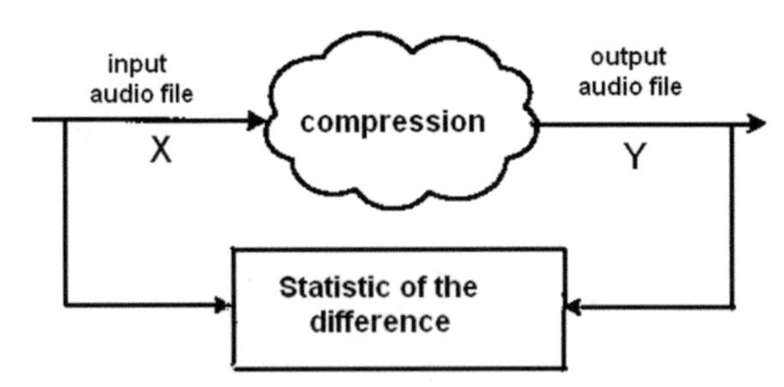

input: $\mathbf{X} = x_1, x_2, \ldots, x_i, \ldots$, values of the samples;
output: $\mathbf{Y} = y_1, y_2, \ldots, y_i, \ldots$, values after compression.

We aim after a comprehensive investigation of the embedding technique presented in Fig. 1 to propose a method of choosing its parameters so that to achieve acceptable probability of errors (according to characteristics of the watermark) without perceptible loss of quality of the audio file. The proper choice of error correcting code (ECC) is important both to achieve the aforesaid goals and to assure the maximum possible capacity for embedding. To make the choice we need to know the probability of errors, that is, the error characteristics of the medium channel presented by a rectangular in Fig. 1. Obviously, its error characteristics are function of parameters of embedding techniques and distortion due to compression. On the other hand the proper parameters of embedding techniques themselves are function of characteristic of compression. Hence, we need a good knowledge about characteristics of the inner channel presented in Fig. 2.

Our investigation consists of two parts:

- Collecting knowledge about statistical characteristics of the inner and medium channels presented in the above figures;
- Making conclusions about the most proper embedding parameters and how powerful error correcting codes to be used. (For a detail information about ECC we refer to [4].)

In this paper we describe a part of our study, namely, in the case when key-dependable dither modulation and Haar wavelet transformation are used in the embedding process. There are two reasons to study that case: first, the good properties of aforesaid transformations, and the second, the same embedding process is used in [2], thus, we have a base for comparison of the obtained results. Unfortunately, due to page limitations we have no possibility to present even a short review of numerous significant papers in the area and refer the reader to [2].

The remaining part of the paper is organized as follows. The necessary knowledge about wavelets and dither modulation are given in Sect. 2. In Sect. 3 we describe our investigations and present some results. Section 4 illustrates with examples some of our observations. The last section summarizes our conclusions and recommendations.

2 Preliminaries

2.1 Audio Compression Formats

There is much information about audio compression techniques in the Internet. Herein we give only several basic facts about MP3 and AAC that concern our study.

MP3 format. MP3 is commonly used name for MPEG-1 or MPEG-2 Audio Layer III lossy audio compression. It commonly achieves 75–95% reduction in size compared to CD quality digital audio. The CD digital audio represents music by 2 channels with 44,100 samples per second, each sample presented by 16 bits, that is, by $2 \times 16 \times 44,100 = 1\,411.2$ Kbits/s. MPEG-1 Audio Layer III standard supports

3 sampling frequencies of 32, 44.1 and 48 kHz and 14 bit-rates: 32, 40, 48, 56, 64, 80, 96, 112, 128, 160, 192, 224, 256 and 320 kbit/s. The most used bit-rate is 128 Kbits/s.

Although MP3 is still very popular amongst consumers most state-of-the-art media services use modern ISO-MPEG codecs such as the Advanced Audio Coding family which can deliver more features and a higher audio quality at much lower bit-rates compared to mp3. In a result the developer of MP3 in May 2017 terminated its licensing program and announces the "end" of MP3.

AAC format. Advanced Audio Coding is designed to be the successor of the MP3 format and it generally achieves better sound quality than MP3 at the same bit rate. AAC is the default audio format for the most modern devices and platforms. It supports sampling frequencies from 8 to 96 kHz AAC uses the modified discrete cosine transform together with the window length of 1024 (or 960 points). As a result the length of a file after decompression is a integer multiple of 1024. **In order to compare it with the original file the first and the last block of 1024 samples have to be removed. The original file has to be cut to the largest multiple of 1024 less than its length**.

We have made experiments only with sampling frequency 44.1 KHz and bit-rates 128 Kbits/s.

2.2 Haar Wavelet Transform

Haar wavelet transform is a linear transformation based on multiplication of pairs of coordinates of a target vector (block of samples in our case) by the orthogonal matrix

$$\mathbf{H}_2 = \frac{1}{\sqrt{2}}\begin{pmatrix} 1 & 1 \\ 1 & -1 \end{pmatrix}, \qquad \mathbf{H}_2^{-1} = \mathbf{H}_2.$$

The target vector with length 2^n is separated in consecutive pairs and each one is multiplied by $\mathbf{H}_2$. The first coordinate of the resultant pair is put at the corresponding place among the first 2^{n-1} positions of the output vector. The second coordinate is put at the same place in the last 2^{n-1} positions. The described transformation is referred to as ***Haar wavelet decomposition (HWD) level 1***.

Haar wavelet decomposition ***level 2*** means applying the transformation twice but the second time it is applied only to the first 2^{n-1} coordinates of the output vector of the level 1. And so on, the last transformation in the case of Haar wavelet decomposition ***level k*** is applied only to the first 2^{n-k+1} coordinates of the output of level $k-1$. The output has the structure

$$(*) \qquad [\, cA_k,\ cD_k,\ \ldots,\ cD_3,\ cD_2,\ cD_1 \,],$$

where the length of $|cD_i| = 2^{n-i}$, $i = 1, \ldots, k$, and $|cA_k| = 2^{n-k}$. In analogy to the Fourier transformation the positions in the output vector are often called *sub-bands*. Haar wavelet decomposition is realized by function *wavedec* in Matlab (see [3]).

For our investigation we need to know how an "error" is propagated after applying of inverse transformation: the so called *Haar wavelet reconstruction (HWR)* (function *waverec* in Matlab). More precisely, if a level k Haar reconstruction is applied to a vector with 1 only in one position and zeros anywhere else how many positions are nonzero in the output and what their magnitudes are. Knowing the described above structure (*) of Haar wavelet decomposition level k output and the fact that it preserves energy it is not difficult to conclude that if the input vector is $\mathbf{e}_i = [\underbrace{0, \ldots, 0}_{i-1}, 1, 0, \ldots, 0]$, then

- if $1 \leq i \leq 2^{n-k}$, i.e., i belongs to the subset cA_k then the output vector has 2^k consecutive nonzero coordinates (the i-th such group) of values $(1/\sqrt{2})^k$ and zeros anywhere else.
- if $i \in cD_j$, $j = k, k-1, \ldots, 1$, then 2^j consecutive coordinates are nonzero, first half of them are equal to $(1/\sqrt{2})^j$, the second half are equal to $-(1/\sqrt{2})^j$.

2.3 *Quantization-Based Embedding Algorithms and Dither Modulation*

Quantization-based embedding algorithms are very popular because they are easy to be implemented, assure computational flexibility, high embedding rate and amenability to theoretical analysis. In [1] Chen and Wornell introduced a new quantization-based data hiding algorithm called *Quantization Indexed Modulation (QIM)*. Since then, several variations have been developed. We use a variant of QIM known as *Dither Modulation (DM)* introduced in the same paper [1]. DM's main characteristics are low-complexity by using scalar and uniform quantizers and the possibility of incorporating a private key in the embedding process.

The parameters of the used in our study DM are

- ***Quantizer*** $q(.)$ which is characterized by
  - set Q of points at distance Δ, which is called ***step*** of the quantizer;
  - $q : x \rightarrow q(x)$, where $q(x)$ is the closest to x point of Q.
- Dither (pseudo-random key-dependable) parameters:
  - $v(0)$ uniformly distributed in $[-\Delta/2, \Delta/2]$ pseudo-random sequence;
  - $v(1) = \begin{cases} v(0) + \Delta/2, & \text{if } v(0) < 0 \\ v(0) - \Delta/2, & \text{if } v(0) \geq 0 \end{cases}$.

Embedding Algorithm
Data:

- $\mathbf{x} = x_1, x_2, \ldots, x_i, \ldots$ signal points;
- $\mathbf{m} = m_1, m_2, \ldots, m_i, \ldots$ binary sequence (watermark);

- $\hat{\mathbf{x}} = \hat{x}_1, \hat{x}_2, \ldots, \hat{x}_i, \ldots$ output signal points.

Embedding: $\hat{x}_i = q(x_i + v(m_i)) - v(m_i)$.

Retrieving Algorithm
Data:

- $\mathbf{y} = y_1, y_2, \ldots, y_i, \ldots$, received signal points, which are the noisy $\mathbf{x}$ due to attack or other manipulations
- $\hat{\mathbf{m}} = \hat{m}_1, \hat{m}_2, \ldots, \hat{m}_i, \ldots$ - retrieved watermark.

Retrieving:

$$\hat{m}_i = b \in \{0, 1\} \text{ such that } y_i + v(b) - q(y_i + v(b)) \text{ is minimal.}$$

Remark. To perform the retrieving step it is not necessary to keep the sequences $v(0)$ and $v(1)$, but only the length of embedded sequence and the initial state (the "key" of embedding) of the pseudo-random generator.

3 Description of Investigations and Experiments

In accordance with our intentions given in the Introduction our study past through the following main research stages:

1. Choosing 30 musical files each of duration ≈ 4 minutes from five genres: classical, instrumental, pop, hard rock, soft rock. From our point of view the chosen files represent well the variety in each genre.
2. Determining statistical characteristics of the inner channel, that is, the distortion effect of compression.
3. Making conclusion about the value of Δ, the level of HWD, and in which sub-bands to embed.
4. Determining statistical characteristics of the middle (binary) channel.
5. Taking a decision about parameters of interleaving and error correcting codes.

3.1 Stage 1 and Stage 2

All files are .wav stereo format with sampling rate 44.1 KHz and resolution 16 bits. They are chosen randomly aiming only the files in each genre to differ as much as possible. To process file we use Matlab. Audio files are presented by $N \times 2$ matrices with entries in $[-1, 1]$, where the columns correspond to the left and right stereo channels. In Table 1 are given the mean *mean(abs)*, standard deviation *std(abs)*, and maximum *max(abs)* of absolute values of the elements of corresponding matrices

Table 1 Characteristics of different musical genres

Musical genres	Mean(abs)		Std(abs)		Max(abs)	
	1	2	1	2	1	2
Classic1	0.0626	0.0535	0.0668	0.0557	0.6184	0.5844
Classic2	0.0564	0.0535	0.0546	0.0524	0.6261	0.7410
Instrumental1	0.1004	0.1101	0.1019	0.1074	0.9199	0.9214
Instrumental2	0.1644	0.1519	0.1425	0.1330	0.9989	0.9989
HardRock1	0.0888	0.0890	0.0794	0.0801	0.9756	0.9938
HardRock2	0.1291	0.1194	0.1187	0.1107	0.9039	0.9039
SoftRock1	0.1319	0.1361	0.1266	0.1238	0.9204	0.9206
SoftRock2	0.1610	0.1666	0.1412	0.1458	0.9888	0.9888
Pop1	0.1360	0.1361	0.1295	0.1287	0.9989	0.9989
Pop2	0.1730	0.1782	0.1583	0.1620	0.9989	0.9989

for two files from each genre. The representative files of the genres are chosen to be maximum different from one another.

One can observe the differences in each genre as well as between genres. We would like to attract attention of the reader to the files with classical music. The larger difference with the files of the other genres can be also observed in next results given below.

The next step in our study is to collect some information about the distortion effect of compression, that is, to study the inner channel given in Fig. 2. For that purpose we compress and then decompress each files:

$$\text{file.wav} \Longrightarrow \text{file.aac} \Longrightarrow \text{file-modified.wav}$$

Then we compute and analyze the difference between matrices corresponding to file.wav and file-modified.wav. The obtained results are illustrated with Table 2. The file used for making Table 2 are the same as ones used for Table 1 and it demonstrates the following two facts observed during our experiments:

- The mean M and the standard deviation σ are approximately equal;
- Mutual relations among characteristics of a file, among files and genres in both tables are very similar. Hence, by determining only simple characteristics of the musical file we can make conclusions about distortions of compression (both for AAC and MP3).

One can achieve the aforesaid conclusions by carefully analyzing the methods used for AAC and MP3 compressions but such considerations are out of the range of this paper.

Table 2 The absolute value of differences for 128 Kbits/s AAC; $\alpha = M + \sigma$

Musical genre	Mean M	Std σ	Max	% diff > α	% diff > 2α
Classic1	0.0032	0.0034	0.1074	11.50	1.58
Classic2	0.0022	0.0024	0.0910	12.32	2.00
Instrumental1	0.0168	0.0179	0.3976	13.44	2.25
Instrumental2	0.0195	0.0215	0.5318	12.96	2.12
HardRock1	0.0120	0.0120	0.3629	13.62	1.40
HardRock2	0.0117	0.0117	0.3200	12.78	1.20
SoftRock1	0.0201	0.0198	0.3074	14.01	1.65
SoftRock2	0.0247	0.0245	0.4341	14.64	1.88
Pop1	0.0203	0.0216	0.6197	13.96	2.25
Pop2	0.0250	0.0297	1.1321	11.39	2.20

3.2 Stage 3

Process of embedding carried out in the middle channel consists of the following steps

Step E1. The sequence of samples of both stereo channels (or only of one) is divided into blocks of length $1024 = 2^{10}$ (or 512). If the length of the audio file is not multiple of 1024 we cut it or add zeros at the end.

Step E2. Applying Haar wavelet decomposition level k to each block of 1024 samples. Usually $k = 5$ or 6.

Step E3. Embedding 2^{10-k} bits from the watermark $\mathbf{m}$ into each block using dither modulation method with step Δ as it is described in Sect. 2.3. The embedding is into sub-bands cA_k or cD_k.

Step E4. Applying Haar wavelet reconstruction level k to each modified block of 1024 samples.

Step E5. Reassembling blocks as a file and compressing it into .aac or .mp3 file with bit rate 128 Kbits/s.

The parameters of the embedding process are step Δ and level k. After Stage 2 we have sufficient information to decide how to choose Δ. Depending on the watermark the acceptable probability is less than 10^{-3}, in many cases the upper bound is 10^{-4} or less. The values of the changes in Step 3 belong to $[-\Delta/2, \Delta/2]$ but after Step 4 they result in modifications with values in

$$\left[-\frac{\Delta}{2(\sqrt{2})^k}, \frac{\Delta}{2(\sqrt{2})^k}\right].$$

If the above interval is very small the compression will distort modifications. In such a case the bit error probability (BER) for medium channel is about 1/2 and no error correcting codes can help to recover the watermark. On the other hand if the interval is large it worsens the quality of the musical file and makes embedding too perceptible. Hence the choice of Δ has to assure the balance between two opposing goals: good quality of the modified file and lower BER.

The last two columns of Table 2 gives an idea how to choose Δ. They show for how many percents of the samples the distortion values are greater than α and 2α, respectively. In order the embedded bit to be retrieved the altering of the sample value has to be greater than the distortion value. Therefore to achieve BER about $10^{-2} \div 10^{-1}$ we need

$$\alpha \leq \frac{\Delta}{2(\sqrt{2})^k} \leq 2\alpha \qquad \Longrightarrow \qquad \Delta \in [2\alpha(\sqrt{2})^k,\ 4\alpha(\sqrt{2})^k]$$

For "noisy" musical files and genres Δ can be chosen about the right end of the interval or even greater in order to decrease BER. On the other hand for some files, mainly classical music and some instrumental, Δ must be close or even less than $2\alpha(\sqrt{2})^k$ to assure acceptable quality. But in this case very powerful error correcting codes has to be used which results in lower capacity of embedding. Note that with exception of classical music genre $\alpha \approx 0.035 \div 0.045$.

The above observations confirm the conclusion about classical music in [2].

Where to embed?

Let us first note that embedding into $cD_j,\ j < k$, sub-bands is equivalent to use of HWD level j. Hence the use of HWD level k means that we embed into cA_k or cD_k sub-bands. There is an widespread understanding that modifications to low-frequencies produce audible distortion. In [2] the authors recommend to embed in several sub-bands with highest correlation between amount of noise and average energy of sub-band. It is logically argued but leads to "buzz" effect. The embedding only into several sub-bands modifies only several samples in each block that introduces a periodical signal in the modified file heard as a buzz.

On the contrary, if we embed into all sub-bands of cA_k or cD_k all positions of the time-domain block are altered. The embedded signal correlates with the pseudo random sequences, it is very similar to white Gaussian noise, and it is perceived as a noise of the speaker or computer ventilator by the listener if the noise is enough strong.

Therefore we **recommend embedding into all sub-bands of** cA_k **or** cD_k.

Also we recommend the Haar wavelet transformation to be with level k equals at least 5, but there is no noticeable improvement for larger k.

At the end of this stage we carry out the steps E4 and E5 and produce *file-modified.aac*.

3.3 Stage 4 and Stage 5

To compute BER of the middle channel for the studied file we convert *file-modified.aac* to *file-aftercompression.wav*, retrieve watermark (with errors) following the procedure given below, and compare it with the original.

Table 3 The probability of error (BER)

Musical genre	E(Δ)	Δ = 0.15	Δ = 0.7	Δ = 0.9	Δ = 1.1
Classic1	0.1584	0.0522	0.0000	0.0000	0.0000
Classic2	0.0996	0.0235	0.0000	0.0000	0.0000
Instrumental1	0.7580	0.4232	0.0386	0.0170	0.0083
Instrumental2	0.9322	≈ 0.5	0.0883	0.0430	0.0210
HardRock1	0.5453	≈ 0.5	0.0105	0.0035	0.0013
HardRock2	0.5295	≈ 0.5	0.0105	0.0034	0.0012
SoftRock1	0.9028	≈ 0.5	0.0685	0.0310	0.0150
SoftRock2	1.0881	≈ 0.5	0.0892	0.0441	0.0220
Pop1	0.9481	≈ 0.5	0.0412	0.0182	0.0092
Pop2	1.2488	≈ 0.5	0.1034	0.0646	0.0432

Retrieving the watermark:

Step D1. Delete the first 1024 samples for both stereo channels of *file-aftercompression.wav* and then carry out step E1.

Step D2. Do step E2.

Step D3. For each block retrieve 2^{10-k} bits from the positions cA_k (or cD_k if they are target for embedding) using retrieving algorithm described in Sect. 2.3.

Step D4. Reassemble retrieved pieces of bits into a binary sequence in order to compare with the original watermark.

Table 3 presents the probability of error per bit for the same audio files used for Tables 1 and 2. The results are obtained under the following conditions

- HWD level 5 over blocks of 1024 bits;
- Embedding into $1 \div 32$ sub-bands, i.e., into cA_5;
- 128 Kbits/s ACC compression;
- $E(\Delta) = 4\alpha(\sqrt{2})^5$ is given for comparison with the used values of Δ (given in the top of the columns).

The red color in the table means that the quality of the modified file is bad and unacceptable, the yellow color - medium quality, and the green color - good and acceptable quality. The table as well as experiments with the other files confirm our conclusion in Sect. 3.2 that choosing Δ with values close to E(Δ) we can achieve BER of order 10^{-2}. With not very complex error correcting codes the BER can be decreased to 10^{-4} or less. In some cases there is no need of codes: 10^{-2} turn out to be sufficient the watermark to fulfill its function.

The capacity

If we embed in both channels of the target audio file by using HWD level k and embedding in cA_k or cD_k the amount of embedded bits is

$$(2 \times \text{sampling rate} \times 2^{10-k})/2^{10} = \text{sampling rate}/2^{k-1} \quad \text{bits per second.}$$

If we use ECC with rate R the capacity is

$$\frac{\text{sampling rate}}{2^{k-1}} \times R \quad \textbf{information bits per second}.$$

Even when $R = 1/5$ with $k = 5$ we achieve 551 information bits/s, or 132 300 information bits in the whole file. This amount is equivalent of a 128×128 pixels gray-scale picture.

The errors in the middle channel tend to group in bursts. For example, ≈700 zeros and then an interval of length $\approx 200 \div 400$ with many ones. (Demonstrates behavior of Markov chain.) Hence the use of interleaving and/or burst error correcting codes (e.g., Reed-Solomon codes over a large finite field) is a good decision.

In [2] it is proposed to use a combination of LDPC codes with very long repetition codes. Such an approach decreases the capacity and requires a modification of the algorithm for retrieving watermark (Sect. 2.3). Although there are no limitation in time and computing resources for recovering the watermark it is not necessary to involve very complex ECC design. Its choice is also function of the characteristic of the watermark. In the examples given in the next section the retrieved watermark can be doubtless used for proving ownership although very week codes are used.

4 Examples

In this section we give two examples of embedding a gray-scale picture into a musical audio file. In the both examples we use as a target file "Classical 2" - a piano performance having many silent passages that makes modifications more perceivable. This file is a representative of the most inappropriate for embedding audio files. Although the used error correcting codes are very week the qualities of the both retrieved watermark and marked audio file are very good. These examples demonstrate the importance of proper choice of Δ and embedding place. In both examples $\Delta = 0.08$, which is less than but close to $E(\Delta)$. Example 2 illustrates the influance of interleaving.

Example 1 The embedded watermark is the 102×102 pixels photo of Lenna shown in Fig. 2a. After transformation into a $\{0, 1\}$-sequence of length 83,232, it is repeated 7 times to construct the embedded sequence of 582,624 bits. The repetition is equivalent of use of $[7, 1, 1]$ code with some form of interleaving as far as each bit is repeated after 83,232 bits. The embedding is in $1 \div 32$ sub-bands. The decoding algorithm consists of choosing the symbol (0 or 1) dominated in the codeword. The probability achieved after decoding is **BER = 0.0062** and the retrieved image is shown in Fig. 3b. The quality of the modified audio file is very good - the embedded data is imperceivable.

If the embedding is in $33 \div 64$ sub-bands, i.e. in cD_5, we achieve **BER = 0.0004** (see Fig. 3c), but the quality of the modified audio file is slightly worse. Nonetheless it remains acceptable good.

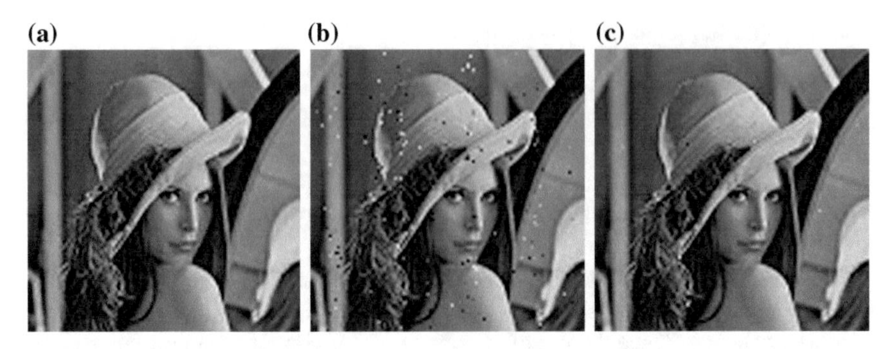

Fig. 3 **a** The original; **b** embedding in 1÷32 sub-bands; **c** in 33÷64 sub-bands

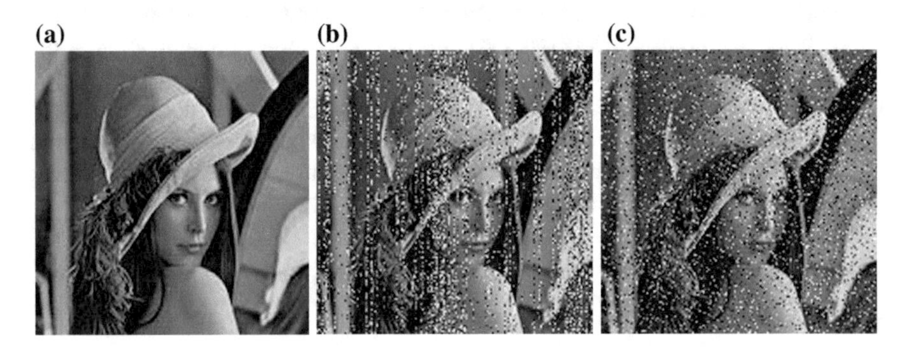

Fig. 4 **a** Original watermark; **b** without interleaving; **c** 177 × 3385 interleaving

Example 2 In this example the target audio file is also Classical2.wav and the watermark is Lenna's photo but with a larger size of 158 × 158 pixels. We use the [15, 5, 7] BCH code with zeros $\beta, \beta^2, \beta^3, \beta^4, \beta^5, \beta^6$, where β is the primitive element of $GF(16)$ with minimal polynomial $x^4 + x + 1$. The generating polynomial of the code is $g(x) = x^{10} + x^8 + x^5 + x^4 + x^2 + x + 1$ and the code can correct up to 3 errors in codeword. Its larger coding rate of 1/3 enables us to increase the amount of embedded data more than two times in comparison with Example 1 but the price we pay is a larger BER and worse quality of the retrieved image.

We performed embedding in two variants: without and with 177 × 3385 interleaving. The values of BER are relatively large 0.1105 and 0.0787, respectively. The original and the retrieved watermarks are shown in Fig. 4.

5 Conclusion

The results of our investigations can be summarized as follows:

- Embedding in all sub-bands of cA_k (or cD_k) improves audio quality.

- The proposed approach to choosing Δ is more practical than the using SNR. The choice of Δ close to $E(\Delta)$ leads to BER in $[10^{-2}, 3.10^{-2}]$. These values are typical for telephone channels and there are many error correcting codes with low complexity designed for such channels.
- The better audio quality requires smaller Δ and more powerful codes to be used. Errors in middle channel demonstrate inclination towards grouping. (More exactly, the probabilistic model of the channel is Markov chain.) Thus, the use of interleaving and burst error correcting codes is recommended.
- The capacity in the case of HWD level k and ECC with rate R is

$$\frac{\text{sampling frequency}}{2^{k-1}} \times R \quad \textbf{information bits per second.}$$

Acknowledgements This work was partially supported by the National Science Fund of Bulgaria under Grant DFNI-I02/8.

References

1. Chen, B., Wornell, G.W.: Quantization index modulation: a class of provably good methods for digital watermarking and information embedding. IEEE Trans. on Inf. Theory **47**(4), 1423–1443 (2001)
2. Martinez-Noriega, R., Nakano, M., Kurkoski, B., Yamaguchi, K.: High payload audio watermarking: toward channel characterization of MP3 compression. J. Inf. Hiding Multimedia Signal Proc **2**(2), 91–107 (2011)
3. MathWorks, https://www.mathworks.com/help/wavelet/index.html
4. Pless, V.S., Huffman, W.C. (eds.): Handbook of Coding Theory. Elsevier, Amsterdam (1998)

Numerical Evaluation of Fokas' Transform Solution of the Heat Equation on the Half-Line

Mohamed Lachaab, Peter R. Turner and Athanassios S. Fokas

Abstract We propose an efficient numerical approach to evaluate the infinite oscillatory integral for the heat equation's transform solution on the half-line that was recently introduced by Fokas. The proposed approach consists of first bounding the tail of the integral by making use of the partitioning-extrapolation method, then applying the trapezoidal rule with repeated Richardson extrapolation (Romberg integration) to estimate the resulting finite oscillatory integral to any desired accuracy. Our computational results show that, in the case of the slow integrand decay, our method improves the estimation of the integral's tail and the solution in general over that obtained by a direct truncation, and the use of Romberg integration reduces the number of function evaluations than that of MATLAB's adaptive subroutine.

1 Introduction

A new method has been recently introduced by Fokas (one of the authors) for solving a large class of PDEs. The new method [3–5], which is referred to as the unified transform, contains the classical methods as special cases and gives solutions to many PDEs where the classical methods cannot. The key to this method is the global relation which combines specified and unknown values of the solution or its derivatives on the boundary. This method, however, arises a problem: the transform solution is

M. Lachaab (✉)
Department of Engineering and Computer Science,
York College of Pennsylvania, York, PA 17403, USA
e-mail: mlachaab@ycp.edu

P. R. Turner
Institute for STEM Education, Clarkson University,
305B Bertrand H. Snell Hall, Potsdam, New York 13699, USA
e-mail: pturner@clarkson.edu

A. S. Fokas
Department of Applied Mathematics and Theoretical Physics,
Cambridge University, Cambridge CB3 0WA, UK
e-mail: T.Fokas@damtp.cam.ac.uk

K. Georgiev et al. (eds.), *Advanced Computing in Industrial Mathematics*,
Studies in Computational Intelligence 793,
https://doi.org/10.1007/978-3-319-97277-0_20

an infinite contour integral in the complex plane that needs to be evaluated numerically. For the heat equation on the half-line, for example, this method constructs the solution $q(x, t)$ as an integral in the complex plane which involves an x-transform of the initial condition and a t-transform of the boundary condition.

In this paper, we propose an efficient numerical evaluation of Fokas' transform solution of the heat equation. Specifically, after a suitable contour's deformation and parameterization, we bound the tail of the integral by partition-extrapolation and then apply the Romberg integration to the resulting finite interval integral. Our numerical approach gives a more accurate solution than that obtained by the simple truncation of the tail used by Flyer and Fokas [3] and [5], especially in the case of a slow decay of the integrand (small x values).

In the following sections, we first review Fokas' transform method of the integral representation of the solution of the heat equation on the half line. This is followed by a derivation of the integral representation of the solution of the heat equation on the real line which is convenient for numerical evaluation. In Sect. 4, we review some existing methods in the literature for estimating oscillatory integrals. In Sect. 5, we present and discuss our numerical estimation (which is the subject matter of this paper). Finally, we conclude the paper.

2 Fokas' Transform Method

We briefly describe Fokas' method in the case of the heat equation on the half-line with a Dirichlet boundary condition. We have:

$$q_t(x, t) = q_{xx}(x, t); \;\; 0 < x < \infty; \;\; 0 < t \leq T \tag{1}$$

with: $q(x, 0) = q_0(x)$, $0 \leq x$, and $q(0, t) = g_0(t)$, $0 \leq t$.

In the new approach, we start by rewriting Eq. (1) as a one-parameter family of PDEs:

$$\left(e^{-ikx+k^2t}q(x, t)\right)_t - \left(e^{-ikx+k^2t}(ikq(x, t) + q_x(x, t)\right)_x = 0, \qquad k \in C \tag{2}$$

We let $\widehat{q}_0(k)$ and $\widetilde{f}_j(k^2, t)$ represent the x-Fourier transform of the initial condition $q_0(x)$ and a t-transform of the boundary condition $g_0(t)$ respectively.

$$\widehat{q}_0(k) = \int_0^\infty e^{-ikx} q_0(x)dx, \qquad \text{Im}k \leq 0,$$

$$\widetilde{f}_j(k^2, t) = \int_0^t e^{k^2 s} \partial_x^j g_0(s)ds, \qquad j = 0, 1.$$

Multiplying the x-Fourier transform of the solution $q(x,t)$ by $e^{k^2 t}$, computing the time evolution of $\widehat{q}(k,t)$, and taking the inverse Fourier transform, the solution of the above heat equation becomes:

$$q(x,t) = \frac{1}{2\pi}\int_{-\infty}^{\infty} e^{ikx-k^2 t}\widehat{q}_0(k)dk - \frac{1}{2\pi}\int_{\infty}^{\infty} e^{ikx-k^2 t}(ik\widetilde{f}_0(k^2,t) + \widetilde{f}_1(k^2,t))dk, \tag{3}$$

By using Cauchy theorem and the analyticity properties of $\widetilde{f}_1(k^2,t)$, the solution becomes:

$$q(x,t) = \frac{1}{2\pi}\int_{-\infty}^{\infty} e^{ikx-k^2 t}\widehat{q}_0(k)dk - \frac{1}{2\pi}\int_{\partial D^+} e^{ikx-k^2 t}\left(2ik\widetilde{f}_0(k^2,t) + \widehat{q}_0(-k)\right)dk, \tag{4}$$

where: ∂D^+ is the positively oriented boundary to the domain: $D^+ = \{k \in C : \arg k \in [\frac{\pi}{4}, \frac{3\pi}{4}]\}$. Assuming zero initial condition ($q_0(x) = 0$), the solution becomes:

$$q(x,t) = -\frac{i}{\pi}\int_{\partial D^+} e^{ikx-k^2 t}k\widetilde{f}_0(k^2,t)dk \tag{5}$$

3 Representation of the Solution over the Real Line

We first evaluate $\widetilde{f}_0(k^2,t)$ in (5), then parameterize the contour, and finally convert the complex integral along the infinite contour to an integral over the real line. We assume that the boundary condition at $x = 0$ is given by $g_0(t) = \sin(\lambda t)$. This choice is made because this function is sufficiently smooth and compatible with the initial condition $q_0(x)$ at the origin, that is, $q_0(0) = g_0(0) = 0$. The t-transform becomes $\widetilde{f}_0(k^2,t) = \int_0^t e^{k^2 s}\sin(\lambda s)ds$ and the integration gives:

$$\widetilde{f}_0(k^2,t) = \frac{(\sin(\lambda t)k^2 - \lambda\cos(\lambda t))e^{k^2 t} + \lambda}{k^4 + \lambda^2}. \tag{6}$$

Therefore:

$$q(x,t) = -\frac{i}{\pi}\int_{\partial D^+} \frac{(\sin(\lambda t)k^3 - \lambda\cos(\lambda t)k + \lambda k e^{-tk^2})e^{ixk}}{k^4 + \lambda^2}dk \tag{7}$$

A suitable path is when integrating along different sectors in the analytic region instead of the angles $\pi/4$ and $3\pi/4$; for example using the angles $c = \pi/5$ and $\pi - c = 4\pi/5$. In this case, the contour of integration can be parameterized as $k(\alpha) = z\alpha$ for $\alpha \in [0,\infty)$ and $k(\alpha) = -\bar{z}\alpha$ for $\alpha \in (-\infty, 0]$ where $z = e^{ic}$ and $\bar{z}$ is the complexe conjugate of z. Using the above parameterization, the integral over the infinite contour can be converted to an integral over the real line as follows:

$$\int_{\partial D^+} f(k)dk = \int_{-\infty}^{0} f(k(\alpha))k^{'}(\alpha)d\alpha + \int_{0}^{\infty} f(k(\alpha))k^{'}(\alpha)d\alpha \tag{8}$$

After applying the contour parameterization and some algebra, the solution in (7) becomes:

$$q(x,t) = -\frac{i}{\pi}\left(\int_{0}^{\infty} f(\alpha,t)e^{izx\alpha}d\alpha - \int_{0}^{\infty} \overline{f(\alpha,t)e^{izx\alpha}}d\alpha\right), \tag{9}$$

where $f(\alpha,t) = \dfrac{z^4\sin(\lambda t)\alpha^3 - \lambda z^2\cos(\lambda t)\alpha + \lambda z^2\alpha e^{-z^2 t\alpha^2}}{z^4\alpha^4 + \lambda^2}$ and $\overline{f(\alpha,t)e^{izx\alpha}}$ is the complex conjugate of $f(\alpha,t)e^{izx\alpha}$. Subtracting the two integrals, we obtain:

$$\begin{aligned} q(x,t) &= -\frac{2}{\pi}\int_{0}^{\infty} Re\left(i * f(\alpha,t)e^{izx\alpha}\right)d\alpha \\ &= \frac{2}{\pi}\int_{0}^{\infty} |f(\alpha,t)|e^{-\sin(\pi/5)x\alpha}\sin\left(\cos(\pi/5)x\alpha + \phi\right)d\alpha, \end{aligned} \tag{10}$$

where $|f(\alpha,t)|$ is the modulus of $f(\alpha,t)$, $\tan\phi = \frac{\mathrm{Im}f(\alpha,t)}{\mathrm{Re}f(\alpha,t)}$, $\mathrm{Re}f(\alpha,t)$ and $\mathrm{Im}f(\alpha,t)$ are the real and imaginary parts of $f(\alpha,t)$. The square of the modulus, real part, and imaginary part of $f(\alpha,t)$ are given by the following formulas:

$$\begin{aligned} |f(\alpha,t)|^2 = \Big(&\sin^2(\lambda t)\alpha^6 - 2\lambda\{\cos(2c)\sin(\lambda t)\cos(\lambda t) - \sin(\lambda t)\cos(\sin(2c)t\alpha^2 \\ &+2c)e^{-\cos(2c)t\alpha^2}\}\alpha^4 + \lambda\{\cos^2(\lambda t) + e^{-2\cos(2c)t\alpha^2} - 2\cos(\lambda t) \\ &\cos(\sin(2c)t\alpha^2)e^{\cos(2c)t\alpha^2}\}\alpha^2\Big) \times \frac{1}{(\alpha^8 + 2\lambda^2\cos(4c)\alpha^4 + \lambda^4)} \end{aligned}$$

$$\begin{aligned} \mathrm{Re}f(\alpha,t) = \Big(&\sin(\lambda t)\alpha^7 - \lambda\{\cos(2c)\cos(\lambda t) - \cos(\sin(2c)t\alpha^2 + 2c)e^{-\cos(2c)t\alpha^2}\} \\ &\alpha^5 + \lambda^2\cos(4c)\sin(\lambda t)\alpha^3 - \lambda^3\{\cos(2c)\cos(\lambda t) - \cos(\sin(2c)t\alpha^2 \\ &-2c)e^{-\cos(2c)t\alpha^2}\}\alpha\Big) \times \frac{1}{(\alpha^8 + 2\lambda^2\cos(4c)\alpha^4 + \lambda^4)} \end{aligned}$$

$$\begin{aligned} \mathrm{Im}f(\alpha,t) = \Big(&\lambda\{\sin(2c)\cos(\lambda t) - \sin(\sin(2c)t\alpha^2 + 2c)e^{-\cos(2c)t\alpha^2}\}\alpha^5 + \lambda^2 \\ &\sin(4c)\sin(\lambda t)\alpha^3 - \lambda^3\{\sin(2c)\cos(\lambda t) + \sin(\sin(2c)t\alpha^2 \\ &-2c)e^{-\cos(2c)t\alpha^2}\}\alpha\Big) \times \frac{1}{(\alpha^8 + 2\lambda^2\cos(4c)\alpha^4 + \lambda^4)}. \end{aligned}$$

Equation (10) shows that the integrand is the product of a damping term, which contains an exponential, and a sine term, which accounts for the oscillatory behavior. The phase function of the oscillating term is nonlinear and has a complicated form, which indicates that the oscillations are irregular. In addition, since the coefficient $-\sin(c)x$ is negative, the exponential term tends to zero as $\alpha \to +\infty$, which shows that the integrand oscillates and decays exponentially in x. Therefore, the integral is an infinite sum of alternating and decreasing terms.

4 Some Methods for Estimating Oscillatory Integrals

The problem of numerical computation of infinite oscillatory integrals has recently attracted the attention of many authors [1, 2, 6, 8, 9, 11] since longman's original paper in 1960. The different approaches that have been proposed to compute such oscillatory infinite integrals mainly depend on the decay nature of the integrand. In the case of a fast decay, a truncation of the integration interval has been justified and the integral is treated as finite using standard integration tool. If the integrand decays so slowly, however, the truncation of the integration range would introduce too large error. Bettess [1] proposed a procedure that is related to the Newton-Cotes method and consists of selecting integration abscissae and then generating Lagrange polynomial for each abscissa. Iserles [6] used the Filon's method to estimate high oscillatory integrals on a finite small interval. The idea is to choose Legendre quadrature nodes in the interval (0, 1) and interpolate the decay function only (rather than the whole integrand) by a quadratic polynomial that can be integrated analytically. This method is only applicable to a finite small integration interval. In our case, the range of integration is infinite.

Lyness [9] used an alternative method based on the partition of the integration interval and integration of the integrand over each sub-interval thereby generating a sequence of alternating series that was treated by Euler transform with forward average operator. The same approach of integration then summation by extrapolation has been used by Blakemore et al. [2] who applied the ε-algorithm and by Sidi [11] who introduced the W-transform as an accelerator. Lucas and Stone [7] compared different extrapolation techniques and endpoint choices in dividing the integral and established the most efficient method for evaluating infinite integrals involving Bessel functions of any order. In this paper, we follow the method proposed by Lyness [9] to bound the tail given our integrand form. After bounding the tail, we estimate the integral on a finite interval by making use of the Romberg integration.

5 Numerical Evaluation of the Transform Solution

Analytical solution of $q(x, t)$ in (10) is impossible given the infinite range of integration and the complexity of the integrand. We therefore resort to the numerical estimation of this integral. Our approach for estimating such integral is first to bound the tail of the integrand and then estimate the integral on a finite interval.

5.1 Step 1: Bounding the Tail of the Integrand

The decay of the integrand in Eq. (10) depends on the exponential term $e^{-\sin(c)x\alpha}$. The smaller the damping x-values the slower the oscillation dies down, and the larger the x-values the quicker the oscillation dies down. Since x takes both small and large values, we use two comparative approaches to compute the tail of the integral. We first start with a truncation estimate of the integration interval, second we adopt Lyness's partition-extrapolation method.

5.1.1 Bounding the Tail by Truncation

From Eq. (10), we have:

$$\left|\int_b^\infty |f(\alpha,t)|\, e^{-\sin(c)x\alpha}\sin(\cos(c)x\alpha+\phi)d\alpha\right| \le 3\int_b^\infty e^{-\sin(c)x\alpha}d\alpha = \frac{3e^{-\sin(c)xb}}{\sin(c)x}, \tag{11}$$

Therefore, given a truncation error $\varepsilon = 3e^{-\sin(c)xb}/\sin(c)x$, we can find the values of $b = -\ln(\sin(c)x\varepsilon/3)/\sin(c)x$. Column 2 of Table 1 gives the values of b for a fixed truncation error level $\varepsilon = 0.5 * 10^{-12}$ and some different values of x between 0.25 and 5. Our choice of the error level $\varepsilon = 0.5 * 10^{-12}$ allows us to evaluate the integral in (10) between 0 and b for an accuracy of twelve figures. Column 2 shows that as x increases the b-values decrease and the interval of integration decreases at a slow logarithmic rate.

It is important to note that using the truncation method the values of b are found using the sum of the absolute value of all terms of the series. This way of finding the b-values works well in the case of rapid decay (large values of x), since the terms of the series become very small quickly. However, if the terms do not go down fast, this sum would be much bigger and the error would be much larger. Therefore, we use another suitable method: the partition-extrapolation method.

5.2 Bounding the Tail by Partition-Extrapolation

As showed by many researchers, a combination of partitioning and extrapolation is the most efficient way to accurately estimate infinite slowly converging integrals with oscillating integrands. Specifically, this approach is based first on dividing the range of integration into an infinite numbers of sub-intervals, secondly approximating the integral over the early finite number of sub-domains using a standard numerical quadrature, and finally estimating the infinite integral by an extrapolation technique of the partial sums. Several different combinations of extrapolation and endpoint choice have been used to evaluate such integrals. In most cases the zeros of the

Table 1 The b-Values given by the truncation and the partition-extrapolation methods

x	Truncation.method $b = -\frac{\ln(\sin(c)x\varepsilon/3)}{\sin(c)x}$	Partition.method $b = \frac{10.41\pi}{\cos(c)x}$	9 pt. Gauss	64 pt. trap rule
0.25	213.2784	161.6974	5.0199E−13	5.0207E−13
0.50	104.2807	80.8487	5.0196E−13	5.0203E−13
0.75	68.6007	53.8991	5.0190E−13	5.0197E−13
1.00	50.9611	40.4243	5.0181E−13	5.0189E−13
1.25	40.4652	32.3395	5.0171E−13	5.0178E−13
1.50	33.5142	26.9496	5.0158E−13	5.0165E−13
1.75	28.5766	23.0996	5.0142E−13	5.0150E−13
2.00	24.8909	20.2122	5.0125E−13	5.0132E−13
2.25	22.0362	17.9664	5.0105E−13	5.0112E−13
2.50	19.7609	16.1697	5.0082E−13	5.0090E−13
2.75	17.9055	14.6998	5.0058E−13	5.0064E−13
3.00	16.3640	13.4748	5.0030E−13	5.0038E−13
3.25	15.0633	12.4383	5.0001E−13	5.0008E−13
3.50	13.9514	11.5498	4.9969E−13	4.9977E−13
3.75	12.9900	10.7798	4.9935E−13	4.9943E−13
4.00	12.1506	10.1061	4.9899E−13	4.9906E−13
4.25	11.4116	9.5116	4.9860E−13	4.9868E−13
4.50	10.7560	8.9832	4.9820E−13	4.9827E−13
4.75	10.1706	8.5104	4.9777E−13	4.9784E−13
5.00	9.6446	8.0849	4.9731E−13	4.9739E−13

integrand are chosen as partition points for convenience. However, finding the exact zeros of (10) is difficult. In fact, these zeros are the roots of the following nonlinear equation:

$$\cos(c)x\alpha + \tan^{-1}\left(\frac{\mathrm{Im}f(\alpha,t)}{\mathrm{Re}f(\alpha,t)}\right) = z\pi, \quad z \in Z \tag{12}$$

or equivalently:

$$\begin{aligned}&\sin(\lambda t)\sin(\cos(c)x\alpha)\alpha^6 - \lambda\cos(\lambda t)\sin(\cos(c)x\alpha - 2c)\alpha^4 + \lambda^2\sin(\lambda t)\\&\sin(\cos(c)x\alpha + 4c)\alpha^2 - \lambda^3\cos(\lambda t)\sin(\cos(c)x\alpha + 2c) - \lambda\sin(\sin(2c)t\alpha^2 - \cos(c)\\&x\alpha + 2c)\alpha^4 e^{-\cos(2c)t\alpha^2} - \lambda^2\sin(\sin(2c)t\alpha^2 - \cos(c)x\alpha - 2c)e^{-\cos(2c)t\alpha^2} = 0,\end{aligned} \tag{13}$$

As stated earlier, we fellow the approach of Lyness which avoids the need to evaluate the exact zeros for the interval endpoints needed for integration. This method is based on the idea that if certain analytic information of the tail of the integrand is available (specifically the distance between consecutive zeros $x_{j+1} - x_j$ tends to a finite limit whose numerical value is available), then any choice of interval endpoints in which $x_{j+1} - x_j$ approach asymptotically this distance is sufficient to the proper working of the partition-extrapolation method.

The exponential term $e^{-\cos(2c)t\alpha^2}$ in (13) tends to 0 as $\alpha \to +\infty$, and both $\sin[\sin(2c)t\alpha^2 - \cos(c)x\alpha \pm 2c]$ are bounded functions, and $\lambda(\sin(\sin(2c)t\alpha^2 - \cos(c)x\alpha + 2c)\alpha^4 e^{-\cos(2c)t\alpha^2} + \lambda^2 \sin(\sin(2c)t\alpha^2 - \cos(c)x\alpha - 2c)e^{-\cos(2c)t\alpha^2})$ tends to 0 as $\alpha \to +\infty$. In addition, as $\alpha \to +\infty$, the term $\sin(\lambda t)\alpha^6 \sin(\cos(c)x\alpha)$ dominates $\lambda \cos(\lambda t)\alpha^4 \sin(\cos(c)x\alpha - 2c)$, $\lambda^2 \sin(\lambda t)\alpha^2 \sin(\cos(c)x\alpha + 4c)$, and $\lambda^3 \cos(\lambda t) \sin(\cos(c)x\alpha + 2c)$. Therefore, for large value of α, the zeros of (13) asymptotically correspond to the zeros of the following equation:

$$\sin(\lambda t) \sin(\cos(c)x\alpha)\alpha^6 = 0, \tag{14}$$

and the oscillations become regular and have an asymptotic period of $2\pi/\cos(c)x$. Therefore, by fixing a lower limit b, the interval of integration $[b, +\infty)$ is partitioned into sub-intervals, $[\alpha_l, \alpha_{l+1}]$ all of equal length, $\pi/\cos(\pi/5)x$, except the first one. Hence, the integral tail can be written as:

$$I = \int_b^\infty f(\alpha)d\alpha = \sum_{l=0}^{\infty} s_l, \tag{15}$$

where $s_0 = \int_b^{\alpha_1} f(\alpha)d\alpha$ and $s_l = \int_{\alpha_l}^{\alpha_{l+1}} f(\alpha)d\alpha$, 1 = 1, 2, ...With this partition, the infinite integral I becomes an infinite sum of finite integrals which is treated using the Euler transform. Each finite integral s_l is approximated using the same standard quadrature rule. A good approximation to I is given by:

$$t_{n,p} = \sum_{j=0}^{r-1} s_j + \frac{1}{2}\sum_{q=0}^{p} M^q s_r + \frac{1}{2}M^p s_r + \sum_{l=r+1}^{n-p-1} M^p s_l + \frac{1}{2}M^p s_{n-p} \tag{16}$$

where M is the forward average operator, defined as $M^p s_r = \frac{1}{2^p}\sum_{m=0}^{p}\binom{p}{m}s_{r+m}$. The constants n and p are the number of terms and columns used in the finite average table. The first group of terms in (16) is simply the sum of the first r terms of the series. The second group is the initial $(p + 1)$ terms of the Euler transformation applied to the series starting with the term s_r. The last part of the sum is the sum of the terms in the column p of the finite average table. This way of re-expressing the sum of n terms is useful in cases where many of the terms $M^p s_j$ are small enough to be neglected. This happens when the sequence s_j alternates in sign and decays slowly.

To estimate the values of b, we modified [10] which is written in FORTRAN 77. All the computations have been done in double precision arithmetic. We used two different quadrature rules: 9 point Gauss-Legendre and 64 point trapezoid rule. For the fixed accuracy level of $0.5 * 10^{-12}$,we have experimented with many values of b and found that the optimal values coincide with $b = 10.41\pi/\cos(c)x$ (values given in column 3 of Table 1). In fact, given this choice of the b values, the sums of the terms by the two quadrature methods, as shown in columns 4 and 5, are approximately equal to the overall accuracy $0.5 * 10^{-12}$. To achieve the requested relative error bound (10^{-16}), we have used seven terms in the extrapolation.

As it is clear from columns 2 and 3 of Table 1, that the b-values given by the partition-extrapolation method are much smaller than that given by the truncation method for small values of x but when x increases, we get close values. This shows that the exponential decay is not the dominant feature when x takes small values and the b-values have been improved by the use of the acceleration technique. However, as expected, this exponential decay is dominant when x takes large values, which result in close estimation truncates by the two methods.

5.3 Step 2: Estimating the Integral on a Finite Interval

After bounding the tail of the integrand, we approximate the integral in (10) on the finite interval [0, b]. We use the Romberg integration method to find an approximation to this integral that is accurate within the fixed tolerance $\varepsilon = 0.5 * 10^{-12}$. The first step in the Romberg process involves obtaining the following recursive trapezoidal rule approximations:

$$R_{k,1} = \frac{1}{2}\left[R_{k-1,1} + h_{k-1}\sum_{i=1}^{2^{k-1}} f(a + (i - \frac{1}{2})h_{k-1})\right], \qquad k = 2, 3, \ldots, n, \tag{17}$$

where $R_{1,1} = \frac{(b-a)}{2}\left[f(a) + f(b)\right]$ and $h_k = \frac{(b-a)}{2^{k-1}}$. The second step applies the Richardson extrapolation procedure to the values $R_{k,1}$ using the following formula:

$$R_{k,j} = \frac{4^{j-1}R_{k,j-1} - R_{k-1,j-1}}{4^{j-1} - 1}, k = 2, 3, \ldots n; j = 2, \ldots, k. \tag{18}$$

The term $R_{k,j}$ is the value of the integral with 2^{k-1} the number of sub-intervals and j the number of Richardon extrapolations. The values with larger j index correspond to successively higher order Newton-Cotes formulas. The Romberg recursion continues until the error $\left|R_{k,k} - R_{k,k-1}\right|$ is within the given tolerance. In our estimation, the maximum number of sub-intervals that is allowed to be reached is 2^{25}. We have initialized the error to be $0.5 * 10^{-10}$(100$*$ error level). Figure 1 represents the solution by the Romberg integration method in the case of an overall accuracy 10^{-12}. Figure 2 shows the error estimation which is smaller than the fixed tolerance ($0.5 * 10^{-12}$) for all fixed values of x and t. In order to improve more the efficiency and speed of estimation, we have computed the number of function evaluations given by Romberg estimation and MATLAB quadl.m subroutine. Figures 3 and 4 display the number of function evaluations by the Romberg integration and MATLAB quadl.m subroutines successively. The first method gives a large difference in the number of functions

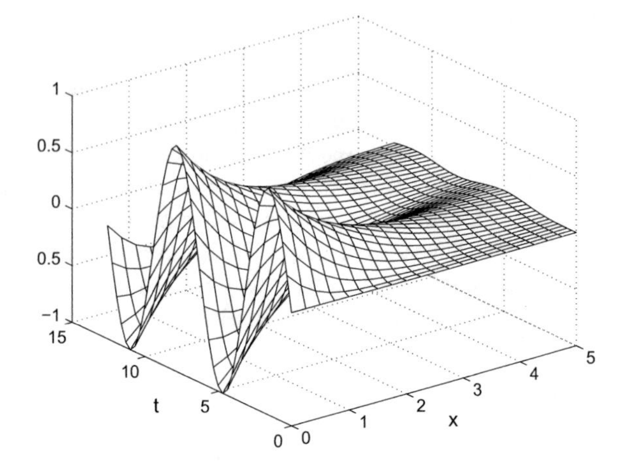

Fig. 1 Solution using extrapolation and Romberg integration: 10^{-12} accuracy

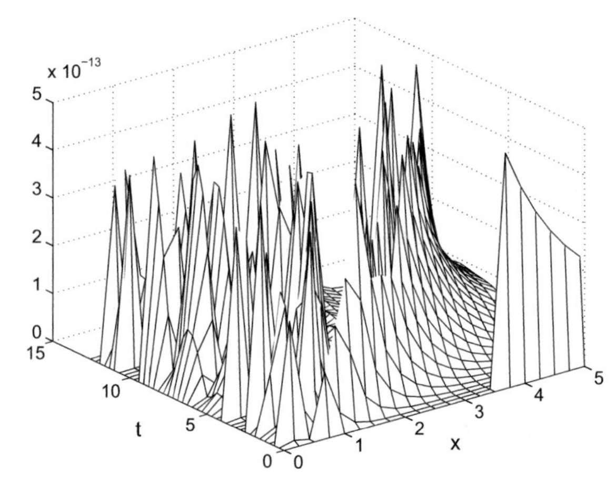

Fig. 2 Error estimation by Romberg integration

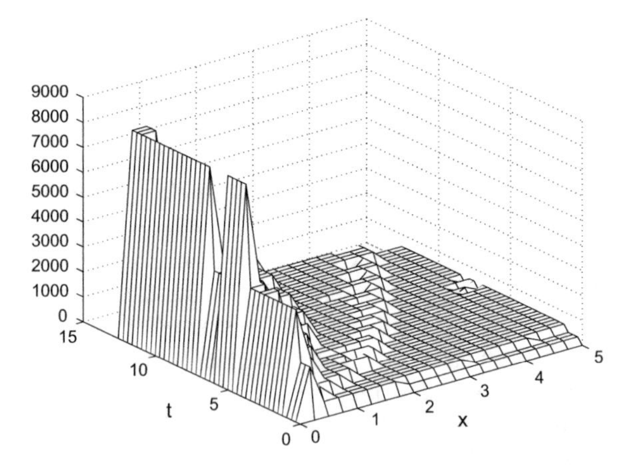

Fig. 3 Number of function evaluations by Romberg integration

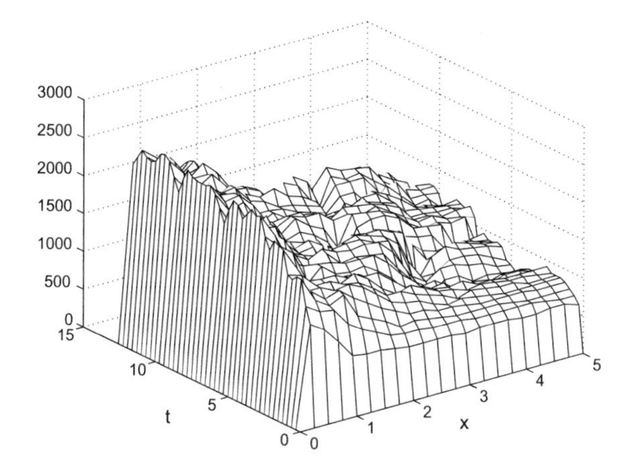

Fig. 4 Number of function evaluations by Matlab subroutine

Table 2 Comparison of the total number of function evaluations by two quadrature methods as the overall accuracy level increases

Overall accuracy	10^{-12}	10^{-13}	10^{-14}	10^{-15}	10^{-16}
Values of b	$10.41\pi/\cos(c)x$	$11\pi/\cos(c)x$	$12\pi/\cos(c)x$	$13\pi/\cos(c)x$	$14\pi/\cos(c)x$
Romberg method	1100736	1524000	1967912	2753312	3723552
quadl.m subroutine	1122300	1599630	2212350	3073770	4163580

evaluations between small and large x values. This difference, however, is small in the case of quadl.m. We observed that this pattern is the same for other overall accuracies less than 10^{-12}. We have also computed the total number of function evaluations in each method for different overall accuracy. Table 2 shows that the quadl.m subroutine uses more function evaluations than that of Richardson recursion for all levels of accuracy considered.

6 Conclusion

We have proposed an efficient numerical approach to approximate the transform solution of the diffusion equation. The approach consists of two steps. First we bound the tail of the integral by making use of a partition-extrapolation method, secondly we estimate the integral over a finite interval by using the Romberg integration. Our estimation result show that, when the integrand decays slowly, the acceleration technique improves the truncation estimate of the integral tail and the solution in general, and the Romberg integration uses less function evaluations than that of MATLAB quadl.m subroutine. One of the limitations of our work is the error that remains not counted for when we applied the same quadrature rule on each finite integral. To better improve the quality of this estimation by extrapolation, we consider to replace the same quadrature rule by some sufficiently accurate algorithm in the future.

References

1. Bettess, P.: Infinite Elements. Penshaw Press (1992)
2. Blakemore, M., et al.: Comparison of some methods for evaluating infinite range oscillatory integrals. J. Comp. Phys. **22**, 352–376 (1976)
3. Flyer S., Fokas A.S.: A hybrid analytical numerical method for solving evolution partial differential equations. In: Proceedings of the R. Soc Lond. A, pp. 1823–1849 (2008)
4. Fokas, A.S.: A Unified Approach to Boundary Value Problems. SIAM (2007)
5. Fokas, A,S, et al.: A semi analytical numerical method for solving evolution and elliptic partial differential equation. J. Comp. Appl. Math, **227**, 59–74 (2009)
6. Iserles, A.: On the numerical quadrature of highly-oscillating integrals i: fourier transforms. IMA J. Num. Anal. **24**,365–391 (2004)
7. Longman, I.M.: A method for the numerical evaluation of finite integrals of oscillatory functions. Math. Comp **14**, 53–59 (1960)
8. Lucas, S.K., Stone, H.A.: Evaluating infinite integrals involving Bessel functions of arbitrary order. J. Comput. Appl. Maths **64**, 217–231 (1995)
9. Lyness, J.N.: Integrating some infinite oscillating tails. J. Comp. Appl. Math **12**, 109–117 (1985)
10. Lyness, J.N., Hines, G.: Algorithm 639: to integrate some infinite oscillating tails. ACM Trans. Mathe Software **12**, 24–25 (1986)
11. Sidi, A.: The numerical evaluation of very oscillatory infinite integrals by extrapolation. Math. Comput. **38**, 517–529 (1982)

Molecular Dynamics Study of the Solution Behaviour of Antimicrobial Peptide Indolicidin

Rositsa Marinova, Peicho Petkov, Nevena Ilieva, Elena Lilkova and Leandar Litov

Abstract Understanding the mechanism of action of antimicrobial peptides (AMP-s) on bacterial cells requires detailed knowledge of how AMPs interact with bacterial membranes. Our hypothesis is that the peptides do not interact with the membrane as monomers, but rather form clusters, that collectively approach the cell and attack the membrane. In this paper we investigate the behavior of the antimicrobial peptide indolicidin in solution, prior to their interaction with the bacterial membrane, by means of coarse grain molecular dynamics simulations (CG-MD). We show that indolicidin in particular and, probably, charged linear AMPs in general tend to aggregate in solution, forming globular amphipathic clusters with a central hydrophobic core. The dependence of the clusters size on the peptide concentration and on the temperature is studied, as well as the influence of the finite size of the simulation box. Our results manifest the investigation of the AMPs behavior in solution prior to membrane impact as an indispensable element in revealing the mechanism of their antimicrobial activity.

R. Marinova (✉) · P. Petkov · L. Litov
Faculty of Physics, Sofia University St. Kl. Ohridski,
5 James Bourchier Blvd., 1164 Sofia, Bulgaria
e-mail: rosie.marinova@gmail.com

P. Petkov
e-mail: peicho@phys.uni-sofia.bg

L. Litov
e-mail: leandar.litov@cern.ch

N. Ilieva · E. Lilkova
Institute of Information and Communication Technologies,
Bulgarian Academy of Sciences, 25A, Acad. G. Bonchev Str., 1113 Sofia, Bulgaria
e-mail: nevena.ilieva@parallel.bas.bg

E. Lilkova
e-mail: elilkova@parallel.bas.bg

K. Georgiev et al. (eds.), *Advanced Computing in Industrial Mathematics*,
Studies in Computational Intelligence 793,
https://doi.org/10.1007/978-3-319-97277-0_21

1 Introduction

The increased emergence of antibiotic-resistant strains of bacteria that renders inefficient the existing traditional antibiotics and the decline in the development of new antibacterial agents [15] prompts considerable efforts to be focused on the development of alternative therapies for the treatment of infectious diseases. The characteristics, broad spectrum and largely nonspecific activity against grampositive and gram-negative bacteria, fungi and viruses [13, 36] of the antimicrobial peptides (AMPs) qualify them as possible candidates for therapeutic alternatives [22] and their potential against multi-drug resistant pathogen infections has attracted a lot of interest [29, 38]. These small (<10kDa) amphiphilic molecules (from a few to several tens of amino acid residues) are classified in four major groups: alpha-helical, beta-stranded, mixed and extended structures [26, 28, 33]. Their antibiotic activity is thought to be based on their cationic and amphiphilic nature, which enables them to interact with negatively charged bacterial surfaces and membranes, thus causing membrane disruption or altering metabolic processes [7]. There are three commonly accepted models of action of antimicrobial peptides: the carpet model, the barrel-stave model and the toroidal-pore model [5, 26, 34]. Two different states of AMPs activity have been identified: the S-state, in which the peptide is surface-bound, and the I-state, in which the peptide is inserted into the membrane bilayer [17]. Most model studies focus on these states represented by configurations of a model membrane and closely placed single peptides or parts of peptides, with possible sequential addition of further peptide monomers [3, 25, 35]. However, the complete understanding at atomic level of the relationship between the peptide structure and the mechanisms of peptide-membrane interaction remains an open question [12].

We propose that certain antimicrobial peptides, in particular extended ones, form clusters prior to their interaction with the target membrane and it is these clusters that exercise antimicrobial action by penetrating through the membrane lipid bilayer or inducing rearrangements in it. Thus, it is of primary importance for forecasting or designing certain modes of antibiotic activity based on AMPs, to study their behavior in water solution for identifying the starting conformation of the peptide-membrane interaction. We exemplify our concept by studying the solution behavior of a particular antibacterial peptide – indolicidin.

Indolicidin (IL) is the shortest antimicrobial peptide (13 aa residues: ILPWKWP WWPWRR-NH2), which is disordered in solution but supposedly forms coils and turn in the course of interaction with the bacterial membrane. It was isolated from the cytoplasmic granules of bovine neutrophils. It is active against Gram-negative and Gram-positive bacteria, causing disruption of the cytoplasmic membrane by channel formation [10]. IL has the highest percentage of tryptophan (39%) and proline (23%) residues of all known proteins [32]. These characteristics make indolicidin both a simple and an interesting model object for investigation of peptide-membrane interactions. We used coarse-grain molecular dynamics (CG-MD) simulations to investigate the process of peptide aggregation in solution, the cluster size dependence on peptide concentration and temperature and the influence of the finite volume of the simulation box on the results.

2 Materials and Methods

In the present study, coarse-grain molecular dynamics simulations were performed with the MD package GROMACS 2016.1 [14]. In contrast to all-atom (AA) simulations, coarse-grain (CG-MD) approaches consider certain groups of atoms as individual system components (beads), thus essentially reducing the number of degrees of freedom and allowing for longer simulated evolution [18, 21]. The protein was parameterized using the MARTINI 22p force field [20, 24]. The treatment of the solvent is decisive for the properties derived from simulation studies. In our simulation protocol, we used a polarizable coarse-grained water model that allows for a reliable modeling of water interactions with charged particles [37]. The model systems were prepared using the Solution option of the Martini Maker of the web-based graphical user interface for generation of molecular simulation system setups, standardizing the usage of common and advanced simulation techniques, CHARMM-GUI [19, 30]. For the starting conformation of the IL peptides in our studies we used as initial structure the PDB [4] entry 1G89 [31].

Sodium cations and chlorine anions were added at a concentration of 0.15 M to neutralize the net charge of each system. All systems were simulated under periodic boundary conditions. The simulations were performed in the isothermal-isobaric ensemble at a constant temperature 310 K maintained by the v-rescale thermostat [6] with a time-constant of 1 ps^{-1} and pressure of 1 bar maintained through a Parrinello-Rahman barostat [27] with a coupling constant of 12 ps^{-1}. The electrostatic interactions were treated using the smooth particle mesh Ewald (PME) algorithm [9] with a cutoff of the direct summation of 1.2 nm. Van der Waals interactions were cut at 1.1 nm. The leapfrog integrator was used with an integration time step of 20 fs. The setup properties and the total simulated time for the different test systems are given in Table 1.

Table 1 Parameters of the test systems

System	I	II	III	IV
Simulation box dimensions [Å^3]	160×160×160	160×160×160	150×150×150	76.7×76.7×76.7
Number of peptides	27	54	54	27
Indolicidin concentration [mM]	10.9	21.9	26.6	99.4
Total simulated time [μs]	1.0	1.0	1.5	1.0

3 Results and Discussion

In protein-membrane interactions, the processes of interest span over time scales often inaccessible to all-atom MD simulations. This justifies the application of coarse-grain approaches (CG), which overcome the time- and lengthscale restrictions by reducing the level of detail in the representation of the system, though at the price of loss of resolution [23]. In order to explore the aggregation properties of IL in water solution, we performed CG-MD simulations at different concentrations of the peptide in the simulation boxes with volumes of the order of several atolitres (10^{-18} L or 10^{-15} mL), determined by the simulation protocol and the computational feasibility of the simulations. A direct in silico reproduction of the typical molar concentrations in the experimental studies – of the order of a few tens of μg/ml – is not possible even for coarsegrained studies, since they correspond to molar concentrations of a few hundredths of mM. For instance, the experimentally determined concentration for indolicidin aggregation is at concentration of 0.026 mM and above 5.2 μM IL exhibits hemolytic activity [1]. Simulations of peptide solutions with such concentrations would require 1000 to 1,000,000 times larger simulation boxes than those used in the present study and thus, unrealistic computational resources. Instead, we adopt proportionally scaled number densities, thus generating representative systems in the concentration range of interest. For calibration, we refer to earlier model studies and available experimental data. Four different concentrations of the peptide at T = 310 K were investigated 10.9, 21.9, 26.6, and 99.4 mM, denoted as systems I–IV, respectively (see 1). In all simulations, in the initial state the peptides were placed randomly in the water bulk. In the course of simulations, a/an aggregation/selforganization process took place, leading to a different number and size of clusters in the final state. Floeck et al. [11] used MD to study small amphiphilic peptides with similar characteristics to indolicidin. They simulated solutions at comparable concentrations and reported a formation of clusters as well.

In system I, with an indolicidin concentration of 10.9 mM, the peptide aggregation leads to the formation of three globular clusters. In system II, six clusters are formed, so the number of clusters increases twice, compared to system I. In system III, with an indolicidin concentration of 26.6 mM, we observed formation of ten clusters at T = 310 K, with an average cluster size of 11 peptide units. In the simulation of system IV, at indolicidin concentration of 99.4 mM, saturation of the solution takes place, with only one cluster formed, containing all peptide monomers. Thus, above a certain concentration threshold (between 99.4 and 26.6 mM), the peptides create a linked filamentary structure. The obtained filament is shown in Fig. 1, in periodic boundary conditions. Between the experimentally determined aggregation concentration threshold of 0.026 mM and the solution saturation threshold, IL peptides in solution form globular amphipathic structures.

For studying the temperature dependence of the aggregation process, a simulation of system III was performed at a higher temperature, T = 330 K, following the same protocol as described in Section II, with a duration of 1.6 μs. System III was chosen as having the highest peptide concentration below the saturation threshold. In this

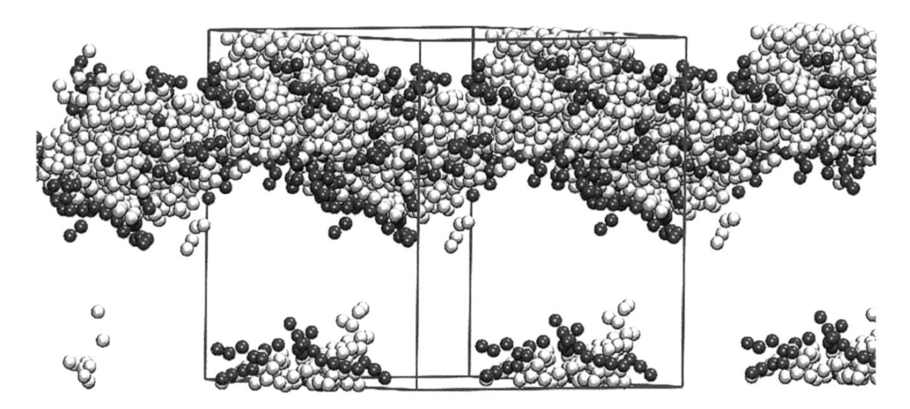

Fig. 1 Solvent saturation at concentration of 99.4 mM: formation of a filamentary cluster, containing all solvated peptides (hydrophobic residues in white, cationic residues – in blue)

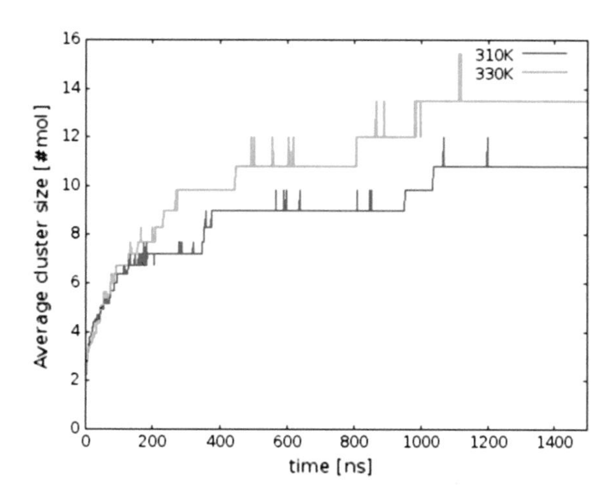

Fig. 2 Average cluster size at IL concentration of 26.6 mM at T = 310 K (red) and T = 330 K (green)

simulation, eight clusters were formed, with an average size of 13 peptide units. At both temperatures, however, the largest cluster contained 23 peptide units, with a similar cluster-formation pattern (Fig. 2).

Such a behavior is to be expected, because of the faster diffusion at higher temperatures and the hydrophobic effect, driving the assembly process. Thus, the increased assembly probability results in formation of larger clusters, clearly less in number because of the preserved simulation parameters. However, the topology of the self-assembled structure remains unchanged.

The entropic part of the cluster-formation free energy is proportional to the temperature, while the electrostatic free energy term is temperature independent. Therefore, the largest cluster should be formed when the entropic and the coulomb free energy terms are equal. Moreover, because of the cluster-formation free energy structure, the larger a cluster, the less stable it will be.

The finite box size in general and periodic boundary conditions in particular might lead to artifacts or distortions in the resulting evolution of the investigated system. In our studies, the observed aggregation may well be considered as an established fact, but the cluster size might be subjected to finite-volume influences. The average cluster size is an important parameter in constructing the initial state for studying the peptide-membrane interaction dynamics. Therefore, in order to eliminate the influence of the finite simulation box and thus, of the periodic boundary conditions employed, on the average cluster size, we performed simulations at a constant concentration of 26.6 mM and variable simulation box volumes in geometric progression: V, 2V, 4V (box sizes $150\times150\times75$ Å^3, $150\times150\times150$ Å^3 and $150\times150\times300$ Å^3 , respectively). The results for the average cluster size in each simulation are presented in Fig. 3. By extrapolating the obtained line, one can estimate the average cluster size for an infinite simulation box. A linear fit with R2 = 0.84 allows to deduce the average cluster size for an infinite simulation box, so with no periodic boundary conditions imposed, being about 14 peptide monomers.

The typical aggregates, formed in our simulations, have an average diameter of 4.5 nm and an average size of 14 peptide units (see Fig. 4). The clusters are characterized by a central hydrophobic core, comprised of proline (Pro) and tryptophan (Trp) residues (the white spheres in the cluster interior in Fig. 4). The globular structures are surrounded by positively charged domains (the blue spheres in Fig. 4), which provides excellent conditions for electrostatic interactions with the hydrophilic lipid heads of bacterial membranes. Our simulations show that in all four systems the hydrophobic-driven aggregation occurs already in the first nanoseconds. Such a swift self-assembly was also observed in [11], where the reported aggregation took place already after 20 ns simulated evolution. Although the noncovalent interactions, involved in the self-assembly process, have a much lower binding energy than the covalent bonds, the so-formed clusters remain very stable during the whole simulations (Fig. 1). In fact, once the clusters are formed, virtually no peptides dissociate from them. Occasionally two clusters merge and then split again, but no peptides are released from the clusters as monomers.

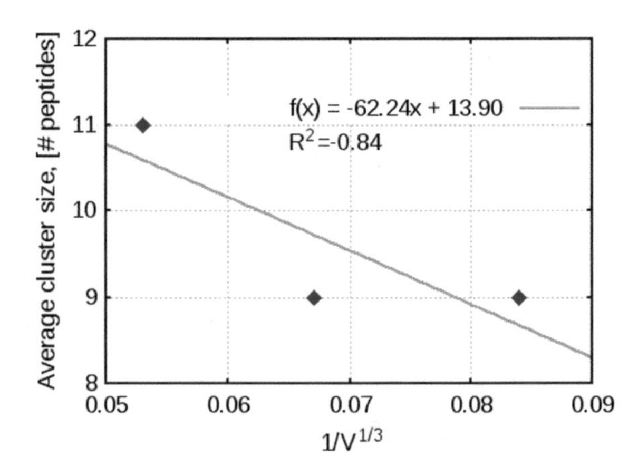

Fig. 3 The average cluster size as a function of the inverse cubic root of the volume of the simulation box

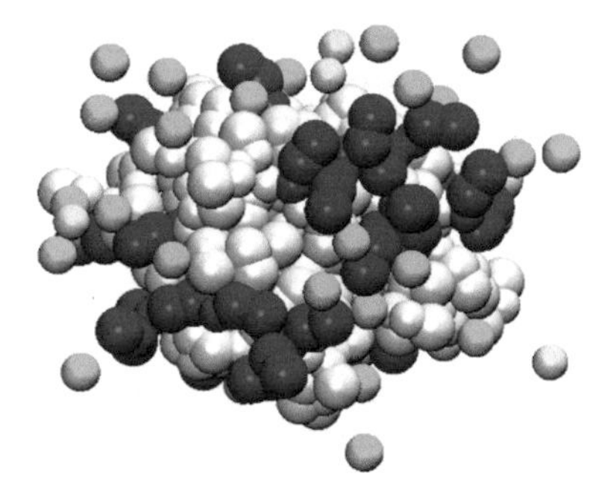

Fig. 4 Typical cluster structure (hydrophobic residues in white and the cationic residues in blue), with ions in the hydration shell (green and yellow)

The results of our simulations are in agreement with the experimental observations in [16] suggesting a globular form for indolicidin in aqueous solution. The specific amphiphilic nature, hydrophobicity and cationic charge were pointed out as decisive features that allow for spontaneous self-assembling of indolicidin monomers in aqueous solution into organized 3D structures (clusters) [8]. In a structure-activity study of indolicidin [2] both its antimicrobial and haemolytic activity were attributed to the tryptophan (Trp) residues at positions 4 and 9. In our simulations, these amino acid residues are buried in the center of the cluster to form its hydrophobic core.

4 Conclusions

In this paper, we report the first results of molecular dynamics studies of the behavior of the antimicrobial peptide indolicidin in water solution, its concentration and temperature dependence and the influence of the periodic boundary conditions employed. In the concentration range from 0.026 mM to an upper threshold, between 26.4 and 99.4 mM, we observe spontaneous aggregation of the IL monomers into globular amphiphilic clusters, characterized by a central hydrophobic core and positively charged residues exposed to the solvent. The self-assembly process is very fast, even at the lowest studied concentration, and the aggregates appear to be very stable over time. By analyzing the dependence of the average cluster size on the simulation box volume, the cluster size in an infinite box and thus, without periodic boundary conditions, was estimated to be about 14 IL monomers within the investigated concentration range. At concentrations above the upper threshold, the peptide solution saturates and the peptide monomers form large filamentary structures.

Our results highlight the importance of studying the AMPs monomers behavior in solution, prior to their contact with the bacterial membrane. The aggregation process, the size and the composition (in case of multicomponent substances) of the so-formed clusters might be a key element in AMPs mode of antimicrobial activity and thus indispensable for its understanding, modeling and modulation. Knowing the proper clusters certain AMPs form will allow for building adequate initial configurations as a prerequisite for modelling the peptide-membrane interactions and design of peptides with predefined properties.

Acknowledgements The simulations were performed on the supercomputer Avitohol at BAS and on the HPC Cluster at the Faculty of Physics of Sofia University St. Kl. Ohridski. This work was supported in part by National Science Fund under Grant DNTS-China-01/9/2014.

References

1. Ahmad, I., Perkins, W.R., Lupan, D.M., Selsted, M.E., Janoff, A.S.: Liposomal entrapment of the neutrophil-derived peptide indolicidin endows it with in vivo antifungal activity. Biochim. Biophys. Acta **1237**(2), 109–114 (1995)
2. Ando, S., Mitsuyasu, K., Soeda, Y., Hidaka, M., Ito, Y., Matsubara, K., Shindo, M., Uchida, Y., Aoyagi, H.: Structure-activity relationship of indolicidin, a Trp-rich antibacterial peptide. J. Pept. Sci. **16**(4), 171–177 (2010)
3. Berglund, N.A., Piggot, T.J., Jefferies, D., Sessions, R.B., Bond, P.J., Khalid, S.: Interaction of the antimicrobial peptide polymyxin B1 with both membranes of E. coli: a molecular dynamics study. PLOS Comput. Biol. **11**(4), 1–17 (2015)
4. Berman, H.M., Westbrook, J., Feng, Z., Gilliland, G., Bhat, T.N., Weissig, H., Shindyalov, I.N., Bourne, P.E.: The protein data bank. Nucleic Acids Res. **28**(1), 235–242 (2000)
5. Brogden, K.A.: Antimicrobial peptides: pore formers or metabolic inhibitors in bacteria? Nat. Rev. Microbiol. **3**, 238–250 (2005)
6. Bussi, G., Donadio, D., Parrinello, M.: Canonical sampling through velocity rescaling. J. Chem. Phys. **126**(1), 014101 (2007). http://www.ncbi.nlm.nih.gov/pubmed/17212484
7. Copolovici, D.M., Langel, K., Eriste, E., Langel, I.: Cell-penetrating peptides: design, synthesis, and applications. ACS Nano **8**(3), 1972–1994 (2014)
8. Dehsorkhi, A., Castelletto, V., Hamley, I.W.: Self-assembling amphiphilic peptides. J. Pept. Sci. **20**(7), 453–467 (2014)
9. Essmann, U., Perera, L., Berkowitz, M.L., Darden, T., Lee, H., Pedersen, L.G.: A smooth particle mesh ewald method. J. Chem. Phys. **103**(19), 8577–8593 (1995)
10. Falla, T.J., Karunaratne, D.N., Hancock, R.E.W.: Mode of action of the antimicrobial peptide indolicidin. J. Biol. Chem. **271**, 19298–19303 (1996)
11. Flöck, D., Rossetti, G., Daidone, I., Amadei, A., Nola, A.D.: Aggregation of small peptides studied by molecular dynamics simulations. Proteins: Struct., Funct., Bioinf. **65**(4), 914–921 (2006)
12. Galdiero, S., Falanga, A., Cantisani, M., Vitiello, M., Morelli, G., Galdiero, M.: Peptide-lipid interactions: experiments and applications. Int. J. Mol. Sci. **14**(9), 18758–18789 (2013)
13. Hancock, R.E.W., Sahl, H.G.: Antimicrobial and host-defense peptides as new anti-infective therapeutic strategies. Nat. Biotechnol. **24**, 1551–1557 (2006)
14. Hess, B., Kutzner, C., van der Spoel, D., Lindahl, E.: Gromacs 4: algorithms for highly efficient, load-balanced, and scalable molecular simulation. J. Chem. Theor. Comput. **4**(3), 435–447 (2008)
15. Högberg, L.D., Heddini, A.: The global need for effective antibiotics: challenges and recent advances. Trends Pharmacol. Sci. **31**(11), 509–515 (2010)
16. Hsu, C.H., Chen, C., Jou, M.L., Lee, A.Y.L., Lin, Y.C., Yu, Y.P., Huang, W.T., Wu, S.H.: Structural and dna-binding studies on the bovine antimicrobial peptide, indolicidin: evidence for multiple conformations involved in binding to membranes and dna. Nucleic Acids Res. **33**(13), 4053–4064 (2005)
17. Huang, H.W.: Molecular mechanism of antimicrobial peptides: the origin of cooperativity. Biochim. Biophys. Acta **1758**(9), 1292–1302 (2006)
18. Inglfsson, H.I., Lopez, C.A., Uusitalo, J.J., de Jong, D.H., Gopal, S.M., Periole, X., Marrink, S.J.: The power of coarse graining in biomolecular simulations. Wiley Interdisc. Rev. Comput. Mol. Sci. **4**(3), 225–248 (2014)

19. Jo, S., Kim, T., Iyer, V.G., Im, W.: Charmm-gui: a web-based graphical user interface for charmm. J. Comput. Chem. **29**(11), 1859–1865 (2008)
20. de Jong, D.H., Singh, G., Bennett, W.F.D., Arnarez, C., Wassenaar, T.A., Schfer, L.V., Periole, X., Tieleman, D.P., Marrink, S.J.: Improved parameters for the martini coarse-grained protein force field. J. Chem. Theor. Comput. **9**(1), 687–697 (2013)
21. Kmiecik, S., Gront, D., Kolinski, M., Wieteska, L., Dawid, A.E., Kolinski, A.: Coarse-grained protein models and their applications. Chem. Rev. **116**(14), 7898–7936 (2016)
22. López-Meza, J.E., Ochoa-Zarzosa, A., Barboza-Corona, J.E., Bideshi, D.K.: Antimicrobial peptides: current and potential applications in biomedical therapies. BioMed. Res. Int. **2015**(367243), 2 (2015)
23. Marrink, S.J., Risselada, H.J., Yefimov, S., Tieleman, D.P., de Vries, A.H.: The martini force field: coarse grained model for biomolecular simulations. J. Phys. Chem. B **111**(27), 7812–7824 (2007)
24. Monticelli, L., Kandasamy, S.K., Periole, X., Larson, R.G., Tieleman, D.P., Marrink, S.J.: The martini coarse-grained force field: extension to proteins. J. Chem. Theor. Comput. **4**(5), 819–834 (2008)
25. Neale, C., Hsu, J., Yip, C., Pomès, R.: Indolicidin binding induces thinning of a lipid bilayer. Biophys. J. **106**(8), L29–L31 (2014)
26. Nguyen, L.T., Haney, E.F., Vogel, H.J.: The expanding scope of antimicrobial peptide structures and their modes of action. Trends Biotechnol. **29**(9), 464–472 (2011)
27. Parrinello, M., Rahman, A.: Polymorphic transitions in single crystals: a new molecular dynamics method. J. Appl. Phys. **52**(12), 7182–7190 (1981)
28. Pasupuleti, M., Schmidtchen, A., Malmsten, M.: Antimicrobial peptides: key components of the innate immune system. Crit. Rev. Biotechnol. **32**(2), 143–171 (2012)
29. Pathan, F.K., Venkata, D.A., Panguluri, S.K.: Recent patents on antimicrobial peptides. Recent Pat. DNA Gene Sequences (DIscontinued) **4**(1), 10–16 (2010)
30. Qi, Y., Inglfsson, H.I., Cheng, X., Lee, J., Marrink, S.J., Im, W.: Charmm-gui martini maker for coarse-grained simulations with the martini force field. J. Chem. Theor. Comput. **11**(9), 4486–4494 (2015)
31. Rozek, A., Friedrich, C.L., Hancock, R.E.W.: Structure of the bovine antimicrobial peptide indolicidin bound to dodecylphosphocholine and sodium dodecyl sulfate micelles. Biochemistry **39**(51), 15765–15774 (2000)
32. Selsted, M., Novotny, M., Morris, W., Tang, Y., Smith, W., Cullor, J.: Indolicidin, a novel bactericidal tridecapeptide amide from neutrophils. J. Biol. Chem. **267**(7), 4292–4295 (1996)
33. Takahashi, D., Shukla, S.K., Prakash, O., Zhang, G.: Structural determinants of host defense peptides for antimicrobial activity and target cell selectivity. Biochimie **92**(9), 1236–1241 (2010)
34. Toke, O.: Antimicrobial peptides: new candidates in the fight against bacterial infections. Peptide Sci. **80**(6), 717–735 (2005)
35. Wang, Y., Chen, C.H., Hu, D., Ulmschneider, M.B., Ulmschneider, J.P.: Spontaneous formation of structurally diverse membrane channel architectures from a single antimicrobial peptide. Nat. Commun. 7(13535) (2016).
36. Yeaman, M.R., Yount, N.Y.: Mechanisms of antimicrobial peptide action and resistance. Pharmacol. Rev. **55**(1), 27–55 (2003)
37. Yesylevskyy, S.O., Schfer, L.V., Sengupta, D., Marrink, S.J.: Polarizable water model for the coarse-grained martini force field. PLOS Comput. Biol. **6**(6), 1–17 (2010)
38. Zhuang, J., Coates, C.J., Zhu, H., Zhu, P., Wu, Z., Xie, L.: Identification of candidate antimicrobial peptides derived from abalone hemocyanin. Dev. Comp. Immunol. **49**(1), 96–102 (2015)

A Functional Expansion and a New Set of Rapidly Convergent Series Involving Zeta Values

Lubomir Markov

Abstract We derive a power series expansion for the function $x \mapsto x \log(\tan x) - x \log x + x - \mathrm{Ti}_2(\tan x)$, and then use it to obtain the expansion of the function $x \mapsto \mathrm{Ti}_2(\tan x)$; as usual, Ti_2 denotes the inverse tangent integral. Two similar expansions involving Legendre's chi-function follow from the above. These representations are used to obtain new rapidly convergent numerical series involving zeta values.

Keywords Riemann zeta function · Inverse tangent integral · Legendre's chi-function · Rapidly convergent zeta series · Ramanujan's series · Catalan's constant

1 Introduction and Preliminaries

The origin of the interesting subject of series involving values of the Riemann zeta function $\zeta(z)$ may be traced back to a classical paper by Euler [5], where he establishes, in equivalent form, the following representation for $\zeta(3)$:

$$\zeta(3) = \frac{\pi^2}{7} - \frac{4\pi^2}{7} \sum_{k=1}^{\infty} \frac{\zeta(2k)}{(2k+1)(2k+2)2^{2k}}.$$

This series has since been rediscovered several times, most recently by Ewell [6], who derives it following a method of Choe [3], which method is itself a modified Euler argument (see [1]). It is clear that the above sum converges much more rapidly than the defining series of $\zeta(3) = \sum_{n=1}^{\infty} \frac{1}{n^3}$. In recent years there has been a substantial amount of research regarding such rapidly convergent representations, and a large number of series have been evaluated in closed form. (See primarily Srivastava's

L. Markov (✉)
Department of Mathematics and CS, Barry University, 11300 N.E. Second Avenue, Miami Shores, FL 33161, USA
e-mail: lmarkov@barry.edu

K. Georgiev et al. (eds.), *Advanced Computing in Industrial Mathematics*, Studies in Computational Intelligence 793,
https://doi.org/10.1007/978-3-319-97277-0_22

expository article [10] and the book by Srivastava and Choi [11], but also the papers [2, 4, 13].) As a rather suggestive example of the progress that has been achieved, one may note the formula

$$\zeta(3) = -\frac{120\pi^2}{1573} \cdot \sum_{k=0}^{\infty} \frac{8576k^2 + 24{,}286k + 17{,}283}{(2k+1)(2k+2)(2k+3)(2k+4)(2k+5)(2k+6)(2k+7)} \frac{\zeta(2k)}{2^{2k}},$$

discovered recently by Srivastava and Tsumura [12, equation (4.6)]. Our main objective is to give four functional expansions (believed to be new), involving the inverse tangent integral Ti_2 and Legendre's chi-function χ_2. We then use these expansions to sum in closed form several rapidly convergent numerical series which contain zeta values at the even positive integers. It has been known since Euler's time (and proven by him) that these values are given by the formula

$$\zeta(2n) = \frac{(-1)^{n+1} B_{2n}}{2(2n)!} (2\pi)^{2n},$$

where B_{2n} are the Bernoulli numbers (see e.g. [11]).

Throughout the paper, we denote the hyperbolic tangent and the inverse hyperbolic tangent by $\mathrm{th}(\cdot)$ and $\mathrm{arth}(\cdot)$, respectively. For the higher transcendental functions we adhere to the now standard notation introduced by Lewin [8]. Recall (see [8] or [11]) the definitions of the inverse tangent integral Ti_2 and Legendre's chi-function χ_2 in terms of the dilogarithm function Li_2:

$$\mathrm{Ti}_2(z) := \frac{1}{2i}\Big[\mathrm{Li}_2(iz) - \mathrm{Li}_2(-iz)\Big],$$

$$\chi_2(z) := \frac{1}{2}\Big[\mathrm{Li}_2(z) - \mathrm{Li}_2(-z)\Big].$$

For $|y| \le 1$ there also hold the power series representations:

$$\mathrm{Ti}_2(y) = \sum_{k=0}^{\infty} (-1)^k \frac{y^{2k+1}}{(2k+1)^2},$$

$$\chi_2(y) = \sum_{k=0}^{\infty} \frac{y^{2k+1}}{(2k+1)^2}.$$

The special value $\mathrm{Ti}_2(1)$ does not seem to be reducible to known constants, and so is taken to be a new constant of Analysis, known as Catalan's constant and denoted by G:

$$G = \frac{1}{1^2} - \frac{1}{3^2} + \frac{1}{5^2} - \frac{1}{7^2} + \cdots = 0.9159655941\cdots.$$

At present, it is not even known whether G is irrational.

We proceed to derive the power series expansion for the function $x \mapsto x \log(\tan x) - x \log x + x - \mathrm{Ti}_2(\tan x)$, which will then imply an expansion for the function $x \mapsto \mathrm{Ti}_2(\tan x)$. These representations may be of some interest, and to the best of our knowledge are new. From them two similar formulas will be obtained, involving Legendre's chi-function χ_2.

2 The Main Result

Theorem 1 *The following series expansions hold true in the unit disk* $\{z \in \mathbb{C} : |z| \le 1\}$:

$$(A)\ \sum_{m=1}^{\infty} \frac{2\left(2^{2m-1}-1\right)}{m(2m+1)} \frac{\zeta(2m)}{\pi^{2m}} x^{2m+1} = x \log(\tan x) - x \log x + x - \mathrm{Ti}_2(\tan x);$$

$$(B)\ \sum_{m=0}^{\infty} \frac{4\left(2^{2m-1}-1\right)}{2m+1} \frac{\zeta(2m)}{\pi^{2m}} x^{2m+1} = \mathrm{Ti}_2(\tan x);$$

$$(C)\ \sum_{m=1}^{\infty} \frac{(-1)^{m-1} 2\left(2^{2m-1}-1\right)}{m(2m+1)} \frac{\zeta(2m)}{\pi^{2m}} x^{2m+1} = \chi_2\big(\mathrm{th}(x)\big) - x \log\big(\mathrm{th}(x)\big) + x \log x - x;$$

$$(D)\ \sum_{m=0}^{\infty} \frac{(-1)^{m} 4\left(2^{2m-1}-1\right)}{2m+1} \frac{\zeta(2m)}{\pi^{2m}} x^{2m+1} = \chi_2\big(\mathrm{th}(x)\big).$$

Proof Initially, let $x \in \left[0, \frac{\pi}{2}\right)$. For such x, Ewell has proven the identity (see [7]):

$$-x \log x + x - \sum_{n=0}^{\infty} \frac{\sin\left[2(2n+1)x\right]}{(2n+1)^2} - 2\sum_{m=1}^{\infty} \frac{\left(2^{2m-1}-1\right)\zeta(2m)}{m(2m+1)\pi^{2m}} x^{2m+1} = $$
$$= i\left\{\frac{\pi}{2}x - \frac{\pi^2}{8} + \sum_{n=0}^{\infty} \frac{\cos\left[2(2n+1)x\right]}{(2n+1)^2}\right\}.$$

This can only be true if both the real and the imaginary parts vanish, i.e. we must have

$$\frac{\pi}{2}x - \frac{\pi^2}{8} + \sum_{n=0}^{\infty} \frac{\cos\left[2(2n+1)x\right]}{(2n+1)^2} = 0; \tag{1}$$

$$-x\log x + x - \sum_{n=0}^{\infty}\frac{\sin\left[2(2n+1)x\right]}{(2n+1)^2} - \\ - 2\sum_{m=1}^{\infty}\frac{\left(2^{2m-1}-1\right)\zeta(2m)}{m(2m+1)\,\pi^{2m}}x^{2m+1} = 0. \tag{2}$$

We recognize the first sum in (2) to be the same as the one that appears in Ramanujan's formula (see [9]):

$$\sum_{n=0}^{\infty}\frac{\sin\left[2(2n+1)x\right]}{(2n+1)^2} = \mathrm{Ti}_2(\tan x) - x\log(\tan x). \tag{3}$$

Substituting (3) in (2) establishes (A). To prove (B), use the known expansion

$$\log(\tan x) = \log x + \sum_{m=1}^{\infty}\frac{2\left(2^{2m-1}-1\right)}{m}\frac{\zeta(2m)}{\pi^{2m}}x^{2m}, \tag{4}$$

and multiply both sides by x:

$$x\log(\tan x) = x\log x + \sum_{m=1}^{\infty}\frac{2\left(2^{2m-1}-1\right)}{m}\frac{\zeta(2m)}{\pi^{2m}}x^{2m+1}. \tag{5}$$

Substituting (5) in (A) gives

$$\begin{aligned}\mathrm{Ti}_2(\tan x) &= x + \sum_{m=1}^{\infty}\frac{2\left(2^{2m-1}-1\right)}{m}\frac{\zeta(2m)}{\pi^{2m}}x^{2m+1} - \sum_{m=1}^{\infty}\frac{2\left(2^{2m-1}-1\right)}{m(2m+1)}\frac{\zeta(2m)}{\pi^{2m}}x^{2m+1} = \\ &= x + \sum_{m=1}^{\infty}\frac{4\left(2^{2m-1}-1\right)}{2m+1}\frac{\zeta(2m)}{\pi^{2m}}x^{2m+1} = \\ &= \sum_{m=0}^{\infty}\frac{4\left(2^{2m-1}-1\right)}{2m+1}\frac{\zeta(2m)}{\pi^{2m}}x^{2m+1}.\end{aligned}$$

This establishes (B). It is easily seen that the domain of validity of (A) and (B) can be extended to the unit disk $\{z \in \mathbb{C} : |z| \le 1\}$ by analytic continuation. Finally, replacing x with ix in (A) and (B) gives (C) and (D), respectively. The proof is complete.

In the next section, we combine Theorem 1 with known values for $\mathrm{Ti}_2(\cdot)$, $\chi_2(\cdot)$ and $\mathrm{th}(\cdot)$ to obtain new rapidly convergent zeta series. A list of exact values is given in the Appendix; the derivations can be found in [8].

3 New Rapidly Convergent Series Involving Zeta Values

Putting $x = \frac{\pi}{4}$, $x = \frac{\pi}{12}$, $x = \frac{5\pi}{12}$ in Theorem 1(B) gives, respectively:

$$\sum_{m=0}^{\infty} \frac{\left(2^{2m-1}-1\right)\zeta(2m)}{2m+1} \frac{1}{4^{2m}} = \frac{G}{\pi}, \tag{6}$$

$$\sum_{m=0}^{\infty} \frac{\left(2^{2m-1}-1\right)\zeta(2m)}{2m+1} \frac{1}{12^{2m}} = \frac{2G}{\pi} + \frac{1}{4}\log(2-\sqrt{3}), \tag{7}$$

$$\sum_{m=0}^{\infty} \frac{\left(2^{2m-1}-1\right)\zeta(2m)}{2m+1} \left(\frac{5}{12}\right)^{2m} = \frac{2G}{5\pi} + \frac{1}{4}\log(2+\sqrt{3}). \tag{8}$$

Putting $x = \frac{\pi}{4}$, $x = \frac{\pi}{12}$, $x = \frac{5\pi}{12}$ in Theorem 1(A) gives, respectively:

$$\sum_{m=1}^{\infty} \frac{\left(2^{2m-1}-1\right)\zeta(2m)}{m(2m+1)} \frac{1}{4^{2m}} = -\frac{2G}{\pi} - \frac{1}{2}\log\left(\frac{\pi}{4e}\right), \tag{9}$$

$$\sum_{m=1}^{\infty} \frac{\left(2^{2m-1}-1\right)\zeta(2m)}{m(2m+1)} \frac{1}{12^{2m}} = -\frac{4G}{\pi} - \frac{1}{2}\log\left(\frac{\pi}{12e}\right), \tag{10}$$

$$\sum_{m=1}^{\infty} \frac{\left(2^{2m-1}-1\right)\zeta(2m)}{m(2m+1)} \left(\frac{5}{12}\right)^{2m} = -\frac{4G}{5\pi} - \frac{1}{2}\log\left(\frac{5\pi}{12e}\right). \tag{11}$$

The next formula follows from the three-term relation for Ti_2 given in the Appendix (see also [8, p. 47, equations 2.40, 2.41]):

Set $x = \frac{\pi}{24}$, $x = \frac{5\pi}{24}$, $x = \frac{\pi}{8}$ in Theorem 1(B); we have

$$\sum_{m=0}^{\infty} \frac{4\left(2^{2m-1}-1\right)\zeta(2m)}{2m+1} \frac{1}{\pi^{2m}}\left[\frac{\pi}{24}\right]^{2m+1} - \sum_{m=0}^{\infty} \frac{4\left(2^{2m-1}-1\right)\zeta(2m)}{2m+1} \frac{1}{\pi^{2m}}\left[\frac{5\pi}{24}\right]^{2m+1} +$$

$$+ \tfrac{2}{3}\sum_{m=0}^{\infty} \frac{4\left(2^{2m-1}-1\right)\zeta(2m)}{2m+1} \frac{1}{\pi^{2m}}\left[\frac{\pi}{8}\right]^{2m+1} =$$

$$= \mathrm{Ti}_2\left(\tan\frac{\pi}{24}\right) - \mathrm{Ti}_2\left(\tan\frac{5\pi}{24}\right) + \tfrac{2}{3}\mathrm{Ti}_2\left(\tan\frac{\pi}{8}\right) =$$

$$= -\frac{\pi}{6}\log\left[\frac{\tan(5\pi/24)}{\tan(\pi/8)}\right] = \frac{\pi}{6}\log\left[\frac{\sqrt{2}-1}{(\sqrt{3}-\sqrt{2})(\sqrt{2}+1)}\right].$$

Combining terms in the three sums and simplifying gives

$$\sum_{m=0}^{\infty} \frac{\left(2^{2m-1}-1\right)\left(5^{2m+1}-2\cdot 3^{2m}-1\right)}{2m+1}\frac{\zeta(2m)}{24^{2m}} = \log\left[(\sqrt{3}-\sqrt{2})(3+2\sqrt{2})\right]. \tag{12}$$

Similarly to the previous derivation, upon setting $x = \dfrac{\pi}{24}, \; x = \dfrac{5\pi}{24}, \; x = \dfrac{\pi}{8}$ in Theorem 1(A) and using again the three-term relation for Ti_2, we obtain after a lengthy but elementary reduction the sum

$$\sum_{m=1}^{\infty} \frac{\left(2^{2m-1}-1\right)\left(5^{2m+1}-2\cdot 3^{2m}-1\right)}{m(2m+1)}\frac{\zeta(2m)}{24^{2m}} = \log\left[\frac{72\mathrm{e}}{5^{5/2}\pi}\right]. \tag{13}$$

Next, we make use of the fact that $\mathrm{th}\big(\log(\phi)\big) = \dfrac{\sqrt{5}}{5}$, where $\phi = \dfrac{\sqrt{5}+1}{2}$ is the golden ratio: put $x = \log(\phi)$ respectively in Theorem 1(C) and (D); this gives the sums

$$\sum_{m=1}^{\infty} \frac{(-1)^{m-1}2\left(2^{2m-1}-1\right)\zeta(2m)}{m(2m+1)}\left[\frac{\log(\phi)}{\pi}\right]^{2m} =$$

$$= \tfrac{1}{\log(\phi)}\chi_2\left(\frac{\sqrt{5}}{5}\right) + \log\left[\frac{\sqrt{5}\cdot\log(\phi)}{\mathrm{e}}\right], \tag{14}$$

$$\sum_{m=0}^{\infty} \frac{(-1)^m 4\left(2^{2m-1}-1\right)\zeta(2m)}{2m+1}\frac{\left[\log(\phi)\right]^{2m+1}}{\pi^{2m}} = \chi_2\left(\frac{\sqrt{5}}{5}\right). \tag{15}$$

Setting $x = \mathrm{arth}(\sqrt{2}-1) = \tfrac{1}{2}\log(\sqrt{2}+1)$, $x = \mathrm{arth}\left(\tfrac{\sqrt{5}-1}{2}\right) = \tfrac{1}{2}\log(\sqrt{5}+2) = \tfrac{3}{2}\log(\phi)$, $x = \mathrm{arth}(\sqrt{5}-2) = \tfrac{1}{2}\log(\phi)$ in Theorem 1(D) gives, respectively:

$$\sum_{m=0}^{\infty} \frac{(-1)^m\left(2^{2m-1}-1\right)\zeta(2m)}{2m+1}\frac{\left[\tfrac{1}{2}\log(\sqrt{2}+1)\right]^{2m+1}}{\pi^{2m}} =$$

$$= \frac{\pi^2}{64} - \frac{1}{16}\log^2(\sqrt{2}-1)\,; \tag{16}$$

$$\sum_{m=0}^{\infty} \frac{(-1)^m\left(2^{2m-1}-1\right)\zeta(2m)}{2m+1}\frac{\left[\tfrac{3}{2}\log(\phi)\right]^{2m+1}}{\pi^{2m}} =$$

$$= \frac{\pi^2}{48} - \frac{3}{16}\log^2\left(\frac{\sqrt{5}-1}{2}\right); \tag{17}$$

$$\sum_{m=0}^{\infty} \frac{(-1)^m (2^{2m-1}-1)\,\zeta(2m)}{2m+1} \frac{\left[\frac{1}{2}\log(\phi)\right]^{2m+1}}{\pi^{2m}} =$$

$$= \frac{\pi^2}{96} - \frac{1}{48}\log^2(\sqrt{5}-2) = \frac{\pi^2}{96} - \frac{3}{16}\log^2\left(\frac{\sqrt{5}-1}{2}\right). \tag{18}$$

Similarly, set $x = \operatorname{arth}(\sqrt{2}-1) = \frac{1}{2}\log(\sqrt{2}+1)$, $x = \operatorname{arth}\left(\frac{\sqrt{5}-1}{2}\right) = \frac{1}{2}\log(\sqrt{5}+2) = \frac{3}{2}\log(\phi)$, $x = \operatorname{arth}(\sqrt{5}-2) = \frac{1}{2}\log(\phi)$ in Theorem 1(C), respectively; after simplification, we obtain:

$$\sum_{m=1}^{\infty} \frac{(-1)^{m-1} 2(2^{2m-1}-1)\,\zeta(2m)}{m(2m+1)} \left[\frac{\log(\sqrt{2}+1)}{2\pi}\right]^{2m} =$$

$$= \frac{\pi^2}{8\log(\sqrt{2}+1)} + \log\left[\frac{(\sqrt{2}+1)^{1/2}\cdot\log(\sqrt{2}+1)}{2\mathrm{e}}\right]; \tag{19}$$

$$\sum_{m=1}^{\infty} \frac{(-1)^{m-1} 2(2^{2m-1}-1)\,\zeta(2m)}{m(2m+1)} \left[\frac{3\log(\phi)}{2\pi}\right]^{2m} =$$

$$= \frac{\pi^2}{18\log(\phi)} + \log\left[\frac{3\sqrt{\phi}\cdot\log(\phi)}{2\mathrm{e}}\right]; \tag{20}$$

$$\sum_{m=1}^{\infty} \frac{(-1)^{m-1} 2(2^{2m-1}-1)\,\zeta(2m)}{m(2m+1)} \left[\frac{\log(\phi)}{2\pi}\right]^{2m} =$$

$$= \frac{\pi^2}{12\log(\phi)} + \log\left[\frac{\phi^{3/2}\cdot\log(\phi)}{2\mathrm{e}}\right]. \tag{21}$$

Finally we mention that further representations can be obtained via transformations of the above series. For example, subtracting (18) from (17) gives

$$\sum_{m=0}^{\infty} \frac{(-1)^m (2^{2m-1}-1)}{2m+1} \frac{\zeta(2m)}{\pi^{2m}} \left\{\left[\frac{3}{2}\log(\phi)\right]^{2m+1} - \left[\frac{1}{2}\log(\phi)\right]^{2m+1}\right\} =$$

$$= \frac{\pi^2}{96}, \tag{22}$$

and subtracting (21) from (20) gives

$$\sum_{m=1}^{\infty} \frac{(-1)^{m-1} 2(2^{2m-1}-1)(3^{2m}-1)\,\zeta(2m)}{m(2m+1)} \left[\frac{\log(\phi)}{2\pi}\right]^{2m} =$$

$$= \log\left[\frac{3}{\phi}\right] - \frac{\pi^2}{36\log(\phi)}. \tag{23}$$

4 An Interesting Approximation

While performing a numerical verification of (15), we noticed that

$$2 \cdot \chi_2\left(\frac{\sqrt{5}}{5}\right) \approx G,$$

correct to four decimals:

$$2 \cdot \chi_2\left(\frac{\sqrt{5}}{5}\right) = 0.91590089\ldots, \text{whereas } G = -i \cdot \chi_2(i) = 0.91596559\ldots.$$

5 Appendix: Exact Values

The following three values are known for $\mathrm{Ti}_2(\cdot)$:

$$\begin{aligned}
&\mathrm{Ti}_2(1) = G; && \left[\tan\left(\frac{\pi}{4}\right) = 1\right]\\
&3\mathrm{Ti}_2(2-\sqrt{3}) = 2\mathrm{Ti}_2(1) + \frac{\pi}{4}\log(2-\sqrt{3}); && \left[\tan\left(\frac{\pi}{12}\right) = 2-\sqrt{3}\right]\\
&3\mathrm{Ti}_2(2+\sqrt{3}) = 2\mathrm{Ti}_2(1) + \frac{5\pi}{4}\log(2+\sqrt{3}); && \left[\tan\left(\frac{5\pi}{12}\right) = 2+\sqrt{3}\right]
\end{aligned}$$

In addition, there holds the three-term relation:

$$\mathrm{Ti}_2\left(\tan\frac{\pi}{24}\right) - \mathrm{Ti}_2\left(\tan\frac{5\pi}{24}\right) + \tfrac{2}{3}\mathrm{Ti}_2\left(\tan\frac{\pi}{8}\right) = -\frac{\pi}{6}\log\left[\frac{\tan(5\pi/24)}{\tan(\pi/8)}\right],$$

or equivalently

$$\mathrm{Ti}_2\left(\frac{\sqrt{3}-\sqrt{2}}{\sqrt{2}+1}\right) - \mathrm{Ti}_2\left(\frac{\sqrt{3}-\sqrt{2}}{\sqrt{2}-1}\right) + \tfrac{2}{3}\mathrm{Ti}_2\left(\sqrt{2}-1\right) = \frac{\pi}{6}\log\left[\frac{\sqrt{2}-1}{(\sqrt{3}-\sqrt{2})(\sqrt{2}+1)}\right].$$

The following values are known for $\chi_2(\cdot)$:

$$\chi_2\left(\tan\frac{\pi}{8}\right) = \frac{\pi^2}{16} - \frac{1}{4}\log^2\left(\tan\frac{\pi}{8}\right),$$

or equivalently

$$\chi_2\left(\sqrt{2}-1\right) = \frac{\pi^2}{16} - \frac{1}{4}\log^2\left(\sqrt{2}-1\right);$$
$$\chi_2\left(\frac{\sqrt{5}-1}{2}\right) = \frac{\pi^2}{12} - \frac{3}{4}\log^2\left(\frac{\sqrt{5}-1}{2}\right);$$
$$\chi_2\left(\sqrt{5}-2\right) = \frac{\pi^2}{24} - \frac{1}{12}\log^2\left(\sqrt{5}-2\right).$$

References

1. Ayoub, R.: Euler and the Zeta function. Amer. Math. Monthly **81**, 1067–1086 (1974)
2. Chen, M.-P., Srivastava, H.M.: Some families of series representations for the Riemann $\zeta(3)$. Result. Math. **33**, 179–197 (1998)
3. Choe, R.: An elementary proof of $\sum_{n=1}^{\infty}\frac{1}{n^2}=\frac{\pi^2}{6}$. Amer. Math. Monthly **94**, 662–663 (1987)
4. Dąbrowski, A.: A note on the values of the Riemann Zeta Function at positive odd integers. Nieuw Arch. Wisk. **14**, 199–207 (1996)
5. Euler, L.: Exercitationes analyticae. Novi Comment. Acad. Sci. Imp. Petropol. **17**, 173–204 (1772)
6. Ewell, J.: A new series representation for $\zeta(3)$. Amer. Math. Monthly **97**, 219–220 (1990)
7. Ewell, J.: On the zeta function values $\zeta(2k+1)$, $k=1,2,\ldots$. Rocky Mountain J. Math. **25**, 1003–1012 (1995)
8. Lewin, L.: Polylogarithms and Associated Functions. Elsevier (North-Holland), New York/London/Amsterdam (1981)
9. Ramanujan, S.: On the integral $\int_0^x \frac{\tan^{-1}t}{t}dt$. J. Indian Math. Soc. **7**, 93–96 (1915)
10. Srivastava, H.M.: Some properties and results involving the zeta and associated functions. Funct. Anal. Approx. Comput. **7**, 89–133 (2015)
11. Srivastava, H.M., Choi, J.: Series Associated with the Zeta and Related Functions. Kluwer Academic Publishers, Dordrecht/Boston/London (2001)
12. Srivastava, H.M., Tsumura, H.: Inductive construction of rapidly convergent series representations for $\zeta(2n+1)$. Internat. J. Comput. Math. **80**, 1161–1173 (2003)
13. Zhang, N.-Y., Williams, K.S.: Some series representations of $\zeta(2n+1)$. Rocky Mountain J. Math. **23**, 1581–1591 (1993)

Simulated Annealing Method for Metal Nanoparticle Structures Optimization

Vladimir Myasnichenko, Leoneed Kirilov, Rossen Mikhov, Stefka Fidanova and Nikolay Sdobnyakov

Abstract The goal of this paper is to develop an efficient method to search for metal and bimetal nanoparticle structures with the lowest possible potential energy. This is a global optimization problem. In computational complexity theory, global optimization problems are NP-hard, meaning that they cannot be solved in polynomial time. Because of the severe difficulty of finding the global minimum, the simulated annealing algorithm was selected as main strategy. At the first step we use the lattice Monte Carlo method with different lattices. Then we relax the resulting nanoparticle structures at low temperature within molecular dynamics, choosing one of them as approximation of the global minimum. The numerical solution of an optimal cluster structure of Ag (200) shows the efficiency of the proposed method.

Keywords Nanostructures configurations · Global optimization · Hybrid approach · Simulated annealing · Monte Carlo method

1 Introduction

At present, it is already obvious that the structure of nanoclusters plays a decisive role in the study of the thermodynamic characteristics determined in the course of phase transitions (melting/crystallization). The description of the mechanisms of formation and dynamics of changes in the internal structure of nanoparticles can allow predicting the properties of these nanoparticles. Despite the modern development of the experimental base and theoretical approaches, certain problems in the study of structural characteristics, including the search for stable configurations, the descrip-

---

V. Myasnichenko (✉) · N. Sdobnyakov
Tver State University, Tver, Russia
e-mail: viplabs@yandex.ru

L. Kirilov · R. Mikhov · S. Fidanova
Institute of Information and Communication Technologies—Bulgarian Academy of Sciences, Sofia, Bulgaria
e-mail: l_kirilov_8@abv.bg

K. Georgiev et al. (eds.), *Advanced Computing in Industrial Mathematics*,
Studies in Computational Intelligence 793,
https://doi.org/10.1007/978-3-319-97277-0_23

tion of the criteria for thermal stability, and also in the study of the evolution of the structure, depending on the size, are far from being solved.

Further, we note that to date the stability/instability problem of nanoparticles does not even have a clear formulation and, accordingly, a comprehensive solution, despite its unconditional importance from the fundamental and applied points of view. Some aspects of the thermal stability of metallic nanomaterials were discussed in the review [14]. The concept of mechanical stability/instability of nanoparticles, due to the magnitude of volume fluctuations, was proposed in [20].

The purpose of this work is to reveal the regularities and conditions for the formation of the most stable free globular metal nanoclusters in the range of sizes.

2 Available Approaches

Randomly searching through all possible minima on a potential energy surface is impractical. Instead, computer algorithms were written to search for minima that are believed to be close to the global minimum (GM). In other words, the GM was approximated using time-efficient optimization strategies. The difficulty of global optimization of atomic clusters is even greater for nanoalloys, where permutation of unalike atoms likewise leads to many more isomers [14, 20].

A comparison between the two search techniques, molecular dynamics and basin-hopping [34, 35] Monte Carlo, has been performed in [30]. It was found that the basin-hopping algorithm is more efficient than a molecular dynamics minimization approach in the investigation of the most stable conformations for metal clusters.

The convergence of a class of Metropolis-type Markov-chain annealing algorithms for global optimization of a smooth function was established in [6]. Metropolis Monte Carlo with a simulated annealing (SA) procedure was used to search for GM configurations (up to 3000 atoms) of transition metals nanoparticles in [24].

The investigating of nanoparticle structures is very important for creating new metals with predefined specific properties. The basic step of investigating metal clusters is to locate their stable geometrical structures. Potential functions play an important role in computer simulation of metal structures [22]. There exist different approaches: the Gupta potential function, for example, has been widely applied to study face-centered cubic structures [4], hexagonal closed packed structures [18], transition metals and their alloy clusters [13]. The stable geometrical structure of six metallic clusters has been optimized by a dynamic lattice search method [40].

Various methods have been developed to solve these kinds of problems. In [26, 35, 36], methods for finding the minimal potential energy in large molecules are proposed. In [16, 17], the basin-hopping algorithm is applied for global optimization to find the global minima of the Lennard-Jones cluster. In [37] is proposed a global optimization analysis of water clusters. Other approaches are fast annealing evolutionary algorithms [1, 2], random tunneling algorithms [15, 32], genetic algorithms (GAs) [8, 11, 39], simulated annealing approaches [21, 29, 41]. Due to the large number of parameters to be optimized, a local search method is necessary to improve the optimization algorithm and to speed-up the search process [31]. Some authors apply the limited memory quasi-Newton method [19]. In [1, 2], binary clustering using dynamic lattice is applied. In [15] is proposed an adaptive immune optimization algorithm. A Parallel random tunneling algorithm for structural optimization is proposed in [32]. A stochastic algorithm for global optimization of Lennard-Jones clustering is applied in [16]. In [23] there is a review of the complex functional materials and existing methods to determine their atomic structure.

Successful optimization methods were developed to solve the problem in the last decades. Examples include variants of the basin-hopping algorithm [7], minimum hopping algorithm [15], random tunneling algorithm, adaptive immune optimization algorithm [3], heuristic algorithm with the surface and interior operators [33], high-efficiency differential evolutionary algorithm [28]. A cluster surface smoothing method that can quickly locate the minimum of the funnels in the potential energy surface is designed in [25].

In [28] is studied the implementation of the basin-hopping algorithm for the optimization of the potential energy of nanoclusters that combines two different types of moves, namely shape-changing moves and exchange moves. The approach is demonstrated on nanoalloys of 400 atoms composed by Au and Rh or by Au and Cu. A review of the new applications of GAs in materials science and in related fields (solid state physics and chemistry, crystallography, production and engineering) is presented in [25] with representative examples. In [27], a global optimization problem of determining the n-atom cluster configuration that yields the minimum Lennard-Jones potential energy is studied by a genetic algorithm combined with a stochastic search procedure on icosahedral lattices. In [5], the physical aspects of the global optimization of the geometry of atomic clusters are studied. Namely, the structural principles that determine the nature of the lowest energy structure, the physical reasons why some clusters are especially difficult to optimize, and how the basin-hopping transformation of the potential energy surface enables these difficult clusters to be optimized.

The silver clusters have become a hot topic in clusters science, because of their application in catalysis, photography, electronic materials and production of metal materials [12]. Therefore the structure of silver clusters needs to be optimized according the potential function which gives the stable structure of the metal.

3 New Approach

Because of the severe difficulty of finding the GM, the simulated annealing (SA) algorithm was selected as main strategy. At the first step we use the lattice Monte Carlo method with different lattices. Then we relax the resulting nanoparticle structures at low temperature within molecular dynamics, choosing one of them as an approximation of the global minimum.

The interaction of atoms is modeled by the Gupta potential (see [4]), with a maximum radius of interaction R_{cut}:

$$E = \sum_i \left(\sum_{j \neq i} E_{ij}(a, b) - \sqrt{\sum_{j \neq i} B_{ij}(a, b)} \right) \quad (1)$$

$$E_{ij}(a, b) = A_{ab} \exp\left(-p_{ab} \left(\frac{r_{ij}}{r_{0,ab}} - 1 \right) \right) \quad (2)$$

$$B_{ij}(a, b) = \xi_{ab}^2 \exp\left(-2q_{ab} \left(\frac{r_{ij}}{r_{0,ab}} - 1 \right) \right), \quad (3)$$

where i ranges over all atoms; j ranges over all atoms other than i but within distance R_{cut} from i; a and b represent the species of the atoms i and j; $E_{ij}(a, b)$ is the repulsive component of the potential due to the atoms i and j; $B_{ij}(a, b)$ is the binding component of the potential due to the atoms i and j; r_{ij} is the distance between the atoms; $r_{0,ab}$, A_{ab}, p_{ab}, ξ_{ab}, q_{ab} are parameters that depend only on the species of the atoms. R_{cut} is the maximum distance beyond which the interaction is assumed to be zero.

The algorithm takes as input a list of nodes, given by their Cartesian coordinates. Each node can either be empty or contain an atom. The distance between adjacent nodes in the input data may slightly vary (within 15%). We keep the nodes in two arrays `N` and `A`, where `N` gives the index of a node into `A` and `A` gives the index of a node into `N`. `N` is sorted in input order, while `A` is sorted to begin with the atoms and end with the holes. This allows us to query information about nodes and atoms, select atoms at random, add, remove and move atoms around, all in constant time.

Before starting the main cycle, we do some pre-computation, inspired by the approach discussed in [23]:

Step A. For each node i, compute the list of nodes j within distance R_{cut} from it. As a sub-list, remember the list of immediate neighbors of i.

Step B. For each node i, each node j within distance R_{cut} from i, and each combination of a and b, use (2) and (3) to compute the values of $E_{ij}(a, b)$ and $B_{ij}(a, b)$. From this point on, we can forget about Cartesian coordinates and work exclusively with indexes and these pre-computed values, according to (1).

Step C. For each atom i, compute $\sum_{j \neq i (r_{ij} \leq R_{\text{cut}})} B_{ij}(a, b)$. This value is used to simplify Step 5. It is remembered and kept updated throughout the algorithm (namely, when placing and moving atoms).

After pre-computation, a number of atoms are placed in the empty nodes, initially at random. Then starts the main cycle, which consists of the following steps:

Step 1. Check the exit criteria. The algorithm stops when either the requested number of iterations is exceeded, or the system has reached equilibrium. Equilibrium is declared whenever we have a number of iterations without significant decrease of the energy.

Step 2. Adjust the temperature. This check is performed once every several thousand iterations and the temperature changes according to the following formula:

$$T = \max\{1, T_0 + s\Delta T\}, \tag{4}$$

where T_0 and ΔT are constants, and s is the iteration number.

Step 3. Choose an atom at random.

Step 4. Choose a neighboring empty node at random. This can be done in constant time, owing to the pre-computations of Step A. If there are no empty neighbors, return to Step 1.

Step 5. Calculate the potential energy difference for the atom moving into the selected empty node. This can be done very efficiently, owing to the pre-computed energy values of Step B. We need to iterate over the R_{cut}-vicinities of the atom and of the empty node, but not over vicinities-of-vicinities, owing to the pre-computations of Step C. The updated energies are remembered at this point, in case the atom will jump into the empty node.

Step 6. If the energy would not increase, perform the jump and return to Step 1.

Step 7. Otherwise, calculate the jump probability $P = \exp(-\Delta E/kT)$, and generate a random number p $(0 \leq p < 1)$.

Step 8. If the number is smaller than the probability, perform the jump, otherwise do nothing. Either way, return to Step 1

The flowchart of the method is given in Fig. 1.

Comments:

(1) To date, a sufficiently large number of works have been devoted to the solution of the problem of metallic potentials development [4, 9]. Usually they employ experimental data on the energy of atomic cohesion. At present, the Gupta potential is used most frequently in the calculation of the properties of metallic nanoclusters [10, 38]. Like other metallic potentials, it is many-body and depends on local density. Here we apply a variant of the Gupta potential in the form of Cleri-Rosato [4].

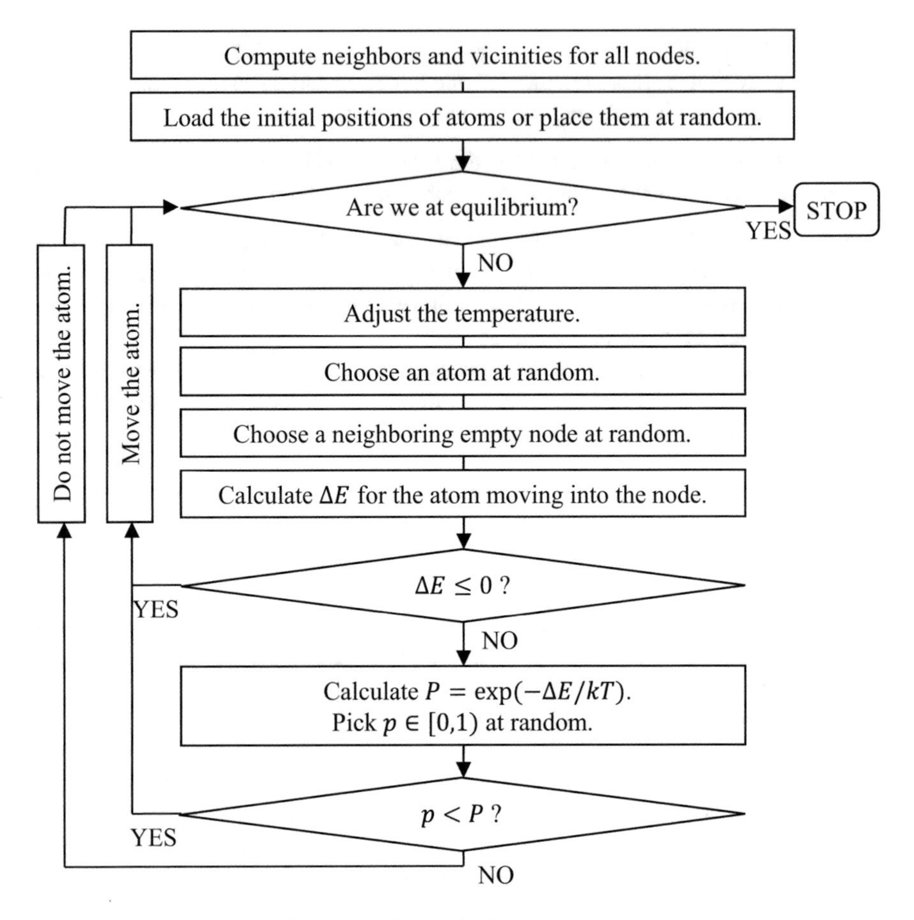

Fig. 1 Flowchart of the simulated annealing method

(2) There is no temperature effect increasing the potential energy because comparison of the resulting energy occurs after cooling by the Molecular Dynamics method down to 0.01 K.
(3) Only the running time of the computer experiment is limited. It is understood that the maximum number of atoms per cluster can be several tens of thousands.

4 Results

We present the optimal cluster structure of Ag (200) (silver 200 atoms) on a decahedral lattice of 432 nodes. At the beginning 130 nodes are filled already with Ag

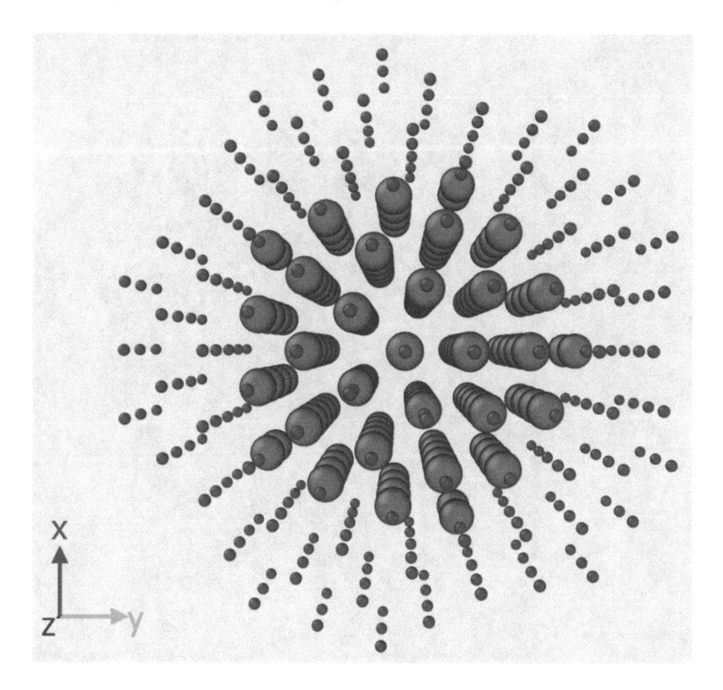

Fig. 2 Ag130_initial: a perspective projection for the initial configuration containing the filled 130 atom nodes

atoms. The optimization is calculated by our program Tsuyoyama, written in C, implementing the above method.

On Fig. 2 is presented a perspective projection for the initial configuration. Figures 3, 4 and 5 visualize the minimum obtained by our method.

The form of such global minimum was first published in [12].

Two series of experiments were conducted for the same Ag200 cluster.

Further we show:

(1) The influence of the number of filled nodes (in the initial configuration) on the resultant energy of the cluster—see Fig. 6. Averaged over 10 experiments.
(2) The effect of the stretching coefficient (identical for all three coordinate axes) for a lattice with 130 filled atoms—see Fig. 7. Averaged over 10 experiments.

The following comments can be made:

(1) It is optimal to begin with an initial filling/initialization of 130 atoms.

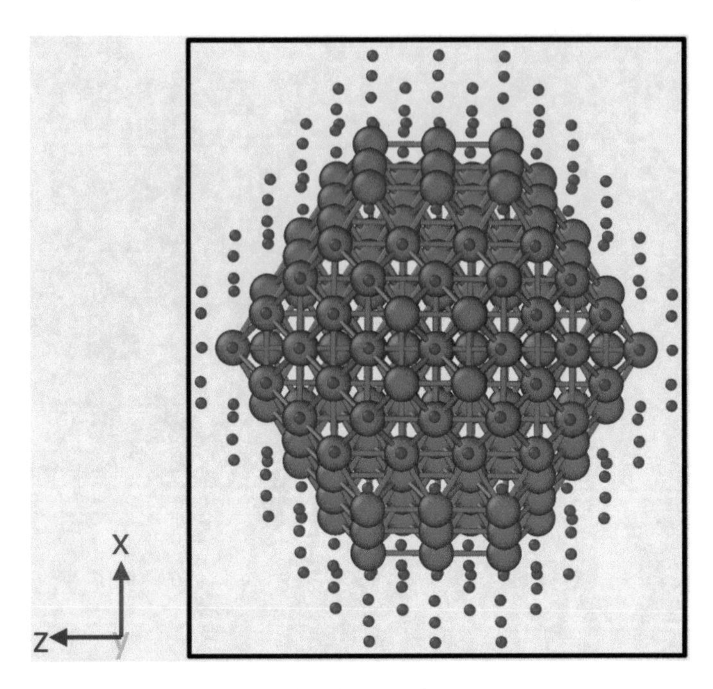

Fig. 3 Ag200 global minimum—projection along *Y* axis

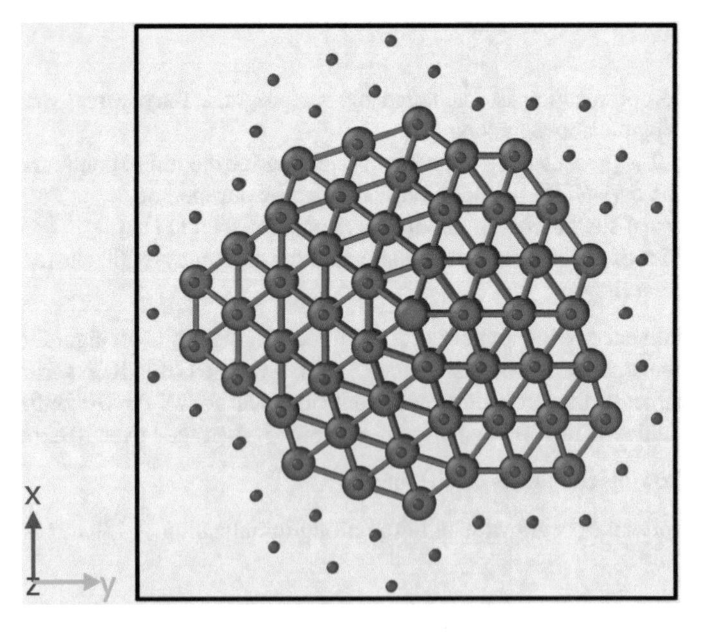

Fig. 4 Ag200 global minimum—projection along *Z* axis

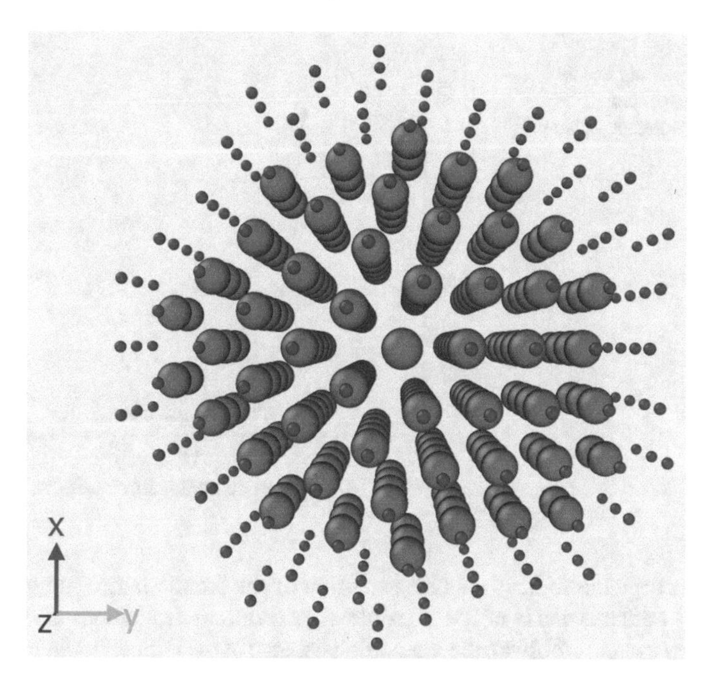

Fig. 5 Ag200 global minimum—perspective projection

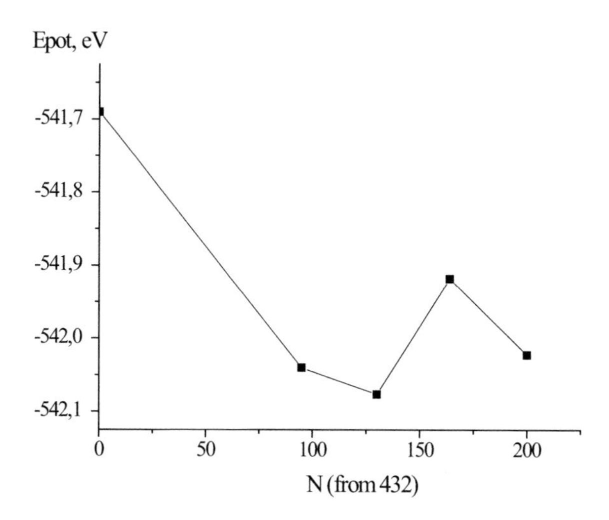

Fig. 6 Influence of the number of pre-filled lattice nodes on the total resulting potential energy of the Ag200 cluster

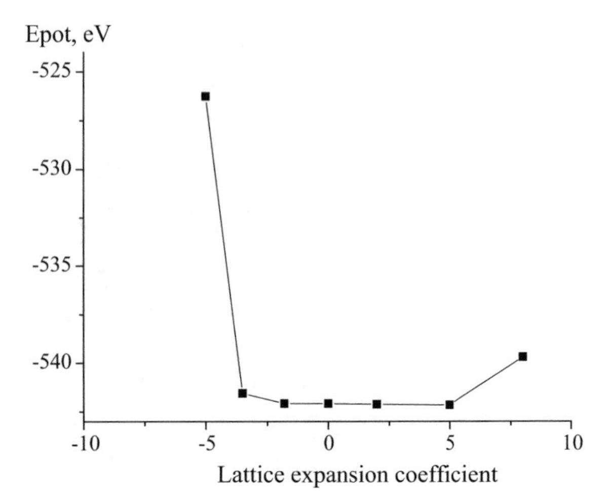

Fig. 7 Effect of the isotropic stretching on the total resulting potential energy of the Ag200 cluster

(2) It is acceptable to have a small extension of the atomic lattice, up to 5%, but lattice compression is not permissible. As a result, in simulations of binary and ternary systems (alloys), the interstitial distance of the lattice should be selected for metals with a large atomic radius.

The parameters for this sample run are as follow:

- Maximum number of iterations: 16,590,000;
- Starting temperature: $T_0 = 2500$ K;
- Step for changing the temperature $\Delta T = -0.00015$ K;

The temperature was changed according to (4) once every 5000 iterations. The final temperature (at iteration 16,590,000) was 12.25 K.

5 Conclusion

An efficient method for searching for metal and bimetal nanoparticle structures with the lowest possible potential energy is presented in the paper. The simulated annealing (SA) algorithm was selected as a main strategy. At the first step we use the lattice Monte Carlo method with different lattices. Then we relax the resulting nanoparticle structures at low temperature within molecular dynamics, choosing one of them as approximation of global minimum.

The numerical solution of a real optimal cluster structure of Ag (200) is discussed in details.

Acknowledgements The research presented here is partially supported by the Russian Foundation for Basic Research (project No. 17-53-04010 and project No. 18-38-00571) and the Bulgarian National Scientific Fund under the grant DFNI-DN 12/5 "Efficient Stochastic Methods and Algorithms for Large Scale Problems".

References

1. Cai, W.S., Shao, X.G.: A fast annealing evolutionary algorithm for global optimization. J. Comput. Chem. **23**(4), 427–435 (2002)
2. Cai, W.S., Feng, Y., Shao, X.G., Pan, Z.X.: Optimization of Lennard-Jones atomic clusters. J. Mol. Struct. (Theochem) **579**(1), 229–234 (2002)
3. Cheng, L.J., Cai, W.S., Shao, X.G.: A connectivity table for cluster similarity checking in the evolutionary optimization method. Chem. Phys. Lett. **389**(4), 309–314 (2004)
4. Cleri, F., Rosato, V.: Tight-binding potentials for transition metals and alloys. Phys. Rev. B **48**, 22–33 (1993)
5. Doye, J.P.K.: Physical perspectives on the global optimization of atomic clusters. In: Pintér, J.D. (eds) Global Optimization. Nonconvex Optimization and Its Applications, vol. 85, pp. 103–139. Springer, Boston, MA (2006)
6. Gelfand, S.B., Mitter, S.K.: Metropolis-type annealing algorithms for global optimization in {R}^d. SIAM J. Control Optim. **31**(1), 111–131 (1993)
7. Goedecker, S.: Minima hopping: An efficient search method for the global minimum of the potential energy surface of complex molecular systems. J. Chem. Phys. **120**(21), 9911–9917 (2004)
8. Gregurick, S.K., Alexander, M.H., Hartke, B.: Global geometry optimization of (Ar)n and B(Ar)n clusters using a modified genetic algorithm. J. Chem. Phys. **104**(7), 2684–2691 (1996)
9. Guevara, J., Llois, A.M., Weissmann, M.: Model potential based on tight-binding total-energy calculations for transition-metal systems. Phys. Rev. B: Condens. Matter. **52**(15), 11509–11516 (1995)
10. Gupta, R.P.: Lattice relaxation at a metal surface. Phys. Rev. B: Condens. Matter. **23**(12), 6265–6270 (1981)
11. Hartke, B.: Global cluster geometry optimization by a phenotype algorithm with Niches: Location of elusive minima, and low-order scaling with cluster size. J. Comput. Chem. **20**(16), 1752–1759 (1999)
12. Huang, Wenqi, Lai, Xiangjing, Ruchu, Xu: Structural optimization of silver clusters from Ag141 to Ag310 using a modified dynamic lattice searching method with constructed core. Chem. Phys. Lett. **507**(1), 199–202 (2011)
13. Husic, B.E., Schebarchov, D., Wales, D.J.: Impurity effects on solid–solid transitions in atomic clusters. NANO **8**, 18326–18340 (2016)
14. Jellinek, J., Krissinel, E.B.: NinAlm alloy clusters: analysis of structural forms and their energy ordering. Chem. Phys. Lett. **258**(1–2), 283–292 (1996)
15. Jiang, H.Y., Cai, W.S., Shao, X.G.: A random tunneling algorithm for the structural optimization problem. Phys. Chem. Chem. Phys. **4**(19), 4782–4788 (2002)
16. Leary, R.H.: Global Optimization on Funneling Landscapes. J. Global Optim. **18**(4), 367–383 (2000)

17. Leary, R.H., Doye, J.P.K.: Tetrahedral global minimum for the 98-atom Lennard-Jones cluster. Phys. Rev. E **60**(6), R6320–R6322 (1999)
18. Li, X.J., Fu, J., Qin, Y., Hao, S.Z., Zhao, J.J.: Gupta potentials for five HCP rare earth metals. Comput. Mater. Sci. **112**, 75–79 (2016)
19. Liu, D.C., Nocedal, J.: On the limited memory BFGS method for large scale optimization. Math. Prog. B **45**(1), 503–528 (1989)
20. Lloyd, L.D., Johnston, R.L., Salhi, S., Wilson, N.T.: Theoretical investigation of isomer stability in platinum-palladium nanoalloy clusters. J. Mater. Chem. **14**(11), 1691–1704 (2004)
21. Ma, J.P., Straub, J.E.J.: Simulated annealing using the classical density distribution. Chem. Phys. **101**(1), 533–541 (1994)
22. Michaelian, K., Rendón, N., Garzón, I.L.: Structure and energetics of Ni, Ag, and Au nanoclusters. Phys. Rev. B **60**, 2000–2010 (1999)
23. Myshlavtsev, A.V., Stishenko, P.V.: Modification of the metropolis algorithm for modeling metallic nanoparticles. Omsk scientific newspaper No 1(107) 21–25 (in Russian) (2012)
24. Myshlyavtsev, A.V., Stishenko, P.V., Svalova, A.I.: A systematic computational study of the structure crossover and coordination number distribution of metallic nanoparticles. Phys. Chem. Chem. Phys. **19**(27), 17895–17903 (2017)
25. Paszkowicz, W.: Genetic algorithms, a nature-inspired tool: a survey of applications in materials science and related fields: part II. Mater. Manuf. Process. **28**, 708–725 (2013). https://doi.org/10.1080/10426914.2012.746707
26. Pillardy, J., Liwo, A., Scheraga, H.A.: An efficient deformation-based global optimization method (Self-Consistent Basin-to-Deformed-Basin Mapping (SCBDBM)). Application to Lennard-Jones Atomic Clusters. J. Phys. Chem. A **103**(46), 9370–9377 (1999)
27. Romero, D., Barrón, C., Gómez, S.: The optimal geometry of Lennard-Jones clusters: 148–309. Comput. Phys. Commun. **123**(1999), 87–96 (1999)
28. Rossi, G., Ferrando, R.: Combining shape-changing with exchange moves in the optimization of nanoalloys. Comput. Theor. Chem. (in press) (2017)
29. Schelstraete, S., Verschelde, H.J.: Finding minimum-energy configurations of Lennard-Jones clusters using an effective potential. Phys. Chem. A **101**(3), 310–315 (1997)
30. Sebetci, A., Güvenç, Z.B.: Global minima for free Pt_N clusters (N = 22–56): a comparison between the searches with a molecular dynamics approach and a basin-hopping algorithm. Eur. Phys. J. D **30**(1), 71–79 (2004)
31. Shao, X.G., Cheng, L.J., Cai, W.S.: A dynamic lattice searching method for fast optimization of Lennard-Jones clusters. J. Comput. Chem. **25**(14), 1693–1698 (2004)
32. Shao, X.G., Jiang, H.Y., Cai, W.S.: Parallel random tunneling algorithm for structural optimization of Lennard-Jones clusters up to N = 330. J. Chem. Inf. Comput. Sci. **44**(1), 193–199 (2004)
33. Takeuchi, H.: Clever and efficient method for searching optimal geometries of Lennard-Jones clusters. J. Chem. Inf. Model. **46**(5), 2066–2070 (2006)
34. Wales, D.J.: Global optimization of clusters, crystals, and biomolecules. Science 1999, Science **285**(5432), 1368–1372 (1999)
35. Wales, D.J., Doye, J.P.K.: Global optimization by basin-hopping and the lowest energy structures of Lennard-Jones clusters containing up to 110 atoms: condensed matter; atomic and molecular clusters. J. Phys. Chem. A **101**(28), 5111–5116 (1997)
36. Wales, D.J., Scheraga, H.A.: Global optimization of clusters, crystals, and biomolecules. Science 1999, Science **285**(5432), 1368–1372 (1999)
37. White, R.P., Mayne, H.R.: An investigation of two approaches to basin hopping minimization for atomic and molecular clusters. Chem. Phys. Lett. **287**(5–6), 463–468 (1998)
38. Wilson, N.T., Johnson, R.L.: A theoretical study of atom ordering in copper–gold nanoalloy clusters. J. Mater. Chem. **12**(10), 2913–2922 (2002)

39. Wolf, M.D., Landman, U.: Genetic algorithms for structural cluster optimization. J. Phys. Chem. A **102**(30), 6129–6137 (1998)
40. Xia, W., Sun, Y.: Stable structures and potential energy surface of the metallic clusters: Ni, Cu, Ag, Au, Pd, and Pt. J. Nanopart. Res. **19**, 201 (2017). https://doi.org/10.1007/s11051-017-3907-6
41. Xue, G.L.: Improvement on the northby algorithm for molecular conformation: Better solutions. J. Global Optim. **4**(4), 425–440 (1994)

Orientation Selectivity Tuning of a Spike Timing Neural Network Model of the First Layer of the Human Visual Cortex

Simona Nedelcheva and Petia Koprinkova-Hristova

Abstract The paper deals with the influence of some parameters determining spatial structure of a spike timing neural network model of the first layer of the human visual cortex on its orientation selectivity. For this aim the model was implemented in NEST simulator and a recently proposed approach for spatial structure design of the orientation columns in its recurrent layer was adopted. The aim was to tune the model to recognize spacial orientation of moving through the visual field stimuli with different size and orientation. The values of the parameters defining columns position and thickness as well as the photo-receptors size and variance of the LGN neurons receptive fields were determined in dependence on the stimuli characteristics. The obtained results showed that bigger size stimuli were detected by wider receptive fields while orientation of smaller stimuli was properly recognized by thicker orientation columns.

1 Introduction

The neuroscience was interested in functioning of visual system of living creatures for centuries. There are numerous neurophysiological findings from in-vivo experiments with mammals (cats, rats, monkeys etc.) as well from human brain investigations [7, 12, 14]. By far there there exist a commonly accepted hierarchical model of the visual system [3, 5]. Its firs layer starts with common for all creatures light sensor - the eye. The sensory cells in the retina convert a light stimulus into an electrical signal that is passed to the next processing layer in the brain - thalamus - trough the optic nerve. Here is observed a structure of neurons called lateral geniculate nucleus (LGN) that acts as a relay station between the retina and the next layer - primary

S. Nedelcheva (✉) · P. Koprinkova-Hristova
Institute of Information and Communication Technologies, Bulgarian Academy of Sciences, Acad. G. Bonchev str. bl.25A, 1113 Sofia, Bulgaria
e-mail: croft883@gmail.com

P. Koprinkova-Hristova
e-mail: pkoprinkova@bas.bg

K. Georgiev et al. (eds.), *Advanced Computing in Industrial Mathematics*, Studies in Computational Intelligence 793,
https://doi.org/10.1007/978-3-319-97277-0_24

visual cortex (V1). It is also discovered that individual neurons in these three layers (retina, LGN, and V1) receive and process information from spatially closer regions from the previous layer (or visual field for the retina layer). These regions, specific for each type of neurons, are called receptive fields. In the retina there are two types of receptive fields in dependence on whether corresponding neuron firing is enhanced by positive or negative difference between light intensity at a given spot and its surrounding. Since in awake animals eyes are constantly moving, the neurons in visual processing layers react strongly to sudden transitions in the level of image illumination, i.e. they accumulate temporal as well as spatial information from the observed world. LGN neurons have similar receptive fields like the corresponding to them retinal cells. They respond best to circular spots of light surrounded by darkness or dark spots surrounded by light. Neurons in V1 layer respond best to elongated light or dark bars or to boundaries between light and dark regions. Hence due to the shape of their receptive fields neurons in primary visual cortex are directionally selective, i.e. they respond more strongly to stimuli moving in one direction than in the other.

Besides the inter-layer connections it is well known that lateral connections between spatially closer neurons from the same layer also exist. In dependence on whether a given connection enhances or depresses the receiving neuron activity, connections were considered to be exciting or inhibiting. There are several topological structures of the primary visual system [4] among which [10, 14] is commonly accepted. It is based on in-vivo experimental results reported in [14] and since then numerous researchers adopted this structure.

Recent investigations of the spatial structure of visual cortex in higher mammals (e.g. macaques or cats) revealed that it has columnar structure, i.e. neurons with identical directional selectivity are spatially grouped into columns. The map presenting such a spatial organization of a layer of neurons is called "orientation map". It was observed that orientation maps in mammalian brain have "pinwheel-structure". Different computational models were developed in attempt to create such maps through a process of self-organization or other techniques [1, 6, 9]. But it still remains not completely clear how these maps emerged and function. In [13] a new model of orientation selectivity is proposed in attempt to explain the emergence of orientation maps. It is able to generate spatial patterns of orientation selectivity closer to the maps found in cats or monkeys.

One more thing that complicates brain models further, is that individual neurons of mammalian brain are nonlinear processing units. There exits variety of models revealing neural cells functioning [8]. Among them spike timing models were considered as closer to reality and biologically plausible approximation of the complex bio-chemical reactions that are engines of neurons functionality. In order to support development of biologically plausible brain models an open source library called NEST simulator was developed [11]. It allows simulation of hierarchical structures of neurons with complex dynamics and variety of dynamical connections.

Here we adopted the model structure proposed in [10, 14]. Since the orientation sensitivities of neurons in that model were randomly selected, we upgraded it using approach for orientation columns generation from [13]. In literature most simulation experiments tested orientation selectivity of such models using sinus grating stimulation since it was reported as a proper signal for this aim [3]. However, such kind of stimuli are not natural. Since we aim to extend further our model to mimic human eye movements as reaction to natural stimulation, we tested the orientation selectivity of the model generating a moving rectangular shape with different size and orientation. Our aim was to select combination of parameters able to distinguish between two possible orientation of the stimuli. The values of the parameters defining columns position and thickness as well as the sizes of photo-receptors, LGN and V1 neurons receptive fields were chosen so that the average spike frequency of all neurons in the output layer with corresponding to the stimuli orientation is higher than that of groups of neurons with different orientation selectivity.

First experiments showed that bigger size stimuli were detected by wider receptive fields of V1 neurons with respect to LGN layer while orientation of smaller stimuli was properly recognized by thicker orientation columns in V1 layer. Our next aim will be to tune also receptive fields of laminar connections.

The rest of the paper is organized as follows: Sect. 2 describes the model structure and its parameters; Sect. 3 describes the simulation investigations and the obtained results; the paper finishes with concluding remarks and directions for future work.

2 Model Structure and Parameters

The model structure, adopted from [10], is shown on Fig. 1. It consists of three layers of neurons following the hierarchical structure of early visual system: retina receptors, LGN and V1. Neurons in each layer are positioned on two-dimensional regular grids.

The structures of photo-receptors/LGN and V1 layer are showed on Fig. 2.

According to Fig. 1 we have two layers of photo-receptors/LGN neurons. In our model each photo-receptor/LGN layer consists of 100 cells placed on a regular gird of 10×10 size. Both layers have identical positions of ON and OFF receptors placed in reverse order, as it is demonstrated on the first two plots of Fig. 3. The number of retina and LGN neurons is the same since each photo-receptor cell is connected to its corresponding LGN neuron. Their positions are relative to the visual scene.

The V1 layer consists of totally 1000 neurons, separated into four groups - two exciting (E1 and E2) and two inhibiting (I1 and I2) populations. According to [10, 14] the ratio exiting/inhibiting neurons should be 4/1. Hence in our model the size of each excitatory population is 400 neurons while the size of each inhibitory population is 100 neurons. All neurons of the V1 layer are placed on regular grids of size 20×20 and 10×10 neurons respectively for excitatory and inhibitory groups. These grids are overlapped at the same plane of the V1 model layer, as it is shown on Fig. 3, last plot. Thus the inhibiting neurons are dispersed among bigger group of exciting neurons. The next section describes the used neuron models for each layer.

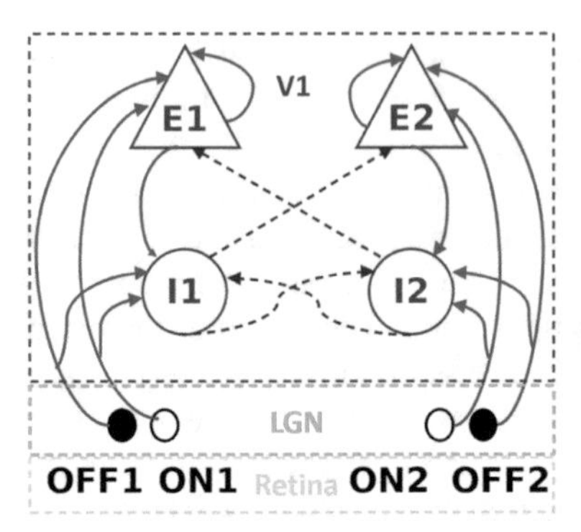

Fig. 1 Model structure. It consists of a layer of retina photo-receptors (yellow rectangle), a layer of LGN neurons (green rectangle) and a layer of primary visual cortex V1 (violet rectangle). The V1 layer is separated into four types of neuron populations: inhibitory (I1 and I2) and excitatory (E1 and E2) neurons. The red solid lines represent excitatory connections while the blue dashed-the inhibitory connections between the corresponding groups of neurons

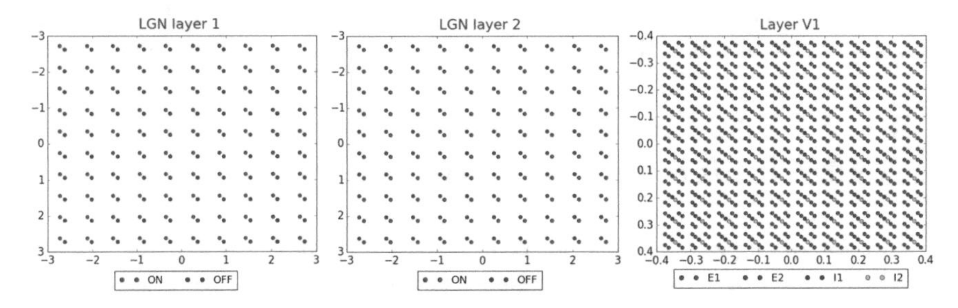

Fig. 2 Topologies of model layers. The photo-receptors/LGN layers consist of neurons with ON receptive fields, denoted by red dots and OFF receptive fields denoted by blue dots. V1 layer contains two excitatory neuron populations marked by red (E1) and pink (E2) dots and two inhibitory neuron populations, marked by blue (I1) and cyan (I2) dots

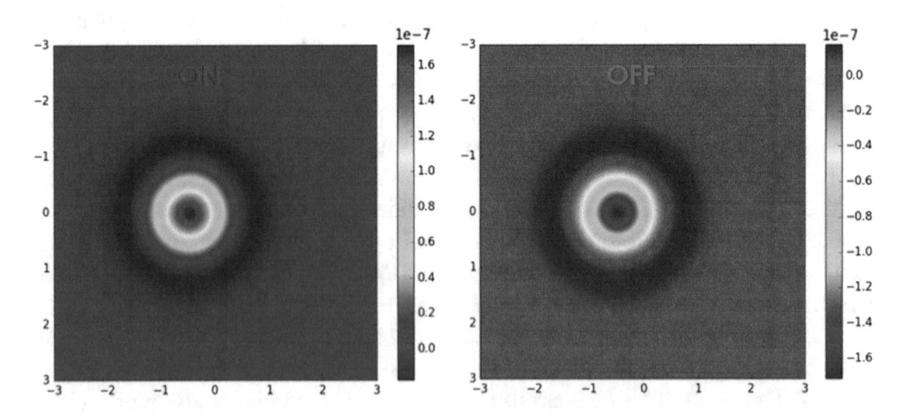

Fig. 3 On-center off surround (ON) and off-center on-surround (OFF) receptive filters of retina cells

2.1 Neuron Models

As [10] proposed, we used the model of photo-receptors from [14] to transform the visual stimuli to continuous electric current while for both LGN and V1 layers we used leaky integrate-and-fire neurons whose output signal has discrete form of spike trains. However, in contrast to [10], where identical spike-timing models were used for both LGN and V1 neurons, we decided to chose the different models from the NEST 2.12.0 library [11].

2.1.1 Retina Receptors

Following models in [14] the photo-receptors in retina are modeled as spatio-temporal filter that is convolved with visual stimuli. The spatial component of the filter has center-surround form. It is defined as difference of two Gaussian functions as in [10, 16]:

$$RF = Center - Surround \tag{1}$$

$$Center = \frac{a_c}{{\sigma_c}^2} \exp - \frac{R}{{\sigma_c}^2} \tag{2}$$

$$Surround = \frac{a_s}{{\sigma_s}^2} \exp - \frac{R}{{\sigma_s}^2} \tag{3}$$

$$R = \sqrt{(x - x_{RF})^2 + (y - y_{RF})^2} \tag{4}$$

Here a_c and a_s are gain parameters of center/surround Gaussians with standard deviations σ_c and σ_s respectively. The size of the receptive field can be varied by multiplication of a scaling factor rf_{scale} both variances. x and y denote the coordinates in two-dimensional space of dots from visual stimuli and receptive field respectively. Figure 3 presents examples of ON and OFF receptive fields of the spatial component of the retina receptors.

The temporal component has bi-phasic profile determined by difference of two Gamma functions [10, 14, 16]. The convolution of the spatio-temporal kernel with the moving visual stimuli transforms the images to electrical signal that is generating current for the corresponding LGN neuron.

2.1.2 LGN Neurons

For the LGNs we used the proposed in [2] model whose parameters were determined from in-vivo experiments. The models equations are as follows:

$$C\frac{dV}{dt} = -G_L\left(V - V_L\right) - G_E\left(t\right)\left(V - V_E\right) - G_I\left(t\right)\left(V - V_I\right) - G_A\left(t\right)\left(V - V_A\right) \tag{6}$$

$$G_X = \sum_j g_X\left(t - t_j\right) H\left(t - t_j\right) \tag{7}$$

$$g_X\left(t\right) = \bar{g_X}\frac{t}{\tau_X}\exp - \frac{t - \tau_X}{\tau_X} \tag{8}$$

Here C is capacity of neuron cell membrane, V is membrane potential, G_L is leakage conductance, G_E and G_I are the conductances of total excitatory and inhibitory synaptic inputs respectively, G_A is the conductance of the potassium-mediated AHP channel and V_L, V_E, V_I and V_A are the corresponding reversal potentials. The time-dependent conductances (denoted here by G_X where X stands for E, I and A respectively) are described by Eqs. (7) and (8), where t_j denotes the time of event (release of a neurotransmitter into corresponding synapse), τ_X is duration of the event, H is Heviside step function and the alpha function g_X is defined by Eq. (8) to achieve its maximum $\bar{g_X}$ at the moment $t_j = \tau_X$. The excitation input to LGN neuron comes from retina photo-receptors in the form of generating current produced by the first layer of the model.

2.1.3 V1 Neurons

For this layer we chose the same model as in [10], proposed in [15]. The model equations are:

$$C\frac{dV}{dt} = -G_{rest}\left(V\left(t\right) - V_{rest}\right) + I_{syn} \tag{9}$$

$$I_{syn}\left(i\right) = \sum_j A_{ij} y_{ij}\left(t\right) \tag{10}$$

Here C is capacity of neuron cell membrane, V is membrane potential, G_{rest} is the membrane conductance at resting state V_{rest} and I_{syn} is the synaptic current that is modeled as sum of postsynaptic currents from all neurons j connected to a given neuron i according to Eq. (10). The parameter A_{ij} determines the absolute strength of the synaptic connection. The factor y_{ij} describes the contribution to a synaptic current of neuron i of the postsynaptic currents from neurons j that is determined by a dynamic system of equations from [15].

2.2 Connection Weights

2.2.1 LGN to V1 Connections

As in [10, 14], neurons from V1 layer have elongated receptive fields defined by a Gabor probablity function [16]:

$$Gabor = \exp -\frac{X^2 + \gamma Y^2}{2\sigma^2} \cos (2\pi \omega X + \varphi) \tag{11}$$

$$X = x \cos \theta + y \sin \theta \tag{12}$$

$$Y = -x \sin \theta + y \cos \theta \tag{13}$$

Here x and y denote LGN neurons coordinates; σ is standard deviation of the Gaussian envelop, ω is the frequency of the sinusoidal factor, φ is phase offset, θ is orientation and γ is aspect ratio. Figure 4 presents two examples of a V1 neuron Gabor filters generating the corresponding values of weight connections to the LGN layer.

Each neuron from V1 layer has its own orientation and phase parameters that determine its orientation selectivity. In our model these parameters were determined by an approach for generation of naturalistic orientation maps [13] that is explained further in the paper. The values of frequency and standard deviation determine the size and shape of the corresponding receptive field. They were subject of tuning too.

2.2.2 Lateral Connections in V1

The lateral connections between the four groups of neurons within the first layer of visual cortex have structure presented on Fig. 1. Their absolute values are determined on the basis of neurons correlations by their positions, phases and orientations. For this aim Gabor correlation was used. The sign of a connection weight depends on

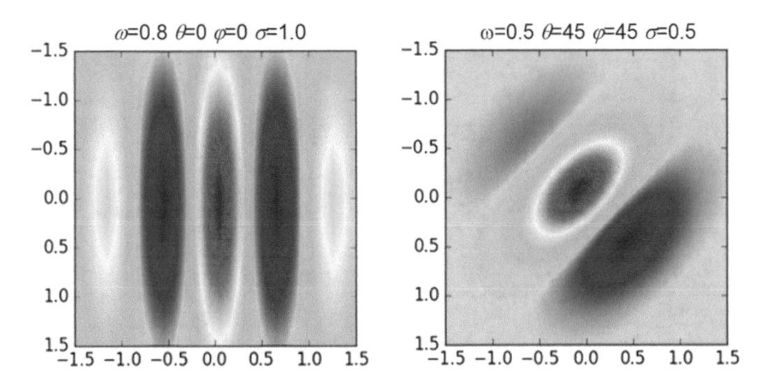

Fig. 4 Gabor receptive fields for two different sets of parameters. Aspect ratio is $\gamma = 1$

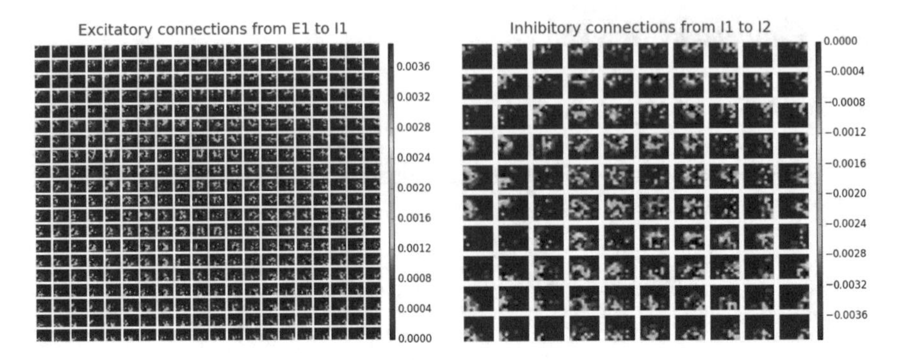

Fig. 5 Lateral connections within V1 layer. Left - excitatory connections between E1 and I1 groups of neurons; right - inhibitory connections between I1 and I2 groups of neurons

whether it is excitatory (positive) or inhibitory (negative). Besides, as in [10], neurons form inhibitory populations connect preferentially to neurons having a receptive field phase difference of around 180°.

In our model we defined frequencies and standard deviations of Gabor filters for lateral connections so as to obtain approximately circular receptive fields for all neurons in the layer. An example of obtained lateral connections is shown on Fig. 5.

2.3 *Orientation Map*

The maps containing the orientations and phases of all neurons within a layer define its columnar structure. In [10] these maps were generated randomly. However, as it was observed by neurological investigations, the orientation maps in mammalian brains have "pinwheel-structure". Among the proposed approaches for mathematical design of such a structure, [13] is a relatively new and easily implemented one. That is why we used it to design orientation and phase maps in our model.

The algorithm from [13] is the following:

$$\theta, \varphi = \arg\max IN(\theta, \varphi) \tag{14}$$

$$IN(\theta, \varphi) = R_{V1}\left(r_{V1}^{j}\right) * g_{\theta,\varphi} \tag{15}$$

$$g_{\theta,\varphi}(x, y) = I_{mean} + I_{mod}\sin\left(\frac{2\pi}{\lambda}(x\sin\theta + y\cos\theta) + \varphi\right),\ I_{mean} = I_{mod} \tag{16}$$

$$R_{V1}\left(r_{V1}^{j}\right) = \sum_{i} G\left(r_{V1}^{j} - r_{col}^{j}, \sigma_{col}\right) R_{col}\left(r_{col}^{i}\right) \tag{17}$$

$$R_{col}(r_{col}) = G(r_{col}, \sigma_{col}) \tag{18}$$

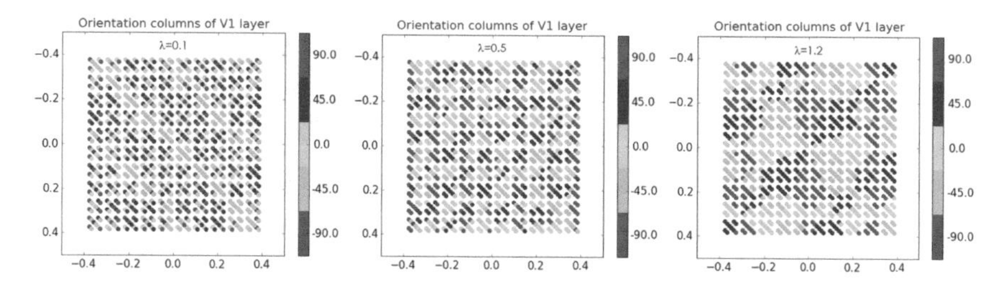

Fig. 6 Orientation maps obtained using different values of parameter λ. Different colors denote different orientations θ of the neurons placed at their spatial position in V1 layer

Here σ_{col} is the column standard deviation; centers of the columns $r_{col} = (x_{col}, y_{col})$ are randomly positioned near by the neurons of V1 layer $r_{V1} = (x_{V1}, y_{V1})$. The idea is to simulate presentation of a sinus grating stimulus g at each position with coordinates x and y on the visual field having defined spatial frequency λ. *IN* determines the input to the given neuron from V1. By varying values of orientations θ and phases φ of the stimulus we select for each neuron its own orientation and phase giving maximal response modulation. G denotes the column receptive field function (in this case Gaussian centered at the corresponding position r). The stimulus intensity is denoted by I; I_{mean} and I_{mod} are the mean and modulation of I; $I_{mean} = I_{mod}$ for stimulus with maximal contrast.

We investigated the influence of the parameter defining spatial frequency of the grating stimulus (λ) on the positions and shapes of the obtained orientation columns. Figure 6 shows three different spatial structures obtained by varying this parameter.

We used five different orientations and phases in the range $[-90, +90]$ and $[-180, +180]$ respectively but the approach is applicable to any number of orientations and phases.

From Fig. 6 is obvious that with increase of λ the number of spatially closer neurons with identical orientation increases while its smallest value resulted in orientation columns composed of only few neighbor neurons. Hence the grating stimuli with smaller λ generated thinner orientation columns while those of bigger λ yielded thicker orientations columns.

3 Simulation Results

3.1 Test Stimuli

In order to test orientation selectivity of our model we generated two types of stimuli consisting of moving bars with two different orientations $-45°$ and $90°$ - shown on Fig. 7.

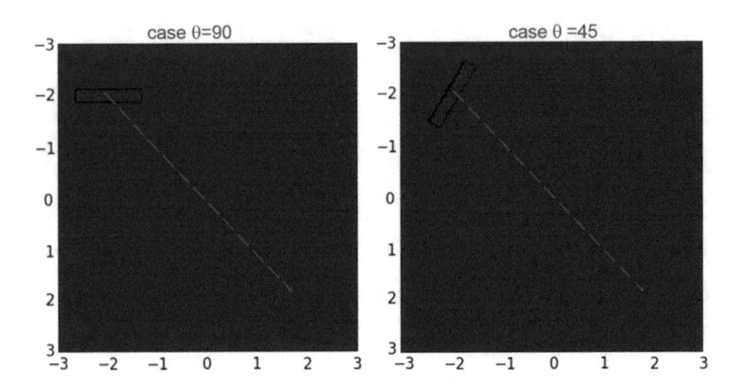

Fig. 7 Test stimuli. The dashed red arrow shows direction of movement of the red bar that is placed at its starting position. The luminance of the background in blue shows that it is lower that the luminance of the bar

Table 1 Parameters, for which maximum firing rate of the V1 neurons having orientation corresponding to the presented stimulus was achieved

Bar size (in dots)	11×1	25×1	25×3
rf_{scale}	0.5	0.5	0.25
λ	1.2	0.25	0.25
σ	1	3	4
ω	0.5	0.5	0.5

Beside the bar orientation, we changed its size. Three different bar sizes (given in Table 1 in next section) were simulated.

3.2 Results

The model parameters under investigation were: seven values of the spatial frequency λ (0.1, 0.15, 0.2, 0.25, 0.5, 0.8 and 1.2) of the sinus grating stimulus used for orientation maps generation, three values of the scaling factor rf_{scale} (1, 0.5 and 0.25) of the standard deviations σ_c and σ_s of the Gaussian receptive fields of the retina, three values of the standard deviation σ (1, 3 and 4) and two values of sinusoidal factor frequency ω (0.5 and 0.25) of the Gabor probability receptive fields of V1 connections with the LGN layer.

Figure 8 represents an example of our model output - spike trains together with spike frequencies of all neurons in V1 layer - for a given set of parameters (case 2 in Table 1 in the next section) for both orientations of the bar in the test stimuli.

Different colors denote spikes of neurons with corresponding orientations (the color scheme is identical to that in orientation maps above).

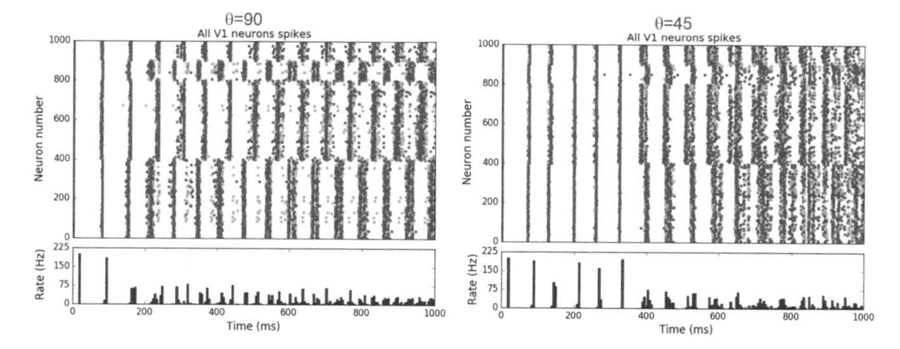

Fig. 8 Spike trains for two tested orientations of a bar stimulus with height of 25 dots and width of 1 dot. The color scheme is the same as on Fig. 6

Although it is obvious that the spike trains on Fig. 8 differ, it is hard to distinguish both stimuli orientations from it. Since the luminosity of the background is lower that that of the bar in the stimulus, we suppose that the firing rate of the population of V1 neurons having the same orientation as the bar should be higher than that of all other neuron groups having different orientations. That is why we tested the mean spike frequency of the five orientation groups of neurons and chose those parameters of the model that result in highest spike frequency for both orientations of the same size bar stimuli. For different bar sizes we received different parameter values that is expected result.

Table 1 summarizes the best parameters for all three bar sizes.

As it was expected, bigger bars needed bigger values of the receptive fields standard deviations of the V1 layer receptive fields, i.e σ increases with bar size. However, for the thicker bar (width = 3 dots) obviously smaller center/surround receptive fields of the retina photo-receptors/LGN neurons were better ($rf_{scale} = 0.25$). Concerning the Gabor sinusoidal factor frequency ω, its best value remains the same for all test stimuli. Concerning the spatial frequency λ of the sinus grating stimulus used for generation of orientation maps, it appears that bars with bigger height were discovered by thinner orientation columns while the smaller bar was recognized by the orientation map having thickest orientation columns among the considered options.

4 Conclusions

The present investigations showed that, although in [13] it was reported that value of the spatial frequency λ of the sinus grating stimulus used for generating of the orientation map of V1 layer is not so important, in practice it leads to different columnar structure and hence different sensitivity of the primary visual cortex to the size and orientation of the visual stimuli.

The size of the photo-receptors in the retina is obviously important parameter, but it should be also dependent on the density of receptor cells in the layer. Since in our model we simulated layers with smaller number of cells that those reported in [10, 14], we suppose that in future we have to test also the dependence of the receptors size defining parameters on the cell density.

As it was expected, the variance of the V1 cells receptive field is strongly dependent on the size of the stimulus we want to detect. Hence in future we should tune our model in dependence on the visual stimuli characteristics. Since in future experiments we'll test our model on widely used in neuro-psychological tests moving dot stimuli similar to those in [5], we should map that parameter value to the size of dots and stimuli generation peculiarities.

The model has also parameters that were fixed in this study - those of the lateral connections receptive fields. They will be subject of further simulation investigations.

Acknowledgements The reported work is a part of and was supported by the project DN02/3/2016 "Modelling of voluntary saccadic eye movements during decision making" funded by the Bulgarian Science Fund.

References

1. Bednar, J.A., Choe, Y., De Paula, J., Miikkulainen, R., Provost, J., Tversky, T.: Modeling cortical maps with Topographica. Neurocomputing **58–60**, 1129–1135 (2004)
2. Casti, A., Hayot, F., Xiao, Y., Kaplan, E.: A simple model of retina-LGN transmission. J. Comput. Neurosci. **24**, 235–252 (2008)
3. Dayan, P., Abbott, L.F.: Theoretical Neuroscience: Computational and Mathematical Modeling of Neural Systems. The MIT Press, Cambridge, Massachusetts (2001)
4. Fregnac, Y., Bathellier, B.: Cortical correlates of low-level perception: from neural circuits to percepts. Neuron **88**, 110–126 (2015)
5. Grossberg, S., Pilly, P.K.: Temporal dynamics of decision-making during motion perception in the visual cortex. In: CAS/CNS Technical Report (2008). Boston University Libraries OpenBU
6. Hansel, D., van Vreeswijk, C.: The mechanism of orientation selectivity in primary visual cortex without a functional map. J. Neurosci. **32**(12), 4049–4064 (2012)
7. Hubel, D.H., Wiesel, T.N.: Receptive fields, binocular interaction and functional architecture in the cats visual cortex. J. Physiol. **160**, 106–154 (1962)
8. Izhikevich, E.M.: Dynamical Systems in Neuroscience: The Geometry of Excitability and Bursting. The MIT Press, Cambridge, Massachusetts (2007)
9. Keil, W., Wolf, F.: Coverage, continuity, and visual cortical architecture. Neural Syst. Circuits **1** (2011). https://doi.org/10.1186/2042-1001-1-1
10. Kremkow, J., Perrinet, L.U., Monier, C., Alonso, J.-M., Aertsen, A., Frgnac, Y., Masson, G.S.: Push-pull receptive field organization and synaptic depression: mechanisms for reliably encoding naturalistic stimuli in V1. Front. Neural Circuits **10** (2016). https://doi.org/10.3389/fncir.2016.00037
11. Kunkel, S. et al.: NEST 2.12.0. Zenodo (2017) https://doi.org/10.5281/zenodo.259534
12. Rust, N.C., Schwartz, O., et al.: Spatiotemporal elements of macaque V1 receptive fields. Neuron **46**(6), 945–56 (2005)
13. Sadeh, S., Rotter, S.: Statistics and geometry of orientation selectivity in primary visual cortex. Biol. Cybern. **108**, 631–653 (2014)

14. Troyer, T.W., Krukowski, A.E., Priebe, N.J., Miller, K.D.: Contrast invariant orientation tuning in cat visual cortex: thalamocortical input tuning and correlation-based intracortical connectivity. J. Neurosci. **18**, 5908–5927 (1998)
15. Tsodyks, M., Uziel, A., Markram, H.: Synchrony generation in recurrent networks with frequency-dependent synapses. J. Neurosci. **20**(RC50), 1–5 (2000)
16. http://www.opensourcebrain.org/projects/111

Gaussian Model Deformation of an Abdominal Aortic Aneurysm Caused by the Interaction between the Wall Elasticity and the Average Blood Pressure

N. Nikolov, S. Tabakova and St. Radev

Abstract The deformation behaviour of an elastic tube possessing a Gaussian defected profile is discussed depending on the inlet pulsatile fluid flow, as an opportunity to imitate the interaction between abdominal aortic aneurysm (AAA) and blood flow. The numerical simulations are performed using the fluid-structure interaction (FSI) of ANSYS software. Three deformation regimes of the tube wall are established depending on the wall elasticity and average fluid pressure. Some stable and unstable deformed shapes of the defected part are found together with the corresponding distributions of displacements and von Misses stresses in the wall zone.

1 Introduction

The initiation and growth of malformations (aneurysms) in the abdominal aorta (AA) are some of the reasons for human mortality. This important human organ possesses such mechanical characteristics, which are difficult for in vivo measurements. The work and the state of all other organs and systems, keeping the metabolism, depend on the optimal function of the aorta. Therefore the in silico (numerical simulations) study of the aortic mechanical behaviour, depending on its elastic properties and the blood flow parameters, becomes a powerful tool to predict the deviations on its optimal functioning. Such optimal functions are generally expressed by the aortic work in compliance with the heart pulses. That is closely related to the elasticity and geometrical form of the aorta as well as the blood flow parameters, and definitely to the interaction between the two continuous media: fluid and structure. The proper combinations of aortic elasticity and blood flow pressure play an important part in the

N. Nikolov · S. Tabakova (✉) · St. Radev
Institute of Mechanics, BAS, Acad. G. Bontchev str., bl. 4, 1113 Sofia, Bulgaria
e-mail: stabakova@gmail.com

N. Nikolov
e-mail: n.nikolov@imbm.bas.bg

St. Radev
e-mail: stradev@imbm.bas.bg

K. Georgiev et al. (eds.), *Advanced Computing in Industrial Mathematics*,
Studies in Computational Intelligence 793,
https://doi.org/10.1007/978-3-319-97277-0_25

compliance/stiffness of the cardiovascular system, and the undesirable declinations could even lead to dissections or ruptures.

The blood flow can be modeled using different models: analytical, asymptotic or computational, which can assume the artery walls either as rigid or deforming (elastic, viscoelastic, etc.) [1]. The blood flow can be considered either as incompressible Newtonian or as non-Newtonian and in rare cases as compressible. The non-Newtonian flow consideration is usually significant for the smaller blood vessels, while for the larger ones the Newtonian viscosity model is sufficient [1, 2]. The models including wall deformation dependence on the blood flow oscillations are more complicated and computer resources consuming [1] and they enter into the so called fluid-structure interaction (FSI) models.

The abdominal aorta aneurysms (AAA) have been studied both from experimental and theoretical point of view by many authors, for example [3–8]. Some of them use artificial geometrical models of AAA, others rely on computer tomography scanned images of real AAA. However, the parametric (artificial) geometrical models are easier for analysis, since they can give a proper insight into the interplay between the different parameters entering the FSI simulations. In the present work such a model is proposed performed by the software ANSYS (MECHANICAL and FLUENT) to model a pulsatile blood flow in a cylindrical elastic tube with an aneurysm part (defected part) described by a Gaussian profile. It must be emphasized that in this work, the considered regimes of the average pressure and elasticity modulus cover a broader spectrum of values than the clinically measured ones. This is done for a deeper parametrical investigation of the elastic wall behaviour during its interaction with the oscillatory blood flow at a prescribed gradient of the oscillatory pressure. During the numerical simulations, an appearance of instability has been established in the pulsating displacements, strains and stresses of the AA wall.

2 Problem Statement

The AA is modeled as a straight circular tube with a total length of 1.12 m, wall thickness 1.5 mm and radius $R = 0.015$ m. The much longer tube than its radius has been chosen in order to obtain a developed flow inside the tube as it has been stated in [2]. A Cartesian coordinate system $Oxyz$ is introduced, where the Oz axis is along the tube axis. The center of the coordinate system O coincides with the center of the geometrical defect as a model aneurysm, whose mechanical behaviour is a subject of the present study. This defect is modeled by the Gaussian shape function [2] and has a total length 0.12 m:

$$r(z) = R + H \exp\left(-\frac{z^2}{2W^2}\right), \quad (1)$$

where $r(z)$ is the current radius of the defect (model aneurysm), H is its height and W is its width at the middle of the height. In the next, it is assumed that the defect is previously created and developed up to its form (1) because of changes in the healthy artery properties. In the numerical simulation the geometrical parameters are accepted as equal, e.g., $H = W = R$.

The tube deformation process depending on the pulsatile fluid flow is modeled applying the fluid structure interaction (FSI) approach of the ANSYS software under the following assumptions. A four-fold symmetry of an isotropic homogeneous tube imitating aorta is assumed; the inlet front surface of the tube is motionless; the outlet front surface is free to move; the outside lateral surface of the tube is free of structural loads; the inside lateral surface is subjected to the pulsatile fluid pressure through its definition as a fluid-solid interface, incorporated in the System Coupling module of the FSI; the wall Poissons ratio is 0.45; the artery wall density is $1060\,\mathrm{kg/m^3}$. The blood is assumed incompressible with constant density $\rho = 1060\,\mathrm{kg/m^3}$ and Newtonian with constant viscosity $\mu = 0.00345\,\mathrm{Pa\ s}$.

The elastic wall dynamics has been solved by the ANSYS/MECHANICAL module using a mesh of 99132 finite elements, while the fluid flow calculations have been performed by the ANSYS/FLUENT on a mesh of 9776 finite volumes. As a test, the results for a rigid tube are confirmed by previous our results obtained in [2] for Womersley number $\alpha = R\sqrt{\frac{n\rho}{\mu}} = 22.72$, where $n = 2.4\pi$ is the pulse frequency of flow oscillations.

Further, three cases of different wall elasticity, i.e., with different Young's modulus E, are considered: $E = 500\,\mathrm{kPa}$, $E = 1000\,\mathrm{kPa}$ and $E = 1500\,\mathrm{kPa}$.

The task, as it is formulated, is solved in a time interval of 6 s at time step of 0.0833 s. Following the approaches accepted in our previous work concerning flows of carotid arteries [2], the cross-section mean velocity based on the analytical solution in a straight infinitely long tube is used as an inlet boundary condition:

$$\bar{u} = 0.41776 \sin(nt + 0.06421) \tag{2}$$

This solution is based on a prescribed gradient of the oscillatory pressure amplitude 3.52 kPa/m, which corresponds to 23.62 mm mercury column per meter. Then, the pressure is expected to consist of a steady part p_{out} and an oscillatory part. The outlet boundary condition p_{out} (an analogue to the average blood pressure) appears as a resistance of the working part in the cardiovascular system against the heart pumping function. Different values of the constant pressure p_{out} in the range of 0–20 kPa are used in the simulations.

3 Results for the Elastic Wall Dynamics

The aortic stress-strain state results will be given for 8 points belonging to the middle of the arterial wall in Oyz plane and possessing z-coordinates as follows: point 1, $z = 0$ m at the center of the aneurysm; points 2 and 3, at $z = -/+0.56$ m, inlet/outlet, respectively; points 4 and 5, at $z = -/+0.06$ m, respectively; point 6 at $z = -0.04$ m;

point 7 at $z = -0.02$ m and point 8 at $z = -0.01$ m. The fluid flow functions of velocity and pressure will be given along the z-axis, in 8 points with the same z-coordinates as above.

A similar problem of a pulsatile flow created by a given oscillatory pressure gradient in a rigid carotid artery with a model aneurysm has been treated in [9] and [10]. However, in this case, it occurs that the different values of the outlet pressure p_{out} does not influence the flow velocity, which corresponds always to zero outlet pressure, $p_{out} = 0$.

3.1 Elastic Modulus $E \geq 1500\,kPa$

In Fig. 1a the total pressure (oscillatory and steady average pressure) distribution caused by the oscillatory velocity given by (2) at $p_{out} = 0$ and $E = 1500$ kPa is presented at the eight points 1–8 placed along the tube z-axis. The elastic modulus $E = 1500$ kP corresponds to that elasticity, at which the tube begins visibly to deform by the oscillatory pressure gradient. The AA wall starts to play its function of compression/tension in accordance with the heart beats in order to decrease the pressure towards the arteries, capillaries, tissues. The inlet pressure pulsations are in the interval [−3.5 kPa, 3.3 kPa], while the pressure at the inlet of the defected profile (AAA, point 4) is in the interval [−/+1.75 kPa]. This is connected only with the effect of the wall elasticity, since at the same $p_{out} = 0$, for the rigid wall case, the latter pulsations are around [−/+1.5 kPa]. The total mesh displacements of the wall are in the interval of [−/+0.06 mm], symmetrical for all the pulses during the time of calculations, as shown in Fig. 1b. The defected profile starts to pulsate around its initial shape, given by Eq. (1). It can be summarized that the observed deformation process is a result only of the oscillatory pressure gradient. Here, we would like to note that all the numerical simulations started from rest, i.e., from zeroth initial state, which means that the at least 1–1.5 s are necessary for the development of the fluid-structure interaction solution.

Increasing the average pressure p_{out} to 8 kPa at elasticity modulus $E = 1500$ kPa a tube extension is created in the interval of 0.75/1.025 mm (as shown in Fig. 2a) and a pulsatile inlet pressure in the interval [4.6 kPa, 11.25 kPa]. The wall is loaded by a biaxial tension, the profile changes its initial Gaussian shape (Fig. 2b), which is still not very evident. The total mesh displacements pulsate with an amplitude of 0.125/0.125/0.15 mm (Fig. 2a), respectively at the non-defected part, at the center of the defect and at the entry part of the defect, points 4, 1 and 7.

The further increase of the average pressure $p_{out} = 13.5$ kPa at the same elasticity $E = 1500$ kPa leads to a stably pulsations of the pressure at the tube inlet in the interval [10.5 kPa, 16.5 kPa], which corresponds to a blood pressure of 124.66/78.9 mm mercury column (known as a healthy blood pressure for humans [4]). A shorter pressure interval at the inlet of the defected profile (point 4) is established: [12 kPa, 15 kPa]. The total mesh displacements are pulsating with an amplitude 0.13/0.15/0.3 mm

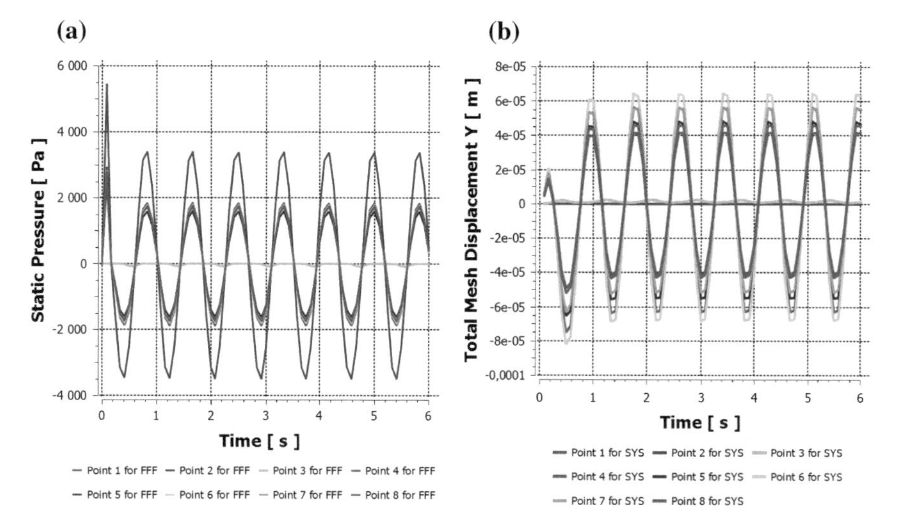

Fig. 1 Time evolution for elastic wall with $E = 1500\,\text{kPa}$ and outlet pressure $p_{out} = 0\,\text{kPa}$ of: **a** the fluid pressure at points 1–8 in the fluid zone (FFF); **b** the total mesh displacement at points 1–8 in the wall zone (SYS)

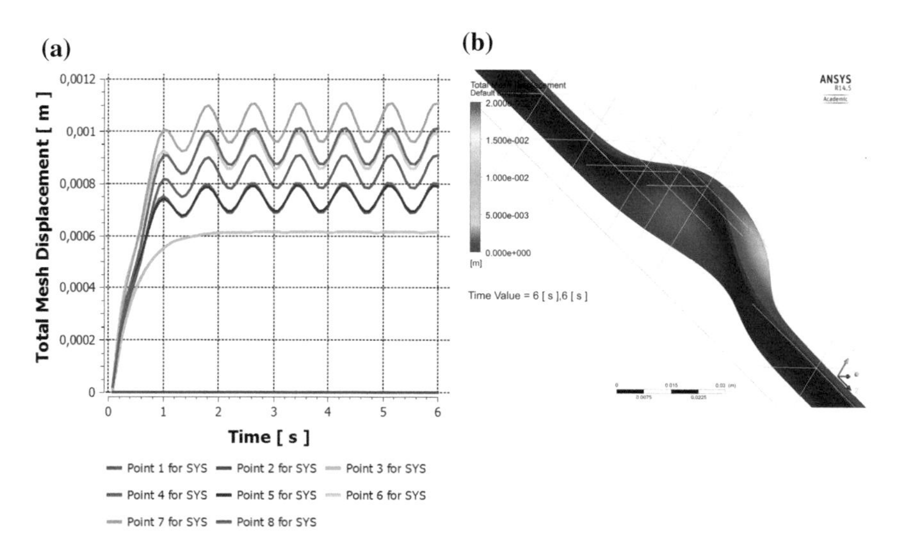

Fig. 2 Time evolution for elastic wall with $E = 1500\,\text{kPa}$ and outlet pressure $p_{out} = 8\,\text{kPa}$ of: **a** the total mesh displacements at points 1–8 in the wall zone (SYS); **b** the total mesh displacements in the 1/4 of the wall zone (SYS)

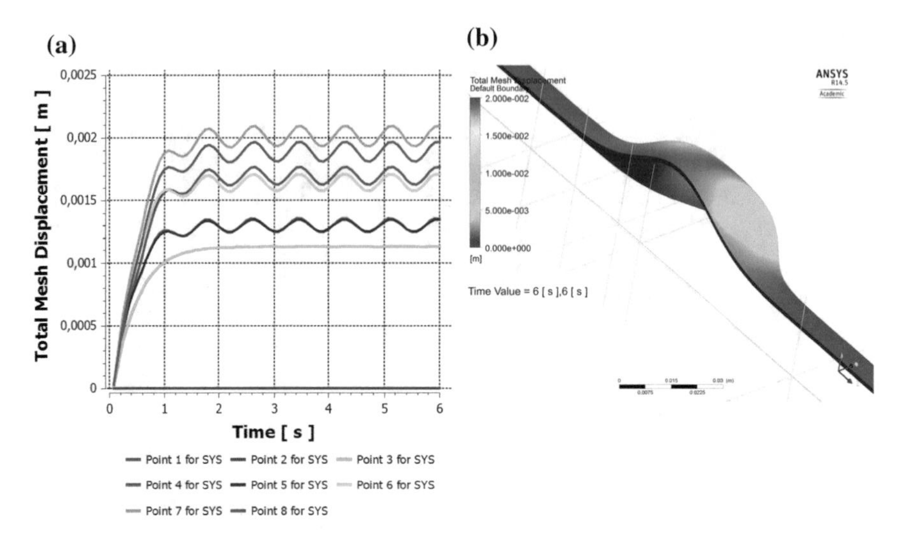

Fig. 3 Time evolution for elastic wall with $E = 1500\,\text{kPa}$ and outlet pressure $p_{out} = 13.5\,\text{kPa}$ of: **a** the total mesh displacements at points 1–8 in the wall zone (SYS); **b** the total mesh displacements in the 1/4 of the wall zone (SYS)

(points 4, 1 and 7), as shown in Fig. 3a. The initial defected shape is essentially deformed: elongated and flattered towards a cylindrical form (Fig. 3b).

The simulations with average pressure $p_{out} \geq 20\,\text{kPa}$ lead to a more elongated defected part of the tube with a bigger volume than the initial one. However, the lower and upper limits of the oscillatory pressure approach to each other and the wall stresses increase significantly. These cases have no realistic analogues in the clinically obtained data, but have been studied here to understand better the different scenarios of the FSI processes.

If the elasticity modulus is augmented to $E \geq 2000\,\text{kPa}$, the wall "solidifies", its deformation decreases substantially and does not interact with the fluid flow, i.e., there is no FSI. Therefore, it can be concluded, that only walls with elastic modulus $E < 2000\,\text{kPa}$ are of any interest for the study of the coupled problem - elasticity/ fluid flow.

3.2 Elastic Modulus $E < 1500\,kPa$

The lower elastic modulus $E = 1000\,\text{kPa}$ makes the wall more deformable. In Fig. 4a the inlet pressure pulsates in the interval $11\,\text{kPa}/16\,\text{kPa}$ (correspondent to diastolic/systolic pressure $83/120$ mm mercury column at $p_{out} = 13.5\,\text{kPa}$. The pulsations of the total mesh displacements are in the interval $0.15/0.25/0.3$ mm, respectively at the points of the non-defected/defected part, that can be seen in Fig. 4b. A better impression of the displacements can be obtained from their distribution in the 1/4 of the wall zone in Fig. 5a. The same tendency is observed, but with a higher

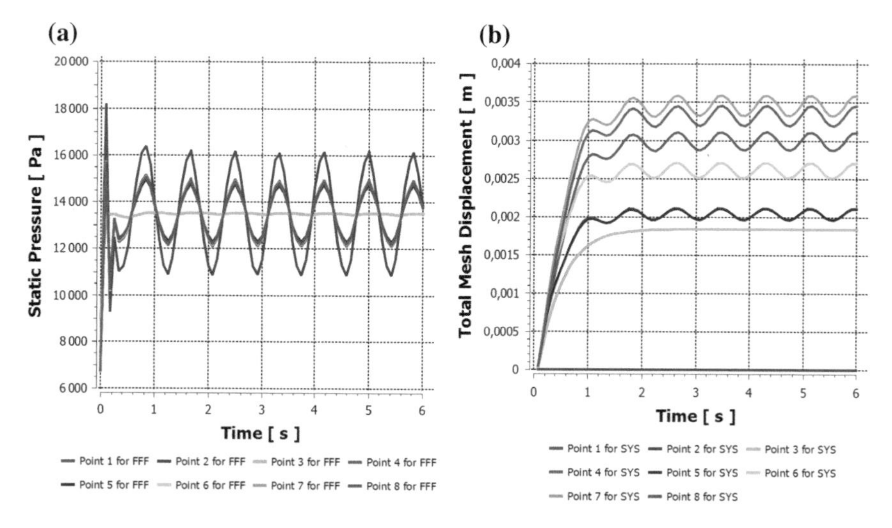

Fig. 4 Time evolution for elastic wall with $E = 1000\,\text{kPa}$ and outlet pressure $p_{out} = 13.5\,\text{kPa}$ of: **a** the fluid pressure at points 1–8 in the fluid zone (FFF); **b** the total mesh displacement at points 1–8 in the wall zone (SYS)

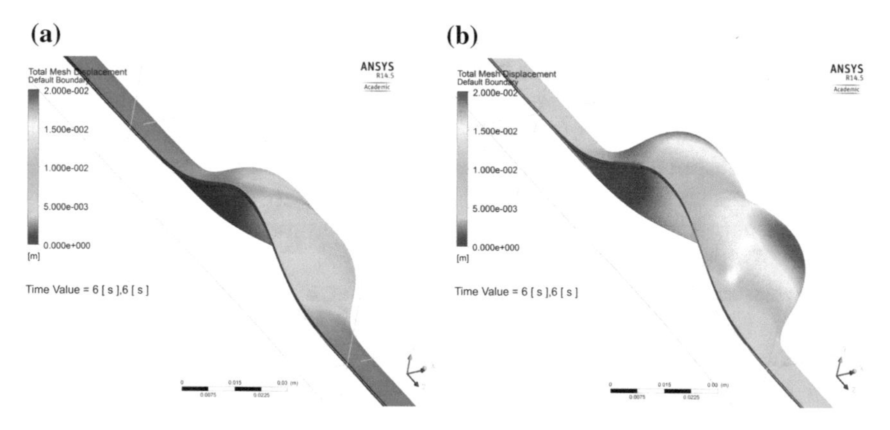

Fig. 5 The total mesh displacements in the 1/4 of the wall zone (SYS) at $p_{out} = 13.5\,\text{kPa}$ and: **a** $E = 1000\,\text{kPa}$; **b** $E = 500\,\text{kPa}$

concentration of the deformations at $p_{out} = 20\,\text{kPa}$. The stable shape, pulsating with respect to the defected part of the tube increases its deformation, as seen on Fig. 5a at time of 6 s.

The elasticity increase, which means decrease of the elastic modulus E to 500 kPa leads to a higher modification of the defected profile at the same average pressure $p_{out} = 13.5\,\text{kPa}$ as discussed above. The inlet pressure is in the interval 12 kPa/15 kPa (correspondent to diastolic/systolic pressure 90 mm/113 mm mercury column, i.e., the pressure difference is less than that for $E = 1000\,\text{kPa}$. However, the higher elasticity makes the wall more deformable and the deformations are

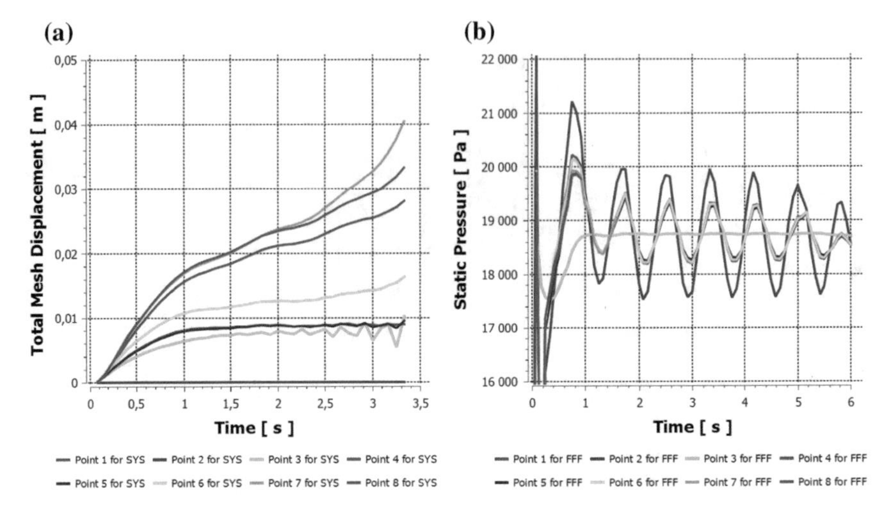

Fig. 6 Time evolution for elastic wall with $E = 500$ kPa of: **a** the total mesh displacements at points 1–8 in the wall zone (SYS) at $p_{out} = 20$ kPa; **b** the fluid pressure at points 1–8 in fluid zone (FFF) at $p_{out} = 18.5$ kPa

substantially higher. In Fig. 5b the total mesh displacements in the 1/4 of the wall zone are presented. It is well seen that the maximum displacements are of order 0.02 m, while for $E = 1000$ kPa (Fig. 5a), they are 0.0087 m.

At the increase of the average pressure to $p_{out} = 20$ kPa the tube ruptures abruptly for 3.41 s by concentration of the deformations in the defected part without pulsations, as seen in Fig. 6a. This process can be defined as a unstably non-pulsating non-reversible deforming regime.

The obtained here results for the displacements at $E = 500$ kPa are in good correspondence with the presented ones in [5], where AA wall material is modeled by the hyperelastic Mooney-Rivlin model.

3.3 Critical Dynamic Regimes at $E = 500$ kPa

The comparison between the images in Figs. 3b and 5a, b give us the opportunity to follow the different stability shapes of the defected profile (model aneurysm) at one and the same average pressure $p_{out} = 13.5$ kPa (101 mm mercury column) with respect to the different wall elasticity. At this average pressure the system performs in stably pulsating deforming regimes. At $E = 1000$ kPa and $E = 1500$ kPa, when increasing the average pressure to 20 kPa the shape change tendences are preserved, as the amplitude of the inlet pulsating pressure is decreased negligibly at the higher elasticity. With the further increase of the elasticity to the elasticity modulus $E \leq 500$ kPa the deformations begin to concentrate to the defected part, that increases its volume till reaching the state of rupture. Then, the deformation regime becomes unstably non-pulsating non-reversible deformation regime.

Decreasing the average pressure to $p_{out} = 18.5$ kPa, the amplitude of the induced inlet pressure decreases considerably, as its two upper and lower limits tend to the average pressure of the system, which is seen in Fig. 6b. At this case the volume of the defected part starts to increase in steps and after the 6th s the tube is destroyed. The von Misses stresses and the total mesh displacements of the wall are increasing with around 1 mm per pulse reaching 60–65 mm at the rupture are shown in Fig. 7a, b, respectively. Thus an unstably-pulsating reversible deforming regime is observed, which means that the process is stepwise in time and can be stabilized to a stably pulsating deformation regime, if the average pressure is decreased to $p_{out} \leq 17.5$ kPa, as shown in Fig. 8a, b. From Figs. 7a and 8a, it can be concluded that the von Misses stresses over 220 kPa in the undefected part lead to tube rupture in the defected part.

Comparing these results, it can be speculated that the interval of the average pressure [17.5 kPa, 18.5 kPa] at $E = 500$ kPa is a limiting one for the stable work of the AA tube, as above this pressure the defected part increases its volume and ruptures. At this elasticity modulus of the wall, the upper limit of the inlet pressure is in the interval of 140–150 mm mercury column, which is accepted as the maximum upper limit of a healthy blood pressure in the physiological sense for humans.

In some references [7], the lowest possible wall strength is bounded to 212 kPa based on similar values which are measured in [8]. If the obtained here results for the von Mises stresses in the undefected tube areas (before point 4, and after point 5) will be followed, it will be seen that stress state in these areas is within the same values or lower for each value of p_{out}. An increase of the stress above this value of 212 kPa

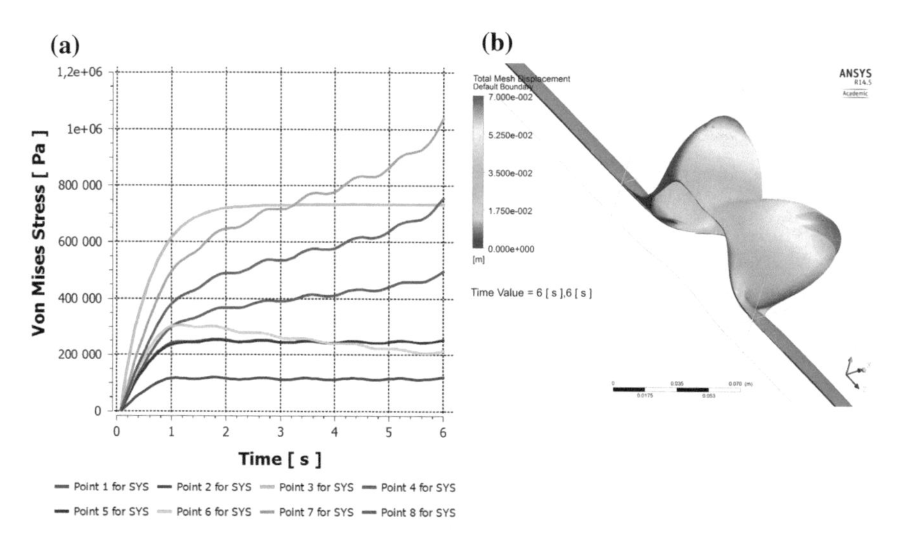

Fig. 7 Wall dynamics at $E = 500$ kPa and $p_{out} = 18.5$ kPa: **a** time evolution of the von Misses stresses at points 1–8 in the wall zone (SYS); **b** total mesh displacements in the 1/4 of the wall zone (SYS) at time 6s

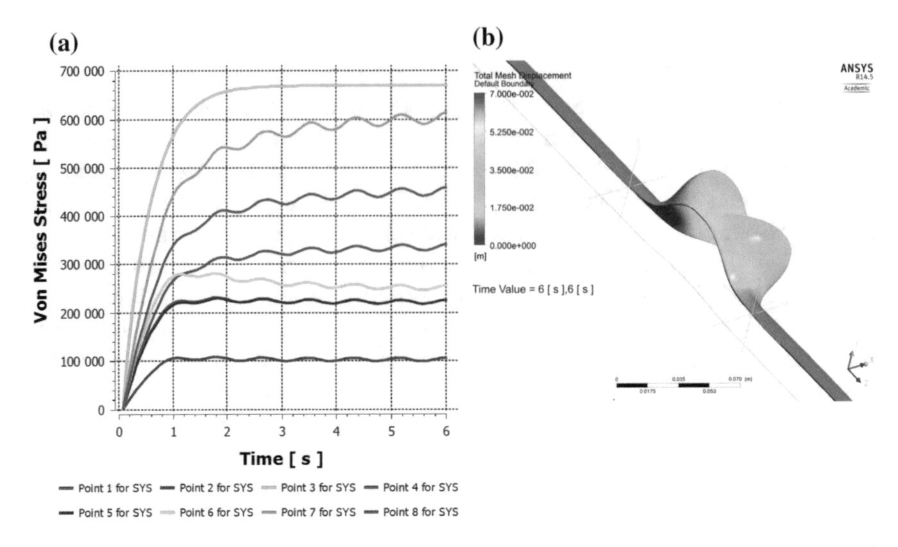

Fig. 8 Wall dynamics at $E = 500$ kPa and $p_{out} = 17.5$ kPa: **a** time evolution of the von Misses stresses at points 1–8 in the wall zone (SYS); **b** total mesh displacements in the 1/4 of the wall zone (SYS) at time 6s

appears in the defected areas only (points 1, 6–8). This circumstance suggests that the appearance of the geometrical defect is the reason for the tube rupture. Otherwise, the tube will work under stress state similar or lower than 210–220 kPa.

4 Conclusion

In the present work, the fluid structure interaction method (FSI, of ANSYS software) is applied to model the pulsatile blood flow in a cylindrical elastic tube with a defected part described by the Gaussian profile (1). The numerical experiments as formulated are imitating the FSI between abdominal aortic walls with Young's modulus of 500, 1000 and 1500 kPa defected geometrically by an abdominal aortic aneurysm (AAA) and average blood pressure in range of 0–20 kPa. Depending on the pulsatile fluid flow, different pulsatile stress strain states are established along the tube length and tube thickness.

Under the combination of the different aortic wall elasticity with different average blood pressure three basic regimes of the development of the Gaussian AAA profile are observed: stably-pulsating deforming regime (see Figs. 2, 3, 4, 5 and 8), with constantly pulsating stress and strain in the defected part of the tube leading to stably pulsating AAA volume around a constant profile in dependence on the initial one; unstably pulsating reversible deforming regime (see Figs. 6b and 7) with increasing pulsating-reversible stress and strain in the defected part leading to pulsating growth of the AAA volume (here there is a possibility to do not achieve AAA rupture under conditions of smooth decrease in the average constant blood pressure); unstably non-pulsating non-reversible deforming regime (see Fig. 6a) with non reversible linearly increasing stress and strain in the defected part leading to sharp increase of the AAA volume and rupture (here there is no possibility to reverse the system in a smooth stable pulsatile regime).

In our previous works treating the carotid artery aneurysms [9, 10], similar dynamic regimes have been established. However, the values of the elastic modulus and average pressure, at which the correspondent system is stable, are completely different, as well as the result values of the different mechanical functions.

In the real case, the real forms of the AAA volumes deviate from the observed model forms here, at least because of the materials anisotropy and non-homogeneity in the boundary conditions. The model forms established here together with the corresponding stress strain states, boundary conditions and the mechanical characteristics of the tube wall and fluid flow define a mathematical space in which the clinically measured data determinate a subspace optimal for keeping the live functions. Corresponding interpretations of the obtained numerical data are available from this point of view.

Acknowledgements The authors have been partially supported for this research by the National Science Fund of Bulgarian Ministry of Education and Research: Grant DFNI-I02/3.

References

1. Morris, P.D., Narracott, A., von Tengg-Kobligk, H., Soto, D.A.S., Hsiao, S., Lungu, A., Evans, P., Bressloff, N.W., Lawford, P.V., Hose, D.R., Gunn, J.P.: Computational fluid dynamics modelling in cardiovascular medicine. Heart **102**(1), 18–28 (2016)
2. Tabakova, S., Raynov, P., Nikolov, N., Radev, St.: Newtonian and non-Newtonian pulsatile blood flow in arteries with model aneurysms. In: Georgiev, K. et al. (eds.) Advanced Computing in Industrial Mathematics, Studies in Computational Intelligence, vol. 681, pp. 187–197. Springer, Berlin (2017)
3. Di Martino, E.S., Guadagni, G., Fumero, A., Ballerini, G., Spirito, R., Biglioli, P., Redaelli, A.: Fluidstructure interaction within realistic three dimensional models of the aneurysmatic aorta as a guidance to assess the risk of rupture of the aneurysm. Ann. Biomed. Eng. **23**, 647655 (2001)
4. Lamata, P., Pitcher, A., Krittian, S., Nordsletten, D., Bissell, M., Cassar, T., Barker, A.J., Markl, M., Neubauer, S., Smith, N.P.: Aortic relative pressure components derived from four-dimensional flow cardiovascular resonance. Magn. Reson. Med. **72**, 1162–1169 (2014)
5. Jozsa, T.I., Paal, G.: Boundary conditions for flow simulations of abdominal aortic aneurysms. Int. J. Heat Fluid Flow. **50**, 342–351 (2014)
6. Benim, A.C., Nahavandi, A., Assmann, A., Schubert, D., Feindt, P., Suh, S.H.: Simulation of blood flow in human aorta with emphasis on the outlet boundary conditions. Appl. Math. Model. **35**, 3175–3188 (2011)
7. Maier, A., Gee, M.W., Reeps, C., Pongratz, J., Eckstein, H.-H., Wall, W.A.: A Comparison of diameter, wall stress, and rupture potential index for abdominal aortic aneurysm rupture risk prediction. Ann. Biomed. Eng. **38**(10), 31243134 (2010)

8. Vande Geest, J.P., Wang, D.H.J., Wisniewski, S., Makaroun, M., Vorp, D.: Towards a noninvasive method for determination of patient-specific wall strength distribution in abdominal aortic aneurysms. Ann. Biomed. Eng. **34**(7), 10981106 (2006)
9. Nikolov N., Radev, St., Tabakova, S.: Pulsatile blood flow in elastic artery with model aneurysm. In: AIP Conference Proceedings, **1906**, 050005 (2017)
10. Nikolov, N., Tabakova, S., Radev, St.: Blood flow and elasticity of carotid artery. Compt. Rend. de l′Acad. Bulg. des Sci. **70**(11), 1579–1584 (2017)

EM Estimation of the Parameters in Latent Mallows' Models

Nikolay I. Nikolov and Eugenia Stoimenova

Abstract Mallows' models are often convenient initial tool for analyzing a set of rank data. They capture the main structure of the data with only one parameter and could be the basis for further research. However, it is usually unrealistic to expect a one-parameter model to reveal all features of the data. One possible generalization of these models could be made by assuming that there are several latent groups in the population. In this paper, we propose an algorithm to find maximum likelihood estimates of the unknown parameters of Latent Mallows' models by making use of the EM algorithm. As an application of the considered estimation algorithm, a comparison between models based on different metrics is made via simulation study.

1 Introduction

The need of analyzing rank data arises in a wide range of fields such as psychology, biology, sociology, educational testing and economics. A full ranking of N items simply assigns a complete ordering to the items: first choice, second choice, ..., N-th choice. Thus any such ranking vector can be considered as an element π of the permutation group $\mathbf{S}_N$, which is generated by the first N positive integers. Typically, the notation

$$\pi^{-1} = \langle \pi^{-1}(1), \pi^{-1}(2), \ldots, \pi^{-1}(N) \rangle$$

is used for the ordering vector $\pi^{-1} \in \mathbf{S}_N$ which corresponds to listing the various items in their ranked order.

There are various probability methods for modelling rank data. One of the most popular was first proposed by Mallows [7] and is based on metrics on $\mathbf{S}_N$. The

N. I. Nikolov (✉) · E. Stoimenova
Institute of Mathematics and Informatics, Bulgarian Academy of Sciences,
Acad. G.Bontchev str., block 8, 1113 Sofia, Bulgaria
e-mail: n.nikolov@math.bas.bg

E. Stoimenova
e-mail: jeni@parallel.bas.bg

K. Georgiev et al. (eds.), *Advanced Computing in Industrial Mathematics*,
Studies in Computational Intelligence 793,
https://doi.org/10.1007/978-3-319-97277-0_26

classical Mallows' models assign larger probabilities for rankings that are *close* to a *modal* ranking π_0. Some methods and techniques for model fitting and finding the maximum likelihood estimates (MLE's) of the unknown parameters can be found in [8]. An asymptotic computable approximation of the MLE's, in the case of general exponential families, is proposed by Mukherjee in [10].

For most rank data sets, the one-parameter models are not rich enough to reveal the full structure in the data. Their important generalizations with an easy interpretation are the Latent Mallows' models, which assume that the population is divided into several latent groups. In this case we observe the full rankings, but not from which group they arose.

The goal of our study is to construct an EM algorithm to find the MLE's of the unknown parameters in Laten Mallows' models. In Sect. 2, Mallows' models and their relation to metrics on permutations are presented. Description of the proposed EM algorithm and some convergence properties are given in Sect. 3. In Sect. 4, the described algorithm is used to make a comparison between eight simulated models.

2 Mallows' Models

The probability mass function of the classical Mallows' models is given by

$$\mathbf{P}_{\theta,\pi_0}(\pi) = e^{\theta d(\pi,\pi_0)-\psi(\theta)} \quad \text{for } \pi \in \mathbf{S}_N, \tag{1}$$

where θ is a real parameter ($\theta \in \mathbf{R}$), $d(\cdot,\cdot)$ is a metric on $\mathbf{S}_N$, π_0 is a fixed ranking and $\psi(\theta)$ is a normalizing constant. When $\theta > 0$, π_0 is a *modal* ranking, for $\theta < 0$, π_0 is an *antimode*, and $\mathbf{P}_{\theta,\pi_0}$ is the uniform distribution for $\theta = 0$. Usually, the metric $d(\cdot,\cdot)$ is chosen in advance and the only unknown parameters of the model are θ and π_0, since $\psi(\cdot)$ is a constant which depends on $d(\cdot,\cdot)$, θ and π_0. More generally, Diaconis [5] suggests models with $d(\cdot,\cdot)$ being a discrepancy function. All distances used in this study are metrics on $\mathbf{S}_N$ and are given in Sect. 2.1. The special cases, when $d(\cdot,\cdot)$ is Kendall's tau and Spearman's rho, are first investigated by Mallows [7]. Later on, models based on Cayley's distance and Hamming distance are considered by Fligner and Verducci [6]. For these models, there are standard methods for finding the exact MLE's of the unknown parameters θ and π_0.

2.1 Metrics on $\mathbf{S}_N$

Deza and Huang [4] considered the following metrics on $\mathbf{S}_N$ which are widely used in applied scientific and statistical problems.

$d_F(\pi,\sigma) = \sum_{i=1}^{N} \lvert\pi(i) - \sigma(i)\rvert$	Spearman's footrule
$d_R(\pi,\sigma) = \left(\sum_{i=1}^{N} (\pi(i) - \sigma(i))^2\right)^{1/2}$	Spearman's rho
$d_M(\pi,\sigma) = \max_{1\le i\le N} \lvert\pi(i) - \sigma(i)\rvert$	Chebyshev metric
$d_K(\pi,\sigma) = \#\{(i,j) : 1 \le i \le N,\ 1 \le j \le N,\ \pi(i) < \pi(j),\ \sigma(i) > \sigma(j)\}$	Kendall's tau
$d_C(\pi,\sigma) = N$ minus the number of cycles in $\sigma\pi^{-1}$	Cayley's distance
$d_U(\pi,\sigma) = N$ minus the length of longest increasing subsequence in $\sigma\pi^{-1}$	Ulam's distance
$d_H(\pi,\sigma) = \#\{i \in \{1,2,\ldots,N\} : \pi(i) \ne \sigma(i)\}$	Hamming distance
$d_L(\pi,\sigma) = \sum_{i=1}^{N} \min(\lvert\pi(i) - \sigma(i)\rvert, N - \lvert\pi(i) - \sigma(i)\rvert)$	Lee distance

It is not difficult to show that the listed metrics possess the following important property.

Definition 1 The metric d on $\mathbf{S}_N$ is called right-invariant, if and only if $d(\pi,\sigma) = d(\pi \circ \tau, \sigma \circ \tau)$ for all $\pi, \sigma, \tau \in \mathbf{S}_N$.

Critchlow [1] pointed that the right-invariance of a metric is a necessary requirement since it means that the distance between rankings does not depend on the labelling of the items. More statistical properties of the presented metrics can be found in [1, 5, 8]. If $d(\cdot,\cdot)$ is right-invariant, then the distribution of $d(\pi,\pi_0)$ for uniformly chosen permutation $\pi \sim Uniform(\mathbf{S}_N)$ does not depend on π_0 and model (1) could be significantly simplified. In this situation, it is easy to prove that $\psi(\theta) = \log(N!\,m(\theta))$, where $m(t)$ is the moment generating function of $d(\pi,\pi_0)$.

2.2 *Latent Models*

One effective approach to extend the classical Mallows' models is to assume that there are K latent groups, $G_1, G_2, \ldots, G_K$ in the population and that the distributions of the rankings within each group can be described by model (1), i.e.

$$\mathbf{P}_{\theta_j,\pi_{0,j}}(\pi) = \mathrm{e}^{\theta_j d(\pi,\pi_{0,j}) - \psi(\theta_j)}, \quad \pi \in \mathbf{S}_N,$$

where θ_j is a real parameter and $\pi_{0,j}$ is the *modal* ranking in group G_j, for $j = 1, 2, \ldots, K$. Then the overall density for this Latent Mallows' model is

$$\mathbf{P}_{\boldsymbol{\theta},\mathbf{p},\boldsymbol{\pi}_0}(\pi) = \sum_{j=1}^{K} p_j \mathrm{e}^{\theta_j d(\pi,\pi_{0,j}) - \psi(\theta_j)}, \quad \pi \in \mathbf{S}_N, \tag{2}$$

where $\boldsymbol{\theta} = (\theta_1, \theta_2, \ldots, \theta_K)$, $\boldsymbol{\pi_0} = (\pi_{0,1}, \pi_{0,2}, \ldots, \pi_{0,K})$ and $\mathbf{p} = (p_1, p_2, \ldots, p_K)$ are vectors of unknown parameters. Since the probability p_j represents the proportion of the population in group G_j, for $j = 1, 2, \ldots, K$, the elements of $\mathbf{p}$ sum up to 1, i.e. $\sum_{j=1}^{K} p_j = 1$. More detailed description of model (2) can be found in [8]. In the next subsection, some of the statistical properties of these models are considered.

2.3 Statistical Inference

Let $\boldsymbol{\pi}^* = \left(\pi^1, \pi^2, \ldots, \pi^n\right)$ be a sample of n complete rankings and $\ell(\boldsymbol{\theta}, \mathbf{p}, \boldsymbol{\pi}_0, \boldsymbol{\pi}^*)$ be the loglikelihood function of model (2). Then for testing the hypothesis of uniform model ($\boldsymbol{\theta} = \mathbf{0}$) against the alternative that $\boldsymbol{\theta} \neq \mathbf{0}$, Marden [8] considered the likelihood ratio statistic,

$$LRS = 2\left[\ell(\hat{\boldsymbol{\theta}}, \hat{\mathbf{p}}, \hat{\boldsymbol{\pi}}_0, \boldsymbol{\pi}^*) - \ell(\mathbf{0}, \hat{\mathbf{p}}, \hat{\boldsymbol{\pi}}_0, \boldsymbol{\pi}^*)\right],$$

where $\left(\hat{\boldsymbol{\theta}}, \hat{\mathbf{p}}, \hat{\boldsymbol{\pi}}_0\right)$ are the MLE's of $(\boldsymbol{\theta}, \mathbf{p}, \boldsymbol{\pi_0})$.

Let $k(\pi)$ be the number of observations that are equal to a given permutation $\pi \in \mathbf{S}_N$. Then the empirical probability for π is $\frac{k(\pi)}{n}$ and a quantity, that measures the total nonuniformity of the data, could be defined as

$$TNU = 2\sum_{\pi \in \mathbf{S}_N} k(\pi)\left[\log\left(\frac{k(\pi)}{n}\right) - \log\left(\frac{1}{N!}\right)\right].$$

Similarly to the multiple correlation coefficient in linear regression, Marden [8] considered the coefficient

$$R^2 = \frac{LRS}{TNU}, \tag{3}$$

which can be used to measure the percentage of nonuniformity in the data that is explained by the fitted model. When $R^2 = 1$ the model exactly fits the data, and $R^2 = 0$ if it performs no better than the uniform model.

3 EM Estimation

It is not possible to estimate the unknown parameters $\boldsymbol{\theta}$, $\mathbf{p}$ and $\boldsymbol{\pi}_0$ in model (2) directly. However, the Expectation-Maximization (EM) algorithm proposed by Dempster et al. [3] can be applied. A complete description of the concept of the EM iteration procedure is given in [9]. Croon and Luijkx [2] considered similar algorithm in the case of Latent Models when $\boldsymbol{\pi}_0$ is known or can be approximated by other methods, for example *K-means clustering*.

3.1 Classical EM Algorithm

The aim of the algorithm is to find the expected value of the group loglikelihood function $\ell(\boldsymbol{\theta}, \mathbf{p}, \boldsymbol{\pi}_0, \boldsymbol{\pi}^*, G)$ for given initial approximations of $\boldsymbol{\theta}$, $\mathbf{p}$ and $\boldsymbol{\pi}_0$ (*E-step*). This expectation is usually denoted by

$$Q^{(t)}\left(\boldsymbol{\theta}, \mathbf{p}, \boldsymbol{\pi}_0, \boldsymbol{\pi}^*\right) = \mathbf{E}_{G|\boldsymbol{\theta}^{(t)}, \mathbf{p}^{(t)}, \boldsymbol{\pi}_0{}^{(t)}, \boldsymbol{\pi}^*}\left[\ell(\boldsymbol{\theta}, \mathbf{p}, \boldsymbol{\pi}_0, \boldsymbol{\pi}^*, G)\right]$$

for some initial values $\boldsymbol{\theta}^{(t)}$, $\mathbf{p}^{(t)}$ and $\boldsymbol{\pi}_0{}^{(t)}$. From (2) it follows that

$$Q^{(t)}\left(\boldsymbol{\theta}, \mathbf{p}, \boldsymbol{\pi}_0, \boldsymbol{\pi}^*\right) = \sum_{i=1}^{n}\sum_{j=1}^{K} \frac{p_j^{(t)} P_{\theta_j^{(t)}, \pi_{0,j}^{(t)}}\left(\pi^i\right)}{\sum_{s=1}^{K} p_s^{(t)} P_{\theta_s^{(t)}, \pi_{0,s}^{(t)}}\left(\pi^i\right)} \left[\log(p_j) + \theta_j d\left(\pi^i, \pi_{0,j}\right) - \psi(\theta_j)\right].$$

The next step is to maximize $Q^{(t)}\left(\boldsymbol{\theta}, \mathbf{p}, \boldsymbol{\pi}_0, \boldsymbol{\pi}^*\right)$ with respect to $\boldsymbol{\theta}$, $\mathbf{p}$ and $\boldsymbol{\pi}_0$ (*M-step*), i.e.

$$\left(\boldsymbol{\theta}^{(t+1)}, \mathbf{p}^{(t+1)}, \boldsymbol{\pi}_0{}^{(t+1)}\right) = \underset{(\boldsymbol{\theta}, \mathbf{p}, \boldsymbol{\pi}_0)}{\operatorname{argmax}} \left\{Q^{(t)}\left(\boldsymbol{\theta}, \mathbf{p}, \boldsymbol{\pi}_0, \boldsymbol{\pi}^*\right)\right\}. \qquad (4)$$

The optimal solution of (4) with respect to $\mathbf{p}$ is

$$p_j^{(t+1)} = \frac{1}{n}\sum_{i=1}^{n} \frac{p_j^{(t)} P_{\theta_j^{(t)}, \pi_{0,j}^{(t)}}\left(\pi^i\right)}{\sum_{s=1}^{K} p_s^{(t)} P_{\theta_s^{(t)}, \pi_{0,s}^{(t)}}\left(\pi^i\right)}, \quad \text{for } j = 1, 2, \ldots, K, \qquad (5)$$

which is independent of the values of $\boldsymbol{\theta}$ and $\boldsymbol{\pi}_0$.

The optimal value θ_j, for $j = 1, 2, \ldots, K$, is the solution of the equation

$$\sum_{i=1}^{n} \frac{p_j^{(t)} P_{\theta_j^{(t)}, \pi_{0,j}^{(t)}} \left(\pi^i\right)}{\sum_{s=1}^{K} p_s^{(t)} P_{\theta_s^{(t)}, \pi_{0,s}^{(t)}} \left(\pi^i\right)} \left[d\left(\pi^i, \pi_{0,j}\right) - \psi'(\theta_j)\right] = 0, \tag{6}$$

which depends on $\pi_{0,j}$. Therefore, the values of $\boldsymbol{\theta} = (\theta_1, \theta_2, \ldots, \theta_K)$ should be calculated from (6) for every possible choice of $\boldsymbol{\pi_0} = (\pi_{0,1}, \pi_{0,2}, \ldots, \pi_{0,K})$, where $\pi_{0,j} \in \mathbf{S_N}$ for $j = 1, 2, \ldots, K$ and $\pi_{0,j} \neq \pi_{0,s}$ for $j \neq s$. Then the optimal solution $\left(\boldsymbol{\theta}^{(t+1)}, \boldsymbol{\pi_0}^{(t+1)}\right)$ is the pair $(\boldsymbol{\theta}, \boldsymbol{\pi_0})$ that maximizes $Q^{(t)}\left(\boldsymbol{\theta}, \mathbf{p}^{(t+1)}, \boldsymbol{\pi_0}, \boldsymbol{\pi}^*\right)$.

After $\boldsymbol{\theta}^{(t+1)}$, $\mathbf{p}^{(t+1)}$ and $\boldsymbol{\pi_0}^{(t+1)}$ are obtained, they are substituted as initial approximations in the *E-step* for calculating the new values of $Q^{(t+1)}\left(\boldsymbol{\theta}, \mathbf{p}, \boldsymbol{\pi_0}, \boldsymbol{\pi}^*\right)$ and so on. This procedure continues until some optimal criteria are met, for example the change of the likelihood function is relatively small or a prefixed number of iterations is reached.

The monotonicity and convergence of the EM algorithm are proven in the general case in Chapter 3 of [9]. However, the convergence rate strongly depends on the initial values $\boldsymbol{\theta}^{(0)}$, $\mathbf{p}^{(0)}$ and $\boldsymbol{\pi_0}^{(0)}$. It looks reasonable to assume that all elements of $\mathbf{p}$ are equal, i.e. $p_j^{(0)} = \frac{1}{K}$ for $j = 1, 2, \ldots, K$. The initial point $\boldsymbol{\pi_0}^{(0)}$ could be taken as a combination of permutations in $\mathbf{S_N}$ for which the empirical probability is large (*modal* rankings) or close to zero (*antimodes*). From the empirical experience it seems that $\theta_j^{(0)} = \frac{1}{2}$ is a good initial approximation, when the corresponding ranking $\pi_{0,j}^{(0)}$ is *modal*, and $\theta_j^{(0)} = -\frac{1}{2}$ when $\pi_{0,j}^{(0)}$ is an *antimode*.

3.2 Generalized EM Algorithm

Since there are $\frac{N!}{(N-K)!}$ possible choices for the values of $\boldsymbol{\pi_0}$ and it is necessary to use some numerical method, for example Newton-Raphson method, to find the corresponding values of $\boldsymbol{\theta}^{(t+1)} = (\theta_1^{(t+1)}, \theta_2^{(t+1)}, \ldots, \theta_K^{(t+1)})$, the algorithm described in the previous subsection requires solving $\frac{K(N!)}{(N-K)!}$ equations of the form (6) at each iteration. To simplify the complexity of the procedure, a generalized version of the EM algorithm can be applied. Condition (4) for $\left(\boldsymbol{\theta}^{(t+1)}, \mathbf{p}^{(t+1)}, \boldsymbol{\pi_0}^{(t+1)}\right)$ could be relaxed and replaced by

$$Q^{(t)}\left(\boldsymbol{\theta}^{(t+1)}, \mathbf{p}^{(t+1)}, \boldsymbol{\pi_0}^{(t+1)}, \boldsymbol{\pi}^*\right) \geq Q^{(t)}\left(\boldsymbol{\theta}^{(t)}, \mathbf{p}^{(t)}, \boldsymbol{\pi_0}^{(t)}, \boldsymbol{\pi}^*\right). \tag{7}$$

It can be shown, see [9, p. 78], that (7) is sufficient to ensure the monotonicity of the algorithm, i.e. the likelihood is not decreased after an EM iteration. Thus $\boldsymbol{\pi_0}^{(t+1)} = (\pi_{0,1}^{(t+1)}, \pi_{0,2}^{(t+1)}, \ldots, \pi_{0,K}^{(t+1)})$ can be defined as

$$\pi_{0,j}^{(t+1)} = \underset{\pi \in \mathbf{S_N}}{\operatorname{argmax}} \left\{ \sum_{i=1}^{n} \frac{p_j^{(t)} P_{\theta_j^{(t)}, \pi_{0,j}^{(t)}}\left(\pi^i\right)}{\sum_{s=1}^{K} p_s^{(t)} P_{\theta_s^{(t)}, \pi_{0,s}^{(t)}}\left(\pi^i\right)} \theta_j^{(t)} d\left(\pi^i, \pi\right) \right\}, \text{ for } j = 1, 2, \ldots, K, \tag{8}$$

where if $\pi_{0,j}^{(t+1)}$ is in $\left\{\pi_{0,j+1}^{(t)}, \pi_{0,j+2}^{(t)}, \ldots, \pi_{0,K}^{(t)}\right\}$ or $\left\{\pi_{0,1}^{(t+1)}, \pi_{0,2}^{(t+1)}, \ldots, \pi_{0,j-1}^{(t+1)}\right\}$, then $\pi_{0,j}^{(t+1)} = \pi_{0,j}^{(t)}$. The corresponding value of $\theta_j^{(t+1)}$ can be found as the solution of

$$\sum_{i=1}^{n} \frac{p_j^{(t)} P_{\theta_j^{(t)}, \pi_{0,j}^{(t)}}\left(\pi^i\right)}{\sum_{s=1}^{K} p_s^{(t)} P_{\theta_s^{(t)}, \pi_{0,s}^{(t)}}\left(\pi^i\right)} \left[d\left(\pi^i, \pi_{0,j}^{(t+1)}\right) - \psi'(\theta_j^{(t+1)})\right] = 0, \text{ for } j = 1, 2, \ldots, K. \tag{9}$$

Proposition 1 *Let* $\mathbf{p}^{(t+1)}$, ${\boldsymbol{\pi}_0}^{(t+1)}$ *and* $\boldsymbol{\theta}^{(t+1)}$ *are given by (5), (8) and (9) respectively. Then condition (7) holds.*

Proof Since $\mathbf{p}^{(t+1)}$ is the solution of (4) and is independent of $(\boldsymbol{\theta}, \boldsymbol{\pi}_0)$,

$$Q^{(t)}\left(\boldsymbol{\theta}^{(t)}, \mathbf{p}^{(t+1)}, {\boldsymbol{\pi}_0}^{(t)}, \boldsymbol{\pi}^*\right) \geq Q^{(t)}\left(\boldsymbol{\theta}^{(t)}, \mathbf{p}^{(t)}, {\boldsymbol{\pi}_0}^{(t)}, \boldsymbol{\pi}^*\right).$$

From (8) and the definition of $Q^{(t)}$ it follows that

$${\boldsymbol{\pi}_0}^{(t+1)} = \underset{\boldsymbol{\pi}_0}{\operatorname{argmax}} \left\{Q^{(t)}\left(\boldsymbol{\theta}^{(t)}, \mathbf{p}, \boldsymbol{\pi}_0, \boldsymbol{\pi}^*\right)\right\},$$

for every vector $\mathbf{p}$. Thus,

$$Q^{(t)}\left(\boldsymbol{\theta}^{(t)}, \mathbf{p}^{(t+1)}, {\boldsymbol{\pi}_0}^{(t+1)}, \boldsymbol{\pi}^*\right) \geq Q^{(t)}\left(\boldsymbol{\theta}^{(t)}, \mathbf{p}^{(t+1)}, {\boldsymbol{\pi}_0}^{(t)}, \boldsymbol{\pi}^*\right).$$

Finally,

$$Q^{(t)}\left(\boldsymbol{\theta}^{(t+1)}, \mathbf{p}^{(t+1)}, {\boldsymbol{\pi}_0}^{(t+1)}, \boldsymbol{\pi}^*\right) \geq Q^{(t)}\left(\boldsymbol{\theta}^{(t)}, \mathbf{p}^{(t+1)}, {\boldsymbol{\pi}_0}^{(t+1)}, \boldsymbol{\pi}^*\right),$$

since from (9) we have that $\boldsymbol{\theta}^{(t+1)}$ is the local maximum of $Q^{(t)}$ when $\boldsymbol{\pi}_0 = {\boldsymbol{\pi}_0}^{(t+1)}$.□

Since condition (7) is sufficient only for convergence of the algorithm to a local maximum of the likelihood function, the resulting points $\hat{\boldsymbol{\theta}}$, $\hat{\mathbf{p}}$ and $\hat{\boldsymbol{\pi}}_0$ could differ from the actual MLE's. When the parameter $\boldsymbol{\pi}_0$ is fixed there is a convergence of the EM sequence to a stationary point, see [2]. Therefore, it is recommended to run the generalized EM procedures several times with different initial approximations of $\boldsymbol{\pi}_0$. One possibility is to simulate various values from the set of initial points ${\boldsymbol{\pi}_0}^{(0)}$ described in Sect. 3.1.

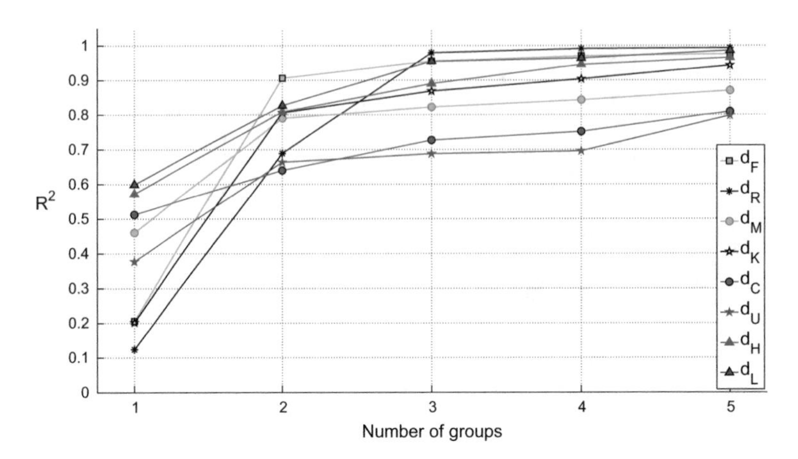

Fig. 1 The average percentage of nonuniformity R^2 for models based on different metrics

4 Simulation Study

In this study, a comparison between Latent Mallows' models based on different metrics is made. The comparison is constructed from 800 Monte Carlo simulations of data samples with size $n = 1000$ from model (2) for $N = 4$, i.e. 100 data samples for each of the listed eight metrics in Sect. 2.1. The used theoretical parameters are: $K = 3$, $\mathbf{p} = \left(\frac{1}{3}, \frac{1}{3}, \frac{1}{3}\right)$, $\pi_{0,1} = \langle 1, 2, 3, 4\rangle$, $\pi_{0,2} = \langle 4, 3, 2, 1\rangle$, $\pi_{0,3} = \langle 3, 1, 4, 2\rangle$ and $\boldsymbol{\theta} = \left(-1, -\frac{1}{2}, \frac{1}{4}\right)$.

By applying the proposed EM algorithm in Sect. 3.2, the values of the percentage of nonuniformity R^2, defined by (3), are estimated for every sample and in cases when the number of the underlying groups K is from 1 to 5. The average values of R^2 for each metric and for each K are presented on Fig. 1. As it is expected, there is a significant improvement in the goodness of fit of the models when $K \geq 2$. Models base on d_R, d_F and d_L explain most of the nonuniformity of the data for $K \geq 3$. Even when $K = 1$, models based on d_H and d_L perform well for the chosen theoretical parameters $\mathbf{p}$, $\boldsymbol{\pi}_0$ and $\boldsymbol{\theta}$.

Acknowledgements The work of the first author was partly supported by the National Science Fund of Bulgaria under Grant DFNI-I02/19 and by the Support Program of Bulgarian Academy of Sciences for Young Researchers under Grant 17-95/2017. The work of the second author was supported by the National Science Fund of Bulgaria under Grant DH02-13.

References

1. Critchlow, D.E.: Metric Methods for Analyzing Partially Ranked Data. Lecture Notes in Statistics, No. 34. Springer, New York (1985)

2. Croon, M.A., Luijkx, R.: Latent Structure Models for Ranking Data. In: Fligner M.A., Verducci J.S. (eds.) Probability Models and Statistical Analyses for Ranking Data. Lecture Notes in Statistics, vol. 80. Springer, New York (1993)
3. Dempster, A.P., Laird, N.M., Rubin, D.B.: Maximum likelihood from incomplete data via the EM algorithm. Journal of the Royal Statistical Society **39**, 1–38 (1977)
4. Deza, M., Huang, T.: Metrics on permutations, a survey. J. Comb. Inf. Syst. Sci. **23**, 173–185 (1998)
5. Diaconis, P.: Group Representations in Probability and Statistics. IMS Lecture Notes - Monograph Series, vol. 11. Hayward, Carifornia (1988)
6. Fligner, M., Verducci, T.: Distance based ranking models. J. Roy. Stat. Soc. **48**, 359–369 (1986)
7. Mallows, C.M.: Non-null ranking models. I, Biometrika **44**, 114–130 (1957)
8. Marden, J.I.: Analyzing and modeling rank data. Monographs on Statistics and Applied Probability, No. 64. Chapman & Hall, London (1995)
9. McLachlan, G., Krishnan, T.: The EM Algorithm and Extentions, 2nd edn. Wiley, Hoboken (2008)
10. Mukherjee, S.: Estimation in exponential families on permutations. Ann. Stat. **44**, 853–875 (2016)

Evolution Equation for Propagation of Blood Pressure Waves in an Artery with an Aneurysm

Elena V. Nikolova

Abstract In this study we derive an evolution equation for propagation of perturbed pressure waves in a blood-filled artery with a local dilatation (an aneurysm) in a long-wave approximation. The equation is a version of the perturbed Korteweg-deVries (KdV) equation with variable coefficients. Exact travelling-wave solution of this equation is obtained by the modified method of simplest equation where the differential equation of Riccati is used as simplest equation. Numerical simulations of the obtained solution are performed and discussed from the point of view of arterial disease geometry.

1 Introduction

The theoretical and experimental investigations of pulse wave propagation in large human arteries has a long history [1, 2]. Over the past decades, however, the scientific efforts have been concentrated on theoretical investigations of nonlinear wave propagation in arteries with a variable radius. The question 'How local imperfections appeared in the artery can disturb the blood flow?' is important for understanding the nature and main features of various cardiovascular diseases, such as stenoses and aneurysms. In order to study propagation of small-but-finite amplitude elastic waves in a stenosed artery, Tay and co-authors treated the artery as a homogeneous, isotropic and thin-walled elastic tube with an local stenosis. The blood was modeled as an incompressible inviscid fluid [3], Newtonian fluid with constant viscosity [4], and Newtonian fluid with variable viscosity [5]. Using a specific perturbation method, in a long-wave approximation the authors obtained the forced Korteweg-de Vries (KdV) equation with variable coefficients [3], the forced perturbed KdV equation with variable coefficients [4], and the forced Korteweg-de Vries–Burgers equation with variable coefficients [5] as evolution equations for the perturbed radial wall displacement. The same theoretical frame was used in [6, 7] to examine

E. V. Nikolova (✉)
Institute of Mechanics, Bulgarian Academy of Sciences, Sofia, Bulgaria
e-mail: elena@imbm.bas.bg

K. Georgiev et al. (eds.), *Advanced Computing in Industrial Mathematics*,
Studies in Computational Intelligence 793,
https://doi.org/10.1007/978-3-319-97277-0_27

propagation of small-but-finite amplitude elastic waves in an artery with an aneurysm. In [6, 7] Nikolova et al. considered the injured artery as a straight thin-walled prestretched hyperelastic tube with a local dilatation. The blood was modeled as an incompressible Newtonian fluid. Depending on the blood viscosity effects the authors attained the forced KdV equation with variable coefficients [6] and the general Korteweg-de Vries–Burgers equation with variable coefficients [7] to represent dynamics of the perturbed radial wall displacement. On the other hand evolution equations of a convectional KdV equation type for propagation of perturbed blood pressure waves in a healthy (straight) artery are presented in [8, 9]. The authors considered the arterial wall as an incompressible hyperelastic material and the blood was modeled as an incompressible inviscid fluid. The numerical simulations performed in [8, 9] demonstrate existence of solitary waves whose amplitude and length are modulated by the vessel characteristics. In this paper we shall focus on consideration of the blood flow through an artery with a local dilatation (an aneurysm). The aneurysm is a localized, blood-filled balloon-like bulge in the wall of a blood vessel [10]. In many cases, its rupture causes massive bleeding with associated high mortality. Motivated by investigations in [8, 9] the main goal of this study is to investigate the effects of aneurismal geometry and blood–vessel characteristics on the behaviour of blood pressure wave. For that purpose, we use a reductive perturbation method to obtain the nonlinear evolution equation for perturbed blood pressure. Exact solution of this equation is obtained by using the modified method of simplest equation [11–14]. Recently, this method has been widely used to obtain general and particular solutions of economic, biological and physical models, represented by partial differential equations [15–18]. The paper is organized as follows. A brief description about the derivation of equations governing the blood flow trough a dilated artery is presented in Sect. 2. In Sect. 3 we derive a basic evolution equation in a long-wave approximation. A traveling wave solution of this equation is obtained in Sect. 4. Numerical simulations of the solution are presented in Sect. 5. The main conclusions based on the obtained results are summarized in Sect. 6 of the paper.

2 Mathematical Formulation of the Basic Model

In this section we shall discuss the interaction of the blood with its container. For this purpose we shall consider two types equations which represent (i) the motion of the arterial wall and (ii) the motion of the blood. To model such a complex medium we shall treat the artery as a thin-walled incompressible prestretched hyperelastic tube with a local dilatation. We shall assume the blood to be an incompressible viscous fluid. A brief formulation of the above–mentioned equations follows in the next two subsections.

2.1 Equation of the Tube

It is well-known, that for a healthy human, the systolic pressure is about 120 mm Hg and the diastolic pressure is 80 mm Hg. Thus, the arteries are initially subjected to a mean pressure, which is about 100 mm Hg. Moreover, the elastic arteries are initially prestretched in a longitudinal direction. This feature minimizes its axial deformations during the pressure cycle. Experimental studies show that the longitudinal motion of arteries is very small [19], and it is due mainly to the strong vascular tethering and partly to the predominantly circumferential orientation of elastin and collagen fibers. Like [6, 7], we consider the artery as a straight circularly cylindrical tube with an uniform radius R_0 and a local dilatation which geometry is presented by the function $f^*(z^*)$. We assume that such a tube is subjected to an initial longitudinal stretch λ_z and a uniform (mean) inner pressure $P_0^*(Z)$ which causes relatively high circumferential and longitudinal initial stresses. On the other hand, the pressure deviation in the course of periodic motion of heart is about ±20 mm Hg. Then the dynamical deformation due to this pressure deviation can be assumed to be smaller than the initial deformation. Therefore, the theory of small deformations superimposed on the initial static deformation can be used in studying the wave propagation in such a medium. Under the action of such a variable pressure the position vector of a generic point on the tube can be described by

$$\mathbf{r_0} = [r_0 + f^*(z^*)]\mathbf{e_r} + z^*\mathbf{e_z},\ z^* = \lambda_z Z^* \tag{1}$$

where $\mathbf{e_r}$ and $\mathbf{e_z}$ are unit basic vectors in cylindrical polar coordinates, $\mathbf{r_0}$ is the deformed radius at the origin of the coordinate system, Z^* is the axial coordinate before deformation, z^* is the axial coordinate after static deformation. Upon the initial static deformation, we shall superimpose only a dynamical radial displacement $u^*(z^*, t^*)$, neglecting the contribution of axial displacement because of the experimental observations, given above. Then, the position vector $\mathbf{r}$ of a generic point on the tube is

$$\mathbf{r} = [r_0 + f^*(z^*) + u^*]\mathbf{e_r} + z^*\mathbf{e_z} \tag{2}$$

The arc-lengths along longitudinal and circumferential curves respectively, are:

$$ds_z = \left[1 + \left(f^{*\prime}(z^*) + \frac{\partial u^*}{\partial z^*}\right)^2\right]^{1/2} dz^*,\ ds_\theta = [r_0 + f^*(z^*) + u^*]d\theta \tag{3}$$

In this way, the stretch ratios in longitudinal and circumferential directions in the final configuration are

$$\lambda_1 = \lambda_z \Lambda,\ \lambda_2 = \frac{1}{R_0}(r_0 + f^*(z^*) + u^*) \tag{4}$$

where

$$\Lambda = \left[1 + \left(f^{*\prime}(z^*) + \frac{\partial u^*}{\partial z^*}\right)^2\right]^{1/2} \tag{5}$$

The notation '$\prime$' denotes the differentiation of f^* with respect to z^*. Then, the unit tangent vector **t** along the deformed meridional curve and the unit exterior normal vector **n** to the deformed tube are

$$\mathbf{t} = \frac{(f^{*\prime}(z^*) + \frac{\partial u^*}{\partial z^*})\mathbf{e_r} + \mathbf{e_z}}{\Lambda}, \quad \mathbf{n} = \frac{\mathbf{e_r} - (f^{*\prime}(z^*) + \frac{\partial u^*}{\partial z^*})\mathbf{e_z}}{\Lambda} \tag{6}$$

According to the assumption made about material incompressibility the following restriction holds:

$$h^* = \frac{H}{\lambda_1 \lambda_2} \tag{7}$$

where H and h^* are the arterial wall thicknesses before and after deformation, respectively. For hyperelastic materials, the tensions in longitudinal and circumferential directions have the form:

$$T_1 = \frac{\mu^* H}{\lambda_2}\frac{\partial \Pi}{\partial \lambda_1}, \quad T_2 = \frac{\mu^* H}{\lambda_1}\frac{\partial \Pi}{\partial \lambda_2} \tag{8}$$

where $\mu^* \Pi$ is the strain energy density function of wall material as μ^* is the initial material shear modulus. A detailed analysis of the forces acting on an element of the artery including a free-body diagram can be found in [20, 21]. Finally, according to the second Newton's law, the equation of radial motion of a small tube element placed between the planes $z^* = const$, $z^* + dz^* = const$, $\theta = const$ and $\theta + d\theta = const$ obtains the form:

$$\begin{aligned} -\frac{\mu^*}{\lambda_z}\frac{\partial \Pi}{\partial \lambda_2} + \mu^* R_0 \frac{\partial}{\partial z^*}\left\{\frac{(f^{*\prime}(z^*) + \partial u^*/\partial z^*)}{\Lambda}\frac{\partial \Pi}{\partial \lambda_1}\right\} \\ + \frac{P^*}{H}(r_0 + f^*(z^*) + u^*)\Lambda = \rho_0 \frac{R_0}{\lambda_z}\frac{\partial^2 u^*}{\partial t^{*2}} \end{aligned} \tag{9}$$

where t^* is a time parameter, P^* is the normal reaction force on the inner surface of the tube and ρ_0 is the mass density of the tube material.

2.2 Equation of the fluid

Experimental studies over many years demonstrate that blood behaves as an incompressible non-Newtonian fluid because it consists of a suspension of cell formed elements in a liquid well-known as blood plasma. However, in the larger arteries (with a vessel radius larger than 1 mm) it is plausible to assume that the blood has an approximately constant viscosity, because the vessel diameters are essentially larger than the individual cell diameters. Thus, in such vessels the non-Newtonian behavior becomes insignificant and the blood can be considered as a Newtonian fluid. Here, for our convenience we assume a 'hydraulic approximation' and apply an averaging procedure with respect to the cross–sectional tube area. Then, we obtain

$$\frac{\partial A^*}{\partial t^*} + \frac{\partial}{\partial z^*}(A^* \omega^*) = 0 \tag{10}$$

$$\frac{\partial \omega^*}{\partial t^*} + \omega^* \frac{\partial \omega^*}{\partial z^*} + \frac{\partial p^*}{\partial z^*} - \frac{\mu_f}{\rho_f}\left[\frac{\partial^2 \omega^*}{\partial z^{*2}} - \frac{8\omega^*}{(r_0 + f^*(z^*) + u^*)^2}\right] = 0 \tag{11}$$

In Eqs. (10)– (11) A^* denotes the inner cross-sectional area, i.e., $A^* = \pi r_f^2$ as $r_f = r_0 + f^*(z^*) + u^*$ is the final radius of the tube after deformation, ω^* is the averaged axial fluid velocity, ρ_f is the fluid density and μ_f is the dynamical viscosity of the fluid. Like [6] we assume that the blood flow has a laminar character. This is reflected in the last term in Eq. (11). The substitution of A^* in Eq. (10) leads to

$$2\frac{\partial u^*}{\partial t^*} + 2\omega^*\left[f^{*\prime}(z^*) + \frac{\partial u^*}{\partial z^*}\right] + [r_0 + f^*(z^*) + u^*]\frac{\partial \omega^*}{\partial z^*} = 0 \tag{12}$$

We introduce the following non-dimensional quantities

$$t^* = \left(\frac{R_0}{c_0}\right)t, \quad z^* = R_0 z, \quad u^* = R_0 u, \quad f^*(z^*) = R_0 f(z), \quad \omega^* = c_0 \omega, \tag{13}$$
$$\mu_f = c_0 R_0 \rho_f \nu, \quad P^* = \rho_f c_0^2 p, \quad r_0 = R_0 \lambda_\theta, \quad c_0^2 = \frac{\mu^* H}{\rho_f R_0}, \quad m = \frac{\rho_0 H}{\rho_f R_0}$$

where c_0 is the Moens–Korteweg velocity, ν is the kinematic viscosity of the fluid, λ_θ is the initial stretch ratio in a circumferential direction and m is the wall relative mass. We put (13) in Eqs. (12), (11) and (9), respectively. Thus the final model takes the form:

$$2\frac{\partial u}{\partial t} + 2\omega\left[f'(z) + \frac{\partial u}{\partial z}\right] + [\lambda_\theta + f(z) + u]\frac{\partial \omega}{\partial z} = 0 \tag{14}$$

$$\frac{\partial \omega}{\partial t} + \omega\frac{\partial \omega}{\partial z} + \frac{\partial p}{\partial z} - \nu\left[\frac{\partial^2 \omega}{\partial z^2} - \frac{8\omega}{(\lambda_\theta + f(z) + u)^2}\right] = 0 \tag{15}$$

$$p = \frac{m}{\lambda_z(\lambda_\theta + f(z) + u)} \frac{\partial^2 u}{\partial t^2} + \frac{1}{\lambda_z(\lambda_\theta + f(z) + u)} \frac{\partial \Pi}{\partial \lambda_2} -$$

$$- \frac{1}{(\lambda_\theta + f(z) + u)} \frac{\partial}{\partial z} \left(\frac{f'(z) + \partial u/\partial z}{[1 + (f'(z) + \frac{\partial u}{\partial z})^2]^{1/2}} \right) \frac{\partial \Pi}{\partial \lambda_1} + \frac{4\nu(f'(z) + \partial(u)/\partial(z))\omega}{\lambda_\theta + f(z) + u} \tag{16}$$

Equation (16) is derived by the boundary condition for the normal stresses of the wall and the fluid. The last term in Eq. (16) represents the laminar character of the blood flow.

3 Derivation of an Evolution Equation for Propagation of Perturbed Blood Pressure Waves

In this section we shall use the reductive perturbation method [22] to study the propagation of pressure waves in the fluid-solid structure system, presented by Eqs. (14)–(16). The reductive perturbation method [22] was formulated for the long wave approximation. In the long-wave limit, it is assumed that the variation of the radius along the axial coordinate is small compared with the wave length. As this condition is valid for large arteries, the reductive perturbation method can be applied to study the asymptotic behaviour of dispersive waves in the medium. According to this method an appropriate scale transformation with a perturbation expansion of the dependent variables is introduced. The choice of coordinate transformation (known also as stretching) depends on the dispersion relationship. The dispersion relationship for such systems is derived, e. g., in [20, 21]. According to this relationship the following stretched coordinates are introduced

$$\xi = \varepsilon^{1/2}(z - ct), \quad \tau = \varepsilon^{3/2} z \tag{17}$$

where ε appears in the dispersion relationship. It is a small parameter ($\varepsilon = r/l$, where l is the characteristic wavelength) measuring the weakness of dispersion. In Eq. (17) c is the phase velocity of harmonic wave propagation in the medium in a long-wave approximation. Then, $z = \varepsilon^{-3/2}\tau$, and $f(\varepsilon^{-3/2}\tau) = \chi(\xi, \tau)$. Thus, the variables u, ω and p are functions of the variables (ξ, τ) and the small parameter ε. Taking into account the effect of dilatation, we assume f to be of order of 5/2, i.e.

$$\chi(\xi, \tau) = \varepsilon h(\tau) \tag{18}$$

To account for the effect of viscosity we set:

$$\nu = \varepsilon^{3/2}\overline{\nu} \tag{19}$$

This assumption is valid for laminar flows and it is permissible to model the blood flow in large arteries. Applying the reductive perturbation method we expand the variables u, ω and p in terms of ε as follows:

$$u=\varepsilon u_1+\varepsilon^2 u_2+\cdots,\ \omega=\varepsilon\omega_1+\varepsilon^2\omega_2+\cdots,\ p=p_0+\varepsilon p_1+\varepsilon^2 p_2+\cdots, \tag{20}$$

The other quantities in Eq. (16) are expanded also in terms of the small parameter ε:

$$\lambda_1\cong\lambda_z,\ \lambda_2=\lambda_\theta+\varepsilon(u_1+h(\tau))+\varepsilon^2(u_2+(u_1+h(\tau))^2)+\cdots \tag{21}$$

$$[\lambda_\theta+\varepsilon h(\tau)+u]^{-1}=\frac{1}{\lambda_\theta}\left(1-\frac{(u_1+h(\tau))}{\lambda_\theta}\varepsilon+\left(\frac{u_2}{\lambda_\theta}+\frac{(u_1+h(\tau))^2)}{\lambda_\theta^2}\right)\varepsilon^2-\cdots\right), \tag{22}$$

$$\frac{1}{\lambda_\theta\lambda_z}\frac{\partial\Pi}{\partial\lambda_1}=\gamma_0,\ \frac{1}{\lambda_\theta\lambda_z}\frac{\partial\Pi}{\partial\lambda_2}=\beta_0+\beta_1(u_1+h)\varepsilon+(\beta_1u_2+\beta_2(u_1+h)^2)\varepsilon^2+\cdots,$$

where

$$\beta_0=\frac{1}{\lambda_\theta\lambda_z}\frac{\partial\Pi}{\partial\lambda_\theta},\ \ \beta_1=\frac{1}{\lambda_\theta\lambda_z}\frac{\partial^2\Pi}{\partial\lambda_\theta^2},\ \ \beta_2=\frac{1}{2\lambda_\theta\lambda_z}\frac{\partial^3\Pi}{\partial\lambda_\theta^3} \tag{23}$$

Substituting Eqs. (17)–(23) into Eqs. (14)–(16), we obtain the following equations:

$O(\varepsilon)$ equations

$$-2c\frac{\partial u_1}{\partial\xi}+\lambda_\theta\frac{\partial\omega_1}{\partial\xi}=0,\ -c\frac{\partial\omega_1}{\partial\xi}+\frac{\partial p_1}{\partial\xi}=0,\ p_1=\gamma_1(u_1+h(\tau)) \tag{24}$$

$O(\varepsilon^2)$ equations

$$\begin{gathered}-2c\frac{\partial u_2}{\partial\xi}+2\omega_1\frac{\partial u_1}{\partial\xi}+\lambda_\theta\frac{\partial\omega_2}{\partial\xi}+[u_1+h]\frac{\partial\omega_1}{\partial\xi}+\lambda_\theta\frac{\partial\omega_1}{\partial\tau}=0\\ -c\frac{\partial\omega_2}{\partial\xi}+\omega_1\frac{\partial\omega_1}{\partial\xi}+\frac{\partial p_2}{\partial\xi}+\frac{\partial p_1}{\partial\tau}+\frac{8\overline{\nu}\omega_1}{\lambda_\theta^2}=0\\ p_2=\left(\frac{mc^2}{\lambda_\theta\lambda z}-\gamma_0\right)\frac{\partial^2u_1}{\partial\xi^2}+\gamma_1u_2+\gamma_2u_1^2+2\gamma_2h(\tau)u_1+\gamma_2h^2(\tau)\end{gathered} \tag{25}$$

where

$$\gamma_1=\beta_1-\frac{\beta_0}{\lambda_\theta},\ \ \gamma_2=\beta_2-\frac{\beta_1}{\lambda_\theta} \tag{26}$$

Considering the process in a long tube, the solution of Eq. (24) can be written as

$$u_1=\frac{\lambda_\theta}{2c^2}P(\xi,\tau),\ \ \omega_1=\frac{1}{c}P(\xi,\tau),\ \ p_1=P(\xi,\tau) \tag{27}$$

where $P(\xi, \tau)$ is an unknown function whose governing equation will be obtained later. In order to obtain a non–zero solution of P the relationship $\gamma_1 = \frac{2c^2}{\lambda_\theta}$ (at $h(\tau) \to 0$) must be held. Introducing (27) into Eq. (25) and after some mathematical transformations we obtain:

$$\frac{\partial P}{\partial \tau} + \mu_1 P \frac{\partial P}{\partial \xi} + \mu_2 \frac{\partial^3 P}{\partial \xi^3} + \mu_3 P + \mu_4(\tau) \frac{\partial P}{\partial \xi} = 0 \tag{28}$$

where

$$\mu_1 = \frac{1}{\gamma_1}\left(\frac{5}{2\lambda_\theta} + \frac{\gamma_2}{\gamma_1}\right), \quad \mu_2 = \frac{1}{\gamma_1}\left(\frac{m}{4\lambda_z} - \frac{\gamma_0}{2\gamma_1}\right), \tag{29}$$

$$\mu_3 = 4\overline{\nu}\sqrt{\frac{2}{\gamma_1 \lambda_\theta^5}}, \quad \mu_4(\tau) = \left(\frac{1}{2\lambda_\theta} + \frac{\gamma_2}{\gamma_1}\right) h(\tau)$$

4 Traveling-Wave Solution of the Evolution Equation

In this section we shall derive a travelling wave solution for the variable coefficients evolution equation, presented by Eq. (28). Like [6], we introduce new variables $\overline{\tau} = \tau$, $\overline{\xi} = \xi - \int_0^\tau \mu_4(s) ds$. Then Eq. (28) is reduced to the perturbed Korteweg-deVries equation with constant coefficients:

$$\frac{\partial P}{\partial \overline{\tau}} + \mu_1 P \frac{\partial P}{\partial \overline{\xi}} + \mu_2 \frac{\partial^3 P}{\partial \overline{\xi}^3} + \mu_3 P = 0 \tag{30}$$

The perturbed Korteweg-de Vries equation is analytically unsolvable in the classical meaning. By this reason we shall search for a particular analytical solution of this equation in the form:

$$P(\overline{\xi}, \overline{\tau}) = \phi(\overline{\tau}) \overline{P}(\overline{\xi}, \overline{\tau}) \tag{31}$$

where $\phi(\overline{\tau})$ is the amplitude of P and $\overline{P}(\overline{\xi}, \overline{\tau})$ is its spatial profile. We replace Eq. (31) in Eq. (30) and obtain

$$\left[\frac{\phi'}{\phi} + \mu_3\right] \overline{P} + \left[\frac{\partial \overline{P}}{\partial \overline{\tau}} + \mu_1 \phi \overline{P} \frac{\partial \overline{P}}{\partial \overline{\xi}} + \mu_2 \frac{\partial^3 \overline{P}}{\partial \overline{\xi}^3}\right] = 0 \tag{32}$$

We assume that $\phi(\overline{\tau}) = exp(-\mu_3 \overline{\tau})$. Then $P \to 0$ at $\overline{\tau} \to \infty$, which corresponds to our assumption for no disturbance at $\overline{\tau} \to \infty$. Then Eq. (30) is reduced to the equation

$$\frac{\partial \overline{P}}{\partial \overline{\tau}} + \overline{\mu}_1 \overline{P} \frac{\partial \overline{P}}{\partial \overline{\xi}} + \mu_2 \frac{\partial^3 \overline{P}}{\partial \overline{\xi}} = 0 \tag{33}$$

where $\overline{\mu}_1 = \phi(\overline{\tau})\mu_1$. It is obvious that $\phi(\overline{\tau})$ will have maximum equals one when the coefficient μ_3 is small enough. Then the coefficient $\overline{\mu}_1 \approx \mu_1 \approx$ *constant* (We will show that for a particular case the numerical value of μ_3 is small enough in Sect. 5). Then Eq. (33) is reduced to the convectional Korteweg-de Vries equation. One well–known analytical solution of the convectional Korteweg-de Vries equation is presented in [3, 4, 8]. In [6] the authors derive another analytical solution of the same equation applying the modified method of simplest equation [11, 12]. The ordinary differential equation of Abel of first kind is used as simplest equation in [6]. Now we shall derive a new analytical solution of Eq. (33)(at $\overline{\mu}_1 \approx \mu_1 \approx$ *constant*) applying the same method. According to the methodology of the modified method of simplest equation we introduce the coordinate transformation $\zeta = \overline{\xi} - v^*\overline{\tau}$ where v^* is the traveling wave velocity. Then Eq. (33) is reduced to the following ordinary differential equation:

$$-v^*\frac{d\overline{P}}{d\zeta} + \mu_1\overline{P}\frac{d\overline{P}}{d\zeta} + \mu_2\frac{d^3\overline{P}}{d\zeta^3} = 0 \tag{34}$$

Now we search for a solution of Eq. (34) of kind $\overline{P} = \overline{P}(\zeta) = \sum_{r=0}^{q} a_r g^r$, where $g_\zeta = \sum_{j=0}^{m} b_j g^j$. Here a_r and b_j are parameters, and $g(\zeta)$ is a solution of some ordinary differential equation, referred to as the simplest equation. We replace the proposed solution in Eq. (34) and we balance the highest powers arising in its derivatives. In this way we obtain a balance equation in the form $q = 2m - 2$. We assume that $m = 2$, i.e. the equation of Riccati will play the role of simplest equation. Then

$$\overline{P}(\zeta) = a_0 + a_1 g + a_2 g^2,\ \frac{dg}{d\zeta} = b_0 + b_1 g + b_2 g^2 \tag{35}$$

The relationships among the coefficients of the solution and the coefficients of the model (Eq. (33)) are derived by solving a system of five algebraic equations. Then the solution of the convectional Korteweg-de Vries equation (Eq. (33)) is

$$\overline{P} = -\frac{8b_2b_0\mu_2 + v + \mu_2 b_1^2}{\mu_1} - \frac{12\mu_2 b_1 b_2}{\mu_1}g - \frac{12\mu_2 b_2^2}{\mu_1}g^2 \tag{36}$$

where the solution of the Riccati equation [12] is

$$g(\zeta) = -\frac{b_1}{2b_2} - \frac{\sqrt{b_1^2 - 4b_0b_2}}{2b_2}\tanh\left[\frac{\sqrt{b_1^2 - 4b_0b_2}(\zeta + \zeta_0)}{2}\right] \tag{37}$$

In Eq. (37) b_0, b_1, b_2 are free parameters. Then the solution of the evolution equation (Eq. (28) takes the form:

$$P(\zeta) = exp(-\mu_3\tau)\overline{P},\ \zeta = \xi + v^*\tau - \int_0^\tau \mu_4(s)ds \tag{38}$$

5 Numerical Simulations

Now we shall simulate numerically the obtained solution (38) for the special case of an abdominal aortic aneurysm (AAA). It is obvious that the wave profile of the perturbed blood pressure P (see Eqs. (28)–(29)) depends on the material properties of the arterial wall, on the initial deformations of the wall and on the arterial geometry. In order to see their effect on the wave behaviour of P we need the values of coefficients $\mu_1, \mu_2, \mu_3, \mu_4(\tau)$. For this purpose, the constitutive relationship for tube material must be specified. Here, unlike [3–5], we assume that the arterial wall is an incompressible, anisotropic and hyperelastic material. The mechanical behaviour of such a material can be defined by the strain energy function of Fung for arteries [23]:

$$\Pi = C(e^Q - 1), \quad Q = C_1E_{QQ}^2 + C_2E_{ZZ}^2 + 2C_3E_{QQ}E_{ZZ} \tag{39}$$

where $E_{QQ} = 1/2(\lambda_\theta^2 - 1)$ and $E_{ZZ} = 1/2(\lambda_z^2 - 1)$ are the Green–Lagrange strains in circumferential and axial directions, respectively, and C, C_1, C_2, C_3 are material constants. The numerical values of the material coefficients in (39) are as follows: $C = 2.5$ kPa, $C_1 = 14.5$, $C_2 = 7$, $C_3 = 0.1$. They are derived in [24] from experimental data of human aortic wall segments applying a specific inverse technique. The initial stretch ratios of the aortic wall are $\lambda_z = 1.5$, $\lambda_\theta = 1.2$ [25]. Finally we have to objectify the aneurysm shape. For an idealized AAA, $h(\tau) = \delta exp(\frac{-\tau^2}{2l^2})$, where δ is the aneurysm height, i.e. $\delta = r_{max} - r_0$, and l is the aneurysm length [26]. In order to normalize these geometric quantities, we non-dimensionalize δ by the inlet radius (diameter). Then, the non-dimensional coefficient can be presented by $\delta' = DI - 1$, where $DI = 2r_{max}/2r_0 = D_{max}/D_0$ is a geometric measure of AAA, which is known as a diameter index or a dilatation index [27]. In the same manner, the aneurysm length L is normalized by the maximum aneurysm diameter (D_{max}), i.e. $l' = L/D_{max} = 1/SI$, where SI is a ratio, which is known as a sacular index of AAA [27]. For AAAs, D_{max} varies from 3 to 8.5 cm, and L varies from 5 to 10–12 cm. Taking into account the above given data the numerical values of the coefficients in Eq. (29) are:

$$\mu_1 = 0.0014;\ \mu_2 = -0.35.10^{-5};\ \mu_3 = 1.76.10^{-4};\ \mu_4(\tau) = 5.18h(\tau) \tag{40}$$

They are calculated at $\gamma_0 = 333.3607404$, $\gamma_1 = 4911.521179$, $\gamma_2 = 23394.54967$ and $\nu = 3.28.10^{-6}$ m^2/s.

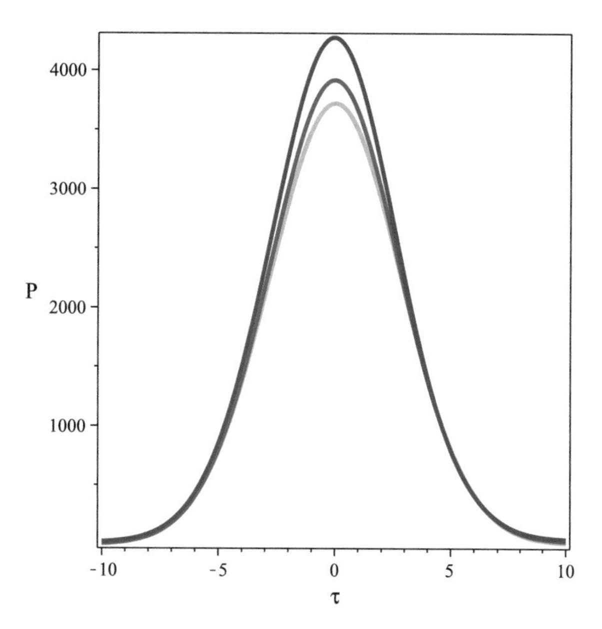

Fig. 1 Variations of the blood pressure perturbation for different values of: $D_{max} = 3\,cm$, $L = 5\,cm$ (green line); $D_{max} = 5\,cm$, $L = 7.5\,cm$ (red line); $D_{max} = 7\,cm$, $L = 10\,cm$ (blue line)

Using these numerical values, the travelling-wave solution of Eq. (28) with respect to the space coordinate τ is plotted in Fig. (1). In all simulations $\xi = 1$, $v^* = 1$, $m = 0.1$. The numerical values of the free parameters in Eq. (37) are $b_0 = 0.03$, $b_1 = 0.001$, $b_2 = -0.03$. They are defined by the symmetry condition at $\tau = 0$ and decay condition at $\tau = +/-\infty$. In more detail Fig. 1 demonstrates the effect of aneurysm geometrical characteristics such as the maximal aneurysm diameter D_{max} and the aneurysmal length L (in particular, DI and SI indexes of AAA) on the wave profile of the pertupbed blood pressure. Taking into account that the healthy aortic diameter D_0 is about 2 cm, various wave profiles of P are obtained for various values of the maximal aneurysm diameter D_{max} and the aneurysm length L. As it is seen from Fig. 1 solitary waves are observed in presence of arterial dilatation. The graph also demonstrates that the amplitude of these waves increases, but the wavelength retains, when the maximal aneurysm diameter and the aneurysm length increase.Translating the observed dimensionless values of the wave amplitudes to the dimensional case, we obtain that the dimensional perturbation of the blood pressure is about 10 kPa for the given data.

6 Conclusions

Modelling the injured artery as a straight thin-walled hyperelastic tube with a local imperfection (an aneurysm), and the blood as a Newtonian fluid we have derived an evolution equation for propagation of perturbed blood pressure in this complex medium. Numerical values of the model parameters are determined for specific mechanical characteristics of the arterial wall and specific aneurismal geometry (typical for an idealized AAA). We have obtained a traveling wave analytical solution of the model evolution equation. The numerical simulations of this solution demonstrate existence of solitary waves, which are symmetric with respect to the aneurysm center. The increasing amplitude of these waves with increasing the maximal diameter and the length of AAA may lead to gradual thinning and weakening of the arterial wall and thereby to its rupture.

Acknowledgements This work was supported by the Bulgarian National Science Fund with grant agreement No. DFNI I 02-3/ 12.12.2014.

References

1. Pedley, T.J.: Fluid Mechanics of Large Blood Vessels. Cambridge University Press, Cambridge (1980)
2. Fung, Y.C.: Biodynamics: Circulation. Springer, New York (1981)
3. Tay, K.G.: Forced Korteweg-de Vries equation in an elastic tube filled with an inviscid fluid. Int. J. Eng. Sci. **44**, 621–632 (2006)
4. Tay, K.G., Ong, C.T., Mohamad, M.N.: Forced perturbed Korteweg-de Vries equation in an elastic tube filled with a viscous fluid. Int. J. Eng. Sci. **45**, 339–349 (2007)
5. Tay, K.G., Demiray, H.: Forced Korteweg-de VriesBurgers equation in an elastic tube filled with a variable viscosity fluid. Chaos Solitons Fractals **38**, 1134–1145 (2008)
6. Nikolova, E.V., Jordanov, I.P., Dimitrova, Z.I., Vitanov, N.K.: Evolution of nonlinear waves in a blood-filled artery with an aneurysm. AIP Conf. Proc. **1895**(1), 070002, 070002-1-070002-8 (2017)
7. Nikolova, E.V., Jordanov, I.P., Dimitrova, Z.I., Vitanov, N.K.: Nonlinear evolution equation for propagation of waves in an artery with an aneurysm: an exact solution obtained by the modified method of simplest equation. In: Advanced Computing in Industrial Mathematics. Studies in Computational Intelligence, 728, pp. 131–144. Springer, Cham (2018)
8. Elgarayhi, A., El-Shewy, E.K., Mahmoud, A.A., Elhakem, A.A.: Propagation of nonlinear pressure waves in blood. Hindawi Publishing Corporation, ISRN Comput. Biol. **2013**(436267), 5 (2013)
9. Abulwafa, E.M., El-Shewy, E.K., Mahmoud, A.A.: Time-fractional effect on pressure waves propagating through a fluid filled circular long elastic tube. Egypt. J. Basic Appl. Sci. **3**, 35–43 (2016)
10. Aneurysm, From Wikipedia, the free encyclopedia, https://en.wikipedia.org/wiki/Aneurysm
11. Vitanov, N.K., Dimitrova, Z.I., Kantz, H.: Modified method of simplest equation and its application to nonlinear PDEs. Appl. Math. Comput. **216**, 2587–2595 (2010)
12. Vitanov, N.K.: Modified method of simplest equation: powerful tool for obtaining exact and approximate traveling-wave solutions of nonlinear PDEs. Commun. Nonlinear Sci. Numer. Simul. **16**, 1176–1185 (2011)
13. Vitanov, N.K.: On modified method of simplest equation for obtaining exact and approximate solutions of nonliear PDEs: the role of simplest equation. Commun. Nonlinear Sci. Numer. Simul. **16**, 4215–4231 (2011)
14. Vitanov, N.K., Dimitrova, Z.I., Vitanov, K.N.: Modified method of simplest equation for obtaining exact analytical solutions of nonlinear partial differential equations: further development of the methodology with applications. Appl. Math. Comput. **269**, 363–378 (2015)
15. Jordanov, I.P.: On the nonlinear waves in (2+ 1)-dimensional population systems. Comptes rendus de lAcadmie bulgare des Sciences **61**(3), 307–314 (2008)

16. Vitanov, N.K., Dimitrova, Z.I.: Application of the method of simplest equation for obtaining exact traveling-wave solutions for two classes ofmodelPDEs from ecology and population dynamics. Commun. Nonlinear Sci. Numer. Simul. **15**, 2836–2845 (2010)
17. Jordanov, I., Nikolova, E.: On nonlinear waves in the spatio-temporal dynamics of interacting populations. J. Theor. Appl. Mech. **43**(2), 69–76 (2013)
18. Nikolova, E., Jordanov, I., Vitanov, N.K.: On nonlinear dynamics of the STAT5a signaling protein. Biomath **3**(1), Article ID: 1404131 (2014)
19. Patel, P.J., Greenfield, J.C., Fry, D.L.: In vivo pressure length radius relationship in certain blood vessels in man and dog. In: Attinger, E.O. (ed.) Pulsatile Blood Flow, p. 277. McGraw-Hill, New York (1964)
20. Demiray, H.: Wave propagation though a viscous fluid contained in a prestressed thin elastic tube. Inf. J. Eng. Sci. **30**, 1607–1620 (1992)
21. Demiray, H.: Waves in fluid-filled elastic tubes with a stenosis: variable coefficients KdV equations. J. Comput. Appl. Math. **202**, 328–338 (2005)
22. Jeffrey, A., Kawahara, T.: Asymptotic Methods in Nonlinear Wave Theory. Pitman, Boston (1981)
23. Fung, Y.: Biomechanics: Mechanical Properties of Living Tissues. Springer, New York (1993)
24. Avril, S., Badel, P., Duprey, A.: Anisotropic and hyperelastic identification of in vitro human arteries from full-field optical measurements. J. Biomech. **43**, 2978–2985 (2010)
25. Kassab, G.S.: Biomechanics of the cardiovascular system: the aorta as an illustratory example. J. R. Soc. Interface **3**(11), 719740 (2006)
26. Gopalakrishnan, S.S., Benot, P., Biesheuvel, A.: Dynamics of pulsatile flow through model abdominal aortic aneurysms. J. Fluid Mech. **758**, 150–179 (2014)
27. Raut, S.S., Chandra, S., Shum, J., Finol, E.A.: The role of geometric and biomechanical factors in abdominal aortic aneurysm rupture risk assessment. Ann. Biomed. Eng. **41**, 1459–1477 (2013)

The Derivatives of Polynomials Represented as a Linear Combination of Finite Differences with Step One

Inna Nikolova

Abstract In this paper we solve the following problem: Given a polynomial represent its derivatives as a linear combination of finite differences with step one. As a corollary we receive the solution of the problem: Given a polynomial in the basis $(-x)_k$ represent its derivatives again in the basis $(-x)_k$.

1 Introduction

For polynomials, orthogonal w.r.t. to a discrete weight, supported on finite or infinite subset of consecutive integers it is true that

$$Tp_n(x) = A_n(x)p_{n-1}(x) - B_n(x)p_n(x)\,, \tag{1}$$

where T is the operator of finite difference with step one forward [4, 5] or backwards [6]. Of interest for the physics of electrostatics is the case when T is the operator of taking first derivative of the polynomial [3]. From (1) it readily follows that this relation is also true when T is the operator Δ^k or ∇^k. One way to solve the problem with the first derivative of the polynomial is to represent the derivative as a linear combination of finite differences. This is a particular case of the problem we solve here.

I. Nikolova (✉)
Institute of Information and Communication Technologies,
Blvd. Tzarigradko shose 125, Bl 2, 1113 Sofia, Bulgaria
e-mail: inna.petrova.nikolova@gmail.com

K. Georgiev et al. (eds.), *Advanced Computing in Industrial Mathematics*,
Studies in Computational Intelligence 793,
https://doi.org/10.1007/978-3-319-97277-0_28

2 Main Result

$$p_n(x)^{(j)} = \sum_{l=j}^{n}(-1)^{l-j}\frac{j!}{l!}\sigma_{l-j}(1,2,\ldots,l-1)\Delta^l p_n(x) \tag{2}$$
$$= \sum_{l=j}^{n}(-1)^{l-j}\frac{j!}{l!}\begin{bmatrix} l \\ j \end{bmatrix}\Delta^l p_n(x),$$

where by $\begin{bmatrix} l \\ j \end{bmatrix}$ are denoted the unsigned Stirling numbers of first kind.
Proof of the result.
We use the following approach. Represent the values of the polynomial in the points $x+k$ in terms of the forward difference operators with step one. Then represent the values of the polynomial in the same points in terms of the exact Taylor expansion and thus form a linear system of equations with the derivatives considered as unknowns and the finite differences are given. The solution of the linear system is not easy since its matrix is a Vandermonde matrix. We solve it in terms of the elementary symmetric functions of the first n integers, which are some times called Stirling numbers of first kind. However we prefer the notation of the sigma functions since during the solution we use symmetric functions of several numbers which are not necessarily consecutive integers. Let us start with the representation of the values $p_n(x+k)$, $k=1,\ldots,n$ in terms of differences .

$$\Delta p_n(x) := p_n(x+1) - p_n(x)$$

Therefore

$$p_n(x+1) = p_n(x) + \Delta p_n(x)$$

Suppose that for some k holds

$$p_n(x+k) = p_n(x) + \binom{k}{1}\Delta p_n(x) + \binom{k}{2}\Delta^2 p_n(x) + \cdots + \binom{k}{k}\Delta^k p_n(x).$$

From the induction supposition follows that

$$p_n(x+k+1) = p_n(x+1) + \binom{k}{1}\Delta p_n(x+1) + \binom{k}{2}\Delta^2 p_n(x+1)$$
$$+ \cdots + \binom{k}{k}\Delta^k p_n(x+1).$$

Moreover from the definition of Δ^k follows that

$$\Delta^k p_n(x+1) = \Delta^{k+1} p_n(x) + \Delta^k p_n(x) .$$

After some elementary series manipulations and using the properties of the binomial coefficients we finish the induction step. From the Taylor expansion of the polynomial we can write that

$$\begin{aligned}
p_n(x+1) &= p_n(x) + p_n'(x)\frac{1}{1!} + p_n(x)''\frac{1^2}{2!} + p_n^{(3)}(x)\frac{1^3}{3!} + \cdots + p_n^{(n)}(x)\frac{1^n}{n!} \\
p_n(x+2) &= p_n(x) + p_n'(x)\frac{2}{1!} + p_n''(x)\frac{2^2}{2!} + p_n^{(3)}(x)\frac{2^3}{3!} + \cdots + p_n^{(n)}(x)\frac{2^n}{n!} \\
&\vdots \\
p_n(x+k) &= p_n(x) + p_n'(x)\frac{k^1}{1!} + p_n''(x)\frac{k^2}{2!} + p_n^{(3)}(x)\frac{k^3}{3!} + \cdots + p_n^{(n)}(x)\frac{k^n}{n!} \\
&\vdots \\
p_n(x+n) &= p_n(x) + p_n'(x)\frac{n}{1!} + p_n''(x)\frac{n^2}{2!} + p_n^{(3)}(x)\frac{n^3}{3!} + \cdots + p_n^{(n)}(x)\frac{n^n}{n!}
\end{aligned}$$

After equating the two different expansions for $p_n(x+k)$, $k = 1, 2, \ldots, n$, one receives the following linear system of equations:

$$\begin{pmatrix}
\frac{1}{1!} & \frac{1^2}{2!} & \frac{1^3}{3!} & \cdots & \frac{1^n}{n!} \\
\frac{2}{1!} & \frac{2^2}{2!} & \frac{2^3}{3!} & \cdots & \frac{2^n}{n!} \\
\vdots & & & & \\
\frac{k}{1!} & \frac{k^2}{2!} & \frac{k^3}{3!} & \cdots & \frac{k^n}{n!} \\
\vdots & & & & \\
\frac{n}{1!} & \frac{n^2}{2!} & \frac{n^3}{3!} & \cdots & \frac{n^n}{n!}
\end{pmatrix}_{n\times n}
\begin{pmatrix}
p_n'(x) \\ p_n''(x) \\ \vdots \\ p_n^{(k)}(x) \\ \vdots \\ p_n^{(n)}(x)
\end{pmatrix}_{n\times 1} =$$

$$= \begin{pmatrix}
\binom{1}{1}\Delta p_n(x) & & & & \\
\binom{2}{1}\Delta p_n(x) + \binom{2}{2}\Delta^2 p_n(x) & & & & \\
\binom{3}{1}\Delta p_n(x) + \binom{3}{2}\Delta^2 p_n(x) + & & & & \\
\vdots \qquad \vdots \qquad \vdots & & & & \\
\binom{k}{1}\Delta p_n(x) + \binom{k}{2}\Delta^2 p_n(x) + \cdots + \binom{k}{k}\Delta^k p_n(x) & & & & \\
\vdots \qquad \vdots \qquad \vdots & & & & \\
\binom{n}{1}\Delta p_n(x) + \binom{n}{2}\Delta^2 p_n(x) + \cdots + \binom{n}{k}\Delta^k p_n(x) + \cdots + \binom{n}{n}\Delta^n p_n(x)
\end{pmatrix}_{n\times 1}$$

Let us solve the above system using Cramer's rule. One can see that the determinant of the main matrix of the system is equal to one. We take out the multiples of $\frac{1}{i!}$ from each column to get multiples for each finite difference of the polynomial. Thus we receive the following formula for the derivatives:

$$p_n^{(j)}(x) = \frac{1}{1!2!\dots(j-1)!(j+1)!\dots n!} \sum_{l=1}^{n} D_{n,l,j} \Delta^l p_n(x),$$

where

$$D_{n,l,j} = \begin{vmatrix} 1^1 & \dots 1^{j-1} & 0 & 1^{j+1} & \dots 1^n \\ 2^1 & \dots 2^{j-1} & 0 & 2^{j+1} & \dots 2^n \\ \vdots & \vdots & \vdots & \vdots & \vdots \\ j^1 & \dots j^{j-1} & 0 & j^{j+1} & \dots j^n \\ \vdots & \vdots & \vdots & \vdots & \vdots \\ (l-1)^1 & \dots (l-1)^{j-1} & 0 & (l-1)^{j+1} & \dots (l-1)^n \\ l^1 & \dots l^{j-1} & \binom{l}{l} & l^{j+1} & \dots l^n \\ (l+1)^1 & \dots (l+1)^{j-1} & \binom{l+1}{l} & (l+1)^{j+1} & \dots (l+1)^n \\ \vdots & \vdots & \vdots & \vdots & \vdots \\ n^1 & \dots n^{j-1} & \binom{n}{l} & n^{j+1} & \dots n^n \end{vmatrix}$$

There are actually two cases $l < j$ and $l \geq j$. The first case is trivial $D_{n,l,j} = 0$. So we pay attention to the second one. But

$$\begin{aligned} \binom{l}{l} &= \frac{l!}{0!l!} = \frac{(1)_l}{l!} \\ \binom{l+1}{l} &= \frac{(l+1)!}{1!l!} = \frac{(2)_l}{l!} \\ \vdots &= \vdots \\ \binom{n}{l} &= \frac{n!}{l!(n-l)!} = \frac{n(n-1)\dots(n-l+1)}{l!} = \frac{(n-l+1)_l}{l!} \end{aligned}$$

Therefore

$$D_{n,l,j} = \frac{1}{l!}\begin{vmatrix} 1^1 & \dots 1^{j-1} & 0 & 1^{j+1} & \dots 1^n \\ 2^1 & \dots 2^{j-1} & 0 & 2^{j+1} & \dots 2^n \\ \vdots & \vdots & \vdots & \vdots & \vdots \\ j^1 & \dots j^{j-1} & 0 & j^{j+1} & \dots j^n \\ \vdots & \vdots & \vdots & \vdots & \vdots \\ (l-1)^1 & \dots (l-1)^{j-1} & 0 & (l-1)^{j+1} & \dots (l-1)^n \\ l^1 & \dots l^{j-1} & (1)_l & l^{j+1} & \dots l^n \\ (l+1)^1 & \dots (l+1)^{j-1} & (2)_l & (l+1)^{j+1} & \dots (l+1)^n \\ \vdots & \vdots & \vdots & \vdots & \vdots \\ n^1 & \dots n^{j-1} & (n-l+1)_l & n^{j+1} & \dots n^n \end{vmatrix}_{n \times n}$$

Start solving the above determinant in a way like solving Vandermonde determinant. The jump of the degrees by two in the j-th column creates some problems. Let us subtract the $(k-1)$-th column from the k-th, $k = 2, 3, \dots j, j+2, j+3 \dots n$ and subtract the $(j-1)$-th column from $(j+1)$-th. Taking into account that

$$a^{k+1} - a^k = a^k(a-1) = (a-1)_2 a^{k-1}$$

and

$$2^{j+1} - 2^{j-1} = 2^{j-1}(2-1)(2+1) = (1)_2 2^{j-2}(1+2) = (1)_2 2^{j-2}\sigma_1(1,2)$$

$$\vdots$$

$$(l-1)^{j+1} - (l-1)^{j-1} = (l-1)^{j-1}[(l-1)^2 - 1] = (l-2)_2(l-1)^{j-2}\sigma_1(1, l-1)$$

Here by $\sigma_k(x_1, x_2, \dots, x_m)$ we denote the elementary symmetric functions of k-th order of the elements x_1 , x_2 , ... x_m i.e.

$$\sigma_k(x_1, x_2, \dots, x_m) = x_1 \times x_2 \times \dots \times x_k + \dots + x_{m-k+1} \times \dots \times x_m \quad m \geq k$$

After making zeros in the first row and expanding the determinant we take out $\sigma(1,2)$ from the corresponding column to make the new first row without sigma functions. We get

$$D_{n,l,j} = (1)_1 \frac{\sigma_1(1,2)}{l!} \times$$

$$
\times \left| \begin{array}{lllll}
(1)_2 & \dots (1)_2 2^{j-3} & 0 & (1)_2 2^{j-2} \\
(2)_2 & \dots (2)_2 3^{j-3} & 0 & (2)_2 3^{j-2} \frac{\sigma_1(1,3)}{\sigma_1(1,2)} \\
\vdots & \vdots & \vdots & \vdots \\
(j-1)_2 & \dots (j-1)_2 j^{j-3} & 0 & (j-1)_2 j^{j-2} \frac{\sigma_1(1,j)}{\sigma_1(1,2)} \\
\vdots & \vdots & & \vdots \\
(l-2)_2 & \dots (l-2)_2 (l-1)^{j-3} & 0 & (l-2)_2 (l-1)^{j-2} \frac{\sigma_1(1,l-1)}{\sigma_1(1,2)} \\
(l-1)_2 & \dots (l-1)_2 l^{j-3} & (1)_l & (l-1)_2 l^{j-2} \frac{\sigma_1(1,l)}{\sigma_1(1,2)} \\
(l)_2 & \dots (l)_2 (l+1)^{j-3} & (2)_l & (l)_2 (l+1)^{j-2} \frac{\sigma_1(1,l+1)}{\sigma_1(1,2)} \\
\vdots & \vdots & & \\
(n-1)_2 & \dots (n-1)_2 n^{j-3} & (n-l+1)_l & (n-1)_2 n^{j-2} \frac{\sigma_1(1,n)}{\sigma_1(1,2)}
\end{array} \right.
$$

$$
\left. \begin{array}{llll}
(1)_2 2^j & (1)_2 2^{j+1} & \cdots (1)_2 2^{n-2} \\
(2)_2 3^j & (2)_2 3^{j+1} & \cdots (2)_2 3^{n-2} \\
\vdots & \vdots & \vdots \\
(j-1)_2 j^j & (j-1)_2 j^{j+1} & \cdots (j-1)_2 j^{n-2} \\
\vdots & \vdots & \vdots \\
(l-2)_2 (l-1)^j & (l-2)_2 (l-1)^{j+1} & \cdots (l-2)_2 (l-1)^{n-2} \\
(l-1)_2 l^j & (l-1)_2 l^{j+1} & \cdots (l-1)_2 l^{n-2} \\
(l)_2 (l+1)^j & (l)_2 (l+1)^{j+1} & \cdots (l)_2 (l+1)^{n-2} \\
\vdots & \vdots & \vdots \\
(n-1)_2 n^j & (n-1)_2 n^{j+1} & \cdots (n-1)_2 n^{n-2}
\end{array} \right|_{(n-1)\times(n-1)}
$$

Step two. Multiply the first column by 2 and subtract it from the second column, Multiply the second column by 2 and subtract it from the 3-th column and so on to the column $(j-3)$. Multiply the column $(j-2)$ by 2^2 and subtract it from column j. We do not use or change column $(j-1)$. It already has zero in the first row. Then proceed using two adjacent columns to make zeros at the first row. We use the following identities.

$$
(j-1)_2 j - 2(j-1)_2 = (j-2)_3 \,,
$$

$$
\begin{aligned}
&(j-1)_2 j^{j-2} \frac{\sigma_1(1,j)}{\sigma_1(1,2)} - 2(j-1)_2\, j^{j-3} = (j-1)_2 j^{j-3} \left[j \frac{\sigma_1(1,j)}{\sigma_1(1,2)} - 2 \right] \\
&= (j-1)_2 j^{j-3} \left[\frac{j(1+j) - 2(1+2)}{\sigma_1(1,2)} \right] = (j-1)_2 j^{j-3} \left[\frac{j + j^2 - 2 - 2^2}{\sigma_1(1,2)} \right] \\
&= (j-1)_2 j^{j-3} \frac{(j-2) + (j-2)(j+2)}{\sigma_1(1,2)} \\
&= (j-1)_2 j^{j-3} (j-2) \frac{1+2+j}{\sigma_1(1,2)} = (j-2)_3 j^{j-3} \frac{\sigma_1(1,2,j)}{\sigma_1(1,2)}
\end{aligned}
$$

$$(j-1)_2 j^j - 2^2 (j-1)_2 j^{j-2} \frac{\sigma_1(1,j)}{\sigma_1(1,2)} = (j-1)_2 j^{j-2} \left[j^2 - 2^2 \frac{\sigma_1(1,j)}{\sigma_1(1,2)} \right]$$
$$= (j-1)_2 j^{j-2} \left[j^2 - 2^2 \frac{\sigma_1(1,j)}{\sigma_1(1,2)} \right]$$
$$= \frac{(j-1)_2}{\sigma_1(1,2)} j^{j-2} \left[j^2 \sigma_1(1,2) - 2^2 \sigma_1(1,j) \right]$$
$$= \frac{(j-1)_2}{\sigma_1(1,2)} j^{j-2} \left[(j-2)(j+2) + 2j(j-2) \right]$$
$$= \frac{(j-2)_3}{\sigma_1 1, 2} j^{j-2} \left[1 \times 2 + 1 \times j + 2 \times j \right] = \frac{(j-2)_3 j^{j-2} \sigma_2(1,2,j)}{\sigma_1(1,2)}$$

After that we expand the determinant using the first row and take out $\sigma_1(1:3)$ and $\sigma_2(1:3)$ from the corresponding columns to leave the new first row without sigma functions. In the next rows we get fractions of sigma functions. Due to a lack of space we will not write the resulting matrices of corresponding two parts of step two.

The step $(j-1)$ again consist of two parts. The first part is to make zeros in the fist row. In order to do that we are using the following identities:

$$(2)_{j-1} j \frac{\sigma_1(1:j-2,j)}{\sigma_1(1:j-1)} - (j-1)(2)_{j-1} = \frac{(1)_j \sigma_1(1:j)}{\sigma_1(1:j-1)}$$

$$(2)_{j-1} j^2 \frac{\sigma_2(1:j-2,j)}{\sigma_2(1:j-2,j-1)} - (j-1)(2)_{j-1} j \frac{\sigma_1(1:j-2,j)}{\sigma_1(1:j-1)}$$
$$= \frac{(2)_{j-1} j}{\sigma_1(1:j-1)\sigma_2(1:j-1)} \left[j\sigma_2(1:j-2,j)\sigma_1(1:j-2,j-1) \right.$$
$$\left. -(j-1)\sigma_2(1:j-2,j-1)\sigma_1(1:j-2,j) \right]$$
$$= \frac{(2)_{j-1} j}{\sigma_1(1:j-1)\sigma_2(1:j-1)}$$
$$\{j[\sigma_2(1:j-2) + j\sigma_1(1:j-2)][\sigma_1(1:j-2) + (j-1)]$$
$$- (j-1) \left[\sigma_2(1:j-2) + (j-1)\sigma_1(1:j-2) \right] \left[\sigma_1(1:j-2) + j \right] \}$$
$$= \frac{(2)_{j-1} j}{\sigma_1(1:j-1)\sigma_2(1:j-1)} \{ j \left[\sigma_2(1:j-2)\sigma_1(1:j-2) + j\sigma_1^2(1:j-2) + \right.$$
$$\left. + (j-1)\sigma_2(1:j-2) + (j-1)j\sigma_1(1:j-2) \right]$$
$$-(j-1) \left[\sigma_2(1:j-2)\sigma_1(1:j-2) + (j-1)\sigma_1^2(1:j-2) \right.$$
$$\left. + j\sigma_2(1:j-2) + (j-1)j\sigma_1(1:j-2) \right] \}$$
$$= \frac{(2)_{j-1} j}{\sigma_1(1:j-1)\sigma_2(1:j-1)} \left[j\sigma_2(1:j-2)\sigma_1(1:j-2) + j^2\sigma_1^2(1:j-2) \right.$$
$$+(j-1)j\sigma_2(1:j-2) + (j-1)j^2\sigma_1(1:j-2)$$
$$-(j-1)\sigma_2(1:j-2)\sigma_1(1:j-2) - (j-1)^2\sigma_1^2(1:j-2)$$

$$- (j-1)j\sigma_2(1:j-2) - (j-1)^2 j\sigma_1(1:j-2)\big]$$

$$= \frac{(1)_j j\sigma_1(1:j-2)}{\sigma_1(1:j-1)\sigma_2(1:j-1)}\big[\sigma_2(1:j-2) + ((j-1)+j)\sigma_1(1:j-2) + (j-1)j\big]$$

$$= \frac{(1)_j j\sigma_1(1:j-2)\sigma_2(1:j)}{\sigma_1(1:j-1)\sigma_2(1:j-1)}$$

$$(2)_{j-1}j^{j-3}\left[j\frac{\sigma_{j-2}(1:j-2,j)}{\sigma_{j-2}(1:j-1)} - (j-1)\frac{\sigma_{j-3}(1:j-2,j)}{\sigma_{j-3}(1:j-1)}\right]$$

$$= \frac{(2)_{j-1}j^{j-3}}{\sigma_{j-3}(1:j-1)\sigma_{j-2}(1:j-1)} \times$$

$$\times \big[j\sigma_{j-2}(1:j-2,j)\sigma_{j-3}(1;j-2,j-1)$$

$$- (j-1)\sigma_{j-2}(1:j-2,j-1)\sigma_{j-3}(1:j-2,j)\big]$$

$$= \frac{(2)_{j-1}j^{j-3}}{\sigma_{j-3}(1:j-1)\sigma_{j-2}(1:j-1)}\{$$

$$j[\sigma_{j-2}(1:j-2)\sigma_{j-3}(1:j-2) + j\sigma_{j-3}^2(1:j-2) +$$

$$(j-1)\sigma_{j-2}(1:j-2)\sigma_{j-4}(1:j-2) + (j-1)j\sigma_{j-3}(1:j-2)\sigma_{j-4}(1:j-2)]$$

$$-(j-1)[\sigma_{j-2}(1:j-2)\sigma_{j-3}(1:j-2) + (j-1)\sigma_{j-3}(1:j-2)^2 +$$

$$j\sigma_{j-2}(1:j-2)\sigma_{j-4}(1:j-2) + (j-1)j\sigma_{j-3}(1:j-2)\sigma_{j-4}(1:j-2)]\big\}$$

$$= \frac{(2)_{j-1}j^{j-3}}{\sigma_{j-3}(1:j-1)\sigma_{j-2}(1:j-1)}[$$

$$j\sigma_{j-2}(1:j-2)\sigma_{j-3}(1:j-2) + j^2\sigma_{j-3}^2(1:j-2) +$$

$$(j-1)j\sigma_{j-2}(1:j-2)\sigma_{j-4}(1:j-2) + (j-1)j^2\sigma_{j-3}(1:j-2)\sigma_{j-4}(1:j-2)$$

$$-(j-1)\sigma_{j-2}(1:j-2)\sigma_{j-3}(1:j-2) - (j-1)^2\sigma_{j-3}^2(1:j-2)$$

$$-(j-1)j\sigma_{j-2}(1:j-2)\sigma_{j-4}(1:j-2)$$

$$-(j-1)^2 j\sigma_{j-3}(1:j-2)\sigma_{j-4}(1:j-2)$$

$$= \frac{(2)_{j-1}j^{j-3}\sigma_{j-3}(1:j-2)}{\sigma_{j-3}(1:j-1)\sigma_{j-2}(1:j-1)}$$

$$\big[1.\sigma_{j-2}(1:j-2) + [(j-1)+j]\sigma_{j-3}(1:j-2) + (j-1)j\sigma_{j-4}(1:j-2)\big]$$

$$= \frac{(1)_j j^{j-3}\sigma_{j-3}(1:j-2)\sigma_{j-2}(1:j)}{\sigma_{j-3}(1:j-1)\sigma_{j-2}(1:j-1)}$$

and

$$(2)_{j-1}j^j - (j-1)^2(2)_{j-1}j^{j-2}\frac{\sigma_{j-2}(1:j-2,j)}{\sigma_{j-2}(1:j-1)}$$

$$
\begin{aligned}
&= \frac{(2)_{j-1} j^{j-2}}{\sigma_{j-2}(1:j-1)} \left[j^2 \sigma_{j-2}(1:j-1) - (j-1)^2 \sigma_{j-2}(1:j-2,j) \right] \\
&= \frac{(2)_{j-1} j^{j-2}}{\sigma_{j-2}(1:j-1)} \Big[j^2 [\sigma_{j-2}(1:j-2) + (j-1)\sigma_{j-3}(1:j-2) \\
&\quad - (j-1)^2 [\sigma_{j-2}(1:j-2) + j\sigma_{j-3}(1:j-2)] \Big] \\
&= \frac{(2)_{j-1} j^{j-2}}{\sigma_{j-2}(1:j-1)} \left[[(j-1)+j]\sigma_{j-2}(1:j-2) + (j-1)j\sigma_{j-3}(1:j-2) \right] \\
&= \frac{(1)_j j^{j-2} \sigma_{j-1}(1:j)}{\sigma_{j-2}(1:j-1)}
\end{aligned}
$$

The second part is to expand the determinant using the first row and factor out multiples of the columns containing sigma functions. Note the tremendous cancellation in the numerator and the denominator in the factors in front of the determinant. So after step $(j-1)$ we obtain the following expression:

$$
D_{n,l,j} = \frac{1}{l!}(1)_1(1)_2(1)_3 \dots (1)_{j-1} \frac{\sigma_1(1:j)\dots\sigma_{j-1}(1:j)}{\sigma_1(1:j-1)\dots\sigma_{j-2}(1:j-1)} \times
$$

$$
\times \begin{vmatrix}
0 & (1)_j & (1)_j j & \cdots & (1)_j j^{j-2} & (1)_j j^j & \cdots & (1)_j j^{n-j} \\
0 & (2)_j \frac{\sigma_1(1:j-1,j+1)}{\sigma_1(1:j)} & (2)_j (j+1) \frac{\sigma_2(1:j-1,j+1)}{\sigma_2(1:j)} & \cdots & (2)_j (j+1)^{j-2} \frac{\sigma_{j-1}(1:j-1,j+1)}{\sigma_{j-1}(1:j)} & (2)_j (j+1)^j & \cdots & (2)_j (j+1)^{n-j} \\
\vdots & \vdots & \vdots & & \vdots & \vdots & & \vdots \\
0 & (l-j)_j \frac{\sigma_1(1:j-1,l-1)}{\sigma_1(1:j)} & (l-j)_j (l-1) \frac{\sigma_2(1:j-1,l-1)}{\sigma_2(1:j)} & \cdots & (l-j)_j (l-1)^{j-2} \frac{\sigma_{j-1}(1:j-1,l-1)}{\sigma_{j-1}(1:j)} & (l-j)_j (l-1)^j & \cdots & (l-j)_j (l-1)^{n-j} \\
(1)_l & (l-j+1)_j \frac{\sigma_1(1:j-1,l)}{\sigma_1(1:j)} & (l-j+1)_j l \frac{\sigma_2(1:j-1,l)}{\sigma_2(1:j)} & \cdots & (l-j+1)_j l^{j-2} \frac{\sigma_{j-1}(1:j-1,l)}{\sigma_{j-1}(1:j)} & (l-j+1) l^j & \cdots & (l-j+1)_j l^{n-j} \\
(2)_l & (l-j+2)_j \frac{\sigma_1(1:j-1,l+1)}{\sigma_1(1:j)} & (l-j+2)_j (l+1) \frac{\sigma_2(1:j-1,l+1)}{\sigma_2(1:j)} & \cdots & (l-j+2)_j (l+1)^{j-2} \frac{\sigma_{j-1}(1:j-1,l+1)}{\sigma_{j-1}(1:j)} & (l-j+2)_j (l+1)^j & \cdots & (l-j+2)_j (l+1)^{n-j} \\
\vdots & \vdots & \vdots & & \vdots & \vdots & & \vdots \\
(n-l+1)_l & (n-j+1)_j \frac{\sigma_1(1:j-1,n)}{\sigma_1(1:j)} & (n-j+1)_j n \frac{\sigma_2(1:j-1,n)}{\sigma_2(1:j)} & \cdots & (n-j+1)_j n^{j-2} \frac{\sigma_{j-1}(1:j-1,n)}{\sigma_{j-1}(1:j)} & (n-j+1)_j n^j & \cdots & (n-j+1)_j n^{n-j}
\end{vmatrix}
$$

The dimension of the determinant is $(n-j+1)\times(n-j+1)$. At the next step we do not use the first column. Start the process of making zeros from using the second column to make zero at the third. We use the same properties of the elementary symmetric functions and again split the process in two parts. One can see that the factors obtained here start already from σ_2. At the next step they will start from σ_3.

$$D_{n,l,j}=\frac{1}{l!}(-1)^1(1)_1(1)_2\dots(1)_j\frac{\sigma_2(1:j+1)\sigma_3(1:j+1)\dots\sigma_j(1:j+1)}{\sigma_2(1:j)\sigma_3(1:j)\dots\sigma_{j-1}(1:j)}\times$$

$$\times\begin{vmatrix}
0 & (1)_{j+1} & (1)_{j+1}(j+1) & \dots & (1)_{j+1}(j+1)^{j-2} & (1)_{j+1}(j+1)^j & \cdots (1)_{j+1}(j+1)^{n-j-1}\\
0 & (2)_{j+1}\frac{\sigma_2(1:j,j+2)}{\sigma_2(1:j+1)} & (2)_{j+1}(j+2)\frac{\sigma_3(1:j,j+2)}{\sigma_3(1:j+1)} & \dots & (2)_{j+1}(j+2)^{j-2}\frac{\sigma_j(1:j,j+2)}{\sigma_j(1:j+1)} & (2)_{j+1}(j+2)^j & \cdots (2)_{j+1}(j+2)^{n-j-1}\\
\vdots & \vdots & \vdots & \vdots & \vdots & \vdots & \vdots\\
0 & (l-j-1)_{j+1}\frac{\sigma_2(1:j,l-1)}{\sigma_2(1:j+1)} & (l-j-1)_{j+1}(l-1)\frac{\sigma_3(1:j,l-1)}{\sigma_3(1:j+1)} & \dots & (l-j-1)_{j+1}(l-1)^{j-2}\frac{\sigma_j(1:j,l-1)}{\sigma_j(1:j+1)} & (l-j-1)_{j+1}(l-1)^j & \cdots (l-j-1)_{j+1}(l-1)^{n-j-1}\\
(1)_l & (l-j)_{j+1}\frac{\sigma_2(1:j,l)}{\sigma_2(1:j+1)} & (l-j)_{j+1}l\frac{\sigma_3(1:j,l)}{\sigma_3(1:j+1)} & \dots & (l-j)_{j+1}l^{j-2}\frac{\sigma_j(1:j,l)}{\sigma_j}(1:j+1) & (l-j)_{j+1}l^j & \cdots (l-j)_{j+1}l^{n-j-1}\\
(2)_l & (l-j+1)_{j+1}\frac{\sigma_2(1:j,l+1)}{\sigma_2(1:j+1)} & (l-j+1)_{j+1}(l+1)\frac{\sigma_3(1:j,l+1)}{\sigma_3(1:j+1)} & \dots & (l-j+1)_{j+1}(l+1)^{j-2}\frac{\sigma_j(1:j,l+1)}{\sigma_j(1:j+1)} & (l-j+1)_{j+1}(l+1)^j & \cdots (l-j+1)_{j+1}(l+1)^{n-j-1}\\
\vdots & \vdots & \vdots & \vdots & \vdots & \vdots & \vdots\\
(n-l+1)_l & (n-j)_{j+1}\frac{\sigma_2(1:j,n)}{\sigma_2(1:j+1)} & (n-j)_{j+1}n\frac{\sigma_3(1:j,n)}{\sigma_3(1:j+1)} & \dots & (n-j)_{j+1}n^{j-2}\frac{\sigma_j(1:j,n)}{\sigma_j(1:j+1)} & (n-j)_{j+1}n^j & \cdots (n-j)_{j+1}n^{n-j-1}
\end{vmatrix}$$

The dimension of the determinant is $(n-j)\times(n-j)$ Expand the determinant and factor out the elementary symmetric functions to obtain at the end of the step $(l-1)$

$$D_{n,l,j}=\frac{1}{l!}(-1)^{l-j}(1)_1(1)_2\dots(1)_{l-1}\times$$

$$\times\frac{\sigma_{l-j+1}(1:l)\sigma_{l-j+2}(1:l)\dots\sigma_{l-2}(1:l)\sigma_{l-1}(1:l)}{\sigma_{l-j+1}(1:l-1)\sigma_{l-j+2}(1:l-1)\dots\sigma_{l-2}(1:l-1)}\times$$

$$
\left| \begin{array}{lllll}
(1)_l & (1)_l & (1)_l l & (1)_l l^2 & \cdots \\
(2)_l & (2)_l \frac{\sigma_{l-j+1}(1:l-1,l+1)}{\sigma_{l-j+1}(1:l)} & (2)_l (l+1) \frac{\sigma_{l-j+2}(1:l-1,l+1)}{\sigma_{l-j+2}(1:l)} & (2)_l (l+1)^2 \frac{\sigma_{l-j+3}(1:l-1,l+1)}{\sigma_{l-j+3}(1:l)} & \cdots \\
(3)_l & (3)_l \frac{\sigma_{l-j+1}(1:l-1,l+2)}{\sigma_{l-j+1}(1:l)} & (3)_l (l+2) \frac{\sigma_{l-j+2}(1:l-1,l+2)}{\sigma_{l-j+2}(1:l)} & (3)_l (l+2)^2 \frac{\sigma_{l-j+3}(1:l-1,l+2)}{\sigma_{l-j+3}(1:l)} & \cdots \\
\vdots & \vdots & \vdots & & \vdots \\
(n-l+1)_l & (n-l+1)_l \frac{\sigma_{l-j+1}(1:l-1,n)}{\sigma_{l-j+1}(1:l)} & (n-l+1)_l n \frac{\sigma_{l-j+2}(1:l-1,n)}{\sigma_{l-j+2}(1:l)} & (n-l+1)_l n^2 \frac{\sigma_{l-j+3}(1:l-1,n)}{\sigma_{l-j+3}(1:l)} & \cdots
\end{array} \right.
$$

$$
\left. \begin{array}{llll}
(1)_l l^{j-3} & (1)_l l^{j-2} & (1)_l l^j & \cdots (1)_l l^{n-l} \\
(2)_l (l+1)^{j-3} \frac{\sigma_{l-2}(1:l-1,l+1)}{\sigma_{l-2}(1:l)} & (2)_l (l+1)^{j-2} \frac{\sigma_{l-1}(1:l-1,l+1)}{\sigma_{l-1}(1:l)} & (2)_l (l+1)^j & \cdots (2)_l (l+1)^{n-l} \\
(3)_l (l+2)^{j-3} \frac{\sigma_{l-2}(1:l-1,l+2)}{\sigma_{l-2}(1:l)} & (3)_l (l+2)^{j-2} \frac{\sigma_{l-1}(1:l-1,l+2)}{\sigma_{l-1}(1:l)} & (3)_l (l+2)^j & \cdots (3)_l (l+2)^{n-l} \\
\vdots & \vdots & \vdots & \vdots \\
(n-l+1)_l n^{j-3} \frac{\sigma_{l-2}(1:l-1,n)}{\sigma_{l-2}(1:l)} & (n-l+1)_l n^{j-2} \frac{\sigma_{l-1}(1:l-1,n)}{\sigma_{l-1}(1:l)} & (n-l+1)_l n^j & \cdots (n-l+1)_l n^{n-l}
\end{array} \right|
$$

The dimension of the determinant is $(n-l+1) \times (n-l+1)$ At the beginning of step l we use new identities. They are

$$
\begin{aligned}
&(2)_l \frac{\sigma_{l-j+1}(1:l-1,l+1)}{\sigma_{l-j+1}(1:l)} - (2)_l \\
&= \frac{(2)_l}{\sigma_{l-j+1}(1:l)} \big[\sigma_{l-j+1}(1:l-1) + (l+1)\sigma_{l-j}(1:l-1) \\
&\quad - \sigma_{l-j+1}(1:l-1) - l\sigma_{l-j}(1:l-1) \big] \\
&= (1)_{l+1} \frac{\sigma_{l-j}(1:l-1)}{\sigma_{l-j+1}(1:l)}
\end{aligned}
$$

$$
\begin{aligned}
&(3)_l \frac{\sigma_{l-j+1}(1:l-1,l+2)}{\sigma_{l-j+1}(1:l)} - (3)_l \\
&= (3)_l \frac{1}{\sigma_{l-j+1}(1:l)} \big[\sigma_{l-j+1}(1:l-1,l+2) \\
&\quad - \sigma_{l-j+1}(1:l-1,l) \big] \\
&= \frac{(3)_l}{\sigma_{l-j+1}(1:l)} \big[\sigma_{l-j+1}(1:l-1) + (l+2)\sigma_{l-j}(1:l-1) \\
&\quad - \sigma_{l-j+1}(1:l-1) - l\sigma_{l-j}(1:l-1) \big] \\
&= \frac{(2)_{l+1}\sigma_{l-j}(1:l-1)}{\sigma_{l-j+1}(1:l)}
\end{aligned}
$$

$$
\begin{aligned}
&(n-l+1)_l \frac{\sigma_{l-j+1}(1:l-1,n)}{\sigma_{l-j+1}(1:l)} - (n-l+1)_l \\
&= (n-l)_{l+1} \frac{\sigma_{l-j}(1:l-1)}{\sigma_{l-j+1}(1:l)}
\end{aligned}
$$

So at the beginning of step l we obtain that

$$
\begin{aligned}
D_{n,l,j} = {} & (-1)^{l-j}(1)_1(1)_2(1)_3 \ldots (1)_{l-1}(1)_l \\
& \times \frac{\sigma_{l-j+1}(1:l)\sigma_{l-j+2}(1:l)\ldots\sigma_{l-1}(1:l)}{\sigma_{l-j+1}(1:l-1)\sigma_{l-j+2}(1:l-1)\ldots\sigma_{l-2}(1:l-1)} \\
& \times \frac{\sigma_{l-j}(1:l-1)\sigma_{l-j+1}(1:l-1)\ldots\sigma_{l-2}(1:l-1)}{\sigma^2_{l-j+1}(1:l)\ldots\sigma^2_{l-1}(1:l)} \times \\
& \times \sigma_{l-j+2}(1:l+1)\ldots\sigma_l(1:l+1)
\end{aligned}
$$

$$
\begin{vmatrix}
(1)_{l+1} & (1)_{l+1} & (1)_{l+1}(l+1) & \cdots & (1)_{l+1}(l+1)^{j-2} & (1)_{l+1}(l+1)^j & \cdots & (1)_{l+1}(l+1)^{n-l-1} \\
(2)_{l+1} & (2)_{l+1}\frac{\sigma_{l-j+2}(1:l,l+2)}{\sigma_{l-j+2}(1:l+1)} & (2)_{l+1}(l+2)\frac{\sigma_{l-j+3}(1:l,l+2)}{\sigma_{l-j+3}(1:l+1)} & \cdots & (2)_{l+1}(l+2)^{j-2} & (2)_{l+1}(l+2)^j & \cdots & (2)_{l+1}(l+2)^{n-l-1} \\
(3)_{l+1} & (3)_{l+1}\frac{\sigma_{l-j+2}(1:l,l+3)}{\sigma_{l-j+2}(1:l+1)} & (3)_{l+1}(l+3)\frac{\sigma_{l-j+3}(1:l,l+3)}{\sigma_{l-j+3}(1:l+1)} & \cdots & (3)_{l+1}(l+3)^{j-2}\frac{\sigma_l(1:l,l+3)}{\sigma_l(1:l+1)} & (3)_{l+1}(l+3)^j & \cdots & (3)_{l+1}(l+3)^{n-l-1} \\
\vdots & \vdots & \vdots & & \vdots & \vdots & & \vdots \\
(n-l)_{l+1} & (n-l)_{l+1}\frac{\sigma_{l-j+2}(1:l,n)}{\sigma_{l-j+2}(1:l+1)} & (n-l)_{l+1}n\frac{\sigma_{l-j+3}(1:l,n)}{\sigma_{l-j+3}(1:l+1)} & \cdots & (n-l)_{l+1}n^{j-2}\frac{\sigma_l(1:l,n)}{\sigma_l(1:l=1)} & (n-l)_{l+1}n^j & \cdots & (n-l)_{l+1}n^{n-l-1}
\end{vmatrix}
$$

Note that after reappearing the factor $\sigma_{l-j}(1:l-1)$ did not cancel and from now on it stays till the end. After step $(n-j)$ we get:

$$
\begin{aligned}
D_{n,l,j} = {} & (-1)^{l-j}\sigma_{l-j}(1:l-1)(1)_1(1)_2(1)_3\ldots(1)_{l+1}\ldots(1)_{n-j} \\
& \times \frac{\sigma_{n-2j+2}(1:n-j+1)\ldots\sigma_{n-j}(1:n-j+1)}{\sigma_{n-2j+1}(1:n-j)\ldots\sigma_{n-j-1}(1:n-j)} \times
\end{aligned}
$$

$$
\begin{vmatrix}
(1)_{n-j+1} & (1)_{n-j+1} & (1)_{n-j+1}(n-j+1) \\
(2)_{n-j+1} & (2)_{n-j+1}\frac{\sigma_{n-2j+2}(1:n-j,n-j+2)}{\sigma_{n-2j+2}(1:n-j+1)} & (2)_{n-j+1}(n-j+2)\frac{\sigma_{n-2j+3}(1:n-j,n-j+2)}{\sigma_{n-2j+3}(1:n-j+1)} \\
\vdots & \vdots & \vdots \\
(j)_{n-j+1} & (j)_{n-j+1}\frac{\sigma_{n-2j+2}(1:n-j,n)}{\sigma_{n-2j+2}(1:n-j+1)} & (j)_{n-j+1}n\frac{\sigma_{n-2j+3}(1:n-j,n)}{\sigma_{n-2j+3}(1:n-j+1)}
\end{vmatrix}
$$

$$\begin{vmatrix} \cdots (1)_{n-j+1}(n-j+1)^{j-2} \\ \cdots (2)_{n-j+1}(n-j+2)^{j-2}\frac{\sigma_{n-j}(1:n-j,n-j+2)}{\sigma_{n-j}(1:n-j+1)} \\ \vdots \\ \cdots (j)_{n-j+1}n^{j-2}\frac{\sigma_{n-j}(1:n-j,n)}{\sigma_{n-j}(1:n-j+1)} \end{vmatrix}$$

We proceed in the same manner. One can easily see that the number of the elementary symmetric functions in the numerator and in the denominator decreases by one at each step since there is no longer jump in the degrees by two in the first row. Using the properties of elementary symmetric functions similar to those, described for step j, but with different coefficients we get that:

$$= (-1)^{l-j}\frac{\sigma_{l-j}(1:l-1)}{l!}(1)_1(1)_2\dots(1)_{n-2}\frac{\sigma_{n-j}(1:n-1)}{\sigma_{n-j-1}(1:n-2)}\times$$

$$\begin{vmatrix} (1)_{n-1} & (1)_{n-1} \\ (2)_{n-1} & (2)_{n-1}\frac{\sigma_{n-j}(1:n-2,n)}{\sigma_{n-j}(1:n-1)} \end{vmatrix}_{2\times 2}$$

$$= (-1)^{l-j}\frac{\sigma_{l-j}(1:l-1)}{l!}(1)_1(1)_2\dots(1)_{n-2}\frac{\sigma_{n-j}(1:n-1)}{\sigma_{n-j-1}(1:n-2)}\times$$

$$\begin{vmatrix} (1)_{n-1} & 0 \\ (2)_{n-1} & (1)_n\frac{\sigma_{n-j-1}(1:n-2)}{\sigma_{n-j}(1:n-1)} \end{vmatrix}_{2\times 2}$$

$$= (-1)^{l-j}\frac{\sigma_{l-j}(1:l-1)}{l!}(1)_1(1)_2\dots(1)_n$$

Therefore (2) holds.This is the end of the proof.

3 Corollary

If we are given a polynomial in the basis $(-x)_k = (-x)(-x+1)\dots(-x+k-1)$ due to the fact that

$$\Delta^l(-x)_k = (-1)^l k(k-1)\dots(k-l+1)(-x)_{k-l}$$

one can easily compute the derivatives, represented again in the basis $(-x)_k$.

References

1. Stirling numbers of the first kind. From Wikipedia, the free encyclopedia. https://en.wikipedia.org/wiki/Stirling_numbers_of_the_first_kind
2. Abramowitz, M., Stegun, I. (eds.).: 24.1.3. Stirling numbers of the first kind. Handbook of Mathematical Functions with Formulas, Graphs, and Mathematical Tables, 9th edn, New York, Dover, p. 824 (1972)
3. Szego, G.: Orthogonal Polynomials. American Mathematical Society, Providence (1939)
4. Ismail, M.E.H., Nikolova, I., Simeonov, P.: Difference equations and Discriminants for discrete orthogonal polynomials. Ramanujan J. **8**(4), 475–502 (2005). https://doi.org/10.1007/s11139-005-0276-z, ISSN: 1572-9303
5. Nikolova, I.: Difference equations for hypergeometric polynomials from the Askey scheme. Some resultants. Discriminants. Methods Appl. Anal. **11**(1), 001–014 (2004)
6. Nikolova, I.: Differences backwards and Askey scheme of hypergeometric polynomials. East J. Approximations **14**(4), 485–495 (2008)

Investigation of As-cast Light Alloys by Selected Homogenization Techniques with Microstructure Effects

Ludmila Parashkevova and Pedro Egizabal

Abstract In the present contribution, upgrading the findings of previous works, [1, 2], new models are proposed for evaluation of effective mechanical properties of light alloys regarded as multiphase composites. These models are aimed to improve the mechanical properties predictions of two groups light alloys: die cast Mg alloys AZ and metal foams with closed cells. The presented models are variants of Mean Field Homogenization (MFH) approach and Differential Homogenization Method, (DHM), both accounting for microstructure size effects. They are appropriate for composite structures where the content of non- matrix phases is predominant. The microstructure - properties relationships for AZ Mg alloy with Continuous (C) and Discontinuous (D) intermetallic phase precipitations are investigated applying MFH and DHM approaches. The basic distinction between cases (C) and (D) consists of different arrangement and volume fraction of harder intermetallic phase Mg17Al12. The type of the microstructure observed depends mainly on the applied cooling regime and chemical composition of the alloy. The elastic-plastic properties predictions for both types of microstructure topology are compared and discussed. The elastic behavior of foams with closed pores is simulated applying DHM, where the size sensitive variant of Mori-Tanaka scheme is used as a basic 'dilute case' procedure. The method is developed to closed form solutions of corresponding system of differential equations. The results obtained by means of the size-sensitive DHM are compared with experimental data for aluminium and glass foams taken from the literature.

L. Parashkevova (✉)
Institute of Mechanics, Bulgarian Academy of Sciences, Sofia, Bulgaria
e-mail: lusy@imbm.bas.bg

P. Egizabal
TECNALIA, Foundry and steelmaking department, Donostia, Spain
e-mail: pedro.egizabal@tecnalia.com

K. Georgiev et al. (eds.), *Advanced Computing in Industrial Mathematics*,
Studies in Computational Intelligence 793,
https://doi.org/10.1007/978-3-319-97277-0_29

1 Introduction

In recent years the development of enhanced technologies for preparing of composites with higher volume fractions of secondary phases shows a significant increase. Such materials are alloys with specific phase morphology, foams, concretes, etc. Special attention has been attracted to light alloys and other materials of metal or ceramic matrix [3–5]. In the work of Dunant and co. [6], several numerical methods are compared to assess the properties and suitability to model microstructures with large phase properties contrast like concretes. Recently a comparative review has been published in [7] discussing on most elastic modulusporosity relationships suggested so far.

This paper is aimed to elucidate some problems of metal material strengthening/ softening caused by the presence of high volume fraction of incorporated inclusions. Upgrading the findings of previous works [1, 2], new models are proposed for evaluation of effective mechanical properties of light alloys regarded as multiphase composites. The approach is applied for analysis of two types of light alloys.

The first type are magnesium (Mg) based AZ alloys with aluminum (Al) content (1–10 wt%).The major advantage, which is keeping the attention on Mg alloys is their lower weight saving of about 40% compared to steel and cast iron and 20% compared to aluminum for the same component performance, [3]. The main representative of the AZ family is the AZ91 alloy. It contains a higher percentage of aluminium (around 9 wt%) than the other Mg - based alloys and about 0.7 wt% zinc. The morphology of the alloy with so-called "continuous" precipitations of intermetallics Mg17Al12 /type (CP)/ is characterized by hard particles (10–20 volume % for AZ91) embedded into the softer matrix. The second microstructure morphology of AZ alloys displays so-called "discontinuous" type of precipitating (DP) and corresponds to a composite with low volume fraction (10–20 volume % for AZ91) of harder intermetallic matrix phase, which contains (80–90 volume % for AZ91) inclusions of softer material alpha Mg. Images of such microstructure can be found in [8], for instance. The schematic patters of both kinds of γ phase distribution are presented on Figs. 1 and 2. The type of microstructure depends mainly on the applied cooling regime and chemical composition of the alloy. As far as in the most AZ alloys Al is supersaturated in the melt, the precipitation type, rate and the amount of intermetallic phase are controlled by the thermal-mechanical manufacturing conditions. In [9] the behavior of AZ91 of type (CP) in non-elastic state was investigated assuming that the total hardening response of the alloy is a sum of contributions of different hardening modes, as Hall-Petch, Orowan, solid solution, work hardening.

The second type of materials studded in this paper are foams containing considerable amount of closed pores. Metal foams are materials of high porosity which demonstrate a multitude of distinctive thermo-mechanical and physical properties, [10]. For their very low specific weights and thus high specific stiffness, they are able to absorb significant amount of deformation energy while guaranteeing other properties such as high fire and heat resistance, noise attenuation and shielding of

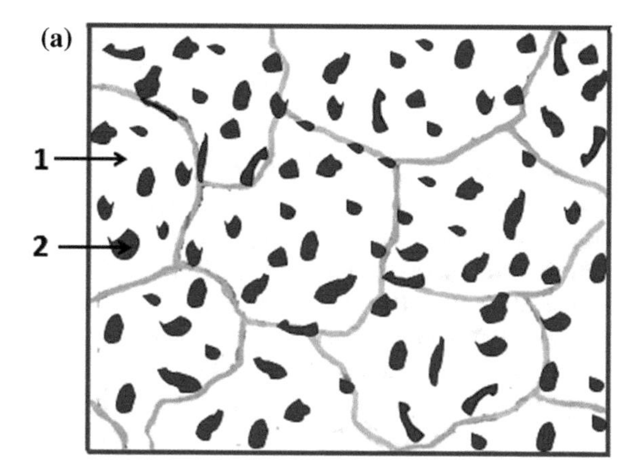

Fig. 1 Simplified patterns of AZ91 alloy's morphology: **a** With "continuous" presipitations; 1- alpha Mg, 2- intermetallic phase

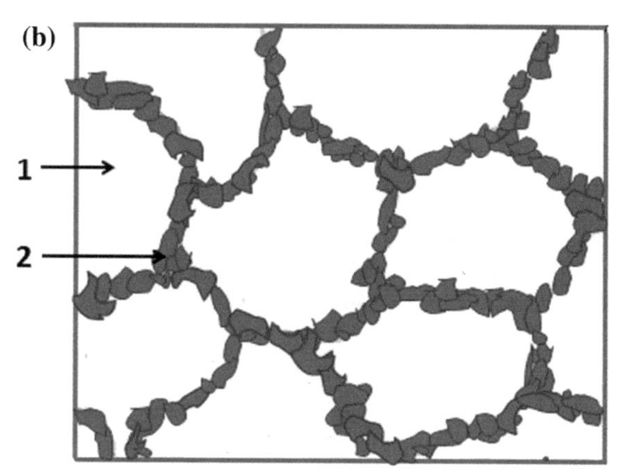

Fig. 2 Simplified patterns of of AZ91 alloy's morphology: **b** with "discontinuous" precipitations; 1- alpha Mg, 2- intermetallic phase

electromagnetic devices, see [7]. In this contribution some new assessments of elastic properties - pore size relationship are presented.

2 Modelling

The overall properties of a multiphase composite at higher concentration of embedded phases are estimated applying a multi step homogenization procedure. On the first step the Representative Volume Element (RVE) of the composite with matrix and n_f phases is resolved into a set of n pseudo grains, following the concept of [11]. Each pseudo grain is a two - phase composite containing a part of the matrix and all inclusions with a particular size and elastic properties. All pseudo grains keep one and the same volume fraction C_{sum}, equal to the total one. The new size sensitive variants of DEM homogenization are presented herein accounting for high volume fractions of non-matrix constituents. These variants are developed for the metal composites considered - AZ alloys with discontinuous intermetallic precipitations and closed cell metal foams but they could be applied to any materials of high porosity or concretes as well. The proper choice of a homogenization method at the first step is a key point for the overall properties predictions of entire model of the composite structure. On the second step the RVE's assemblage of the already homogeneous Cauchy - type pseudo-grains has to be subjected to the final homogenization performing any scheme, symmetric with respect to constituents different isotropic pseudo grains.

For all composites considered in this paper the initial matrix is regarded as isotropic micropolar centrosymmetric material, and the inclusions - as isotropic Cauchy material. For composites with higher volume fraction of matrix phase as alloys of type (CP) the homogenization approach presented in [1] is applyed. For composites with low volume fraction of matrix phase as alloys of type (DP) or closed cell foams a new variant of DHM is discussed in the next sections.

2.1 Differential Approach

The Differential Effective Medium (DEM) approach requires an explicit homogenization scheme for a dilute inclusion problem to be postulated at the beginning. Let the equations (1) represent such a dilute inclusion model for the bulk and shear moduli of two phase composite, [2]:

$$K = F_K\left(K_m, K_i, c, G_m, G_i\right) \quad , \quad G = F_G\left(K_m, K_i, c, G_m, G_i\right) \tag{1}$$

It is suggested that functions F_K and F_G satisfy the requirements of Taylor's theorem and are expanded in Maclaurin series with respect of concentration c. Then:

$$\begin{aligned} K\left(c\right) &= F_K\left(0\right) + \frac{dF_K}{dc}\left(0\right)c + O_K\left(c^2\right) \ , \\ G\left(c\right) &= F_G\left(0\right) + \frac{dF_G}{dc}\left(0\right)c + O_G\left(c^2\right) \end{aligned} \tag{2}$$

where $F_K\left(0\right) = K_m$ and $F_G\left(0\right) = G_m$.

Suppose that RVE of the composite material of volume V contains c volume fraction of inclusions and $\phi = (1-c)$ volume fraction of matrix. Let one remove from the RVE an infinitesimal volume element dV. In this volume element the matrix part is $dV_m = (1-c)dV$. Next a new portion of inclusions is added to the composite replacing the differential volume element by the same small volume consisting of inclusions only. Than, the volume fraction of the matrix will change as follows:

$$\phi + d\phi = \frac{V_m - dV_m}{V - dV + dV} = \phi - (1-c)\frac{dV}{V} \Longrightarrow \frac{dc}{1-c} = \frac{dV}{V} \ . \tag{3}$$

The main idea of the DEM is to regard the new inclusion enriched material as a composite consisting of matrix of equivalent material and inclusions' phase with a very low volume fraction dV/V embedded in it. For such composite the dilute hypothesis is valid and the linear parts of formulae (2) are suitable for evaluation of the overall elastic properties. The linear system (2) is rewritten in the following form: the moduli $K,\ G$ are expressed as $\left(K_{eq}+dK_{eq}\right),\left(G_{eq}+dG_{eq}\right)$ and the matrix index "m" is replaced by matrix index "eq" (equivalent):

$$dK_{eq}=\frac{dF_K\left(K_{eq},K_i,G_{eq},G_i\right)}{dc}|_{c=o}\frac{dc}{1-c},$$
$$dG_{eq}=\frac{dF_G\left(K_{eq},K_i,G_{eq},G_i\right)}{dc}|_{c=o}\frac{dc}{1-c}. \quad (4)$$

2.2 Size - Sensitive Differential Approach

It is worth noting that the system of differential equations (4) can be obtained for any dilute homogenization (1) which allows Taylor series expansion. Hereafter the updated size sensitive Mori-Tanaka homogenization will be applied [12–14], as a representative of (1). For the RVE of the composite with spherical inclusions entirely surrounded by the matrix the moduli of the particular pseudo-grain are obtained by means of equations:

$$K(c)=K_m+\frac{C_{sum}K_m\left(K_i-K_m\right)}{C_m a_m\left(K_i-K_m\right)+K_m},\ a_m=\frac{3K_m}{3K_m+4G_m},$$
$$G(c)=G_m+\frac{C_{sum}G_m\left(G_i-G_m\right)}{C_m\left(b_m-b_{0i}\right)\left(G_i-G_m\right)+G_m},\ b_m=\frac{6\left(K_m+2G_m\right)}{5\left(3K_m+4G_m\right)} \quad (5)$$

where index "m", "i" stays for the matrix and the inclusions, respectively. Appling the procedures described in Sect. 2.1 one obtains the non-liner system of differential equations:

$$dK_{eq}=\frac{\left(3K_{eq}+4G_{eq}\right)\left(K_i-K_{eq}\right)}{\left(3K_i+4G_{eq}\right)}\frac{dc}{1-c},$$
$$dG_{eq}=\frac{G_{eq}\left(G_i-G_{eq}\right)}{\left(G_i-G_{eq}\right)\left[\frac{6}{5}\frac{\left(K_{eq}+2G_{eq}\right)}{\left(3K_{eq}+4G_{eq}\right)}-b_{0i}\right]+G_{eq}}\frac{dc}{1-c}. \quad (6)$$

$$b_{0i}=\frac{6p}{5(p+1)}R_i\left(\eta_i\right),\quad R_i\left(\eta_i\right)=e^{-\eta_i}\left(\eta_i^{-2}+\eta_i^{-3}\right)\left(\eta_i ch\eta_i-sh\eta_i\right), \quad (7)$$

$$\eta_i=\frac{\sqrt{p}}{p+1}\frac{D_i}{l_0},\quad \left(l_0\right)^2=\frac{\beta_0}{G_0}=\frac{\gamma_0}{\kappa_0},\quad \frac{\kappa_0}{G_0}=\frac{\gamma_0}{\beta_0}=p \quad (8)$$

In (8) the moduli G_0 and β_0 connect symmetric parts of stresses - strains, and couple stress- curvature tensors, respectively, and κ_0 and γ_0 connect antisymmetric parts of corresponding tensors at elastic state of the initial micropolar matrix, $p \geq 0$.

The term b_{0i} appearing in (6) and given by (7) is the only trace of Cosserat properties of the initial matrix through the average Eshelby tensor. The new assumption adopted herein is that the term b_{0i} could be regarded as an internal model parameter which expresses in average manner the sensitivity of the genuine matrix to the presence of spherical inclusions of diameter D_i. The term b_{0i} depends on dimensionless model parameters p and D_i/l_0. For metal based composites the internal length l_0 is often related to the matrix microstructure self-organization, as the grain size, for example. From the DEM's point of view such assumption seems reasonable. DEM supposes that after adding the first small portion of inclusions we deal with equivalent matrix of Cauchy type. Further composite properties change is calculated by the simple linear homogenization procedure of "isolated inclusion" type, which should be executed over and over again, accounting for the amount of inclusions added. Every time the equivalent material, obtained at the previous step has to play the roll of matrix at the next step. After such kind of homogenization (5) the overall material is an ordinary isotropic material without any micropolar features, but its moduli are "inherited" their size sensitivity by means of the term b_{0i}.

It should be remarked that there exists an important difference at assessment of the overall moduli given by Eqs. (5) and (6). In upgraded Mori - Tanaka homogenization, [12, 13], the overall bulk modulus does not depend on the size parameter b_{0i}, but in the present variant of DEM both (bulk and shear) overall moduli are size-sensitive, in spite of the fact that the system (6) has been derived on the base of the system (5).

In Garboczi - Berryman (G-B) model, see [15], the size sensitivity is due to the presence of additional intermediate layer like outer inclusion's shell and such size dependence vanishes if the thickness of the mentioned layer tends to zero. In our model the size sensitivity is dictated by size parameters of matrix and inclusions and has no- zero limits even when the diameters of inclusions are infinitesimally small. In G-B model the intermediate shell and the basic inclusion together have to be homogenized in a new inclusion with larger size and corresponding properties. It is worth to mention that the present model is capable to include G-B model, giving rise of double size sensitivity, one induced by matrix-new inclusions interaction, second - induced by the ratio between sizes of the shell and the original inclusion.

Now we pay special attention to the variant of porous composite with $K_i = 0$. In that case the system of differential equations (6) expressed in terms of Poisson's ratio and Young's modulus can be uncoupled:

$$\frac{2}{3} \frac{\left(7 - 5\nu_{eq}\right) + 15\left(1 - \nu_{eq}\right) b_{0i}}{\left(1 - {\nu_{eq}}^2\right)\left[\left(1 - 5\nu_{eq}\right) - 5\left(1 - \nu_{eq}\right) b_{0i}\right]} d\nu_{eq} = \frac{dc}{1 - c}, \tag{9}$$

$$\frac{dE_{eq}}{E_{eq}} = \frac{1}{2\left(2\nu_{eq} - 1\right)} \left[\frac{3\left(1 - \nu_{eq}\right)}{F_\nu{}'\left(eq\nu, 0\right)} + 4\right] d\nu_{eq}\,. \tag{10}$$

where:

$$F_{\nu}{}'(eqv,0) = \frac{3}{2}\frac{\left(1-\nu_{eq}{}^{2}\right)\left[\left(1-5\nu_{eq}\right)-5\left(1-\nu_{eq}\right)b_{0i}\right]}{\left(7-5\nu_{eq}\right)+15\left(1-\nu_{eq}\right)b_{0i}}\,. \tag{11}$$

and it is supposed that $F_{\nu}{}'(eqv,0) \neq 0$.

If $F_{\nu}{}'(eqv,0) = 0$ Eqs. (9), (10) are transformed to the system:

$$d\nu_{eq} = 0 \;\; ; \quad \frac{dE_{eq}}{E_{eq}} = \frac{3\left(1-\nu_m{}^*\right)}{2\left(2\nu_m{}^*-1\right)}\frac{dc}{1-c}\,. \tag{12}$$

where: $\nu_m{}^* = (1-5b_{0i})/5(1-b_{0i})$.

2.3 Limits of Size Effect

For the case of spherical inclusions limiting values of b_{0i}, which serve as bounds for assessment of the size effects have been presented in [2]. Here we will recall that cases.

Case I: $b_{0i} = 0$, $p = 0$ Case I describes those type of matrix inclusions interactions when the size parameter D_i of the embedded phase is much bigger than the internal parameter l_0 related to the matrix microstructure. In other words, the microstructure of the matrix is so fine, that the matrix does not "feel" the presence of the coarse added phase. Case I, of course, also corresponds to the classical solution of DEM when size effects are not considered at all, see ([16]).

Case II: $b_{0i} = 1/5$, $p = 1$. This case represents the limiting case when the diameter of inclusions tends to zero and shear modulus G_0 of the initial matrix is of the same order than Cosserat modulus κ_0. Expressed by b_{0i} this case is situated just in the middle between Case I and Case III.

Case III: $b_{0i} = 2/5$, $p \to \infty$. Theoretically at such ratio between mentioned Cauchy and Cosserat moduli the material of the initial matrix is the most sensitive to the presence of particles or pores. This case also corresponds to the variant when $G_0 \ll \kappa_0$. Case III manifests the limited "resistance" of the matrix when the very small alien objects display themselves as obstacles blocking the matrix to develop its own structures with size $l_0 \gg D_i$.

2.4 Closed form Solutions

Here the closed form analytical solutions for cases which will be used in real composite properties simulations are presented.

(2.4.1) $K_i = K_m$. Thus, $dK_{eq} = 0$ and the solution of the system (6) is:

$$K_{eq} = K_i = K_m = const\ ,\ \ G_{eq}\,(0) = G_m\ ,\ \ G_{eq}\,(1) \to G_i$$
$$\left(\frac{G_m}{G_{eq}}\right)^{(2/5-b_{0i})}\left(\frac{G_i - G_{eq}}{G_i - G_m}\right)\left(\frac{3K_i + 4G_m}{3K_i + 4G_{eq}}\right)^{1/5} = 1 - c\,; \tag{13}$$

(2.4.2) $K_i = 0$ - inclusians are spherical pores.

$$\left(\frac{1-\nu_m}{1-\nu_{eq}}\right)^{1/6}\left(\frac{1+\nu_m}{1+\nu_{eq}}\right)^{\frac{2+5b_{0i}}{3-5b_{0i}}}\left(\frac{{\nu_m}^* - \nu_{eq}}{{\nu_m}^* - \nu_m}\right)^{\frac{5(3+5b_{0i})}{6(3-5b_{0i})}} = 1 - c \tag{14}$$

$$E_{eq} = E_m\left(\frac{1+\nu_m}{1+\nu_{eq}(c)}\right)^{\frac{2+5b_{0i}}{3-5b_{0i}}}\left(\frac{{\nu_m}^* - \nu_{eq}(c)}{{\nu_m}^* - \nu_m}\right)^{\frac{5}{3-b_{0i}}}\ ,\quad {\nu_m}^* = \frac{1-5b_{0i}}{5(1-b_{0i})} \tag{15}$$

(2.4.3) $\nu_m = {\nu_m}^*$. Than the system (12) has the solution:

$$\nu_{eq}(c) = {\nu_m}^* = const\ ,\quad {E_{eq}}^* = E_m\,(1-c)^{\frac{3(1-{\nu_m}^*)}{2(1-2{\nu_m}^*)}} = E_m\,(1-c)^{\frac{6}{3+5b_{0i}}}\ . \tag{16}$$

For a porous composite with a matrix with arbitrary Young's modulus E_m, Poisson's ratio $\nu_m = {\nu_m}^*\ , -1/3 \le {\nu_m}^* \le 1/5,\quad 0 \le c \le 1:$

$$\frac{{E_{eq}}^*(c)}{E_m} = \frac{{K_{eq}}^*(c)}{{K_m}^*} = \frac{{G_{eq}}^*(c)}{{G_m}^*}\ ; \tag{17}$$

$$\frac{{G_{eq}}^*(c)}{{K_{eq}}^*(c)} = const\ ,\quad \frac{{E_{eq}}^*(c)}{{G_{eq}}^*(c)} = const\ ,\quad \frac{{E_{eq}}^*(c)}{{K_{eq}}^*(c)} = const\ . \tag{18}$$

As the above ratios depend on ${\nu_m}^*$ only, at $b_{0i} = \frac{3}{35}$ $\quad \frac{{G_{eq}}^*(c)}{{K_{eq}}^*(c)}|_{{\nu_m}^*=1/8} = 1$.

3 Numerical Simulations and Results

3.1 Mechanical Properties of AZ Mg Alloys

Our aim is to investigate the differences predicted by two models suggested for AZ alloys with (CP) and (DP) microstructure, respectively. To this end the volume fractions of constituents are kept the same in both cases, the inclusions always are considered spherical. The speculation of equality of the bulk moduli of α^* and γ phases is supported too.

For AZ alloy with (CP) microstructure the models, developed in [1] is adopted. For (DP) type of alloy the elastic properties are calculated by means of Eqs. (13), see Table 1. The initial yield stress of (DP) composite is estimated on the suggestion that both phases undergo equal average strain, [17], up to yielding point, so than:

Table 1 Input parameters for simulation of AZ alloy as a composite

Constituent	E	ν	K	κ	σ_{p0}	l_0	D_i
	GPa	–	GPa	GPa	MPa	μm	μm
$\alpha^* Mg$ as matrix at (CP)	44	0.33	44.4	16.5	Eq. (20)	100	–
$Mg17Al12$ as inclusion (CP)	71.94	0.23	44.4	–	286	–	10
$Mg17Al12$ as matrix (DP)	71.94	0.23	44.4	29.24	286	10	–
$\alpha^* Mg$ as inclusion at (DP)	44	0.33	44.4	–	Eq. (20)	–	100

$$\sigma_{p0}^{\alpha^*}\left[1+C_\gamma\left(E_\gamma/E_{\alpha^*}-1\right)\right] \le \sigma_{p0}^{(DP)} \le \sigma_{p0}^{\alpha^*}\left[1+C_\gamma\left(\sigma_{p0}^{\gamma}/\sigma_{p0}^{\alpha^*}-1\right)\right] \quad (19)$$

Analyzing experimental data for tensile behavior of different AZ alloys an expression for the yield strength of α^* phase has been suggested in [1] depending on the amount of Al in the alpha phase:

$$\sigma_{p0}^{\alpha^*} = S_0\left(1+S\,C_{Al}^{wt\alpha}\right), \quad S_0 = 56\text{ MPa},\ S = 17 \quad (20)$$

The data for hardness of α^*Mg phase and γ intermetallic phase of AZ61 alloy ($H_{\alpha^*} = 1.35GPa$, $H_\gamma = 4.3$GPa), [18], obtained through nanoindentation are used to estimate the initial yield stress of Mg17Al12. It is supposed that the ratio between hardness values is the same as the ratio between the yield limits of both phases: $\frac{\sigma_{p0}^{\alpha^*}}{\sigma_{p0}^{\gamma}} \approx \frac{H_{\alpha^*}}{H_\gamma}$. Applying the above relation and Eq. (20) at $C_{Al}^{wt\alpha} = 3.5\,wt\%$ one gets $\sigma_{p0}^{\gamma} = 286$ MPa. The mechanical properties of γ phase may be considered constant for all AZ alloys as far as this secondary compound has chemically stabile structure wherever, but the properties of α^* phase as a solid solution may vary. According to our opinion, the distribution of Al between alpha and gamma phases is the most important factor bringing influence on mechanical properties of α^* phase and on volume fraction of γ phase in a mutual way. In the present study the elastic properties of α^* phase in all AZ are considered constant, but plastic behaviour depends on "residual" Al as it is seen in Eq. (20). Supposing AZ alloy as a two-phase composite, the volume fraction of γ phase obeys the relation, [1]:

$$C_\gamma^{vol\,AZ} \cong C_{\gamma\,max}^{vol\,AZ}(Al)\,\frac{T_{solvus}-T}{T_{solidus}-T}, \quad C_{\gamma\,max}^{vol\,AZ} = \frac{1+a_1-\sqrt{(1+a_1)^2-4a_1a_2}}{2a_1}, \quad (21)$$

where $a_1 = 1-\rho^{Mg}/\rho^{AZ}$, $a_2 = C_{Al}^{wt\,AZ}/C_{Al}^{wt\gamma}$, $\rho^{(.)}$ stays for density, $C_{(..)}^{wt\,(.)}$, $C_{(..)}^{vol\,(.)}$ stays for mass (volume) fraction of the component (..) in (.). The temperature depending Eq. (21) is just an assumption yet. The simulations provided show that the elastic properties of AZ alloy do not depend on the morphology of Mg17Al12, see Fig. 3. The plastic responce is more complicated. If the softer α^* Mg phase is reaching the plastic state first, corresponding to the left inequality of (19), the initial yield limit of (DP) alloy is slightly higher than those of (CP) alloy and as illustrated by black

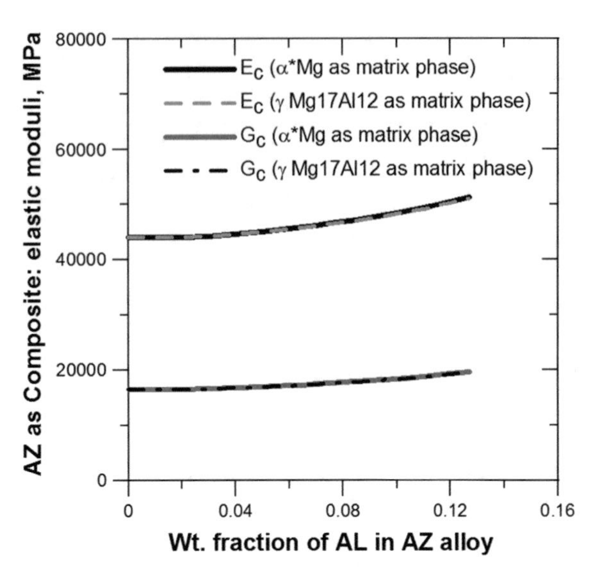

Fig. 3 Comparison between **a** Continuous and **b** Discontinuous Mg17Al12 in AZ alloys. Influence on overall elastic properties of composite

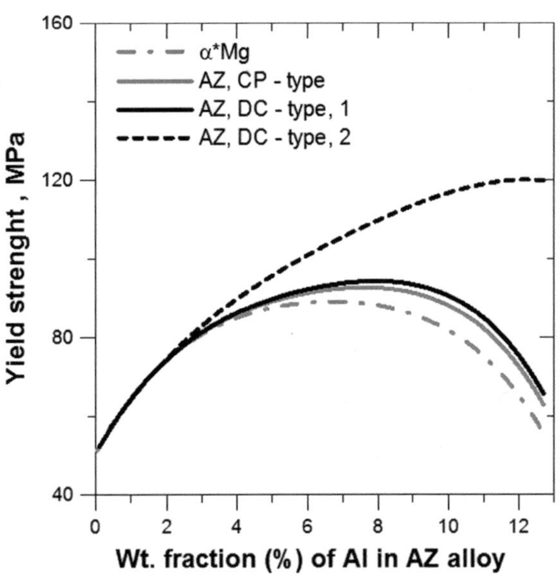

Fig. 4 Comparison between **a** Continuous and **b** Discontinuous Mg17Al12 in AZ alloys. Influence on initial yield stress of composite

curve on Fig. 4. If both constituents enter the plastic regime simultaneously (that is the situation modeled by the right side of (19) the strengthening effect for (DP) composite could be much more pronounced (see black dash curve on Fig. 4), (for (CP) see Fig. 5).

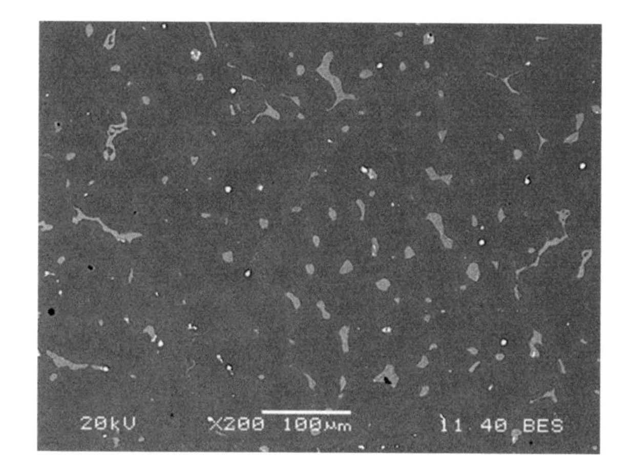

Fig. 5 Microstructure of AZ91 alloy of type (CP). Dark gray areas- alpha Mg matrix; light gray - Mg17Al12; white areas - pores

3.2 *Mechanical Properties of Closed Cell Foams*

The presented size-sensitive model of foams reveals the special role of the Poisson's ratio of the matrix material on the expected elastic deformation of the porous material. On Fig. 6 the different possible ways of composite Poisson's ratio changes depending on the inclusion's size (parameter b_{0i}) and porosity are illustrated. Curves 1–4 are calculated for one and the same initial metal matrix but at different values of b_{0i}. Curves 1, 3, 4 correspond to Cases I, II, III, respectively, which have been discussed in Sect. 2.3. Focusing on curve (3) one can register that at $b_{0i} = 1/5$ if such a metal foam is subjected to axial compression, for instance, at very high porosity state , it would not display any strains on direction perpendicular to compression, as far as its Poisson's ratio should tend to zero.

Predictions for elastic moduli of experimentally investigated foams are provided according to formulae (14), (15) on Figs. 7 and 8. On Fig. 7 the results for relative bulk modulus obtained by means of present model are compared to experimental data for glass foams, presented in [16] and to prediction performed by classical Mori-Tanaka homogenization.

In [19] data for the modulus of elasticity was determined from free vibrations tests. The samples were made of aluminium alloy powder Al 99,7. Available experimental values are recalculated according to the common relation between the relative density and porosity: $c = \rho/\rho_A l$. After that they are plotted on Fig. 8 together with theoretical predictions from (15) for Case I (no size effect) and Case III (maximal size effect). In relative form these results do not depend on matrix Young's modulus but on Poisson's ratio and porosity only. In both type of very different foams -glass and Al-the presented variant of DHM could be considered relevant.

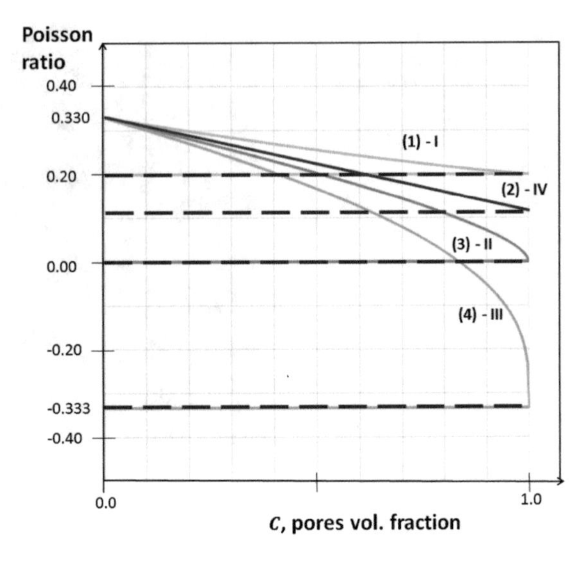

Fig. 6 Influence of b_{0i} on Poisson's ratio of a metal porous composite, (1) $b_{0i} = 0$; (2) $b_{0i} = 1/10$; (3) $b_{0i} = 1/5$; (4) $b_{0i} = 2/5$;

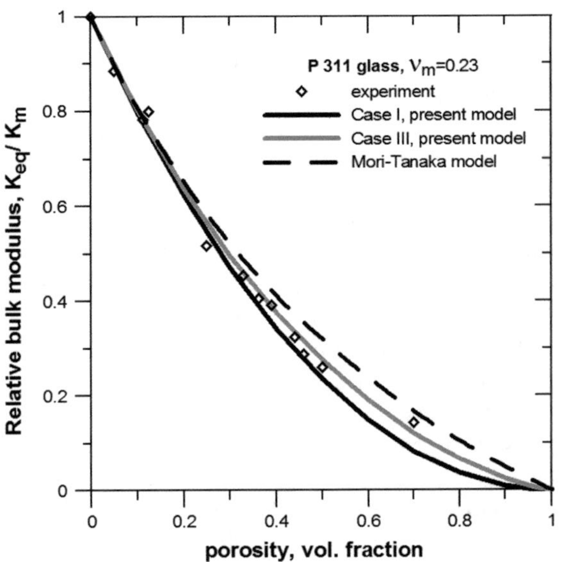

Fig. 7 Relative Bulk modulus of porous P-311 glass (circles), [16], compared to present theory results and classical Mori-Tanaka method

4 Conclusions

The models suggested describe correctly the material behavior of composites, with high volume fraction of multiphase inclusions. The models take into account the microstructure adjustment between the matrix materials and embedded inclusions via internal length properties depending on the size and volume fractions. The models are applicable for metal as well as for non-metal matrix composites. Analytical solutions

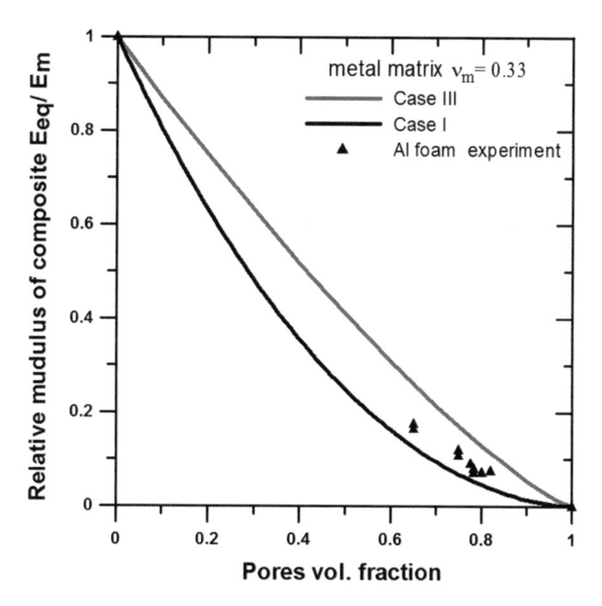

Fig. 8 Relative Young's modulus of a metal porous composite; present theory results and experiments for Al foam, recalculated from [19]

presented in Sect. (2.4) allow some improvement/control of mechanical properties of the end composite to be achieved by a proper combination of size, volume fraction and initial moduli of compounds. The results of the numerical calculations are in accordance with experimental data.

New analytical size- sensitive schemes developed herein could be useful in some modern numerical homogenization approaches which require a choice of an appropriate local analytical homogenization as it was proven, [6], that significant differences were observed in the final values depending on the local scheme chosen.

The model can be developed further for composites with high volume fraction of inclusions of various shapes.

Acknowledgements This research is carrying out in the frame of KMM-VIN - European Virtual Institute. The financial support of BG FSI through the grant DH 07/17/2016 'New approach for structure and properties design of amorphous and nanostructured metallic foams' is gratefully acknowledged.

References

1. Parashkevova, L., Egizabal, P.: Modelling of light Mg and Al based alloys as in situ composites. In: Georgiev, K., Todorov, M., Georgiev, I. (eds.) Advanced Computing in Industrial Mathematics, Studies in Comp. Intelligence, vol. 728, pp. 145–157. Springer, Berlin (2018)
2. Parashkevova, L.: Some considerations on modelling and homogenization of multiphase light alloys. MATEC Web of Conf. **145**, 02008 (2018)
3. Peter, I., Rosso, M.: Light alloys from traditional to innovative technologies. In: Ahmad, Z. (ed.) New Trends in Alloy Development, Characterization and Application, Chapter 1, pp. 3–37. InTech (2015)

4. Lefebvre, LPh, Banhart, J., Dunand, D.: Porous metals and metallic foams: current status and recent developments. Adv. Eng. Mat. **10**, 77–787 (2008)
5. Stanev, L., Kolev, M., Drenchev, B., Drenchev, L.: Application of space holders for obtaining of metallic porous materials part I, production methods. Rev. J. Mat. Sci. Technol. **23**, 303–344 (2015)
6. Dunant, C., Bary, B., Giorla, A., Peniguel, Ch., Sanahuja, J., Toulemonde, Ch., Tran , A.-B., Willot, F., Yvonnet, J.: A critical comparison of several numerical methods for computing effective properties of highly heterogeneous materials. Adv. Eng. Softw. **58**, 1–12 (2013)
7. Koudelka, P., Jirousek, O., Doktor, T., Zlamal, P., Fila, T.: Comparative study on numerical and analytical assessment of elastic properties of metal foams. In: 18th International Conference Engineering Mechanics, May 14–17, 2012, Paper no 218, pp. 691–701. Czech Republic, Svratka (2012)
8. Zhang, M., Kelly, P.M.: Crystallography of Mg17Al12 precipitates in AZ91D alloy. Scr. Mat. **48**, 647–652 (2003)
9. Hutchinson, C.R., Nie, J.-F., Gorsse, S.: Modeling the precipitation processes and strengthening mechanisms in a Mg-Al-(Zn) AZ91 alloy. Metall. Mater. Trans. A **36**, 2093–2105 (2005)
10. Kolken, H.M.A., Zadpoor, A.: Auxetic mechanical metamaterials. RSC Adv. **7**, 5111–5129 (2017)
11. Pierard, O., Friebel, C., Doghri, I.: Mean-field homogenization of multiphase thermo-elastic composites: a general framework and its validation. Compos. Sci. Technol. **64**, 1587–1603 (2004)
12. Hu, G., Liu, X., Lu, T.J.: A variational method for non-linear micropolar composites. J. Alloy Compd. **37**, 407–425 (2005)
13. Parashkevova, L., Bontcheva, N., Babakov, V.: Modelling of size effects on strengthening of multiphase Al based composites. Comp. Mater. Sci. **50**, 527–537 (2010)
14. Parashkevova, L., Bontcheva, N.: Micropolar-based modeling of size effects on stiffness and yield stress of nanoparticles-modified polymer composites. Comp. Mater. Sci. **67**, 303–315 (2013)
15. Garboczi, E.J., Berryman, J.G.: Elastic moduli of a material containing composite inclusions: effective medium theory and finite element computations. Mech. Mat. **33**, 455–470 (2001) brittle materials. J. Mater. Sci. **39**, 3501–35031 (2004)
16. Giordano, S.: Differential schemes for the elastic characterisation of dispersions of randomly oriented ellipsoidse. Eur. J. Mech. A/Solids **22**, 885–902 (2003)
17. Rieger, F.: Work-Hardening of Dual-phase Steel, vol. 7. KIT Scientific Publishing, Karlsruhe (2016)
18. Hay, J., Agee, Ph.: Mapping the mechanical properties of alloyed magnesium (AZ61). In: Hort, N., Mathaudhu, S., Neelameggham, N. (eds.) Magnesium Technology, 142 TMS Annual Meeting. Springer, Berlin (2013)
19. Kovacik, J., Simancik, F.: Aluminium foam - modulus of elasticity and electrical conductivity according to percolation theory. Scripta Mater. **39**, 239–246 (1998)

Performance Analysis of Intel Xeon Phi MICs and Intel Xeon CPUs for Solving Dense Systems of Linear Algebraic Equations: Case Study of Boundary Element Method for Flow Around Airfoils

D. Slavchev and S. Margenov

Abstract The presented analysis is based on simulation of laminar flows around airfoils. The Boundary Element Method is applied for the numerical solution of the related integral equations. The most computationally expensive part of the algorithm is to solve the arising dense system of linear algebraic equations. We examine several libraries that implement some variants of Gaussian elimination. The study is motivated by the recent development of heterogeneous high performance computing architectures, addressing specific issues related to strong scalability of Intel Xeon Phi MICs and Intel Xeon CPUs. The numerical tests are ran on the high performance cluster AVITOHOL at IICT–BAS.

1 Introduction

This work is motivated by the recent development of heterogeneous high performance computing (HPC) architectures. For instance, the energy minimizing requirements result in increased role of accelerators. Intel introduced the Many Integrated Core (MIC) accelerators [1], successfully competing with NVIDIA with respect to price/performance ratio. Although some of the numerical linear libraries and codes were redesigned to be ported successfully to Intel MIC architecture, the optimal convergence could require a careful tuning of parameters and/or significant modifications of the codes, see e.g. [2–4]. Solving systems of linear algebraic equations with dense matrices is one of the most computationally intensive numerical linear algebra problems. This is the topic of this paper where performance analysis of Intel Xeon Phi MICs and Intel Xeon CPUs for solving dense systems of linear equations is presented.

D. Slavchev (✉) · S. Margenov
Institute of Information and Communication Technologies,
Bulgarian Academy of Sciences, Sofia, Bulgaria
e-mail: dimitargslavchev@parallel.bas.bg

S. Margenov
e-mail: margenov@parallel.bas.bg

K. Georgiev et al. (eds.), *Advanced Computing in Industrial Mathematics*,
Studies in Computational Intelligence 793,
https://doi.org/10.1007/978-3-319-97277-0_30

The HPC simulation of laminar flow around airfoils is considered as a test problem. The Boundary Element Method (BEM) is applied for numerical solution of the related integral equation, taking into account the advantages of BEM in the case of exterior domain, see e.g., [5, 6]. The BEM matrix is dense and the Gaussian elimination solver has a complexity of $O(n^3)$, where n is the number of unknowns (degrees of freedom). The Intel Math Kernel Library (MKL) is compared with two other popular libraries: PLASMA (with Atlas or MKL BLAS for CPUs) and MAGMAmic (for the MICs), [7, 8].

The rest of the paper is organized as follows. The used BEM is presented in the next section. The parallel implementation and related scalability tests are discussed and summarized in Sect. 3. Short concluding remarks are provided at the end.

2 Boundary Element Method for Two-Dimensional Problems of Ideal Fluid Flows with Free Boundaries

There are strong advantages of BEM when solving exterior boundary value problems in multiconnected domains with curvilinear boundaries. Based on the approach from [5] we developed an application for a cascade of wings in ideal fluid. The method is based on spline collocation with piecewise linear interpolation, see for more details [9].

Let Ω be a multiconnected domain in $\mathbb{R}$ (typically unbounded in the case of exterior problem) with sufficiently smooth and bounded closed boundary S. We seek the solution of the Laplace equation

$$\nabla^2 \Psi \equiv \frac{\partial^2 \Psi}{\partial x^2} + \frac{\partial^2 \Psi}{\partial y^2} = 0, \tag{1}$$

$$\Psi(P) = -\frac{1}{4\pi} \int_S \gamma(\sigma) \ln\left(r^2(P, Q)\right) d\sigma_Q + \Psi_\infty(P) + C_0, \qquad P \in \Omega, \tag{2}$$

where $r^2(P, Q) = (x - \xi)^2 + (y - \eta)^2$, $P = (x, y)$, $Q = (\xi, \eta)$, $Q \in S$, $d\sigma_Q$ is the measure on S. The first term of the right hand side represents a simple layer (stream function of a vortex layer), where $\gamma(\sigma)$ is the density of the layer. $\Psi_\infty(P)$ is a harmonic function added to the potential of the layer in order to satisfy the condition at infinity for external boundary value problems. The velocity field $\overrightarrow{C} = (u, v)$ is defined by

$$u = \frac{\partial \Psi}{\partial y}, \qquad v = \frac{\partial \Psi}{\partial x}, \tag{3}$$

satisfying the equations

$$u = \frac{1}{2\pi}\int_S \gamma(\sigma)\frac{y-\eta}{r^2}d\sigma, \quad v = -\frac{1}{2\pi}\int_S \gamma(\sigma)\frac{x-\zeta}{r^2}d\sigma. \tag{4}$$

We use also

$$\begin{aligned} \left(\frac{\partial \Psi}{\partial n}\right)_e &= -\frac{\gamma(s)}{2} - \frac{1}{2\pi}\int_S \frac{\cos(r,n_s)}{r}d\sigma, \quad s \in S \\ \left(\frac{\partial \Psi}{\partial n}\right)_i &= \frac{\gamma(s)}{2} - \frac{1}{2\pi}\int_S \frac{\cos(r,n_s)}{r}d\sigma, \quad s \in S \end{aligned} \tag{5}$$

where $\partial\Psi/\partial n$ is the normal derivative on the boundary S and $(f)_{i,e}$ represents the limit of the function f from inside and outside Ω, respectively. Then we obtain the jump of $\partial\Psi/\partial n$ on the boundary S equals the layer's density $\gamma(s)$, that is,

$$\left(\frac{\partial \Psi}{\partial n}\right)_i - \left(\frac{\partial \Psi}{\partial n}\right)_e = \gamma(s), \quad s \in S$$

Let $\overrightarrow{n}$ and $\overrightarrow{t}$ be the normal and the tangential unit vectors to S respectively. Then

$$\overrightarrow{C} = c_n \cdot \overrightarrow{n} + c_t \cdot \overrightarrow{t}$$

and the following relations are satisfied

$$\begin{aligned} \left(\left.\frac{\partial \Psi}{\partial n}\right|_S\right)_{i,e} &= -(c_t)_{i,e} \\ (C_t(P))_e - (C_t(P))_i &= \gamma(P), \quad P \in S. \end{aligned} \tag{6}$$

Finally, the boundary element method boundary condition (6) is written in the form

$$\left|\overrightarrow{C}(P)\right| = |\gamma(P)|, \quad P \in S.$$

2.1 Flow Around Wing Profiles

The considered particular problem concerns the flow around wing profiles. We assume that at infinity the airflow is homogeneous with velocity $\overrightarrow{C}_\infty = (1, 0)$ and S is the contour of the profile/profiles. For a potential flow, the continuity equation leads to the existence of a stream function Ψ satisfying the Laplace's equation (1). The profile/profiles S are impermeable which means that Ψ on S has a constant value and, i.e., the boundary condition $\Psi|_S = K = const$ holds true. In order to satisfy the condition for $\overrightarrow{C}_\infty$, we should choose Ψ_∞ such that

$$\overrightarrow{C}_\infty = \left(\frac{\partial \Psi_\infty}{\partial y}, -\frac{\partial \Psi_\infty}{\partial x}\right)$$

Here, we have

$$\Psi_\infty(P) = \gamma_\infty(P),$$

and the boundary condition takes the form

$$\gamma(P) - \frac{1}{4\pi}\int_S \gamma(\sigma)\ln\left(r^2(P,Q)\right)d\sigma_Q + C = 0.$$

Finally, the Kutta-Joukowsku's condition $\gamma(A) = 0$, is used to obtain a unique solution of the boundary value problem, where A is the point on the sharp tip of the airfoil.

2.2 *Discretization*

The problem (2) leads to a system of linear integral equations in the form

$$(A\gamma)(s) = f(s). \tag{7}$$

We look for a numerical solution

$$\gamma_h(s) = \sum_{i=1}^{n} \gamma_i \varphi_i(s),$$

where $\{\varphi_i(s)\}_{i=1}^n$ is the piecewise linear Lagrangian basis corresponding to the discretization mesh S_h on the wing profiles, and $\gamma_i = \gamma_h(s_i)$, $i = 1 \ldots n$, are the BEM nodal unknowns. The collocation method is applied to the mid points of boundary elements from S_h. Following [6] we obtain the system of linear equations

$$\sum_{i=1}^{n} \gamma_i \psi_{ji} = f_j, \qquad j = 1, 2, \ldots, n, \tag{8}$$

where

$$\psi_{ji} = \psi_i(s_j), \qquad f(s_j) = f_j,$$

$\psi_i(s) = (A\varphi_i)(s)$. Finally, the BEM matrix D has the block form

$$D = \begin{pmatrix} D_{11} & D_{12} & \dots & D_{1m} \\ \vdots & \vdots & D_{lk} & \vdots \\ D_{m1} & D_{m2} & \dots & D_{mm} \end{pmatrix},$$

where m is the number of profiles, and D_{lk} is a block matrix corresponding to the kth profile and the collocation points of the lth profile.

2.3 Test Problem

The geometry of the considered test problem is shown in Fig. 1 (left). The flow around five identical wing profiles arranged in a vertical line is simulated. The profiles are discretized by linear boundary elements. For each boundary element, there is a linear equation from (8) corresponding to the center point (P) and the related $\gamma(P)$. The numerical solution of the exterior boundary value problem is the piecewise function $\gamma(s_i)$, $i = 1, \dots, n$, whose values correspond to the boundary elements s_i, $i = 1, \dots, n$. Roughly speaking the value of $\gamma(s_i)$ represents the "influence" of the boundary element s_i on the flow in the model. The values of $\gamma(s_i)$ for the central airfoil are shown on Fig. 1 (right) against the airfoil itself.

Once the values of $\gamma(s_i)$, $i = 1, \dots, n$ are computed, we obtain the vector field $\vec{C} = (u, v)$ from (3), (4) and (5). The vector norm of $\left|\vec{C}\right|_P$ in a point P is a quantity from which we calculate the air pressure and speed at that point see Fig. 2 (left). Using Euler's method we calculate the streamlines of the vector field $\vec{C}$ Fig. 2 (right).

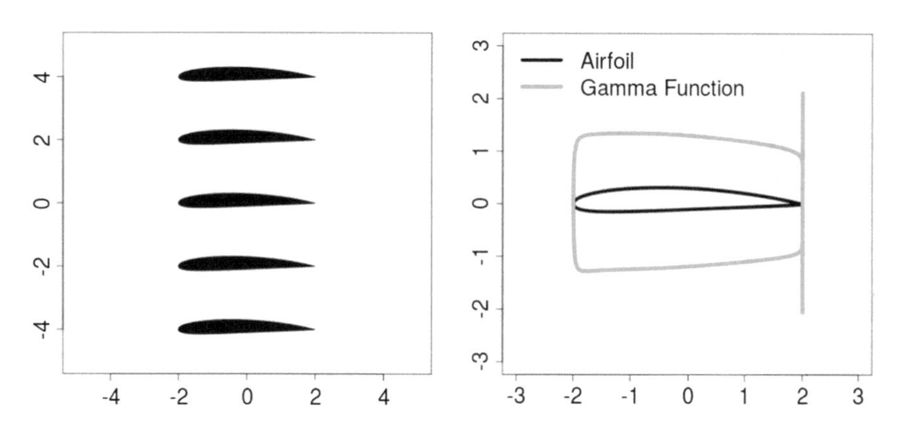

Fig. 1 Wing profiles (left), γ function plotted over the central profile (right)

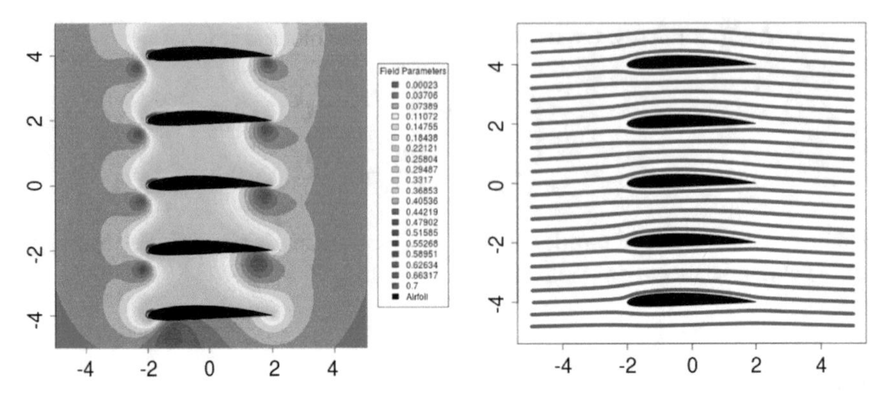

Fig. 2 Velocity field around the wing profiles (left), streamlines (right)

3 Parallel Implementation

The C programming language and the OpenMP standard are used for parallel implementation of the presented algorithm, incorporating several libraries for solution of the arising dense systems of linear algebraic equations. The computational complexity of Gaussian elimination solvers is $O(n^3)$ [10] while the forming of matrix D requires $O(n^2)$ arithmetic operations. The calculation of the airfoils' surface (from given parameters of position, length, angle of attack, etc.) or the lift and drag coefficients (from the solution of the system of equations) are complexity $O(n)$.

On Fig. 3 we present the sequential times to illustrate the comparison between the above mentioned parts of the algorithm. Not surprisingly, beyond a certain size of the matrix, say $n > 4000$, the time needed for the solution of the system of equations is strongly dominating.

The calculation of both the field points and the streamlines are $O(nk)$ complexity processes, where n is the number of points on the profiles and k is the number of points, where we calculate the vector field $\overrightarrow{C}$. It is worth to notice that for larger k, the contribution of computing the field points and the streamlines could be considerably stronger. However, this part of the algorithm is inherently pointwise, i.e. each point of the field/streamline is calculated independently from the rest. This is trivial to parallelize and, what is more important, it is not a bottleneck in the overall parallel efficiency.

In what follows, we focus our attention on the parallel solution of the system of linear equations. The comparative analysis addresses some specific issues related to advanced heterogeneous parallel architectures. The numerical tests are ran on the supercomputer AVITOHOL [11] hosted by IICT-BAS.

AVITOHOL consists of 150 HP Cluster Platform SL250S GEN8 servers with 2 Intel Xeon E 2650 v2 CPUs and 2 Intel Xeon Phi 7120P coprocessors. In this study, we compare different solutions on a single server. One natural choice is the Intel

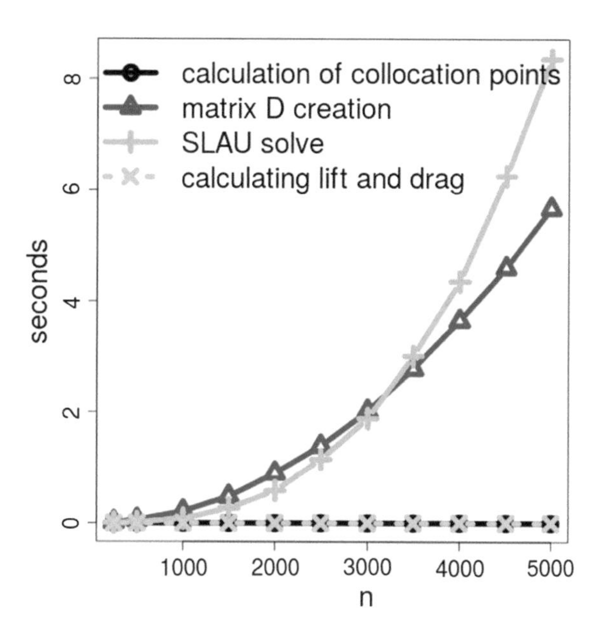

Fig. 3 Comparison of the time needed for the solution of system of linear algebraic equations and the rest main parts of the algorithm

provided Math Kernel Library (MKL) library, which could be used both on either the CPUs or the MICs, or on both simultaneously. The PLASMA (for CPUs) and MAGMAmic (for MICs) libraries are considered as two alternative options.

All numerical tests are ran with double precision using the `dgesv` routine of the corresponding library.

3.1 Intel Xeon CPUs

The parallel efficiency of MKL and PLASMA libraries is examined on the Intel Xeon CPUs where the number of threads is optimized/tuned. The model of the CPUs is Intel Xeon E5-2650v2 8C 2.6 GHz. There are 2 processors with 8 cores each and both are used. Each core allows for 2 threads to run simultaneously on it for a maximum of 32 threads using hyperthreading.

The Parallel Linear Algebra Software for Multicore Architectures (PLASMA) library [12, 13] is a software package for solving problems in dense linear algebra using multicore processors. It uses optimized tile algorithms to solve the system of linear equations.

The efficiency of PLASMA is based on the assumption of highly optimized Basic Linear Algebra Subprograms (BLAS). In this study we have used the MKL BLAS package and the ATLAS BLAS package [8]. The level of parallelism requires that

BLAS subroutines are executed sequentially. We set the `OMP_NUM_THREADS` and `MKL_NUM_THREADS` environment variables to 1 before starting the program and use

```
#pragma omp parallel for ... num_threads(nthreads)
```

to set up the number of threads for the other parallel parts.

For the tests of PLASMA with ATLAS we have compiled the libraries with gcc version 7.2.0. The choice of gcc over the Intel provided icc compiler is because the ATLAS developers have explicitly stated that their library was made with gcc in mind and that they did not provide complete and tested compiler flags for icc. [14]

For the tests of PLASMA with MKL-provided BLAS subroutines, the code was compiled with icc 17.0.2 20170213 and MKL 2017 update 2.

Also, we have examined the MKL library's solve routine. The results are summarized in Fig. 4 and Table 1.

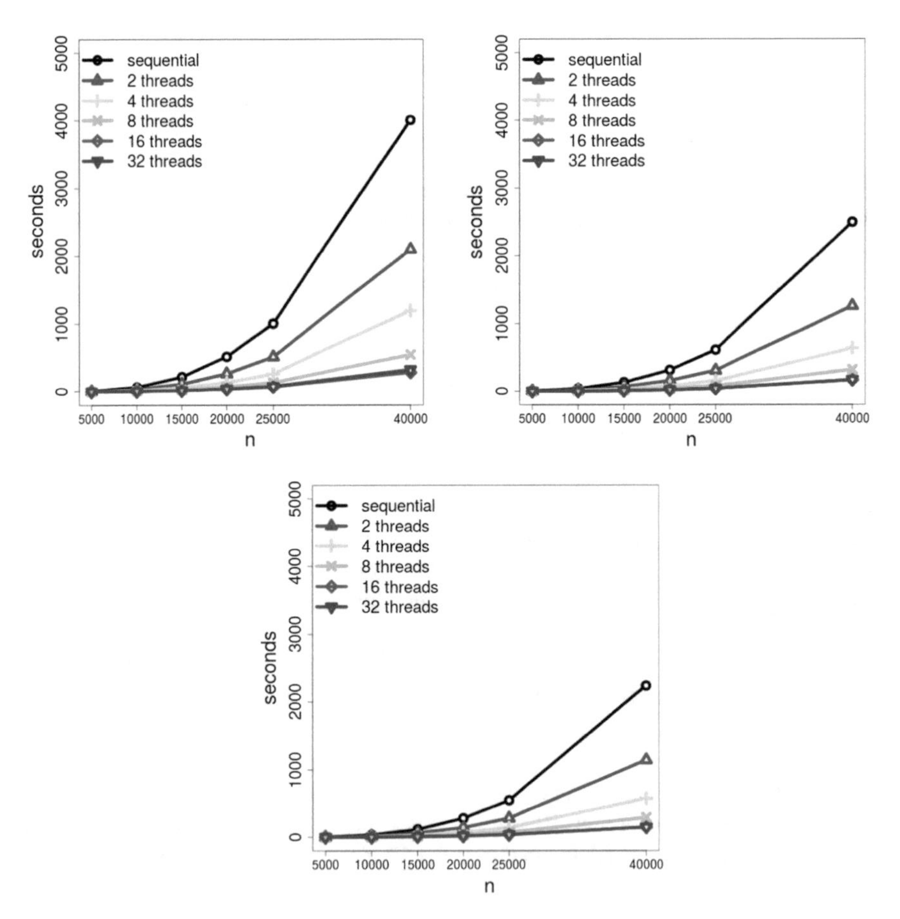

Fig. 4 Comparison of PLASMA with ATLAS (up left), PLASMA with MKL (up right) and MKL (bottom) solvers

Table 1 Comparison of sequential and best parallel times on CPUs

Library used		Plasma + ATLAS		PLASMA + MKL		MKL	
Threads	n	Time [s]	Speedup	Time [s]	Speedup	Time [s]	Speedup
1	5000	8.42	1.00	5.03	1.00	5.30	1.00
16	5000	0.67	12.57	0.47	10.69	0.47	11.26
32	5000	0.88	9.59	0.65	7.76	0.65	8.12
1	40000	4008.76	1.00	2497.12	1.00	2233.93	1.00
16	40000	282.94	14.17	166.41	15.01	147.64	15.13
32	40000	325.17	12.33	169.58	14.73	148.59	15.03

Rather good scalability is observed for all tested libraries up to 16 treads. Using hyperthreading with 2 threads per physical core does not provide any speedup. For the last two settings (libraries), there is a stagnation for 32 threads, while for PLASMA with ATLAS the time even increase substantially.

The parallel efficiency of PLASMA with MKL, and MKL is quite similar, increasing up to 94% for the larger problem, i.e. for $n = 40000$. They outperform more than 1.5 times PLASMA with ATLAS. One possible reason for this could be that gcc compiler doesn't use vectoring as well as the icc.

3.2 *MICs*

The parallel efficiency on Intel Xeon Phi 7120P coprocessors (MICs) is studied in this section. The MICs are designed for massive parallelism and vectorization as required in High Performance Computing. Every MIC has 61 cores and each core can run 4 threads simultaneously for a total of 244 threads. The coprocessors can be used in offload or native mode.

- In native mode the whole program is ran on the coprocessor.
- In offload mode the program is ran on the host (the CPU) and only specific parts are sent to the MICs. In our case MKL and MAGMAmic are taking care for the offloaded regions. Offload mode also needs one core for the offload processes itself and therefore we can use only 60 cores (240 threads). In this study we utilize this mode of operation.

On the MICs we have tested MKL and MAGMAmic MAGMAmic is a port of the MAGMA project to the MIC architecture.

In order to see the scalability of MAGMAmic implementation we use

```
MKL_MIC_ENABLE=1
OFFLOAD_DEVICES=0
MIC_ENV_PREFIX=MIC
MIC_OMP_NUM_THREADS=m
MIC_KMP_AFFINITY="proclist=[1-m],explicit"
```

Where:

- `MKL_MIC_ENABLE` activates the use of the MICs.
- `OFFLOAD_DEVICES` sets the MIC to be used. We consider the case of a single MIC.
- `MIC_KMP_AFFINITY` sets the number of cores (`m`) that we will be used. Note that `m` can be set up to 240, i.e., each physical core can run up to 4 hardware threads simultaneously and each hardware thread is treated as a separate core by the system.
- `MIC_ENV_PREFIX` sets the string to be used for MIC environmental. variables.
- `MIC_OMP_NUM_THREADS` sets the number of OpenMP threads to be used.

For MKL we utilize the compiler-assisted offload option. This means that the same program as the one tested on the CPUs is used. The following environmental variables were set up to instruct MKL to offload the work to the MICs.

```
MKL_MIC_ENABLE=1
OFFLOAD_DEVICES=0
MIC_ENV_PREFIX=MIC
MIC_OMP_NUM_THREADS=m
MIC_KMP_AFFINITY="proclist=[1-m],explicit"
MKL_HOST_WORKDIVISION=0
MKL_MIC_WORKDIVISION=1
MKL_MIC_THRESHOLDS_DGEMM=20,20,20
```

The first five parameters are the same as for MAGMAmic.

- `MKL_HOST_WORKDIVISION` and `MKL_MIC_WORKDIVISION` are two float values that set what part of the work should be done by the host and what part by the MIC. We dedicate the MIC to the whole workload.
- `MKL_MIC_THRESHOLDS_DGEMM` represents the minimal size of the problem that will be offloaded to the MIC instead of being solved on the host CPUs. We have set it to very small numbers to ensure that all tests will be ran on the MIC.

The results from the parallel tests are presented in Fig. 5 and Table 2.

Substantially better performance of MKL is observed. This could be due to some better communication within MKL between the threads. For instance, one can also see that the time for MAGMAmic using 4 threads is worst then the sequential one.

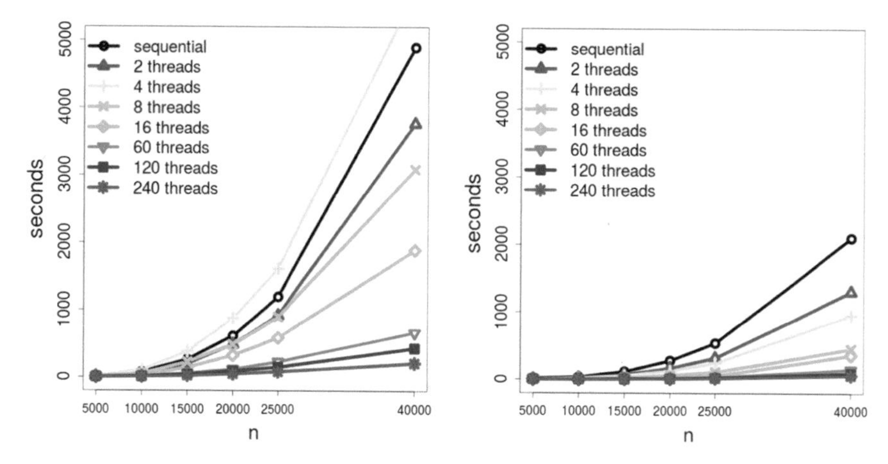

Fig. 5 Comparison of MAGMAmic (left) and MKL (right) on the MIC

Table 2 Comparison of sequential and best parallel times on the MIC

Library used		MAGMAmic		MKL	
Threads	n	Time [s]	Speedup	Time [s]	Speedup
1	5000	11.70	1.00	17.64	1.00
4	5000	14.04	0.83	6.44	2.74
60	5000	5.81	2.01	2.86	6.16
120	5000	6.76	1.73	3.55	4.97
240	5000	5.39	2.17	3.80	4.64
1	40000	4896.49	1.00	2101.93	1.00
4	40000	5805.66	0.84	956.39	2.20
60	40000	665.53	7.36	154.23	13.63
120	40000	432.89	11.31	93.80	22.41
240	40000	208.48	23.49	64.43	32.62

3.3 *Comparison Analysis*

On the MIC, we don't see similar speedups as on the CPUs. However, the large number of cores and the better environment for hyperthreading allow for faster overall execution.

The best parallel performance results for each of the considered packages are presented in Fig. 6. MAGMAmic achieves worst performance then PLASMA with MKL and MKL on the CPUs. This is probably because of some poor thread scheduling on the cores.

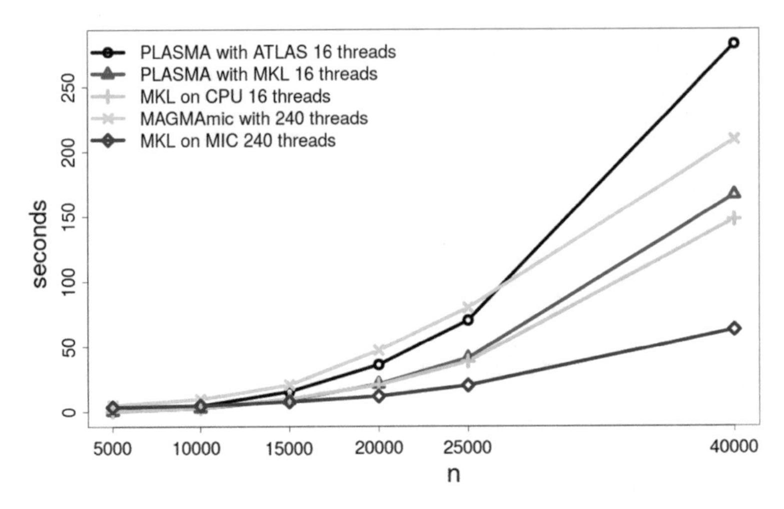

Fig. 6 Comparison of the results of different libraries on CPUs and MIC

The MKL library shows best times when used on the MIC architecture coprocessors. For the larger problem, the achieved speedup is 32. It is important to notice, that the related time of MKL on MIC is almost four time better than the best related time of MIC on CPUs.

4 Concluding Remarks

A comparative performance analysis of Intel Xeon Phi MIC and Intel Xeon CPUs for solving systems of linear equations with dense matrices is presented. The boundary element method is applied for numerical simulation of flows around airfoils, thus defining the benchmarks of the case study. The performance of several implementations based on open access software packages for Intel Xeon E5-2650v2 8C 2.6 GHz CPUs and Intel Xeon Phi 7120P coprocessors is studied. The reported results show a way for efficient solution of real-life large scale problems. For instance, for the biggest test problem (40000 elements) the computing time is dropped from more than an hour (for the sequential run) down to minutes on a single Intel Xeon Phi MIC.

We conclude this work by suggesting two possible future lines of research: (i) Further development of efficient implementations on advanced HPC architectures using several to many heterogeneous nodes; (ii) Analysis of advantages/disadvantages of alternative parallel solvers (e.g., based on low rank hierarchical matrices) for dense linear systems arising in boundary element applications.

Acknowledgements We acknowledge the opportunity to run the numerical tests on the HPC cluster AVITOHOL of the Institute of Information and Communication Technologies, Bulgarian Academy of Sciences [11]. The partial support trough the Bulgarian NSF Grant DN 12/1 is also highly acknowledged.

References

1. Intel xeon phi Coprocessor Instruction Set Architecture Reference Manual (2012)
2. Donfack, S., Dongarra, J., Faverge, M., Gates, M., Kurzak, J., Luszczek, P., Yamazaki, I.: A survey of recent developments in parallel implementations of Gaussian elimination. Concurrency Comput.: Pract. Experience **27**(5), 1292–1309 (2015). https://doi.org/10.1002/cpe.3306
3. Karaivanova, A., Alexandrov, V., Gurov, T., Ivanovska, S.: On the monte carlo matrix computations on intel mic architecture. Cybern. Inf. Technol. **17**(5), 49–59 (2017). DOI 10.1515-2017-0054
4. Stoykov, S., Atanassov, E., Margenov, S.: Efficient sparse matrix-matrix multiplication for computing periodic responses by shooting method on intel xeon phi. In: Application of Mathematics in Technical and Natural Sciences, vol. 1773, p. 110012. AIP (2016). https://doi.org/10.1063/1.4965016
5. Pasheva, V., Lazarov, R.: Boundary element method for 2D problems of ideal fluid flows with free boundaries. Adv. Water Resour. **12**(1), 37–45 (1989). https://doi.org/10.1016/0309-1708(89)90014-6
6. Slavchev, D.: Parallelization of Boundary Element Method for Laplasian Equation. Master's thesis, Technical University, Sofia (2014)
7. Agullo, E., Demmel, J., Dongarra, J., Hadri, B., Kurzak, J., Langou, J., Ltaief, H., Luszczek, P., Tomov, S.: Numerical linear algebra on emerging architectures: the plasma and magma projects. J. Phys.: Conf. Ser. **180**(1), 012,037 (2009)
8. Whaley, R.C.: ATLAS (Automatically Tuned Linear Algebra Software), pp. 95–101. Springer US, Boston, MA (2011). https://doi.org/10.1007/978-0-387-09766-4_85
9. Brebbia, C., Telles, J., Wrobel, L.: Boundary Element Techniques: Theory and Applications in Engineering (1984)
10. Kincaid, D., Cheney, W.: Numer. Anal.: Math. Sci. Comput. Brooks/Cole Publishing Co., Pacific Grove, CA, USA (1991)
11. Avitohol Supercomputer. http://www.hpc.acad.bg/system-1/
12. Buttari, A., Langou, J., Kurzak, J., Dongarra, J.: A class of parallel tiled linear algebra algorithms for multicore architectures. Parallel Comput. **35**(1), 38–53 (2009). https://doi.org/10.1016/j.parco.2008.10.002
13. Dongarra, J., Abalenkovs, M., Abdelfattah, A., Gates, M., Haidar, A., Kurzak, J., Luszczek, P., Tomov, S., Yamazaki, I., YarKhan, A.: Parallel programming models for dense linear algebra on heterogeneous systems. Supercomputing Front. Innovations **2**(4) (2016)
14. Whaley, R.C.: ATLAS Installation Guide (2016). http://math-atlas.sourceforge.net/atlas_install/

Memristor CNNs with Hysteresis

Angela Slavova and Ronald Tetzlaff

Abstract In this paper we first made an overview of memristor computing. Then we presented the dynamics of hysteresis CNN model with memristor synapses (M-HCNN). In order to study the dynamics of the obtained M-HCNN model we applied theory of local activity. In this way we determined the edge of chaos domain in the parameters set for the model under consideration. The processing results do not change under the variations of the memristor weights. This is due to the influence of the binary quantization of the output signals. However, in order to obtain stable solution we need more iterations when some variations arise in the templates. This can reflect in the speed of the M-HCNN performance without change of the quality of the final results.

1 Overview of Memristor Computing

Neuromorphic circuits can be considered for energy efficient computing based on biological principles in future electronic systems. Thereby, memristors are assumed in neuron models and for synapses in several recent investigations in order to overcome the limits of conventional von Neumann architectures by taking these devices as memory elements and as well as devices for computation also in bio-inspired artificial neural networks. Leon Chua has developed the fundamentals of the memistor framework nearly 40 years ago [3] (see Fig. 1). Since 2008, when its existence at the nanoscale was certified at Hewlett-Packard (HP) Labs [7], the memristor has attracted a strong interest from both industry and academia for its central role in the set up of novel integrated circuit (IC) architectures, especially in the design of high-density

A. Slavova (✉)
Institute of Mathematics and Informatics, Bulgarian Academy of Sciences, 1113 Sofia, Bulgaria
e-mail: slavova@math.bas.bg

R. Tetzlaff
Faculty of Electrical and Computer Engineering, Institute of Circuits and Systems, Technische Universität Dresden, Dresden, Germany
e-mail: Ronald.Tetzlaff@tu-dresden.de

K. Georgiev et al. (eds.), *Advanced Computing in Industrial Mathematics*, Studies in Computational Intelligence 793,
https://doi.org/10.1007/978-3-319-97277-0_31

non-volatile memories, programmable analog circuitry, neuromorphic systems, and logic gates [8]. Several applications of memristors such as memory, analog circuits, digital circuits, etc. have been intensively studied. Among these applications, neural network circuits with memristors can be considered very promising, because memristors can be used in synaptic circuits emulating the excitation-dependent change in synaptic coupling factors in neuronal networks.

Memristors introduced by Chua in 1971 [3] exhibit a relationship between flux φ and charge q. The presented equivalent physical examples behave as a non-volatile resistor whose resistance is continuously controlled by the amount of the charge of the flow (current). A general model of memristors is explained in terms of memristance $M(q)$. The dynamics of the memristor is given by:

$$\begin{aligned} i &= g_u(w)v \\ \frac{dw}{dt} &= i, \end{aligned} \tag{1}$$

where v represents the voltage across the memristor; i is the current of the memristor; w is the nominal internal state of the memristor corresponding to the charge flow of the memristor, and $g_u(w)$ is the monotonically non decreasing function with increasing w.

Cellular Nonlinear/Nanoscale Networks (CNN) have been introduced in 1988 by Chua and Yang [1] as a new class of information processing systems which shows important potential applications (Fig. 2). The concept of CNN is based on some aspects of neurobiology and adapted to integrated circuits. CNN are defined as spatial arrangements of locally coupled dynamical systems, referred to as cells. The CNN dynamics are determined by a dynamic law of an isolated cell, by the coupling laws between the cells and by boundary and initial conditions. The cell coupling is confined to the local neighborhood of a cell within a defined sphere of influence. The dynamic law and the coupling laws of a cell are often combined and described by nonlinear ordinary differential- or difference equations (ODE), respectively, referred to as the state equations of cells. Thus a CNN is given by a system of coupled ODE with a very compact representation in the case of translation invariant state equations. Despite of having a compact representation, CNN can show complex dynamics like chaotic behavior, self-organization, and pattern formation or nonlinear oscillation and wave propagation. Furthermore, Reaction-Diffusion Cellular Nonlinear/Nanoscale Networks (RD-CNN) have been applied for modeling complex systems [6]. These networks are not representing a paradigm for complexity only but also establishing novel approaches to information processing by the dynamics of nonlinear complex systems.

The further development of memristor technologies including conventional applications are driven by an obvious market need, like non-volatile high density memory and new nanoelectronic circuits for information processing, will in turn aid novel architecture research. Improvements in device performance in key metrics like switching time, power consumption, on-off ratios, and retention time will make circuit design easier and more amenable to conventional analysis [8]. On the other hand,

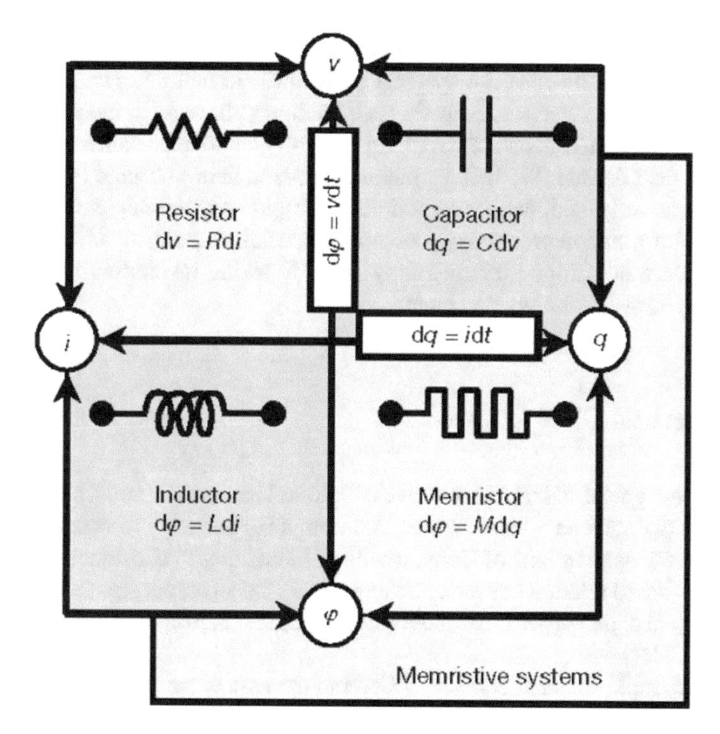

Fig. 1 Memristor - the forth element

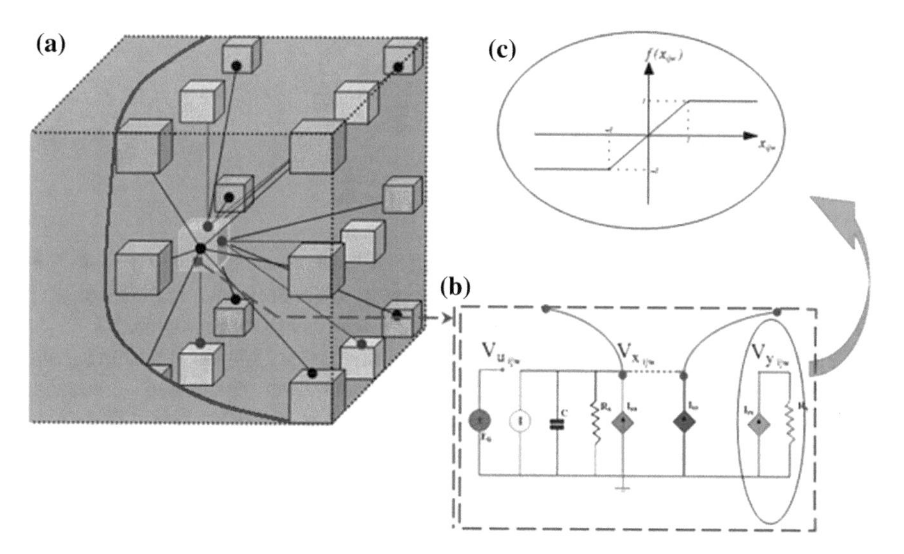

Fig. 2 **a** CNN architecture; **b** cell circuit; **c** output function of CNN

research focused on memory applications only, may tend to overlook the unique advantages offered by material systems with intrinsic internal decay for applications like dynamic synapses. Nonetheless, continued interest in bio-inspired computing will likely make resistive switching technologies an important area of research during the next decade. We believe memristor-based neuromorphic or bio-inspired computing is well positioned to revolutionize computing technologies to handle non-traditional information processing problems. Especially emergent behavior suggests a way forward in artificial intelligent systems by taking inspiration from different complex systems in biology and chemistry.

2 Hysteresis CNN Models

Although the typical CNN [1] does not oscillate and becomes chaotic, one can expect interesting phenomena - bifurcations, and complex dynamics to occur. Moreover, because of the applications of CNN, it will be interesting to consider a special type of memory-based relation between an input signal and an output signal in this circuit. The main goal of this paper is to model and investigate such relation, called hysteresis [10] for a CNN.

As an example we shall consider a CNN in the case when the nonlinearity in the feedback system is allowed to exhibit hysteresis:

$$\frac{dx_{ij}}{dt} = -x_{ij} + \sum_{C(k,l)\in N_r(i,j)} A_{ij,kl} h(x_{kl}) + \sum_{C(k,l)\in N_r(i,j)} B_{ij,kl} u_{kl}, \tag{2}$$

$$1 \leq i \leq M\,,\ 1 \leq j \leq M,$$

x, y and u refer to the state, output and input voltage of a cell $C(i,j)$; A and B, called feedback and control operators, respectively, are usually sparse matrices with a bounded structure containing the template coefficients at proper places.

The hysteresis $h(x_{ij})$ is a real functional determined by an "upper" function h_U and a "lower" function h_L (Fig. 3). The functions h_U and h_L are real valued , piecewise continuous, differentiable functions. Moreover, $h(x_{ij})$ is odd in the sense that

$$h_U(x) = -h_L(-x), \tag{3}$$

for all real numbers $x_{ij} \stackrel{\text{def}}{=} x$. Also $h_U = h_L$ for $|x|$ sufficiently large.

As we have seen in [1] in the typical CNN, the output function is a piecewise-linear function $y_{ij} = \frac{1}{2}(|x_{ij}+1| - |x_{ij}-1|)$, giving a binary- valued output, which property has a lot of applications in pattern recognition and signal processing. Considering HCNN as a memory-based relation we shall look for more interesting dynamic

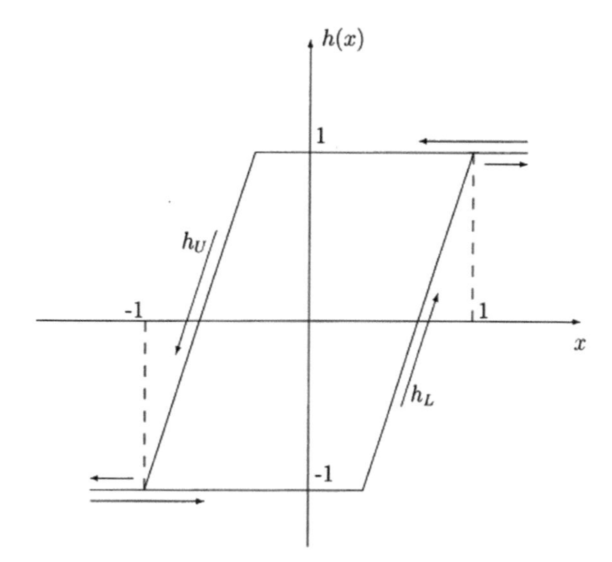

Fig. 3 Hysteresis nonlinearity

behavior of a CNN. Moreover, since the range of dynamics and the connection complexity are independent of the number of processing elements (cells), the implementation is reliable and robust. The easy programming, as well, is based on the geometric aspects of the cloning templates A and B.

Another example of HCNN is one, described by a nonlinear dynamical system of autonomous ODE. We shall generalize the state equation:

$$\frac{dx_{ij}}{dt} = -x_{ij} + \sum_{C(k,l)\in N_r(i,j)} A_{ij,kl} y_{kl} + \\ + \sum_{C(k,l)\in N_r(i,j)} B_{ij,kl} u_{kl} + \sum_{C(k,l)\in N_r(i,j)} \hat{A}(y_{ij}, y_{kl}) + I, \tag{4}$$

$$1 \le i \le M\,,\ 1 \le j \le M,$$

$\hat{A}$ is nonlinear template which is a function of internal states and input signals and we consider it in the form:

$$\hat{A}(y_{ij}, y_{kl}) = \begin{bmatrix} f_{11} & \dots & f_{1n} \\ \vdots & & \vdots \\ f_{m1} & \dots & f_{mn} \end{bmatrix},$$

$m, n \in N_r(i, j)$, functions f_{kl} are defined as

$$f_{kl} = \hat{a}_{kl} y_{ij} y_{kl}^{\alpha},\ \ \alpha \in \mathcal{N}.$$

In this HCNN we shall assume that output equation has its own dynamics where the nonlinearity in the feedback is allowed to exhibit hysteresis:

$$\frac{dy_{ij}}{dt} = -y_{ij} + h(x_{ij}). \tag{5}$$

The hysteresis $h(x_{ij})$ is a real functional defined as in the previous section (Fig. 3).

The above class of CNN can be allowed because they are relatively simple to realize as integrated circuits.

The main difference between CNN and neural networks (NN) is that a CNN has practical dynamic range, which can be calculated by explicit formula. Whereas, general NN often suffers from a severe dynamic range restrictions in the image implementation stage. Let us derive next an estimate for the dynamic range. For HCNN described by (4) and (5), the following proposition provides a foundation for our design:

Proposition 1 *For a HCNN described above by the dynamical equations (4) and (5), all state variables x_{ij} are bounded for all times $t > 0$ and the bound v_{max} can be computed by the formula:*

$$x_{max} = 1 + |I| + max_{1\leq i\leq M, 1\leq j\leq M}[\sum_{C(k,l)\in N_r(i,j)} (|A_{ij,kl}| + \\ +|B_{ij,kl}|) + \sum_{C(k,l)\in N_r(i,j)} max_t|\hat{A}|]. \tag{6}$$

Remark 1 As we mentioned in the previous section neural networks' origin is in neurophysiological models. The idea to develop a model of HCNN has been influenced by the fact that neurobiological studies have demonstrated neurons to exhibit quite complicated dynamical behavior.

3 Memristor HCNN Model

Hysteresis effect can be found in memristive systems. Let us consider the following memristive system:

$$\begin{aligned} y(t) &= g(x, u, t)u(t), \\ \dot{x} &= f(x, u, t) \end{aligned} \tag{7}$$

where the variable $u(t)$ is the input signal, variable $y(t)$ denotes the output signal, and g, f are continuous functions. If the input $u(t)$ is the current $i(t)$, the output $y(t)$ is the voltage $v(t)$, then the hysteresis loop given in Fig. 4 represents the electrical resistance. It can be seen that when we change the slope of the hysteresis curves, a switching between different resistance states appears. It is known that at high

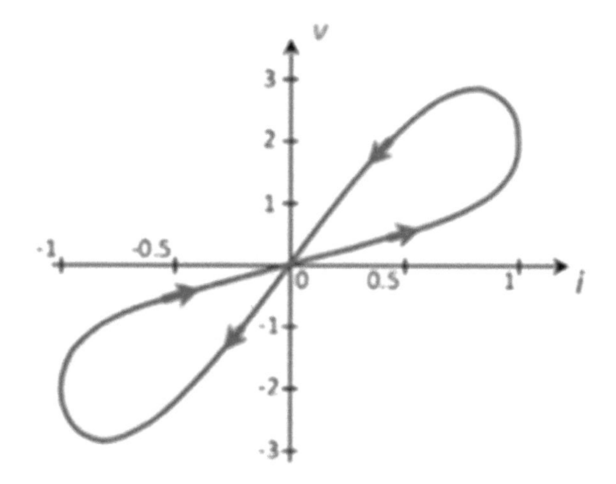

Fig. 4 Hysteresis effect in memristive systems

frequencies, memristive theory predicts the hysteresis effect in high frequencies which means that we obtain a straight line representative of a linear resistor.

Memristors can be encorporated in HCNN architecture in order to implement the synaptic weights (templates) in programmable bridge structure. In this way standard sigmoid output circuit can be eliminated and the size of processing elements is reduced. This leads to the improvement of the speed of computation.

We shall consider in this paper the following memristor HCNN (M-HCNN) model:

$$\frac{dx_{ij}}{dt} = -m(x_{ij}) + bh(x_{ij}) + \sum_{C(k,l)\in N_r(i,j)} A_{k-i,l-j}h(x_{ij}),\ 1 \le i \le M,\ 1 \le j \le M, \tag{8}$$

where b is a constant, the hysteresis functional $h(x_{ij})$ is of the form introduced in the previous section (see Fig. 3), $m(.)$ is the current flowing through the memristor in the form of $m(x_{ij}) = \frac{x_{ij}}{M(t)}$, $M(t)$ is the memristance of memristor state resistor in the memristor bridge circuit which is employed to realize the interactions between the neighboring cells. For the cloning template A we shall choose the following Laplacian template

$$A = \begin{pmatrix} 0 & 1 & 0 \\ 1 & -4 & 1 \\ 0 & 1 & 0 \end{pmatrix}. \tag{9}$$

Because of the usage of the memristor bridge circuits and the memristor state resistor in our M-HCNN model (8), the state x_{ij} can achieve stability itself and thus the output directly, i.e. $y_{ij} = x_{ij}$.

Fig. 5 Computer simulations for $b = -1.5$

The dynamics of an isolated cell is given by [4]

$$\frac{dx_{ij}}{dt} = -m(x_{ij}) + bh(x_{ij}). \tag{10}$$

In the case when $b < -1$, Eq. (10) reduces to a bistable multivibrator. The isolated cell has two stable equilibrium points. When $b > -1$ Eq. (10) reduces to a relaxation oscillator. In the neighborhood of the bifurcation point $b = -1$ the Eq. (8) can generate patterns. Computer simulations for pattern formation are given on Figs. 5 and 6.

It can be observed that Eq. (8) generates some interesting patterns for $b = -1.5$ and $b = -2$ (Figs. 5 and 6). For the simulations forward Euler algorithm with a time step size $\Delta t = 0.01$ is applied.

4 Edge of Chaos in M-HCNN Model

The theory of local activity [2, 5] which will be applied in this paper offers a constructive analytical method. In particular, for M-HCNN model, one can determine the domain of the cell parameters of locally active cells, and thus potentially capable of exhibiting complexity. The central theme of the local activity theory is that emergence and complexity can be rigorously explained by explicit mathematical criteria given to identify a relatively small subset of the local-active parameter region, called the edge of chaos. A locally-active cell kinetic equation can exhibit complex dynam-

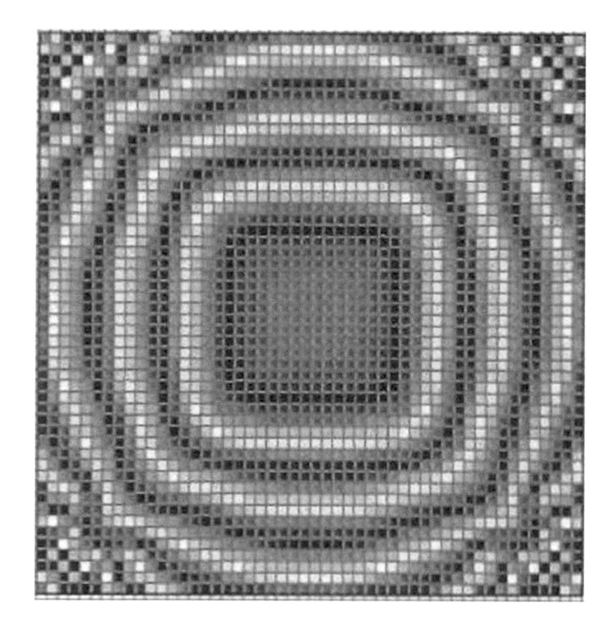

Fig. 6 Computer simulations for $b = -2$

ics such as limit cycles or chaos, even if the cells are uncoupled from each other (by setting all diffusion coefficients to zero). It is not surprising that coupling such cells could give rise to complex spatio-temporal phenomena, such as scroll waves, and spatio-temporal chaos.

We shall rewrite M-HCNN model in its discrete form:

$$\frac{dx_{ij}}{dt} = -m(x_{ij}) + bh(x_{ij}) + A * h(x_{ij}), \tag{11}$$

where A is the Laplacian template (9), $*$ is convolution operator [1].

Next step is to find the equilibrium points of (11). Let us rewrite (11) as follows for simplicity:

$$\frac{dx}{dt} = -m(x) + bh(x) + A * h(x) \equiv F(x), \tag{12}$$

where $x \stackrel{def}{=} x_{ij}$.

Definition 1 An equilibrium point of (12) is a point $x^e \in \mathbf{R}^{M \times M}$ such that [9]

$$F(x^e) = -m(x^e) + bh(x^e) + A * h(x^e) = 0.$$

The associated linear system in the sufficiently small neighborhood of an equilibrium point is

$$\frac{dz}{dt} = DF(x^e)z,$$

where $z = x - x^e$, $DF(x^e) = J$ is the Jacobian matrix of the equilibrium point which can be computed by

$$J_{ps} = \frac{\partial F_p}{\partial x_s}|_{x=x^e}, 1 \leq p, s \leq n, n = M.M.$$

According to local activity theory [2, 5] we shall find the trace $Tr(x^e)$ and the determinant $\Delta(x^e)$ of the Jacobian matrix J in the equilibrium points.

Definition 2 Stable and locally active region $SLAR(x^e)$ at the equilibrium point x^e for the M-HCNN model is such that $Tr(x^e) < 0$ and $\Delta(x^e) > 0$.

Definition 3 *(Complexity)* A spatially continuous or discrete medium made of identical cells which interact with all cells located within a neighborhood (called the sphere of influence), with identical interaction laws is said to manifest complexity if the homogeneous medium can exhibit a nonhomogeneous static or spatio-temporal pattern, under homogeneous initial and boundary conditions.

The initial condition is required to be homogeneous since otherwise, we can consider a system made of only cells which are not coupled to each other, such as a system of reaction-diffusion equations with zero diffusion coefficients. This system can exhibit a non-homogeneous static pattern by choosing the initial condition to correspond to any pattern of cell equilibrium states, assuming each cell has two or more equilibrium states.

In the literature, the so-called edge of chaos (EC) means a region in the parameter space of a dynamical system where complex phenomena and information processing can emerge. We shall try to define more precisely this phenomena till now known only via empirical examples.

Definition 4 M-HCNN (8) is said to be operating on the edge of chaos EC iff there is at least one equilibrium point x^e which is both locally active and stable.

The following theorem holds:

Theorem 1 *M-HCNN model (8) is operating in EC regime iff* $\frac{1}{M} < b < \frac{1}{M} + 4$. *For this parameter set there is at least one equilibrium point which belongs to* $SLAR(x^e)$.

Proof First let us consider the region $|x_{ij}| > 1$ and $|h(x_{ij})| \geq 1$ (see Fig. 3). In this region $|h_U| = |h_L| = h = const$. The Jacobian matrix in the equilibrium point x^e is $J(x^e) = A - \frac{1}{M}$. Then we calculate the trace $Tr(x^e) = -4 - \frac{1}{M} < 0$. Determinant $\Delta(x^e) = 0$. Therefore, in this region the equilibrium point x^e is not locally active and stable.

In the region $|x_{ij}| \leq 1$ and $|h(x_{ij})| \leq 1$ the Jacobian matrix is $J = A * h'(x^e) - \frac{1}{M} + bh'(x^e)$. In this region the trace in the equilibrium point is $Tr(x^e) = -4h'(x^e) - \frac{1}{M} + bh'(x^e)$. The condition $Tr(x^e) < 0$ is satisfied for $b < \frac{1}{M} + 4$. Determinant $\Delta(x^e) = det Ah'(x^e) - \frac{1}{M} + bh'(x^e)$. Therefore, $\Delta(x^e) > 0$ is satisfied for $b > \frac{1}{Mh'(x^e)}$. According to the definition 4 the EC is determined for the following

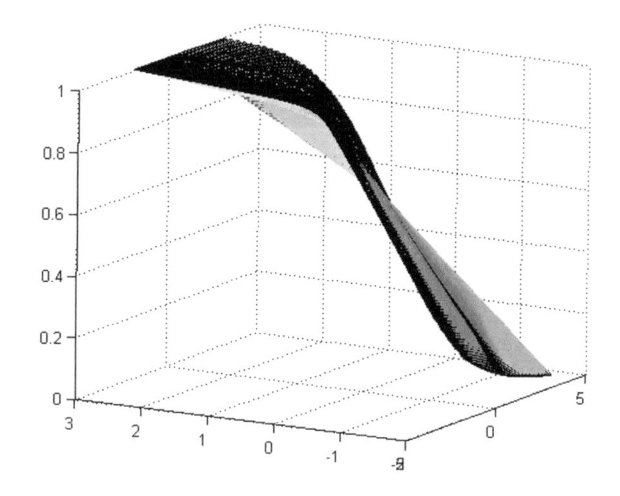

Fig. 7 Edge of chaos domain of M-HCNN model

parameter set $\frac{1}{M} < b < \frac{1}{M} + 4$. According to the local activity theory [2, 5] we have at least one equilibrium point which is both locally active and stable.

Computer simulations show the edge of chaos region for our M-HCNN model (8) on Fig. 7.

Remark 2 It is very important to have circuit model for the physical implementation. Then we can apply results from the classical circuit theory in order to justify the cells local activity. If the cell acts like a source of small signal for at least one equilibrium point then we can say that it is locally active. In this case the cell can inject a net small-signal average power into the passive resistive grids.

5 Conclusion

In this paper we introduce memristor computing as a foundation of future electronic systems. We present two models of HCNN - in the first one the hysteresis nonlinearity is in the feedback circuit, second one allows hysteresis in the output circuit. We develop a novel model of HCNN based on the employment of memristor bridge circuit which realizes the interactions of neighboring cells. In order to study the dynamics of the obtained M-HCNN model (8) we apply theory of local activity. In this way we determine the edge of chaos domain in the parameters set for the model under consideration. We provide two examples of generation of some interesting patterns (Figs. 5 and 6). Memristor bridge circuit leads to the reduction of the size of processing elements, and therefore the architecture of the proposed M-HCNN model is more compact, versatile and suitable for VLSI implementation.

References

1. Chua, L.O., Yang, L.: Cellular neural network: theory and applications. IEEE Trans. CAS. **35**, 1257–1289 (1988)
2. Chua, L.O.: Local activity is the origin of complexity. Int. J. Bifurcat. Chaos **15**(11), 3435–3456 (2005)
3. Chua, L.O.: Memristor the missing circuit element. IEEE Trans. Circuit Theor. **18**(5), 507–519 (1971)
4. Itoh, M., Chua, L.O.: Dynamics of memristor circuits. Int. J. Bif. Chaos **24**(5), 1430015 (2004)
5. Mainzer, K., Chua, L.O.: Local Activity Principle: The Cause of Complexity and Symmetry Breaking. Imperial College Press, London (2013)
6. Slavova, A.: Cellular Neural Networks: Dynamics and Modeling. Kluwer Academic Publishers, Alphen aan den Rijn (2003)
7. Strukov, D.B., Snider, G.S., Stewart, D.R., Williams, R.S.: The missing memristor found. Nat. Lett. **453** (2008). https://doi.org/10.1038/nature06932
8. Tetzlaff, R.: Memristor and Memristive Systems. Springer, Berlin (2014)
9. Vidyasagar, M.: Nonlinear Systems Analysis. Society for Industrial and Applied Mathematics, Philadelphia (2002)
10. Visintin, A.: Models of Hysteresis. Springer, Berlin (1993)

Change Point Analysis as a Tool to Detect Abrupt Cosmic Ray Muons Variations

Assen Tchorbadjieff and Ivo Angelov

Abstract Recently, there have been an increasing number of studies using Big Data. They rely on large data sets of time series to detect artificial or natural patterns in processes of natural sciences and economy. The most possible outcome due to lack of rigid data processing is data contamination with abrupt drifts and regime shifts. They yield either inclusion of undetected errors or missed detection of important observations and events. Possible automatic tools for detection of regime shifts could be delivered from change point statistical methods. However, a major drawback for the most of the currently available change point (CP) methods is the challenge of complex temporal variations in non-stationary natural processes like cosmic rays observed at Earth. This kind of data analysis is applied to experimentally acquired time series from cosmic ray measurements. The observed parameters are muons produced in cosmic ray cascades in atmosphere and acquired in parallel with atmospheric and other meta-data. In this study, we test different approaches for change point detection in compound particle counting process.

1 Introduction

Identifying, quantifying, and understanding the nature of cosmic rays intensity variations at space and Earth atmosphere has been topic for many years in enormous number of different researches. Moreover, the possible consequent impact on climate and natural Earth radioactivity has been the focus of numerous recent research

A. Tchorbadjieff (✉)
Institute of Mathematics and Informatics, Bulgarian Academy of Science, Sofia, Bulgaria
e-mail: atchorbadjieff@math.bas.bg

I. Angelov
South-West University, Blagoevgrad, Bulgaria

I. Angelov
Institute for Nuclear Research and Nuclear Energy, Bulgarian Academy of Science, Sofia, Bulgaria

K. Georgiev et al. (eds.), *Advanced Computing in Industrial Mathematics*, Studies in Computational Intelligence 793,
https://doi.org/10.1007/978-3-319-97277-0_32

of cosmic rays, for more detailed example see the text of Dorman [1]. In most cases, the importance is derived from variations of Galactic Cosmic Rays (GCR) due to Solar perturbations and resulting of Space Weather variability in Space [2]. One of the observed effect on GCR flux are events known as Forbush decreases (FD). They are non-regular in time sudden decrease of particle fluxes which lasts from 3 to 5 days and differ in frequency and intensity [3].

Usually, the in-situ observations of CR are performed on decades-long time series data acquired by neutron and muon detectors, combined with relative atmosphere data. For their detection in large scale an automatic procedures are required for analysis of regime changes in CR variations. For this purpose, different implementations of a change point analysis were tested on available data. It is a statistical hypothesis testing for natural or artificial stochastic shifts in time series. It is mainly popular in financial mathematics, but it also gained a popularity in climatology and environmental science, or specially in detection abrupt changes in time series trend of observations in atmosphere science (see [4, 5]).

For the purpose of our research, variety different models are used for testing and detection of change points in time series with registered FD events. The used data records are acquired from the located at BEO Moussala (2925 m.a.s.l.) muon telescope [6]. The observed time series are intentionally chosen to include data that contain already detected and reported in the past FD events. This enables comparative analysis between independently acquired test results and actual situation. In addition, the data analysis must include related meteorological parameters, such as pressure, following the theoretical connection between intensity of secondary CR particle flux and atmosphere density [7]. Assuming the importance of this natural connection, two main different approaches for implementation are applied for change point analysis. The first one is direct change point analysis on modified univariate muon flux acquired after correction with ambient atmosphere pressure. The second approach is based on detection of regime changes in regression coefficients between muons and pressure. The explanation how theory of change point analysis is applied for both cases is shown in the next paragraph.

2 Change Point Analysis

The first method for regime change detection has been initially introduced as Cumulative Sum (CUSUM) sequential analysis technique by Page in 1954 [8]. The method is control chart scheme for identification of the subsamples and detection of the changes in the parameter value of sequential observations x_i with steps $i = 1, 2, \ldots, n$. The method computes upper C_i^+ and lower C_i^- cumulative statistics with initial values equal to 0 such as:

$$\begin{aligned} C_0^+ &= C_0^- = 0 \\ C_i^+ &= \max(0, C_{i-1}^+ + x_i - k) \\ C_i^- &= \min(0, C_{i-1}^- + x_i - k). \end{aligned} \tag{1}$$

where k represents reference value. For example, when the shift of mean δ is known for i.i.d. and normally distributed x_i, k is equal to $\delta/2$. However, the model performs poorly for unknown δ with values significantly different than expected ones. An regime change is detected when C_i values reaches control limit h. It is a predefined value according average run length (ARL) and usually is proportional to the standard deviation.

2.1 Tests

However, despite the fact that the CUSUM sequence techniques is simple and easy to implement, usually more general statistical tool is required for multiple change point detection. Let $x_1, x_2, \ldots, x_n$ be a sequence of independent random vectors (variables) with any probability distribution functions $F_1, F_2, \ldots, F_n$. For detection of multiple change points the statistical test is run for the following alternative hypotheses H_0 vs H_A:

$$H_0 : F_1 = F_2 = \cdots = F_n \quad (2)$$

$$H_A : F_1 = \cdots = F_{k_1} \neq F_{k_1+1} = \cdots = F_{k_2} \neq F_{k_2+1} = \cdots = F_{k_{q+1}} = \cdots = F_n \quad (3)$$

where $1 < k_1 < k_2 < \cdots < k_q < n$ represents unknown number of changing points q with respective unknown positions $k_1, \ldots, k_q$. When $F_1, F_2, \ldots, F_n$ belongs to common parametric family $F(\theta)$, the null hypothesis is test about population parameters $\theta_i, i = 1, \ldots, n$ and $\theta \in R^n$. Then the test is transformed to [9]:

$$H_0 : \theta_1 = \theta_2 = \cdots = \theta_n = \theta \quad \text{(unknown)} \quad (4)$$

$$H_A : \theta_1 = \cdots = \theta_{k_1} \neq \theta_{k_1+1} = \cdots = \theta_{k_2} \neq \theta_{k_2+1} = \cdots = \theta_{k_{q+1}} = \cdots = \theta_n \quad (5)$$

The computational approach to identify multiple change points is to compute [10]:

$$\min \left[\sum_{i=1}^{q+1} \left[C(x_{k_1,\ldots,k_q}) \right] + \beta f(k) \right] \quad (6)$$

where $C(x_{k_1}, x_{k_2}, \ldots, x_{k_q})$ is a cost function, usually twice negative log-likelihood. The additional part of $f(k)$ is a penalty for avoiding of over-fitting due to data size, number of change points or autocorrelation. There are many different types of proposed penalties. The most easy for implementation cost functions is Minimum AIC Estimate (MAICE) with penalty only on number of breaking points. But, because the penalty for data size is not considered, AIC based models show tendency to over-fit. The criteria with implemented both of penalties on size and number of change points is Schwartz Information Criteria (BIC). Their formulation as selection among K models can be generalized by:

$$\text{AIC}(k) = -2log(L(\theta_k)) + 2k, k = 1, 2, \ldots, K \quad (7)$$

$$\text{SIC}(k) = -2log(L(\theta_k)) + klog(n), k = 1, 2, \ldots, K \quad (8)$$

For the computational work in this paper, the specially dedicated implementation is selected. It is the *changepoint* package [10] available for statistical computation with R [11]. The library delivers many features, which could be used directly for our research. Firstly, the implemented penalties are not constrained only to AIC and BIC, but there is an option for usage of user manually defined penalties. Secondly,the Gamma and Poisson distributions are implemented in addition to CUSUM and Normal. Another positive feature is the available variety of computational optimizations with implementations of splitting algorithms such as binary segmentation [12]; the Segment Neighbourhood [13]; and the PELT [14].

2.2 Using CP in Regression Models

In case of multivariate data, with vector of values y, the change point analysis is over regression model with a non-stochastic (p+1)-vector of predictors $\boldsymbol{x}_i = (1, x_{1i}, \ldots, x_{pi})$ is:

$$y_i = X\beta + \varepsilon_i, \ i = 1, \ldots, n, \quad (9)$$

where $\boldsymbol{\beta}'$ - is a $p + 1$ vector of unknown regression parameters and ε_i are random normally distributed errors with $N(0, \sigma^2)$. The change point analysis is observation about change of regression coefficients due to detected disconnection between the data before and after the point k. Then the statistical test is about the comparison between Null hypothesis of lack of differences in regression coefficients against the alternative of two different models with split point at k [9]:

$$H_0: \ \mu_{y_i} = x_i'\beta \ , \ i = 1, \ldots, n \quad (10)$$

$$H_A: \begin{cases} \mu_{y_i} = x_i'\beta_1, \ i = 1, \ldots, k \\ \mu_{y_i} = x_i'\beta_2, \ i = k+1, \ldots, n \end{cases} \quad (11)$$

where $k = p + 1, \ldots, n - p - 1$ is the location of CP, where β, β_1, β_2 are unknown.

For computational work with regression models a specially dedicated package *strucchange* in R [15] is used. The estimated regression coefficients $\hat{\beta}$ are yielded from Ordinary least squares (OLS). The package provides computations at position k of residuals $\hat{u}_i$ and their recursive values $\tilde{u}_i$ with zero mean and σ^2 under H_0:

$$\hat{u}_i = y_i - x_i'\hat{\beta}^n \quad (12)$$

$$\tilde{u}_i = \frac{y_i - x_i'\hat{\beta}^{(i-1)}}{1 + x_i^T(X^{(i-1)'}X^{(i-1)})^{-1}x_i} \quad (13)$$

The decision for availability of CP is based on different implementations of CUSUM process of residuals. Then, the cumulative sum of standardized residuals $W_n(t)$, is defined as [15]:

$$W_n(t) = \frac{1}{\tilde{\sigma}\sqrt{\eta}} \sum_{i=k+1}^{k+|t\eta|} \tilde{u_i} \tag{14}$$

Thus, under H_0 $W_n \longrightarrow W$ as $n \to \infty$, which is standard Brownian Motion [15]. Similarly, the modified OLS-CUSUM process is equal to [15]

$$W_n^0(t) = \frac{1}{\hat{\sigma}\sqrt{n}} \sum_{i=1}^{nt} \hat{u_i} \tag{15}$$

Then under $H_0 : W^0(t) = W(t) - tW(1)$, or the Brownian Motion starts at 0 and must finish there [15]. The MOSUM method is also implemented in the library, which is a moving sums of residuals.

3 Data

The used data for this research consist with datasets related to already reported and confirmed FD events. They include two periods - the first one is in the middle of February 2011, the second begins in March 1st 2012 and lasts until the end of May. We would refer to them as *Period 1* for year 2011 and *Period 2* in the following text. The time periods with their first and last dates are selected, as it is reported in [16, 17]. The two time series differ in size and number of FD evens which are included. The period of 2011 contains only single short FD event. Conversely, there are registered a chain of overleaped in time FD events and in addition to two other FD events during the *Period 2* (Table 1). Finally, an abrupt shift due to undocumented changes in the measurement procedures is available in data during the *Period 2*. This regime change is produced from higher values in the middle of May 2012. It was removed from the original report [17], but they are included in this work for more precise research. Finally, because the first detected FD during *Period 2* is on March 8th which is in only 7 rows after the first data on March 1st, the additional extended control dataset for daily data starting from Feb. 25th is assumed for verification that the short period before the event does not interfere the results.

The raw data consists of measured in 15 s vertical muon counts from 4 channels and pressure records in 10 min intervals. After both datasets are preprocessed for significant measurement errors they are synchronised in time and merged. For competitive data analysis three different datasets are assumed - two versions with hourly time resolution for periods 1 and 2 and one with a daily averages for events in 2012. The data for every test case are properly aggregated for the two different test scenar-

Table 1 Reported FD events

Start date	Until	Intensity (%)	Number of events	Reporting paper
2011–02–18	2011–02–20	4.5	1	[16][a]
2012–03–08	2012–03–17	6	3	[17]
2012–04–05	2012–04–07	2.5	1	[17]
2012–04–24	2012–04–27	3.5	1	[17]

[a]Note that the results are compared to more sensitive neutron flux, not muons

ios. This difference is required mainly because the very important physical negative dependence of CR particle flux on atmosphere density. For that reason two different type change point tests are assumed for every time period - CP regression models with implicitly included pressure as a predictor and modified muon time series with corrected pressure.

However, corrected data require some additional preparation. For correction are used β_P coefficients from regression models, OLS and Generalized Linear Models (GLM). Then the corrected values for time dependent flux $I(t)$ with averages of I_0 are equal to:

$$I_{corr} = \frac{I - I_0}{I_0} = \beta_P(P(t) - P_0) \tag{16}$$

where $P(t)$ are time relative pressure and P_0 is its average. Usually, the coefficients are computed from previously measured annual data without any significant regime shifts. However, for the current work are used coefficients computed from smaller time intervals occurred exactly before every observed period. The differences in β_P are negligible for the final outcome.

There are two important remarks that should be made for the observed data. Firstly, the overall distribution departure from the normality. The main reason are detected FD events, which skew density function of atmosphere corrected intensity with their lower values (Fig. 1). In general, the overall distribution of data with included FD events usually could not be generalized as Normal, Gamma or exponential for dif-

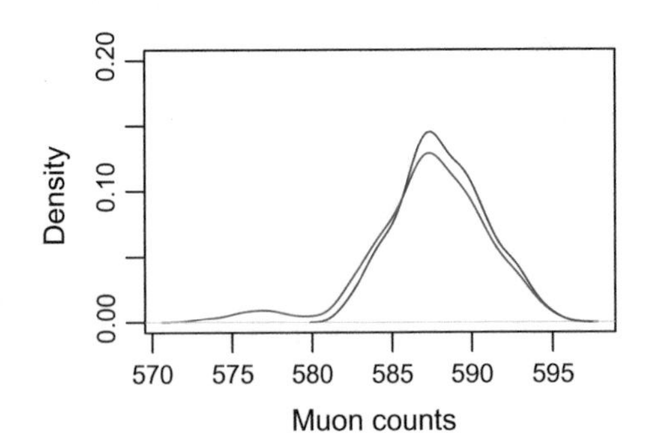

Fig. 1 A combined density plot of corrected with pressure muon data with 1 hour resolution. The selected period is the middle of February 2011, as it is shown in [16]. The density kernel of all data is with red line. The colour of data without reported FD decrease is blue

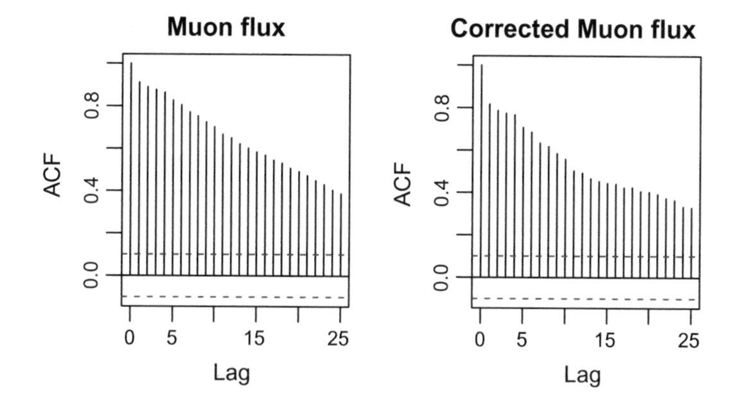

Fig. 2 Autocorrelation function for 1-h muon data (left) and for corrected with pressure values from the same data. The demonstrated period is mid-February 2011, as it shown in [16]

ferent cases. This is confirmed with the results for maximum-likelihood fittings with R function '*fitdistr*'.

Another important characteristic is the non-stationariness of aggregated data, a state common with many other natural processes [4]. The very possible explanation are complicated seasonality and trend with overlapped periods. In cosmic rays intensity, the seasonal periods are connected to rotations of Earth and Sun and daily, 27-daily, annual and 11 years cycles exist. As a result, the unit-root tests unsurprisingly deny stationary and autocorrelation decays very slowly for raw flux and pressure corrected data (see Fig. 2). To compute the optimal lags m it is used Augmented Dickey-Fuller test. All produced results are larger than 2 and the values of lags in every data set are shown in tables of results (Table 2).

4 Applications

All change point testing scenarios initially are run for tests with different penalties and segmentation algorithms without incorporation of autocorrelation properties. They are repeated for all aggregated versions of data. A brief description of implementation and most important results are explained in the following subsections.

4.1 Direct Approach

The function is **'efp'** from *structchange* library is the main method used for the change point analysis of regression coefficients. The models were build starting from the simplest regressive relation between muon counts and pressure. With assumption of periodicity, the regression model is extended with predictors of lagged data according expected periodicities. However, all test failed to produce correct results. Then, the already corrected muon data are run in regression models against their lagged values. The tests are run with pooling function **'breakpoints'**, which run all possible CP cases and select the optional model for minimal BIC or Residual Sum of Squares (RSS). With tests performed on data in 1 hour resolution using OLS-CUSUSM, the exact hour for the FD events with amplitude larger than $>2.5\%$ are detected. However, the numbers of breaking point is very strongly dependent on minimal segment size and usually differ from actual size. All other used variants failed to produce significant results. Some of the results are shown graphically in Fig. 3.

The change point analysis of corrected with pressure muon flux values are also computed with all possible functions **'cpt.mean'**, **'cpt.var'** and **'cpt.meanvar'** from *changepoint* library. Every distinct function is tested with all different options. Firstly, the used distributions were Normal and Gamma. They were run with all possible combinations of available segmentation methods and penalty functions. All results are either over-fitted or not complete, mainly in cases for complicated data from 2012. Some of the results are also shown graphically in Fig. 3.

In general, the proposed tests do not produce successful results. The main reason is the time dependence in measurement data, resulted with large autocorrelation lag for data either equal or shorter than 1 month [4]. Usually, this creates pattern in time series which could be easily confused with regime shift and the risk of false interpretation

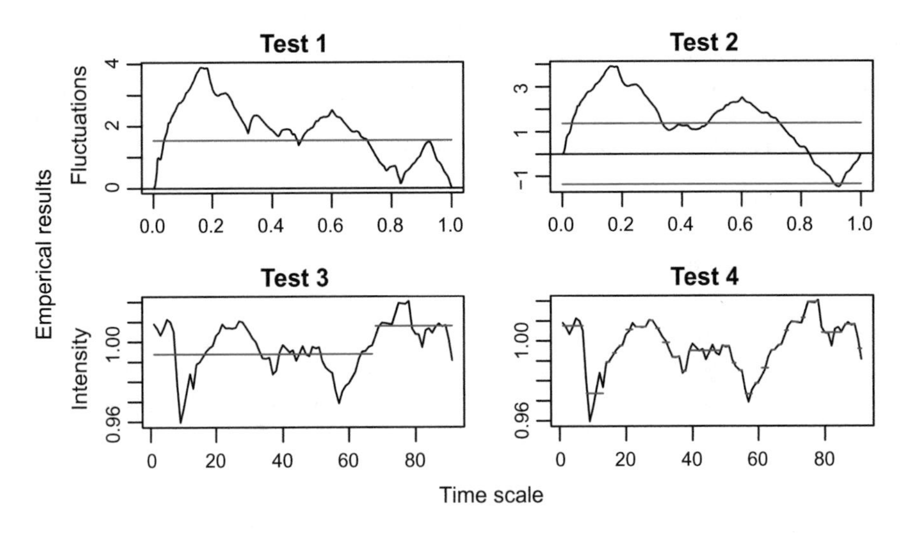

Fig. 3 Graphics show 4 tests detection of change points during the *Period 2*. Test 1 and Test 2 use regression with cumulative sums of standardized residuals for the first and OLS-CUSUM for the second test. Both of them are shown in Brownian motions scale. Tests 3 and 4 compute change point locations over pressure corrected muons reported in [17]. Test 3 and 4 are generated for hypothesis of Normal distribution with AIC penalty. The difference between those test is that for the last one is used PELT for computational acceleration

of ACT tends to increase [18]. However, the FD events represents abrupt shifts with purely random time of arrival, amplitude and the persistent time. Then, the solution for abrupt changes detection is to expand the penalty with incorporation of time dependency in the model.

4.2 Integrating the Autocorrelation

The initial steps for integration of autocorrelation are taken following the models proposed in similar works on environmental data (see [4, 5]). The daily data is computed with different versions of manually pre-computed SIC penalty with m-th order autocorrelation using **'cpt.meanvar'** function from *changepoint* library. The tests confirmed relatively precisely the period with extreme solar activity between March 8th and 19. The FD event in April 25 is also detected, but with a day earlier in comparison to the reported time in [17]. However the disturbances in beginning of April are not detected as separate event, but as a part of larger regime change until April 22. The results are shown in Fig. 4.

However, the tests were not successful in cases of data with hourly time resolution. It is a complex problem because we have data with compound periods incorporated and the proposed autocorrelation integration is on process with not fully removed time dependence. Moreover, the regime shifts of mean and variation usually correlates. Thus, a solution must implement approximation of well known tailed distribution with penalty correction representing the assumption of departure from the main distribution due to additional autocorrelation and regime shift. For this purpose we use a model with fixed penalty that enables the variance V to vary with the mean of distribution:

$$V(\mu) = \phi\mu^s \tag{17}$$

where ϕ is dispersion parameter. This constant ratio between mean and variation imply assumption for Tweedie distribution [19]. This is polymorphic distribution,

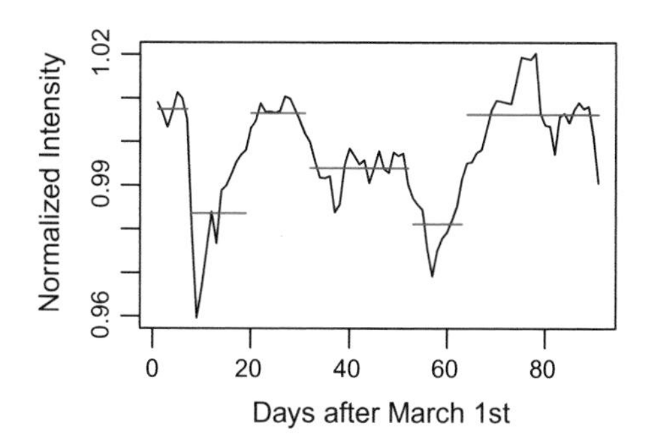

Fig. 4 CP analysis of *Period* 2. The two largest FD events are detected. The change point staring from April 5-th is detected, but the related FD is not distinguished

which with change of s, takes properties of well known distributions. For example, with $s = 2$ it is Gamma, but when $s = 0$ is equal to Norma distribution.

The newly modified model is based on data fitted to Gamma distribution (α, θ), where θ is rate ratio $(1/\mu)$, which inverse FD data on right-tailed. The CP analysis is performed by **'cpt.meanvar'** function with manually selected fixed penalty on variations equal to $\alpha Var(x)$. This, due to properties of Gamma distribution, is equal to $\alpha^2\theta^2 = \mu^2$, or just Tweedie fixed variation for Gamma distribution. Thus, any departure from Gamma may be penalized with $\alpha Var(x)^k$, where the power k enables correction on autocorrelation.The coefficient k is computed as $1 - 1/r$, where r is equal to:

$$r = log(n/m - 1)log(n - m - 2)\frac{1-\rho}{1+\rho}, \tag{18}$$

and n is a number of observed values, m is autocorrelation lag and $\rho < 1$. When the data is stationary, the value of k is equal to 1. When, the lag of autocorrelation is very large, $0 < s < 1$,the range where Tweedie distribution is not defined, thus k must be limited for values above 0.5. The value of $n - m - 2$ is yielded from reordering of the CP model following the assumption that number of CP must be less than n-m-2. The last part represents effective size correction as it is described in [5] with ρ equal to autocorrelation coefficient. Due to very large size n, the current formula is corrected in cases of hourly data as the first part is changed from $log(n/m - 1)$ to $log(n/m^2)$.

The results are obtained from computations performed over the very same datasets for Normally and Gamma distributed statistics with proposed fixed manual penalty. The tests for daily data in 2012, show that proposed model in [4] over-perform any tests with Gamma with fixed penalty. The main disadvantage in Gamma based models are their lesser sensitivity which leads to cut-off last 3 days of the biggest event in March 2012. However, both test with daily data missed the end of FD at April 5th 2012. The summarized results are shown in Table 2.

However, according the tests with hour long time resolution, the models with proposed Gamma statistics with autoregression corrected penalty equal to r in Eq. 18 perform better than all other tests. It is important to be noticed that for events in 2012, the big disturbance in March is split on 2 parts, which exactly corresponds to report

Table 2 Estimated parameters and summarized results

Dataset	n	m	k	Comparison with model of AR(q) correction in [4]
Hourly 2011[a]	382	7	0.63287	Better. Removes non-related CP a day before
Hourly 2012	2137	12	0.4953	Better. The large disturbance in March is slitted
Daily 2012	91	4	0.82696	Worse with March events shorter with 3 days

[a]Note that the data is about more sensitive neutron flux, not muons

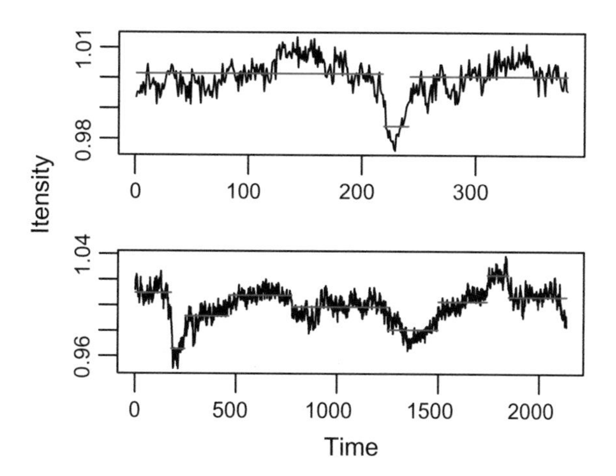

Fig. 5 Detected CP for data with 1 h resolution for both periods. The FD event in February 2012 is shown in figure above. The graphics below shows all events for observed period of 2012. The periods are hours after first hour

[17]. Secondly, the test also detected erroneous regime change in measurements after May 17th, which is shown as a increase of intensity. Thirdly, the begin times the FD events are also correctly detected. However, the sensitivity issue for the end of disturbances on April 5th 2012 remained (Fig. 5).

5 Conclusions

This paper presents first ever work on implementation of change point analysis of automatic detection of Forbush decrease with secondary CR muon flux. The work is tested on two independent in time events with two different approaches to the data. The used CP models, mainly split on dependence of relation between muons and pressure, performed with different success in comparison to already published reports. Most of tested models show weakness, that is partly solved with proposed test modification with incorporation of autoregression. However, some issues remained open without answer or for improvements.

Firstly, the sensitivity remained as issue on particular event on April 5th 2012. But the conclusion on effectiveness of CP analysis on CR muon data may not be drawn firmly. It is important to remember that the Solar disturbances and their impact on CR flux are not fully understood and they are still in research. Secondly, a possible connection with the theory of long memory processes is not investigated in this paper. Any extension to mixture model with implementation of Autoregressive fractionally integrated moving-average models (ARFIMA) could be investigated in future. Thirdly, the automatic procedure could be improved with incorporation of penalty intervals with CROPS option from '*changepoint*' functionalities. Finally, all results and improvements must be done after more extended parallel research of Space Weather and CP analysis on cosmic rays data.

References

1. Dorman, I.: Cosmic rays and space weather: effects on global climate change. Ann. Geophys. **30**, 9–19 (2012)
2. Lilensten, J., et al.: What characterizes planetary space weather. Astron. Astrophys. Rev. 22–79 (2014)
3. Cane, H.: Coronal mass ejections and forbush decreases. Space Sci. Rev. **93**, 55–77 (2000)
4. Beaulieu, C., Chen, J., Sarmiento, J.: Change-point analysis as a tool to detect abrupt climate variations. Phil. Trans. R. Soc. A **370**, 1228–1249 (2012). https://doi.org/10.1098/rsta.2011.0383
5. Siedel, D.J., Lanzante, J.K.: An assessment of three alternatives to linear trends for characterizing global atmospheric temperature changes. J. Geophys. Res. **109** (2004). https://doi.org/10.1029/2003JD004414
6. Angelov, I., Malamova, E., Stamenov, J.: Muon telescopes at basic environmental observatory Moussala and South-West University Blagoevgrad. Sun Geosphere **3**(1), 20–25 (2008)
7. Dorman, L.I.: Cosmic Rays in the Earths Atmosphere and Underground. Springer Netherlands (2004) https://doi.org/10.1007/978-1-4020-2113-8
8. Page, E.S.: Continuous inspection schemes. Biometrika **41**(1–2), 100–114 (1954)
9. Chen, J., Gupta, A.K.: On change point detection and estimation. Commun. Stat. Simul. Comput. **30**(3), 665–697 (2001)
10. Killick, R., Eckley, I.: changepoint: an R package for changepoint analysis. J. Stat. Softw. **58**(3), 1 (2014)
11. R Development Core Team: R: A Language and Environment for Statistical Computing. R Foundation for Statistical Computing, Vienna, Austria (2012)
12. Scott, A.J., Knott, M.: A cluster analysis method for grouping means in the analysis of variance. Biometrics **30**(3), 507–512 (1974)
13. Auger, I.E., Lawrence, C.E.: Algorithms for the optimal identification of segment neighbourhoods. Bull. Math. Biol. **51**(1), 39–54 (1989)
14. Killick, R., Fearnhead, P., Eckley, I.A.: Optimal detection of changepoints with a linear computational cost. JASA **107**(500), 1590–1598 (2012)
15. Zeileis, A.: strucchange: an R package for testing structural change in linear regression models. J. Stat. Softw. **7**, 1 (2002)
16. Abbrescia, M., et al.: Observation of the February 2011 Forbush decrease by the EEE telescopes. Eur. Phys. J. Plus (2011). https://doi.org/10.1140/epjp/i2011-11061-5
17. Tchorbadjieff, A.: Detection of Coronal Mass Ejections (CMEs) in the period of March–May 2012 at Moussala Peak. Proc. Bul. Acad. Sci. **66**(5), 659–666 (2013)
18. Norwood, B., Killick, R.: Long memory and changepoint models: a spectral classification procedure. Stat. Comput. **28**(2), 291–302 (2018)
19. Tweedie, M.C.K.: An index which distinguishes between some important exponential families. In: Statistics: Applications and New Directions. Proceedings of the Indian Statistical Institute Golden Jubilee International Conference. Indian Statistical Institute, Calcutta pp. 579–604 (1984)

Performance Analysis of Effective Methods for Solving Band Matrix SLAEs After Parabolic Nonlinear PDEs

Milena Veneva and Alexander Ayriyan

Abstract This paper presents an experimental performance study of implementations of three different types of algorithms for solving band matrix systems of linear algebraic equations (SLAEs) after parabolic nonlinear partial differential equations – direct, symbolic, and iterative, the former two of which were introduced in Veneva and Ayriyan in Effective methods for solving band SLAEs after parabolic nonlinear PDEs (2018) [3]. An iterative algorithm is presented – the strongly implicit procedure (SIP), also known as the Stone method. This method uses the incomplete LU (ILU(0)) decomposition. An application of the Hotelling-Bodewig iterative algorithm is suggested as a replacement of the standard forward-backward substitutions. The upsides and the downsides of the SIP method are discussed. The complexity of all the investigated methods is presented. Performance analysis of the implementations is done using the high-performance computing (HPC) clusters "HybriLIT" and "Avitohol". To that purpose, the experimental setup and the results from the conducted computations on the individual computer systems are presented and discussed.

1 Introduction

Systems of linear algebraic equations (SLAEs) with pentadiagonal (PD) and tridiagonal (TD) coefficient matrices arise after discretization of partial differential equations (PDEs), using finite difference methods (FDM) or finite element methods (FEM). Methods for numerical solving of SLAEs with such matrices which take into account the band structure of the matrices are needed. The methods known in the literature usually require the matrix to possess special characteristics so as the

M. Veneva (✉) · A. Ayriyan
Laboratory of Information Technologies, Joint Institute for Nuclear Research,
Joliot-Curie 6, 141980 Dubna, Moscow Region, Russia
e-mail: milena.p.veneva@gmail.com

A. Ayriyan
e-mail: ayriyan@jinr.ru

K. Georgiev et al. (eds.), *Advanced Computing in Industrial Mathematics*,
Studies in Computational Intelligence 793,
https://doi.org/10.1007/978-3-319-97277-0_33

method to be stable, e.g. diagonally dominance, positive definiteness, etc. which are not always feasible.

In [1], a finite difference scheme with first-order approximation of a parabolic PDE was built that leads to a TD SLAE with a diagonally dominant coefficient matrix. The system was solved using the Thomas method (see [2]). However, a difference scheme with second-order approximation [3] leads to a matrix which does not have any of the above-mentioned special characteristics. The numerical algorithms for solving multidimensional governing equation, using FDM (e.g. alternating direction implicit (ADI) algorithms (see [4, 5])), ask for a repeated SLAE solution. This explains the importance of the existence of effective methods for the SLAE solution stage.

Two different approaches for solving SLAEs with pentadiagonal and tridiagonal coefficient matrices were explored by us in [3] – diagonal dominantization and symbolic algorithms. These approaches led to five algorithms – numerical algorithms based on LU decomposition (for PD (see [6]) and TD matrices – **NPDM** and **NTDM**), modified numerical algorithm for solving SLAEs with a PD matrix (where the sparsity of the first and the fifth diagonals was taken into account – **MNPDM**), and symbolic algorithms (for PD (see [6]) and TD (see [7]) matrices – **SPDM** and **STDM**). The numerical experiments with the five methods in our previous paper were conducted on a PC (OS: Fedora 25; Processor: Intel Core i7-6700 (3.40 GHz)), using compiler GCC 6.3.1 and optimization -O0. While the direct numerical methods have requirements to the coefficient matrix, the direct symbolic ones only require nonsingularity. Here, we are going to suggest an iterative numerical method which is also not restrictive on the coefficient matrix.

It is a well-known fact that solving problems of the computational linear algebra with sparse matrices is crucial for the effectiveness of most of the programs for computer modelling of processes which are described with the help of differential equations, especially when solving complex multidimensional problems. However, this is exactly how most of the computational science problems look like and hence usually they cannot be modelled on ordinary PCs for a reasonable amount of time. This enforces the usage of supercomputers and clusters for solving such big problems. For example, a numerical solving of a parabolic PDE needs to solve independently (or in parallel) N SLAEs d times at each time step where N is the discretization number, i.e. the matrix dimension, and d is the dimension of the PDE. Thus, it is also important to have an efficient method for serial solving of one band SLAE. Therefore, the aim and the main contribution of this paper is to investigate the performance characteristics of the considered serial methods for band SLAE being executed on modern computer clusters.

The layout of the paper is as follows: in the next section, we introduce the outline of the SIP algorithm. Afterwards, we introduce the experimental setup including the description of the computers used in our experiments and analyze the obtained results.

2 Iterative Approach

An iterative procedure for solving SLAEs with a pentadiagonal coefficient matrix is considered, namely the strongly implicit procedure (SIP) (see [8]), also known as the Stone method. It is an algorithm for solving sparse SLAEs. The method uses the incomplete LU (ILU(0)) decomposition (see [9]) which is an approximation of the exact LU decomposition in the case when a sparse matrix is considered. The idea of ILU(0) is that the zero elements of L and U are chosen to be on the same places as of the initial matrix A. In the case of a pentadiagonal coefficient matrix A, LU is going to be also pentadiagonal, L and U are going to have non-zero elements only on three of their diagonals (main diagonal and two subdiagonals for L; main diagonal and two superdiagonals for U). The Stone method for solving a SLAE of the form $Ax = b$ can be seen in Algorithm 1. There, LU is found using the ILU(0) algorithm suggested in [9]; L and U are extracted using a modification of the Doolittle method (see [4]), namely instead of referencing the matrix A, we reference the already found LU matrix. This way the product of L and U is exactly LU. Every iteration step of the

Algorithm 1 The Stone method for solving a SLAE $Ax = b$

Require: $A, b, L, U, errorMargin$
Ensure: $k, \overrightarrow{x}^{(k)}$
1: $k = 0$ ▷ number of iterations
2: $\overrightarrow{x}^{(k)} = \overrightarrow{0}$ ▷ set an initial guess vector
3: $\overrightarrow{newRHS}^{(k)} = A\overrightarrow{x}^{(k)}$ ▷ new right-hand side (RHS)
4: $\overrightarrow{residual}^{(k)} = \overrightarrow{b} - \overrightarrow{newRHS}^{(k)}$
5: $K = LU - A$
6: **while** $\|\overrightarrow{residual}^{(k)}\|_\infty \geq$ errorMargin **do**
7: $\quad\overrightarrow{newRHS}^{(k)} = K\overrightarrow{x}^{(k)} + \overrightarrow{b}$
8: $\quad$solve $L\overrightarrow{y}^{(k)} = \overrightarrow{newRHS}^{(k)}$
9: $\quad$solve $U\overrightarrow{x}^{(k+1)} = \overrightarrow{y}^{(k)}$
10: $\quad\overrightarrow{residual}^{(k+1)} = \overrightarrow{b} - A\overrightarrow{x}^{(k+1)}$
11: $\quad k{+}{+}$
12: **end while**

Stone method consists of two matrix-vector multiplications with a pentadiagonal matrix, one forward and one backward substitutions with the two triangular matrices of the ILU(0), and two vector additions, i.e. the complexity of the algorithm on every iteration is $31\,N - 36 = O(N)$, where N is the number of rows of the initial matrix.

Remark Instead of using forward and backward substitutions on rows 8–9 of Algorithm 1, one can try to find the inverse matrices of L and U, using a numerical procedure, e.g. the Hotelling-Bodewig iterative algorithm (see [10]). A diagonal matrix can be used as an initial guess for the inverse matrix, as it is suggested in [11]. Since a matrix implementation is going to be very demanding in regards to memory, conduction of computational experiments for a matrix with more than 7×10^3 rows is going to be impossible. For that reason, the algorithms could be redesigned, taking into account the band structure of the data, and so an array implementation could be made. (For the Hotelling-Bodewig iterative algorithm and numerical results from that approach, see Appendix.)

3 Numerical Experiments

Computations were held on the basis of the heterogeneous computing cluster "HybriLIT" at the Laboratory of Information Technologies of the Joint Institute for Nuclear Research in town of science Dubna, Russia and on the cluster computer system "Avitohol" at the Advanced Computing and Data Centre of the Institute of Information and Communication Technologies of the Bulgarian Academy of Sciences in Sofia, Bulgaria.

3.1 Experimental Setup

The direct and iterative numerical algorithms are implemented using `C++`, while the symbolic algorithms are implemented using the `GiNaC` library (version 1.7.2) (see [12]) of `C++`.

The heterogeneous computing cluster "HybriLIT" consists of 13 computational nodes which include two Intel Xeon E5-2695v2 processors (12-core) or two Intel Xeon E5-2695v3 processors (14-core). For more information, visit http://hybrilit.jinr.ru/en. It must be mentioned that for the sake of the performance analysis and the comparison between the computational times only nodes with Intel Xeon E5-2695v2 processors were used.

The supercomputer system "Avitohol" is built with HP Cluster Platform SL250S GEN8. It has two Intel 8-core Intel Xeon E5-2650v2 8C processors each of which runs at 2.6 GHz. For more information, visit http://www.hpc.acad.bg/. "Avitohol" has been part of the TOP500 list (https://www.top500.org) twice – ranking 332nd in June 2015 and 388th in November 2015.

Tables 1 and 2 summarize the basic information about hardware, compilers and libraries used on the two computer systems. The reason why different compilers were used for the numerical and the symbolic methods, respectively, is that the `GiNaC` library does not maintain work with the `Intel` compilers. However, for the numerical methods the `Intel` compilers gave us better results than the `GCC` ones.

Table 1 Information about the available hardware on the two computer systems

Computer system	Processor	Number of processors per node
"HybriLIT"	Intel Xeon E5-2695v2	2
	Intel Xeon E5-2695v3	2
"Avitohol"	Intel Xeon E5-2650v2	2

Table 2 Information about the used software on the two computer systems

Computer system	"HybriLIT"	"Avitohol"
Compiler for the direct and iterative procedures	Intel 2017.2.050 ICPC	Intel 2016.2.181 ICPC
Compiler for the symbolic procedures	GCC 4.9.3	GCC 6.2.0
Needed libraries for the symbolic procedures	GiNaC (1.7.2)	GiNaC (1.7.2)
	CLN (1.3.4)	CLN (1.3.4)
Optimization for the direct and iterative procedures	-O2	-O2
Optimization for the symbolic procedures	-O0	-O0

3.2 Experimental Results

During our experiments wall-clock times were collected using the member function `now()` of the class `std::chrono::high_resolution_clock` which represents the clock with the smallest tick period provided by the implementation; it requires at least standard `c++11` (needs the argument -std=c++11 when compiling). We report the average time from multiple runs. Since the largest supported precision in the `GiNaC` library is **double**, during all the experiments double data type is used. The achieved accuracy during all the numerical experiments is summarized, using infinity norm. The notation is as follows: **NPDM** stands for numerical PD method, **MNPDM** – modified numerical PD method, **SPDM** – symbolic PD method, **NTDM** – numerical TD method, **STDM** – symbolic TD method. The error tolerance used in the iterative method is 10^{-12}. Both the methods comprised in the iterative procedure (ILU(0) and SIP) are implemented using an array representation of the matrices instead of a matrix one.

Remark 1 So as the nonsingularity of the matrices to be checked, a fast symbolic algorithm for calculating the determinant is implemented, using the method suggested in [13]. The complexity of the algorithm is $O(N)$.

Remark 2 The number of needed operations for the Gaussian elimination used so as PD matrices to be transformed into TD ones is $18 + 16\,K$, where K is the number of PD matrix rows with nonzero elements on their second subdiagonal and on their second superdiagonal. Usually, $K \ll N$.

The achieved computational times from solving a SLAE on "HybriLIT" are summarized in Tables 3 and 4. The number of needed iterations for the Stone method is 31.

Tables 5 and 6 sum up the computational times from solving a SLAE on "Avitohol". The number of needed iterations for the Stone method is 31.

Table 3 Results from solving a SLAE on the cluster "HybriLIT" applying direct methods

	Wall-clock time (s)				
N	NPDM	MNPDM	SPDM	NTDM	STDM
10^3	0.0000427	0.0000420	0.1098275	0.0000273	0.0827742
10^4	0.0004310	0.0004270	17.5189275	0.0002693	7.9570979
10^5	0.0041760	0.0040850	5991.4962896	0.0026823	2857.5483843
10^8	2.7946627	2.6662850	–	2.0525187	–
$\max_N \|y - \bar{y}\|_\infty$	2.22×10^{-16}	2.22×10^{-16}	0	2.22×10^{-16}	0

Table 4 Results from solving a SLAE using SIP on the cluster "HybriLIT" applying an iterative method

	Wall-clock time (s)	
N	ILU(0)	SIP
10^2	0.0004090	0.0000847
10^3	0.3493933	0.0007300
10^4	325.6368140	0.0084683
$\max_N \|y - \bar{y}\|_\infty$	–	1.42×10^{-13}

Table 5 Results from solving a SLAE on the supercomputer system "Avitohol" applying direct methods

N	Wall-clock time (s)				
	NPDM	MNPDM	SPDM	NTDM	STDM
10^3	0.0000420	0.0000400	0.1089801	0.0000290	0.0518055
10^4	0.0004234	0.0004110	15.2383414	0.0002610	5.2806483
10^5	0.0040710	0.0039387	2009.6004854	0.0027417	711.9402796
10^8	2.8660797	2.7304760	–	2.1347700	–
$\max_N \|y - \bar{y}\|_\infty$	2.22×10^{-16}	2.22×10^{-16}	0	2.22×10^{-16}	0

Table 6 Results from solving a SLAE using SIP on the cluster "Avitohol" applying an iterative method

N	Wall-clock time (s)	
	ILU(0)	SIP
10^2	0.0011817	0.0001210
10^3	0.5288667	0.0008603
10^4	516.6088950	0.0085333
$\max_N \|y - \bar{y}\|_\infty$	–	1.42×10^{-13}

4 Discussion and Conclusions

Three different approaches for solving a SLAE are compared – direct numerical, direct symbolic, and iterative. The complexity of all the suggested numerical algorithms is $O(N)$ (see Table 7). Since it is unknown what stands behind the symbolic library, evaluating the complexity of the symbolic algorithms is a very complicated task.

Both the achieved computational times and accuracy for the **NPDM** and **SPDM** methods on both the clusters were much better than the ones outlined in [6].

All the experiments with the direct methods gave an accuracy of an order of magnitude of 10^{-16}, while the iterative method gave an accuracy of an order of magnitude of 10^{-13}. Expectedly, the modified version of the numerical method for solving a SLAE with a PD matrix **MNPDM** gave better computational time than the general algorithm **NPDM** in the case of a sparse PD coefficient matrix, since the former method has a lower complexity (usually $K \ll N$, where K is the number of PD matrix rows with nonzero elements on their second subdiagonal and on their second superdiagonal). The fastest numerical algorithm was found to come from the Thomas method. Finally, an iterative algorithm was built – the Stone method. For the needs of the method, additionally, ILU(0) was implemented. An upside of this iterative procedure is that it requires the initial matrix to be nonsingular only. However, this method is not suitable for matrices for which $N > 1 \times 10^5$, since the ILU(0) decomposition of a matrix is computationally demanding on time and memory. Here, likewise the symbolic algorithms, in the case of a piecewise linear

Table 7 Complexity of the investigated methods

Method	NPDM	MNPDM	SPDM	NTDM	STDM	SIP
Complexity	$19N - 29$	$13N + 7K - 14$	–	$9N + 2$	–	$31N - 36$

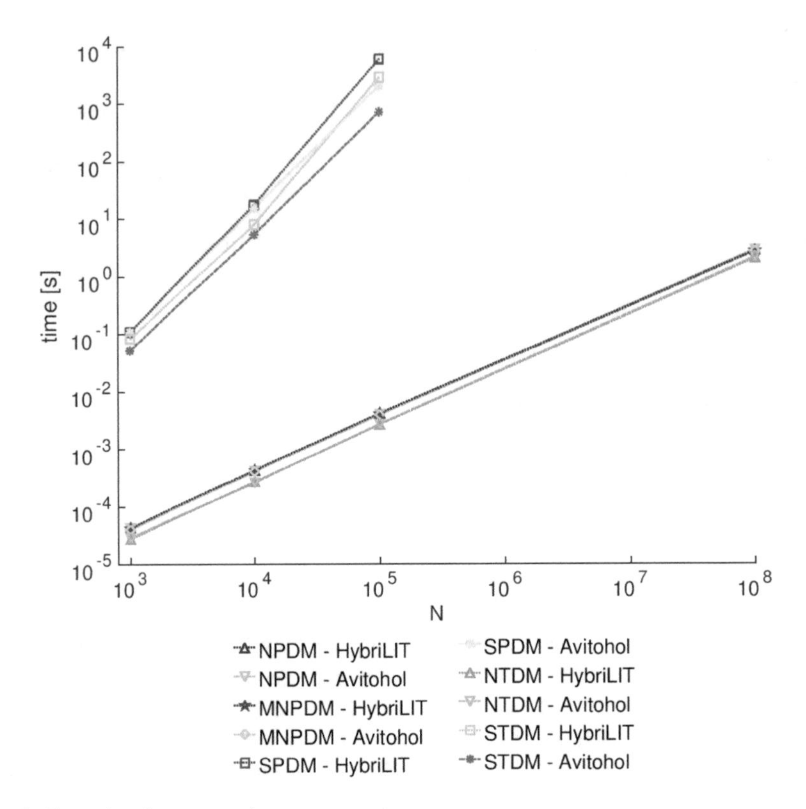

Fig. 1 Execution time comparison

parabolic partial differential equation, they do not add nonlinearity to the right-hand side of the system and hence, there is no need of iterations for the time step to be executed (see [3]). Similarly to the symbolic methods, SIP is not comparable with the numerical algorithms with respect to the required time in the case of a numerical solving of the heat equation when one needs to solve the SLAE many times. Lastly, the obtained accuracy is worse in comparison with any of the other methods. A comparison between the execution times for the direct numerical methods (see Fig. 1) showed only a negligible difference between the two computer systems. However, this was not the case when it comes to the symbolic methods where "Avitohol" performed much better than "HybriLIT". On the other hand, "HybriLIT" behaved better than "Avitohol" with respect to the ILU(0) procedure (see Fig. 2). Only a minimal discrepancy in times was observed for the SIP algorithm.

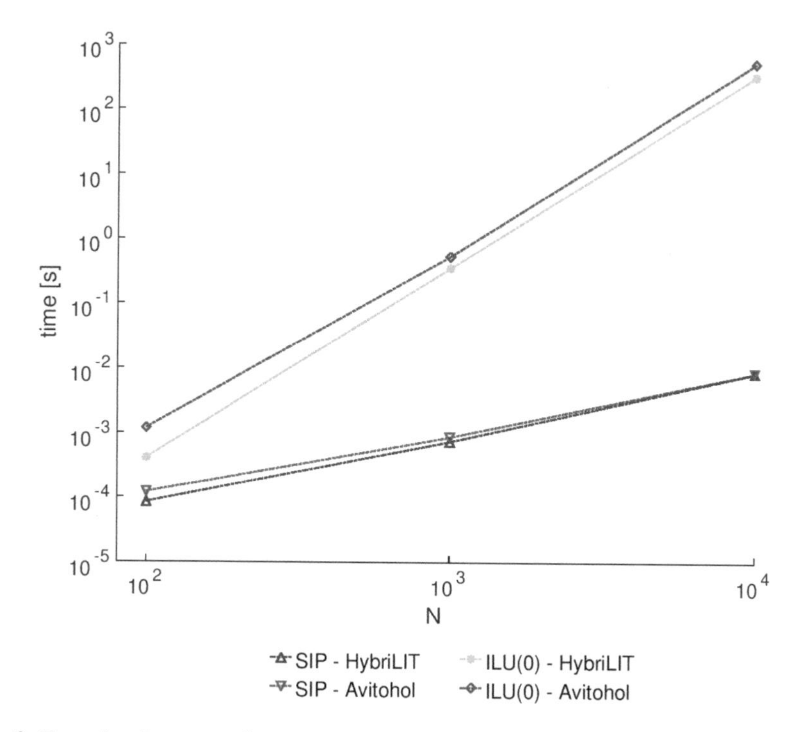

Fig. 2 Execution time comparison

Acknowledgements The authors want to express their gratitude to the Summer Student Program at JINR, Dr. Ján Buša Jr. (JINR), Dr. Andrey Lebedev (GSI/JINR), Assoc. Prof. Ivan Georgiev (IICT & IMI, BAS), the "HybriLIT" team at LIT, JINR, and the "Avitohol" team at the Advanced Computing and Data Centre of IICT, BAS. Computer time grants from LIT, JINR and the Advanced Computing and Data Centre at IICT, BAS are kindly acknowledged. A. Ayriyan thanks the JINR grant No. 17-602-01.

Appendix

The Hotelling-Bodewig iterative algorithm has the form as follows:

$$A_{n+1}^{-1} = A_n^{-1}\,(2\,I - A\,A_n^{-1}), \quad n = 0, 1, \ldots, \tag{1}$$

where I is the identity matrix, A is the matrix whose inverse we are looking for. A_0^{-1} is taken to be of a diagonal form.

The obtained computational times for the ILU(0) method, the Hotelling-Bodewig iterative algorithm and the Stone method, using the heterogeneous cluster "HybriLIT" and the supercomputer system "Avitohol", are summarized in Tables 8, 9, and 10.

Table 8 Results from the ILU(0) method and the numerical method for inverting matrices, using the cluster "HybriLIT"

N	Wall-clock time (s)					
	Matrix implementation			Array implementation		
	ILU(0)	L^{-1}	U^{-1}	ILU(0)	L^{-1}	U^{-1}
10^2	0.0007027	0.0071077	0.0096933	0.0004687	0.0026583	0.0043893
10^3	1.5635590	82.9383600	49.5268260	0.3368320	3.3851580	5.3989410
2×10^3	21.6416160	289.3253220	300.0902740	2.5914510	27.4874390	41.7962950
5×10^3	547.7717120	4835.9211180	6800.0948670	39.5945850	1153.8804500	1606.9331050
7×10^3	1178.6338560	18966.0135900	24345.2476050	108.1988910	3395.9828320	7116.1639450
10^4	–	–	–	314.7906570	10384.6694270	14561.3854660

Table 9 Results from the ILU(0) method and the numerical method for inverting matrices, using the cluster "Avitohol"

N	Wall-clock time (s)					
	Matrix implementation			Array implementation		
	ILU(0)	L^{-1}	U^{-1}	ILU(0)	L^{-1}	U^{-1}
10^2	0.0017620	0.0089710	0.0117817	0.0013103	0.0035383	0.0060527
10^3	1.9317270	85.0694670	76.6738290	0.5320370	4.9676690	6.5183100
2×10^3	27.3982830	299.0649410	370.7769350	4.1901280	32.8010570	51.0338640
5×10^3	495.6995570	5175.7197290	6720.2701160	64.8352820	1227.5802660	1780.3281350
7×10^3	1144.9877790	14829.3973560	22415.4835190	177.4153890	3569.6018970	5279.6295710
10^4	–	–	–	516.4751790	10441.5862030	17833.2337200

Table 10 Results from solving a SLAE using SIP on the clusters "HybriLIT" and "Avitohol"

N	Wall-clock time (s)			
	on "HybriLIT"		on "Avitohol"	
	Matrix implementation	Array implementation	Matrix implementation	Array implementation
	SIP	SIP	SIP	SIP
10^2	0.0005637	0.0001827	0.0006510	0.0002953
10^3	0.0703420	0.0130150	0.0866850	0.0163560
2×10^3	0.3403310	0.0683440	0.3492530	0.0859700
5×10^3	2.3330770	0.5063490	3.7949870	0.5812540
7×10^3	8.7838330	1.1616650	6.3790020	1.2195610
10^4	–	2.0574280	–	2.9845790
$\max_{N} \|y - \bar{y}\|_\infty$	3.13×10^{-14}	3.13×10^{-14}	3.13×10^{-14}	3.13×10^{-14}

The matrix implementations lead to 5, 7, and 34 iterations, respectively for finding L^{-1} and U^{-1}, applying the Hotelling-Bodewig procedure, and for the Stone method while the needed iterations when the array implementations are executed are 5, 6, and 31, respectively. It is expected that inverting L would require less number of iterations, since it is a unit triangular matrix. The achieved accuracy is of an order of magnitude of 10^{-13}, having used an error tolerance 10^{-12}. Comparing the results for the computational times, one can see that the array implementation not only decreased the time needed for the inversion of both the matrices L and U but also it decreases the number of iterations needed so as the matrix U to be inverted. As one can see, the time required for the SIP procedure is also improved by the new implementation approach. One reason being is that the number of iterations is decreased. Overall, the array implementations decrease the computational times with one order of magnitude. Finally, this second approach requires less amount of memory (instead of keeping $N \times N$ matrix, just 5 arrays with length N are stored), which allows experiments with bigger matrices to be conducted. However, this method (even in its array form) is not suitable for too large matrices (with number of rows bigger than 1×10^5), since the evaluation of the inverse of a matrix is computationally demanding on both time and memory. A comparison between the times on the two computer systems showed that overall "HybriLIT" is a bit faster than "Avitohol".

References

1. Ayriyan, A., Buša, J., Jr., Donets, E.E., Grigorian, H., Pribiš, J.: Algorithm and simulation of heat conduction process for design of a thin multilayer technical device. Appl. Therm. Eng. **94**, 151–158 (2016). https://doi.org/10.1016/j.applthermaleng.2015.10.095. (Elsevier)
2. Higham, N.J.: Accuracy and Stability of Numerical Algorithms, 2nd edn, pp. 174–176. SIAM, Philadelphia, PA (2002)

3. Veneva, M., Ayriyan, A.: Effective methods for solving band SLAEs after parabolic nonlinear PDEs. In: AYSS-2017, EPJ Web of Conferences, vol. 177, p. 07004 (2018). https://doi.org/10.1051/epjconf/201817707004, arXiv: 1710.00428v2 [math.NA]
4. Kincaid, D.R., Cheney, E.W.: Numerical Analysis: Mathematics of Scientific Computing, p. 788. American Mathematical Society, Providence, RI (2002)
5. Samarskii, A.A., Goolin, A.: Chislennye Metody, pp. 45–47. Nauka, Moscow (1989). (in Russian)
6. Askar, S.S., Karawia, A.A.: On solving pentadiagonal linear systems via transformations. Math. Probl. Eng. **2015**, 9 (2015). https://doi.org/10.1155/2015/232456. (Hindawi Publishing Corporation)
7. El-Mikkawy, M.: A generalized symbolic Thomas algorithm. Appl. Math. **3**(4), 342–345 (2012). https://doi.org/10.4236/am.2012.34052
8. Stone, H.L.: Iterative solution of implicit approximations of multidimensional partial differential equations. SIAM J. Numer. Anal. **4**(3), 530–538 (1968). https://doi.org/10.1137/0705044
9. Saad, Y.: Iterative Methods for Sparse Linear Systems, 2nd edn, pp. 307–310. SIAM, Philadelphia, PA (2003). https://doi.org/10.1137/1.9780898718003
10. Schulz, G.: Iterative berechnung der reziproken matrix. Zeitschrift fur Angewandte Mathematik und Mechanik **13**, 57–59 (1933). (in German)
11. Soleymani, F.: A rapid numerical algorithm to compute matrix inversion. Int. J. Math. Math. Sci. **2012**, 11 (2012). https://doi.org/10.1155/2012/134653
12. Bauer, C., Frink, A., Kreckel, R.: Introduction to the GiNaC framework for symbolic computation within the C++ programming language. J. Symbolic Comput. **33**, 1–12 (2002). https://doi.org/10.1006/jsco.2001.0494
13. El-Mikkawy, M.: Fast and reliable algorithm for evaluating nth order pentadiagonal determinants. Appl. Math. Comput. **202**(1), 210–215 (2008). (Elsevier)

Statistical Characteristics of a Flow of Substance in a Channel of Network that Contains Three Arms

Nikolay K. Vitanov and Roumen Borisov

Abstract We study the motion of a substance in a channel that is part of a network. The channel has 3 arms and consists of nodes of the network and edges that connect the nodes and form ways for motion of the substance. Stationary regime of the flow of the substance in the channel is discussed and statistical distributions for the amount of substance in the nodes of the channel are obtained. These distributions for each of the three arms of the channel contain as particular case famous long-tail distributions such as Waring distribution, Yule-Simon distribution and Zipf distribution. The obtained results are discussed from the point of view of technological applications of the model (e.g., the motion of the substance is considered to happen in a complex technological system and the obtained analytical relationships for the distribution of the substance in the nodes of the channel represents the distribution of the substance in the corresponding cells of the technological chains). A possible application of the obtained results for description of human migration in migration channels is discussed too.

Keywords Network flow · Distribution of flowing substance · Waring distribution · Zipf distribution

1 Introduction

The studies on the dynamics of complex systems are very intensive in the last decade especially in the area of social dynamics and population dynamics [1–3, 20, 22, 23, 29–37]. Networks and the flows in networks are important part of the structure and processes of many complex systems [9, 11, 19]. Research on network flows has roots in the studies on transportation problems [11] and in the studies on migration flows [10, 13, 14, 38]. Today one uses the methodology from the theory of network

N. K. Vitanov (✉) · R. Borisov
Institute of Mechanics, Bulgarian Academy of Sciences,
Acad. G. Bonchev Str., Bl. 4, 1113 Sofia, Bulgaria
e-mail: vitanov@imbm.bas.bg

K. Georgiev et al. (eds.), *Advanced Computing in Industrial Mathematics*,
Studies in Computational Intelligence 793,
https://doi.org/10.1007/978-3-319-97277-0_34

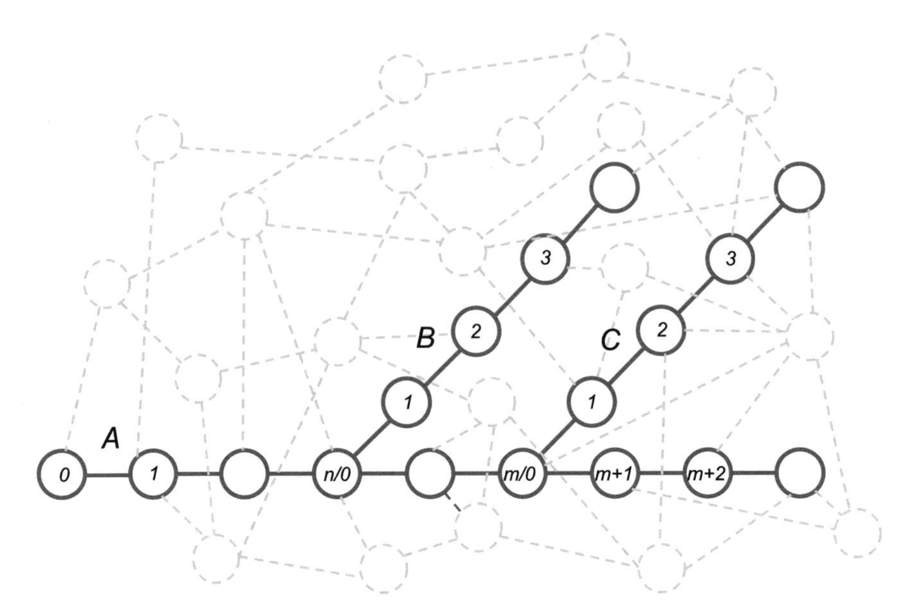

Fig. 1 Part of a network and a channel consisting of three arms - A, B and C. The nodes of the channel are represented by circles and the edges that connect the nodes are represented by solid lines. The main arm (A) splits into two arms at two of the nodes. These nodes are labeled $n/0$ and $m/0$ respectively ($m > n$). The entry nodes of the channels A, B and C are labeled by 0. The nodes of the network that are not a part of the studied channel are connected by edges that are represented by dashed lines. The assumption below is that the lengths of the arms of the channel are infinite

flows [7] to solve problems connected to e.g., minimal cost of the flow, just in time scheduling or electronic route guidance in urban traffic networks [12].

A specific feature of the study below is that we consider a channel of a network and this channel consists of three arms. Several types of channels are possible depending on the mutual arrangement of the arms. In this study we shall consider the situation where the channel consists a main arm (A, Fig. 1) and two other arms connected to the main one in two different nodes ($n/0$ and $m/0$, $m > n$). In this way the channel has a single arm up to the node $n/0$ where the channel splits to two arms (A and B). The first arm A continues up to the node $m/0$ where it also splits to two arms - A and C. Thus the studied channel has three arms: A, B, and C.

There are three special nodes in this channel. The first node of the channel arm A (called also the entry node and labeled by the number 0) is the only node of the channel where the substance may enter the channel from the outside world. After that the substance flows along the channel A up to the node labeled by $n/0$. This is the second special node where the channel splits for the first time and this node (node $n/0$) is the entry node for the second arm B of the channel. The third special node of the network is the node where the channel splits in two arms for the second time (node $m/0$). This node is the entry node for the third arm C of the channel. We assume that the substance moves only in one direction along the channel (from nodes labeled by smaller numbers to nodes labeled by larger numbers). In addition we assume that the substance may quit the channel and may move to the environment. This process will be denoted below as "leakage". As the substance can enter the network only through the entry node of the channel arm A then the "leakage" is possible only in the direction from the channel to the network (and not in the opposite direction).

2 Mathematical Formulation of the Problem

As we have mentioned above the studied channel consists of three arms. The nodes on each arm are connected by edges and each node is connected only to the two neighboring nodes of the arm exclusive for the first node of the channel arm A that is connected only to the neighboring node and the two other special nodes that are connected to three nodes each. We consider each node as a cell (box), i.e., we consider an array of infinite number of cells indexed in succession by non-negative integers. We assume that an amount x^q of some substance is distributed among the cells of the arm q (q can be A, B or C) and this substance can move from one cell to the neighboring cell.

Let x_i^q be the amount of the substance in the i-th cell on the q-th arm. We shall consider channel containing infinite number of nodes in each of its three arms. Then the substance in the q-th arm of the channel is

$$x^q = \sum_{i=0}^{\infty} x_i^q, \quad q = \{A, B, C\}. \tag{1}$$

The fractions $y_i^q = x_i^q / x^q$ can be considered as probability values of distribution of a discrete random variable ζ in the corresponding arm of the channel

$$y_i^q = p^q(\zeta = i), \; i = 0, 1, \ldots \tag{2}$$

The content x_i^q of any cell may change due to the following 4 processes:

1. Some amount s^A of the substance x^A enters the main arm A from the external environment through the 0-th cell;
2. Some amount s^q of the substance x^q enters the q-th arm from the main one through the $n/0$-th or $m/0$-th cell, respectively ($q \in \{B, C\}$);
3. Part f_i^q of x_i^q is transferred from the i-th cell into the $i+1$-th cell of the q-th arm;
4. Part g_i^q of x_i^q leaks out the i-th cell of the q-th arm into the environment;

We assume that the process of the motion of the substance is continuous in the time. Then the process can be modeled mathematically by the system of ordinary differential equations:

$$\frac{dx_0^q}{dt} = s^q - f_0^q - g_0^q;$$
$$\frac{dx_i^q}{dt} = f_{i-1}^q - f_i^q - g_i^q,\ i = 1, 2, \dots, ; q = \{A, B, C\} \tag{3}$$

There are two regimes of functioning of the channel: stationary regime and non-stationary regime. What we shall discuss below is the stationary regime of functioning of the channel. In the stationary regime of the functioning of the channel $dx_i^q/dt = 0$, $i = 0, 1, \dots$. Let us mark the quantities for the stationary case with $*$. Then from Eq. (3) one obtains

$$f_0^{*q} = s^{*q} - g_0^{*q};\ f_i^{*q} = f_{i-1}^{*q} - g_i^{*q}. \tag{4}$$

This result can be written also as

$$f_i^{*q} = s^{*q} - \sum_{j=0}^{i} g_j^{*q} \tag{5}$$

Hence for the stationary case the situation in the each arm is determined by the quantities s^{*q} and g_j^{*q}, $j = 0, 1, \dots$. In this chapter we shall assume the following forms of rules for the motion of substance in Eq. (3) (α_i, β_i, γ_i, σ are constants)

$$s^A = \sigma x_0^A > 0$$
$$s^B = \delta_n^A x_n^A;\ s^C = \delta_m^A x_m^A;\ 1 \geq \delta_{m,n}^A \geq 0$$
$$f_i^q = (\alpha_i^q + \beta_i^q i) x_i^q;\ 1 > \alpha_i^q \geq 0,\ 1 \geq \beta_i^q \geq 0,\ 1 \geq \alpha_i^q + \beta_i^q i \geq 0$$
$$g_i^q = \gamma_i^{*q} x_i^q;\ 1 \geq \gamma_i^{*q} \geq 0 \rightarrow \text{non-uniform leakage in the nodes} \tag{6}$$

γ_i^* is a quantity specific for the present study. $\gamma_i^{*A} = \gamma_i^A + \delta_i^A$ describes the situation with the leakages in the cells on the main arm A. γ_i^A corresponds to leakage from the i-th node of the arm A to the environment. δ_i^A corresponds to leakage from the i-th node of the arm A to the arm B. We shall assume that $\delta_i^A = 0$ for all i except for $i = n$ and $i = m$. This means that in the n-th node in addition to the usual leakage γ_n^A there is a leakage of substance given by the term $\delta_n^A x_n^A$ and this additional leakage supplies the substance that then begins its motion along the second arm of the channel. Furthermore in the m-th node (where the main arm of the channel splits for the second time) in addition to the usual leakage γ_m^A there is additional leakage of substance given by the term $\delta_m^A x_m^A$ and this additional leakage supplies the substance that then begins its motion along the third arm of the channel. In the arms B and C of the channel the leakages are from the corresponding node of the channel to the environment. Then in these arms $\gamma_i^{*B} = \gamma_i^B$ and $\gamma_i^{*C} = \gamma_i^C$. There are no leakages from the arms B and C to the other arms of the channel.

On the basis of all above the model system of differential equations for each arm of the channel becomes

$$\frac{dx_0^q}{dt} = s^q - \alpha_0^q x_0^q - \gamma_0^{*q} x_0^q;$$
$$\frac{dx_i^q}{dt} = [\alpha_{i-1}^q + (i-1)\beta_{i-1}^q]x_{i-1}^q - (\alpha_i^q + i\beta_i^q + \gamma_i^{*q})x_i^q; \quad i = 1, 2, \ldots \qquad (7)$$

Below we shall discuss the situation in which the stationary state is established in the entire channel. Then $dx_0^q/dt = 0$ from the first of the Eq. (7). Hence

$$x_0^q = \frac{s^q}{\alpha_0^q + \gamma_0^{*q}} \quad . \qquad (8)$$

For the main arm A it follows that $\sigma = \alpha_0^A + \gamma_0^A$. This means that x_0^A (the amount of the substance in the 0-th cell of the arm A) is free parameter. For the arm B and C, $dx_0^k/dt = 0$ follows that

$$x_0^{*B} = \frac{\delta_n^A x_n^{*A}}{\alpha_0^B + \gamma_0^B}; \quad x_0^{*C} = \frac{\delta_m^A x_m^{*A}}{\alpha_0^C + \gamma_0^C}. \qquad (9)$$

The solution of Eq. (7) is (see Appendix A)

$$x_i^A = x_i^{*A} + \sum_{j=0}^{i} b_{ij}^A \exp[-(\alpha_j^A + j\beta_j^A + \gamma_j^{*A})t]; \quad i = 1, 2, \ldots$$
$$x_i^q = x_i^{*q} + \sum_{j=0}^{i} b_{ij}^q \exp[-(\alpha_j^q + j\beta_j^q + \gamma_j^{*q})t], i = 1, 2, \ldots; q = \{B, C\} \qquad (10)$$

where x_i^{*q} is the stationary part of the solution. For x_i^{*q} one obtains the relationship (just set $dx^q/dt = 0$ in the second of Eq. (7)

$$x_i^{*q} = \frac{\alpha_{i-1}^q + (i-1)\beta_{i-1}^q}{\alpha_i^q + i\beta_i^q + \gamma_i^{*q}} x_{i-1}^{*q}, \quad i = 1, 2, \ldots; q = \{A, B, C\} \qquad (11)$$

The corresponding relationships for the coefficients b_{ij}^q are $(i = 1, \ldots, q = \{A, B, C\})$:

$$b_{ij}^q = \frac{\alpha_{i-1}^q + (i-1)\beta_{i-1}^q}{(\alpha_i^q - \alpha_j^q) + (i\beta_i^q - j\beta_j^q) + (\gamma_i^{*q} - \gamma_j^{*q})} b_{i-1,j}^q, \; j = 0, 1, \ldots, i-1, \qquad (12)$$

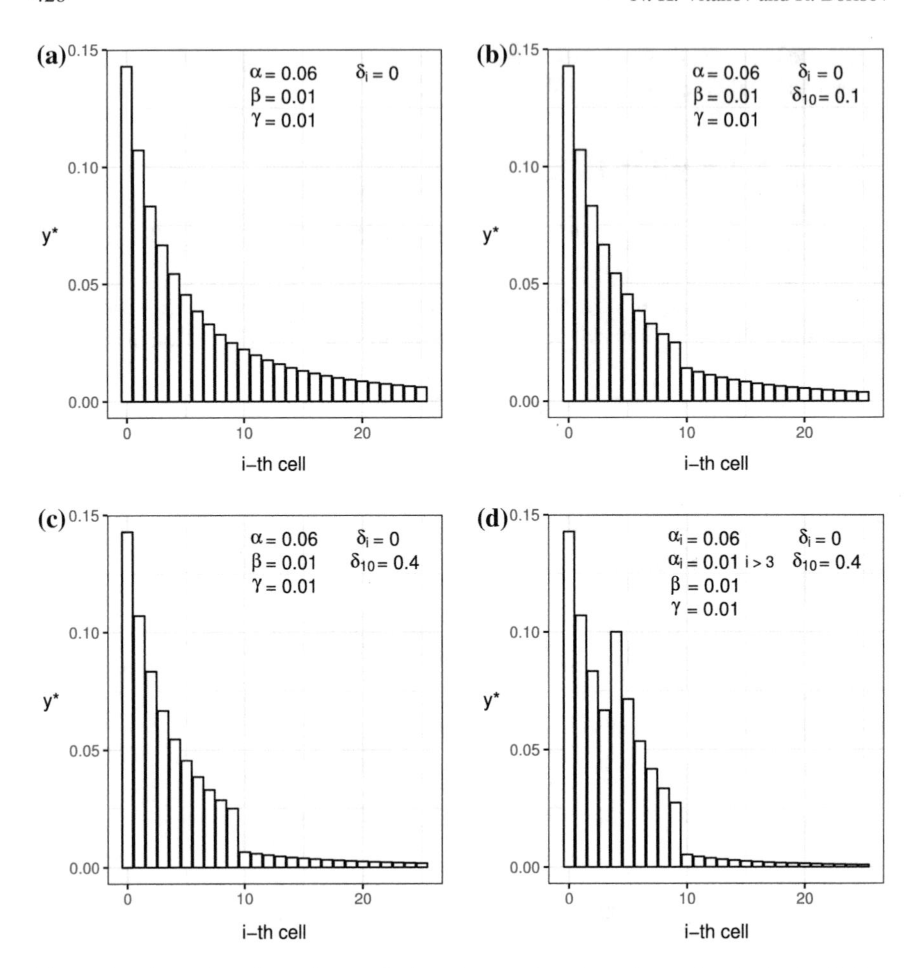

Fig. 2 The distribution of the substance in the main arm of the channel described by the Eq. (14) for various values of the parameters γ_i^*. Values of the parameters $\alpha_i = 0.08$, $\beta_i = 0.01$ and $\gamma_i = 0.01$ are fixed. The figures show the influence of different choices of the parameter $\gamma_i^*(\delta_i)$ on the shape of the distribution. Figure **a** $\delta_i^A = 0$. There is a flow of substance only in the main arm A of the channel. **b** $\delta_i^A = 0$, $\delta_5^A = 0.01$. In addition to the main arm, there is a flow of substance in the second arm B of the channel too. **c** $\delta_i^A = 0$, $\delta_5^A = 0.1$, $\delta_{10}^A = 0.1$. In this case the substance flows in all three arms of the channel. In addition to the main channel, the second and third channels are also present. **d** $\delta_i = 0$, $\delta_5 = 0.1$, $\delta_{10} = 0.4$

From Eq. (11) one obtains

$$x_i^{q*} = \frac{\prod_{j=0}^{i-1}[\alpha^q_{i-j-1} + (i-j-1)\beta^q_{i-j-1}]}{\prod_{j=0}^{i-1}\alpha^q_{i-j} + (i-j)\beta^q_{i-j} + \gamma^{q*}_{i-j}} x_0^{q*} \tag{13}$$

The form of the corresponding stationary distribution $y_i^{q*} = x_i^{q*}/x^{q*}$ (where x^{q*} is the amount of the substance in all of the cells of the arm of the channel) is

$$y_i^{q*} = \frac{\prod\limits_{j=0}^{i-1}[\alpha_{i-j-1}^q + (i-j-1)\beta_{i-j-1}^q]}{\prod\limits_{j=0}^{i-1}[\alpha_{i-j}^q + (i-j)\beta_{i-j}^q + \gamma^{q*}_{i-j}]} y_0^{q*} \tag{14}$$

To the best of our knowledge the distribution presented by Eq. (14) was not discussed up to now outside our research group, i.e. this is a new statistical distribution. Let us write the values of this distribution for the first 5 nodes of the main arm of the channel. We assume that the arm B splits at the 3rd node of the arm A and the arm C splits at the 5th node of the arm A. Thus the values of the distribution for the first 5 nodes of the channel A are

$$\begin{aligned}
y_1^{A^*} &= y_0^{A^*}\frac{\alpha_0^A}{\alpha_1^A+\beta_1^A+\gamma_1^A}\\
y_2^{A^*} &= y_0^{A^*}\frac{\alpha_0^A}{\alpha_1^A+\beta_1^A+\gamma_1^A}\frac{\alpha_1^A+\beta_1^A}{\alpha_2^A+2\beta_2^A+\gamma_2^A}\\
y_3^{A^*} &= y_0^{A^*}\frac{\alpha_0^A}{\alpha_1^A+\beta_1^A+\gamma_1^A}\frac{\alpha_1^A+\beta_1^A}{\alpha_2^A+2\beta_2^A+\gamma_2^A}\frac{\alpha_2^A+2\beta_2^A}{\alpha_3^A+3\beta_3^A+\gamma_3^A+\delta_3}\\
y_4^{A^*} &= y_0^{A^*}\frac{\alpha_0^A}{\alpha_1^A+\beta_1^A+\gamma_1^A}\frac{\alpha_1^A+\beta_1^A}{\alpha_2^A+2\beta_2^A+\gamma_2^A}\frac{\alpha_2^A+2\beta_2^A}{\alpha_3^A+3\beta_3^A+\gamma_3^A+\delta_3}\frac{\alpha_3^A+3\beta_3^A}{\alpha_4^A+4\beta_4^A+\gamma_4^A}\\
y_5^{A^*} &= y_0^{A^*}\frac{\alpha_0^A}{\alpha_1^A+\beta_1^A+\gamma_1^A}\frac{\alpha_1^A+\beta_1^A}{\alpha_2^A+2\beta_2^A+\gamma_2^A}\frac{\alpha_2^A+2\beta_2^A}{\alpha_3^A+3\beta_3^A+\gamma_3^A+\delta_3}\frac{\alpha_3^A+3\beta_3^A}{\alpha_4^A+4\beta_4^A+\gamma_4^A}\times\\
&\quad\frac{\alpha_4^A+4\beta_3^A}{\alpha_5^A+5\beta_4^A+\gamma_5^A+\delta_5}
\end{aligned} \tag{15}$$

Let us show that the distribution described by Eq. (14) contains as particular cases several famous distributions, e.g., Waring distribution, Zipf distribution, and Yule-Simon distribution. In order to do this we consider the particular case when $\beta_i^q \neq 0$ and write x_i^q from Eq. (13) by means of the new notations $k_i^q = \alpha_i^q/\beta_i^q$; $a_i^q = \gamma^{q*}_i/\beta_i^q$; $b_i^q = \beta_{i-1}^q/\beta_i^q$. Let us now consider the particular case where $\alpha_i^q = \alpha^q$ and $\beta_i^q = \beta^q$ for $i = 0, 1, 2, \ldots, q = \{A, B, C\}$. Then from Eq. (14) one obtains

$$x_i^{q*} = \frac{[k^q + (i-1)]!}{(k^q-1)!\prod\limits_{j=1}^{i}(k^q + j + a_j)} x_0^{q*} \tag{16}$$

where $k^q = \alpha^q/\beta^q$ and $a_j^q = \gamma_j^{q*}/\beta^q$. Let us now consider the particular case where $a_0^q = \cdots = a_N^q$. In this case the distribution for $P^q(\zeta = i)$ for $y_i^{q*} = x_i^{q*}/x^{q*}$ is:

$$P^q(\zeta = i) = P^q(\zeta = 0)\frac{(k^q - 1)^{[i]}}{(a^q + k^q)^{[i]}}; \quad k^{q[i]} = \frac{(k^q + i)!}{k^q!}; \; i = 1, 2, \ldots \tag{17}$$

$P^q(\zeta = 0) = y_0^{q*} = x_0^{q*}/x^{q*}$ is the percentage of substance that is located in the first cell of the channel. Let this percentage be $y_0^{q*} = \frac{a^q}{a^q+k^q}$. This case corresponds to the situation where the amount of substance in the first cell is proportional of the amount of substance in the entire channel. In this case Eq. (17) is reduced to:

$$P^q(\zeta = i) = \frac{a^q}{a^q + k^q}\frac{(k^q - 1)^{[i]}}{(a^q + k^q)^{[i]}}; \quad k^{q[i]} = \frac{(k^q + i)!}{k^q!}; \; i = 1, 2, \ldots; q = \{A, B, C\} \tag{18}$$

For all q the distribution (18) is exactly the Waring distribution (probability distribution of non-negative integers named after Edward Waring - the 6th Lucasian professor of Mathematics in Cambridge from the 18th century) [16, 17]. Waring distribution has the form

$$p_l = \rho\frac{\alpha_{(l)}}{(\rho + \alpha)_{(l+1)}}; \; \alpha_{(l)} = \alpha(\alpha + 1)\ldots(\alpha + l - 1) \tag{19}$$

ρ is called the tail parameter as it controls the tail of the Waring distribution. Waring distribution contains various distributions as particular cases. Let $l \to \infty$ Then the Waring distribution is reduced to the frequency form of the Zipf distribution [8] $p_l \approx \frac{1}{l^{(1+\rho)}}$. If $\alpha \to 0$ the Waring distribution is reduced to the Yule-Simon distribution [27] $p(\zeta = l \mid \zeta > 0) = \rho B(\rho + 1, l)$, where B is the beta-function.

3 Discussion

On the basis of the obtained analytical relationships for the distribution of the substance in three arms of the channel we can make numerous conclusions. Let us consider the distribution in the main channel A described by Eq. (14) from the last section (we have to set $q = A$). Below we shall omit the index A keeping in the mind that we are discussing the main arm of the channel. Let us denote as $y_i^{*(1)}$ the distribution of the substance in the cells of the main arm of the channel for the case of lack of second and third arms. Let $y_i^{*(2)}$ be the distribution of the substance in the cells of the main arm of the channel in the case of the presence of second and the absence of the third arm of the channel. Finally, let $y_i^{*(3)}$ be the distribution of the substance in the cells of the main arm of the channel in the case of the presence of both the second and third arms. From the theory in the previous section one easily obtains the relationship

$$\frac{y_i^{*(1)}}{y_i^{*(r)}} = \prod_{j=0}^{i-1} \frac{\alpha_{i-j} + (i-j)\beta_{i-j} + \gamma_{i-j}^*}{\alpha_{i-j} + (i-j)\beta_{i-j} + \gamma_{i-j}} = \prod_{j=0}^{i-1} \left[1 + \frac{\delta_{i-j}}{\alpha_{i-j} + (i-j)\beta_{i-j} + \gamma_{i-j}}\right], \tag{20}$$

where r can be 2 or 3. If there are no second and third arms of the channel then the distribution of the substance in the arm A has a standard form of a long-tail distribution like the distribution shown in Fig. 2a. In the case of presence of additional arms B and C and if $i < n$ then $\delta_i = 0$ and there is no difference between the distribution of the substances in the channel with single arm and in the channel with two arms. The difference arises at the splitting cell $n/0$. As it can be easily calculated for $i \geq n$ and $i < m$ Eq. (20) reduces to

$$\frac{y_i^{*(2)}}{y_i^{*(1)}} = \frac{1}{1 + \frac{\delta_n}{\alpha_n + n\beta_n + \gamma_n}}, \quad m > i \geq n \tag{21}$$

Equation (21) shows clearly that *the presence of the second arm of the channel changes the distribution of the substance in the main arm of the channel*. The "leakage" of the substance to second arm of the channel may reduce much the tail of the distribution of substance in the main arm of the channel. This can appear as a kink in the distribution similar to the kink that can be seen at Fig. 2b.

When $i \geq m$ Eq. (20) reduces to

$$\frac{y_i^{*(3)}}{y_i^{*(1)}} = \left[\frac{1}{1 + \frac{\delta_n}{\alpha_n + n\beta_n + \gamma_n}}\right]\left[\frac{1}{1 + \frac{\delta_m}{\alpha_m + m\beta_m + \gamma_m}}\right], \quad i \geq m \tag{22}$$

Equation (22) shows that the "leakage" of the substance through both the second and the third arms may reduce even more the tail of the distribution of substance in the main arm than only through the second arm. This reduction becomes extremely strong for large values of δ_n and δ_m (Fig. 2c and d). Furthermore, it is easily seen that *the presence of the third arm of the channel does not change the distribution of the substance in the second arm.* We note that the form of the distribution (14) can be different for different values of the parameters of the distribution. One interesting form of the distribution can be observed in Fig. 2d and this form is different than the conventional form of a long-tail discrete distribution shown in Fig. 2a. Thus at Fig. 2 one can obtain visually a further impression that the distribution (14) is more general than the Waring distribution.

4 Concluding Remarks

In this text we have discussed one possible case of motion of substance through a network. Namely our attention was concentrated on the directed motion of substance in a channel of network. Specific feature of the study is that the channel consists of three arms: main arm and two other arms that split from the main arm. We propose a mathematical model of the motion of the substance through the channel and our interest in this study was concentrated on the stationary regime of the motion of the substance through the arms of the channel. The main outcome of the study is the obtained distributions of substance along the cells of the channel. These distributions have a very long tail in the form of the distributions depends on the numerous parameters that regulate the motion of the substance through the channel. Nevertheless we have shown that all of the distributions (i.e. the distributions of the substance along the three arms of the channel) contain as particular case the long-tail discrete Waring distribution which is famous because of the fact that it contains as particular cases the Zipf distribution and the Yule-Simon distribution that are much used in the modeling of complex natural, biological, and social systems.

The model discussed above can be used for study of motion of substance through cells of technological systems. The model can be applied also for study of other system such as channels of human migration or flows in logistic channels. Initial version of the model (for the case of channel containing one arm) was applied for modeling dynamics of citations and publications in a group of researchers [25]. Let us make several notes on the application in the case of human migration as the study of human migration is an actual research topic [4–6, 15, 18, 21, 24, 26, 28, 38–42]. In this case of application of the theory the nodes are the countries that form the migration channel and the edges represents the ways that connect the countries of the channel. Equations (21, 22) show that the additional arms of the channel can be used to decrease the "pressure" of migrants in the direction of the more preferred countries that are relatively away from the initial countries of the channel. This may be done by appropriate increase of the coefficients δ_n and δ_m in Eqs. (21, 22).

The research presented above is connected to the actual topic of motion of different types of substances along the nodes and the edges of various kinds of networks. We intend to continue this research by study of more complicated kinds of channels and by use more sophisticated model that accounts for more kinds of processes that may happen in connection with the studied network flows.

Proof that Eq. (10) is a solution of Eq. (7) for main arm of the channel

Let us consider the first equation from Eq. (7) for main arm. In this case $i = 0$ and Eq. (10) becomes $x_0^A = x_0^{*A} + b_{00}^A \exp[-(\alpha_0^A + \gamma_0^A)t]$. The substitution of the last relationship in the first of the Eq. (7) leads to the relationship

$$0 = (\sigma_0 - \alpha_0^A - \gamma_0^{*A})x_0^* + b_{00}^A \sigma_0 \exp[-(\alpha_0^A + \gamma_0^{*A})t]. \tag{23}$$

Let us assume $\sigma_0 = \alpha_0^A + \gamma_0^{*A}$ and $b_{00}^A = 0$. Then Eq. (10) describes the solution of the first of Eq. (7).

Let us now consider Eq. (7) for the main arm of the channel for $i = 1, 2, \ldots$. Let us fix i and substitute the first of Eq. (10) in the corresponding equation from Eq. (7). The result is

$$\sum_{j=0}^{i-1} \exp[-(\alpha_j^A + j\beta_j^A + \gamma_j^{*A})t]\Big\{ - b_{ij}^A(\alpha_j^A + j\beta_j^A + \gamma_j^{*A}) - \tag{24}$$
$$-b_{i-1,j}^A[\alpha_{i-1}^A + (i-1)\beta_{i-1}^A] + b_{ij}^A(\alpha_i^A + i\beta_i^A + \gamma_i^{*A})\Big\} = 0$$

As it can be seen from Eqs. (12), (24) is satisfied.

References

1. Albert, R., Barabasi, A.-L.: Statistical mechanics of complex networks. Rev. Mod. Phys. **74**, 47–97 (2002)
2. Amaral, L.A.N., Ottino, J.M.: Complex networks. Augmenting and framework for the study of complex systems. Eur. Phys. J. B **38**, 147–162 (2004)
3. Boccaletti, S., Latora, V., Moreno, Y., Chavez, M., Hwang, D.U.: Complex networks: structure and dynamics. Phys. Rep. **424**, 175–308 (2006)
4. Borjas, G.J.: Economic theory and international migration. Int. Migration Rev. **23**, 457–485 (1989)
5. Bracken, I., Bates, J.J.: Analysis of gross migration profiles in England and Wales: some developments in classification. Env. Plann. A **15**, 343–355 (1983)
6. Champion A.G., Bramley G., Fotheringham A.S., Macgill J., Rees P.H.: A migration modelling system to support government decision-making. In: Stillwell, J., Geertman, S. (eds.) Planning support systems in practice, pp. 257–278. Springer Verlag, Berlin (2002)
7. Chan, W.-K.: Theory of nets: flows in networks. Wiley, New York (1990)
8. Chen, W.-C.: On the weak form of the Zipf's law. J. Appl. Probab. **17**, 611–622 (1980)
9. Dorogovtsev, S.N., Mendes, J.F.F.: Evolution of networks. Adv. Phys. **51**, 1079–1187 (2002)
10. Fawcet, J.T.: Networks, linkages, and migration systems. Int. Migration Rev. **23**(1989), 671–680 (1989)
11. Ford Jr., L.D., Fulkerson, D.R.: Flows in Networks. Princeton University Press, Princeton, NJ (1962)
12. Gartner, N.H., Importa, G. (eds.): Urban Traffic Networks. Dynamic Flow Modeling and Control. Springer, Berlin (1995)
13. Gurak D.T., Caces F.: Migration networks and the shaping of migration systems. In: Kitz M.M., Lim L.L., Zlotnik H. (eds.) International Migration Systems: A Global Approach, pp. 150-176. Clarendon Press, Oxford (1992)
14. Harris, J.R., Todaro, M.P.: Migration, unemployment and development: a two-sector analysis. Am Econ. Rev. **60**, 126–142 (1970)
15. Hotelling, H.: A mathematical theory of migration. Env. Plann. **10**, 1223–1239 (1978)
16. Irwin, J.O.: The place of mathematics in medical and biological sciences. J. R. Stat. Soc. **126**, 1–44 (1963)
17. Irwin, J.O.: The generalized waring distribution applied to accident theory. J. R. Stat. Soc. **131**, 205–225 (1968)

18. Lee, E.S.: A theory of migration. Demography **3**, 47–57 (1966)
19. Lu, J., Yu, X., Chen, G., Yu, W. (eds.) Complex Systems and Networks. Dynamics, Controls and Applications. Springer, Berlin (2016)
20. Marsan, G.A., Bellomo, N., Tosin, A.: Complex Systems and Society: Modeling and Simulation. Springer, New York (2013)
21. Massey, D.S., Arango, J., Hugo, G., Kouaougi, A., Pellegrino, A., Edward, Taylor J.: Theories of international migration: a review and appraisal. Populat. Dev. Rev. **19**, 431–466 (1993)
22. Pastor-Satorras, R., Vespignani, A.: Epidemic dynamics and endemic states in complex networks. Phys. Rev. E. **63**, 066117 (2001)
23. Petrov, V., Nikolova, E., Wolkenhauer, O.: Reduction of nonlinear dynamic systems with an application to signal transduction pathways. IET Syst. Biol. **1**, 2–9 (2007)
24. Puu, T.: Hotelling's migration model revisited. Env. Plann. **23**, 1209–1216 (1991)
25. Schubert, A., Glänzel, W.: A dynamic look at a class of skew distributions. A model with scientometric application. Scientometrics **6**, 149–167 (1984)
26. Simon, J.H.: The Economic Consequences of Migration. The University of Michnigan Press, Ann Arbor, MI (1999)
27. Simon, H.A.: On a class of skew distribution functions. Biometrica **42**, 425–440 (1955)
28. Skeldon, R.: Migration and Development: A Global Perspective. Routledge, London (1992)
29. Vitanov, N.K.: Science Dynamics and Research Production: Indicators, Indexes, Statistical Laws and Mathematical Models. Springer, Cham (2016)
30. Vitanov, N.K., Jordanov, I.P., Dimitrova, Z.I.: On nonlinear dynamics of interacting populations: coupled kink waves in a system of two populations. Commun. Nonlinear Sci. Numer. Simul. **14**, 2379–2388 (2009)
31. Vitanov, N.K., Jordanov, I.P., Dimitrova, Z.I.: On nonlinear population waves. Appl. Mathe. Comput. **215**, 2950–2964 (2009)
32. Vitanov, N.K., Dimitrova, Z.I., Ausloos, M.: Verhulst-Lotka-Volterra model of ideological struggle. Phys. A. **389**, 4970–4980 (2010)
33. Vitanov N.K., Ausloos M., Rotundo G.: Discrete model of ideological struggle accounting for migration. Adv. Complex Syst. **15**, Supplement 1, Article number 1250049 (2012)
34. Vitanov, N.K., Ausloos, M.: Knowledge epidemics and population dynamics models for describing idea diffusion. In: Scharnhorst, A., Börner, K., van den Besselaar, P. (eds.) Knowledge Epidemics and Population Dynamics Models for Describing Idea Diffusion, pp. 69–125. Springer, Berlin (2012)
35. Vitanov, N.K., Dimitrova, Z.I., Vitanov, K.N.: Traveling waves and statistical distributions connected to systems of interacting populations. Comput. Mathe. Appl. **66**, 1666–1684 (2013)
36. Vitanov, N.K., Vitanov, K.N.: Population dynamics in presence of state dependent fluctuations. Comput. Mathe. Appl. **68**, 962–971 (2013)
37. Vitanov, N.K., Ausloos, M.: Test of two hypotheses explaining the size of populations in a system of cities. J. Appl. Statist. **42**, 2686–2693 (2015)
38. Vitanov, N.K., Vitanov, K.N.: Box model of migration channels. Mathe. Soc. Sci. **80**, 108–114 (2016)
39. Vitanov, N.K., Vitanov, K.N., Ivanova, T.: Box model of migration in channels of migration networks. In: Georgiev, K., Todorov, M., Georgiev, I. (eds.) Advanced Computing in Industrial Mathematics. Studies in Computational Intelligence, vol. 728, pp. 203–215. Springer, Cham (2018)
40. Vitanov, N.K., Vitanov, K.N.: On the motion of substance in a channel of network and human migration. Phys. A **490**, 1277–1290 (2018)
41. Weidlich, W., Haag, G. (eds.): Interregional Migration. Dynamic Theory and Comparative Analysis. Springer, Berlin (1988)
42. Willekens, F.J.: Probability models of migration: complete and incomplete data. SA J. Demography **7**, 31–43 (1999)

Hermitian and Pseudo-Hermitian Reduction of the GMV Auxiliary System. Spectral Properties of the Recursion Operators

A. B. Yanovski and T. I. Valchev

Abstract We consider simultaneously two different reductions of a Zakharov-Shabat's spectral problem in pole gauge. Using the concept of gauge equivalence, we construct expansions over the eigenfunctions of the Recursion Operators related to the afore-mentioned spectral problem with arbitrary constant asymptotic values of the potential functions. In doing this, we take into account the discrete spectrum of the scattering operator. Having in mind the applications to the theory of the soliton equations associated to the GMV systems, we show how these expansions modify depending on the symmetries of the functions we expand.

Keywords Gauge-equivalent soliton equations · Recursion Operators
Expansions over adjoint solutions

1 Introduction. The GMV System

We are going to study the auxiliary linear problem

$$\tilde{L}^0\psi = (\mathrm{i}\partial_x - \lambda S)\psi = 0, \qquad \lambda \in \mathbb{C}, \qquad S = \begin{pmatrix} 0 & u & v \\ \varepsilon u^* & 0 & 0 \\ v^* & 0 & 0 \end{pmatrix}, \quad \varepsilon = \pm 1 \quad (1)$$

and the theory of expansions over its adjoint solutions. In the above, the potential functions (u, v) are smooth complex valued functions on $x \in \mathbb{R}$ and * stands for the complex conjugation. In addition, u and v satisfy the relations:

A. B. Yanovski (✉)
Department of Mathematics and Applied Mathematics, University of Cape Town,
Cape Town, South Africa
e-mail: Alexandar.Ianovsky@uct.ac.za

T. I. Valchev
Institute of Mathematics and Informatics, Bulgarian Academy of Sciences,
1113 Sofia, Bulgaria
e-mail: tiv@math.bas.bg

K. Georgiev et al. (eds.), *Advanced Computing in Industrial Mathematics*,
Studies in Computational Intelligence 793,
https://doi.org/10.1007/978-3-319-97277-0_35

$$\varepsilon|u|^2 + |v|^2 = 1, \qquad \lim_{x\to\pm\infty} u(x) = u_\pm, \qquad \lim_{x\to\pm\infty} v(x) = v_\pm. \tag{2}$$

We shall call (1) GMV$_\varepsilon$ system or GMV$_\pm$ system.[1] Thus, GMV$_+$ is the original Gerdjikov-Mikhailov-Valchev system [4] obtained after putting $\varepsilon = +1$ in (1).

As demonstrated in [3, 4, 18], the GMV$_\pm$ system arises naturally when one looks for integrable systems whose Lax operators are subject to Mikhailov-type reductions. Indeed, the Mikhailov reduction group G_0 [13, 14] acting on the fundamental solutions of (1) is generated by g_1 and g_2 defined in the following way:

$$\begin{aligned} g_1(\psi)(x,\lambda) &= \left[Q_\varepsilon \psi(x,\lambda^*)^\dagger Q_\varepsilon\right]^{-1}, \qquad Q_\varepsilon = \mathrm{diag}\,(1,\varepsilon,1), \\ g_2(\psi)(x,\lambda) &= H\psi(x,-\lambda)H, \qquad H = \mathrm{diag}\,(-1,1,1) \end{aligned} \tag{3}$$

where ψ is any fundamental solution to (1) and $\dagger$ denotes Hermitian conjugation. Since $g_1g_2 = g_2g_1$ and $g_1^2 = g_2^2 = \mathrm{id}$, we have that $G_0 = \mathbb{Z}_2 \times \mathbb{Z}_2$. Reduction conditions (3) will be called Hermitian when $\varepsilon = 1$ and pseudo-Hermitian when $\varepsilon = -1$. The requirement that G_0 is a reduction group for the soliton equations related to (1) implies that the coefficients of $\tilde{L}^0$ and the coefficients of the A-operators

$$\tilde{A} = \mathrm{i}\partial_t + \sum_{k=0}^{n} \lambda^k \tilde{A}_k, \qquad \tilde{A}_k \in \mathfrak{sl}\,(3,\mathbb{C}), \tag{4}$$

forming $L - A$ pairs for these soliton equations, must satisfy:

$$\begin{aligned} HSH &= -S, \qquad H\tilde{A}_kH = (-1)^k \tilde{A}_k, \\ Q_\varepsilon S^\dagger Q_\varepsilon &= S, \qquad Q_\varepsilon \tilde{A}_k^\dagger Q_\varepsilon = \tilde{A}_k. \end{aligned} \tag{5}$$

It can be checked that S is diagonalizable, indeed, one has

$$g^{-1}Sg = J_0 \tag{6}$$

where

$$g = \frac{1}{\sqrt{2}} \begin{pmatrix} 1 & 0 & -1 \\ \varepsilon u^* & \sqrt{2}v & \varepsilon u^* \\ v^* & -\sqrt{2}u & v^* \end{pmatrix}, \qquad J_0 = \mathrm{diag}\,(1, 0-1). \tag{7}$$

Following [18], we shall write $SU(\varepsilon)$ referring to $SU(3)$ when $\varepsilon = 1$ and $SU(2,1)$ when $\varepsilon = -1$. Similar convention will apply to the corresponding Lie algebras.

Since $g(x) \in SU(\varepsilon)$ (6) means that $S(x)$ belongs to the adjoint representation orbit

$$\mathcal{O}_{J_0}(\mathrm{SU}\,(\varepsilon)) := \{\tilde{X} \in \mathrm{i}\mathfrak{su}\,(\varepsilon) : \tilde{X} = gJ_0g^{-1}, \quad g \in \mathrm{SU}\,(\varepsilon)\}$$

---

[1] A more general system was derived independently by Golubchik and Sokolov [9].

of SU (ε) passing through J_0.

Our approach to the GMV$_\pm$ system will be based on its gauge equivalence to a generalized Zakharov-Shabat auxiliary system (GZS system) on the algebra $\mathfrak{sl}(3, \mathbb{C})$. The auxiliary system

$$L\psi = (\mathrm{i}\partial_x + q(x) - \lambda J)\,\psi = 0, \qquad \lambda \in \mathbb{C} \tag{8}$$

where $q(x)$ and J belong to some irreducible representation of a simple Lie algebra $\mathfrak{g}$ is called generalized Zakharov-Shabat system (for that representation of $\mathfrak{g}$) in canonical gauge. The element J must be such that the kernel of ad_J ($\mathrm{ad}_J(X) \equiv [J, X]$, $X \in \mathfrak{g}$) is a Cartan subalgebra $\mathfrak{h}_J \subset \mathfrak{g}$ while $q(x)$ belongs to the orthogonal complement $\mathfrak{h}_J^{\perp}$ of $\mathfrak{h}_J$ with respect to the Killing form:

$$\langle X, Y\rangle = \mathrm{tr}\,(\mathrm{ad}_X \mathrm{ad}_Y), \qquad X, Y \in \mathfrak{g}\,. \tag{9}$$

It is also assumed that the smooth function $q(x)$ vanishes sufficiently fast as $x \to \pm\infty$. System (8) is gauge equivalent to the system

$$\tilde{L}\tilde{\Psi} = (\mathrm{i}\partial_x - \lambda S(x))\,\tilde{\Psi} = 0, \qquad S(x) \in \mathcal{O}_J(G) \tag{10}$$

where G is the Lie group corresponding to $\mathfrak{g}$. Usually it is also required that

$$\lim_{x\to\pm\infty} S(x) = J$$

where the convergence is sufficiently fast but as we shall see in our case it will be different. The concept of gauge transformation and gauge equivalent auxiliary problems was applied for the first time in the case of the Heisenberg ferromagnet equation [21] and its gauge equivalent — the nonlinear Schrödinger equation. Later, the the integrable hierarchies, the conservation laws and Hamiltonian structures associated with (8) and (10) have extensively been studied by using the so-called gauge-covariant theory of the Recursion Operators related to the GZS systems in canonical and pole gauge [2, 6, 15]. That approach provides a generalization of classical AKNS approach [1]. We recommend the monograph book [5] for further reading.

So for GZS system in pole gauge most of the essential issues could be reformulated from the canonical gauge. The main difficulty is technical — to explicitly express all the quantities depending on q and its derivatives through S and its derivatives. A clear procedure of how to do that is described in [15]. The reader who is interested in that subject can find more details in [16] regarding GZS related to $\mathfrak{sl}(3, \mathbb{C})$ with no reductions imposed and in [20] regarding the geometry of the Recursion Operators for $\mathfrak{sl}(3, \mathbb{C})$ in general position. We also refer to [19] for the case of GMV system.

In the present report we intend to construct expansions over the eigenfunctions of the Recursion Operators related to (1) with arbitrary constant asymptotic values of the potential functions (u, v). In doing this, we are taking into account the whole spectrum of the scattering operator $\tilde{L}^0$, thus extending in a natural way some of the results published in [18]. Next, we are showing how these expansions modify depending on the symmetries of the functions we expand. We would like to stress on the following:

- We shall be dealing with both $\mathrm{GMV}_{\pm}$ systems simultaneously.
- Our approach will be based on the gauge equivalence we mentioned in the above. Consequently, we shall be able to consider general asymptotic conditions — constant limits $\lim_{x\to\pm\infty} u$ and $\lim_{x\to\pm\infty} v$.
- Our point of view on the Recursion Operators when reductions are present is somewhat different from that adopted in [3].
- We show some new algebraic features in the spectral theory.

2 Gauge-Equivalent Systems

As mentioned, our approach to the $\mathrm{GMV}_{\pm}$ system will be based on the fact that it is gauge equivalent to a GZS system on $\mathfrak{sl}(3,\mathbb{C})$ and will follow some of the ideas of [8, 17]. We presented our results for the case of the continuous spectrum in [18]. Here, we shall include also the discrete spectrum. In fact, we have the following basic result:

Theorem 1 *The* $\mathrm{GMV}_{\pm}$ *system is gauge equivalent to a canonical* GZS *linear problem on* $\mathfrak{sl}(3,\mathbb{C})$

$$L^0\psi = (\mathrm{i}\partial_x + q - \lambda J_0)\psi = 0 \tag{11}$$

subject to a Mikhailov reduction group generated by the two elements h_1 *and* h_2*. For a fundamental solution* ψ *of system* (11) *we have:*

$$\begin{aligned} h_1(\psi)(x,\lambda) &= \left[Q_\varepsilon \psi(x,\lambda^*)^\dagger Q_\varepsilon\right]^{-1}, \qquad Q_\varepsilon = diag\,(1,\varepsilon,1), \quad \varepsilon = \pm 1, \\ h_2(\psi)(x,\lambda) &= K\psi(x,-\lambda)K. \end{aligned}$$

Since $h_1^2 = h_2^2 = id$ *and* $h_1h_2 = h_2h_1$ *we have again a* $\mathbb{Z}_2\times\mathbb{Z}_2$ *reduction. In the above*

$$K = \begin{pmatrix} 0 & 0 & 1 \\ 0 & 1 & 0 \\ 1 & 0 & 0 \end{pmatrix}.$$

Proof Indeed, it is enough to put $q = \mathrm{i}\psi_0^{-1}(\psi_0)_x$ where

$$\psi_0 = \exp\left[-\mathrm{i}J'\int_{-\infty}^{x} b(y)\mathrm{d}y\right]g^{-1}.$$

In the above expression $J' = \mathrm{diag}\,(1, -2, 1), b(x) = \mathrm{i}(\varepsilon u u_x^* + v v_x^*)/2$ (note that this expression is real) and g, J_0 are the same as in (7). Then ψ_0 is a solution to (11) for $\lambda = 0$ and GMV$_\varepsilon$ is gauge-equivalent to GZS. □

Also, one gets the following important formulas:

$$K\psi_0 = \psi_0 H, \qquad \psi_0^{-1} K = H\psi_0^{-1}. \tag{12}$$

In order to continue we shall need some simple algebraic facts.

3 Algebraic Preliminaries

The reductions we hsd in the above have clear algebraic meaning: $h(X) = HXH$, $k(X) = KXK$ are obviously involutive automorphisms of the algebra $\mathfrak{sl}\,(3, \mathbb{C})$ and $\sigma_\varepsilon X = -Q_\varepsilon X^\dagger Q_\varepsilon$ defines a complex conjugation of the same algebra. As it is known $\mathfrak{sl}\,(3, \mathbb{C})$ is a simple Lie algebra, the canonical choice for its Cartan subalgebra $\mathfrak{h}$ is the subalgebra of the diagonal matrices. It is also equal to $\mathfrak{h}_J = \ker \mathrm{ad}_J$ where J is any diagonal matrix $\mathrm{diag}\,(\lambda_1, \lambda_2, \lambda_3)$ with distinct λ_i's. In that case we shall call $\mathfrak{h}_J$ *the Cartan subalgebra* and denote it simply by $\mathfrak{h}$ (in particular, we have $\mathfrak{h}_{J_0} = \mathfrak{h}$). More generally, if S is diagonalizable with distinct eigenvalues then $\mathfrak{h}_S = \ker \mathrm{ad}_S$ is also a Cartan subalgebra. We shall denote the projection onto the orthogonal complement $\mathfrak{h}^\perp = \mathfrak{h}_J^\perp$ of $\mathfrak{h} = \mathfrak{h}_J$ (with respect to the Killing form) by $\pi_0 = \pi_J$ when J is diagonal and the projection onto the orthogonal complement $\mathfrak{h}_S^\perp$ of $\mathfrak{h}_S$ by π_S when S is diagonalizable. One can introduce the system of roots Δ, the systems of positive and negative roots $\Delta_\pm$ in a canonical way. The set Δ_+ contains α_1, α_2 and $\alpha_3 = \alpha_1 + \alpha_2$, the Cartan-Weil basis shall be denoted by $E_{\pm\alpha_i}$, H_1, H_2 etc. We use the notation and normalizations used in the well known monograph on semisimple Lie algebras [10]. The matrices H_1, H_2 span $\mathfrak{h}$ and the matrices $E_\alpha, \alpha \in \Delta$ span $\mathfrak{h}^\perp$. The complex conjugation σ_ε defines the real form $\mathfrak{su}\,(3)$ $(\varepsilon = +1)$ or the real form $\mathfrak{su}\,(2, 1)$ $(\varepsilon = -1)$ of $\mathfrak{sl}\,(3, \mathbb{C})$. If we introduce the spaces

$$\tilde{\mathfrak{g}}^{[j]} = \{X : h(X) = (-1)^j X\}, \qquad j = 0; 1 \bmod (2) \tag{13}$$

then we shall have the orthogonal splittings

$$\begin{aligned} &\mathfrak{sl}\,(3, \mathbb{C}) = \tilde{\mathfrak{g}}^{[0]} \oplus \tilde{\mathfrak{g}}^{[1]}, \\ &\mathfrak{su}\,(\varepsilon) = (\tilde{\mathfrak{g}}^{[0]} \cap \mathfrak{su}\,(\varepsilon)) \oplus (\tilde{\mathfrak{g}}^{[1]} \cap \mathfrak{su}\,(\varepsilon)). \end{aligned} \tag{14}$$

In order to explain our results we shall also need the action $\mathscr{K}$ of $k : k(X) = KXK$ on the roots:

$$\mathscr{K}(\pm\alpha_1) = \mp\alpha_2, \qquad \mathscr{K}(\pm\alpha_3) = \mp\alpha_3, \qquad \mathscr{K}(\pm\alpha_2) = \mp\alpha_1$$

so we have $k(E_\alpha) = E_{\mathscr{K}\alpha}$. We also note that we have the following relations which are used in all calculations:

$$h \circ \operatorname{ad}_S = -\operatorname{ad}_S \circ h, \qquad \sigma_\varepsilon \circ \operatorname{ad}_S = -\operatorname{ad}_S \circ \sigma_\varepsilon. \tag{15}$$

Another issue we must discuss is the relation between h from one side and ad_S^{-1} and π_S from the other. Here ad_S^{-1} is defined only on the space $\mathfrak{h}_S^\perp$ but one could extend it as zero on $\mathfrak{h}_S$ which we shall always assume. One obtains that

$$\operatorname{ad}_S^{-1} \circ h = -h \circ \operatorname{ad}_S^{-1}, \qquad \pi_S \circ h = h \circ \pi_S, \tag{16}$$

$$\operatorname{ad}_S^{-1} \circ \sigma_\varepsilon = -\sigma_\varepsilon \circ \operatorname{ad}_S^{-1}, \qquad \pi_S \circ \sigma_\varepsilon = \sigma_\varepsilon \circ \pi_S. \tag{17}$$

4 Recursion Operators of GMV$_\varepsilon$

Recursion Operators (also called Generating Operators or Λ-operators) are theoretical tools that permit:

- To describe the hierarchies of the nonlinear evolution equations (NLEEs) related to the auxiliary linear problems of GZS type (the AKNS approach [1]).
- To describe the hierarchies of conservation laws for these NLEEs.
- To describe the hierarchies of compatible Hamiltonian structures of these NLEEs.
- The expansions over their eigenfunctions permit to interpret the inverse scattering problems for GZS systems as generalized Fourier transforms, **see** [2, 7, 11].
- Recursion operators have important geometric interpretation — the NLEEs could be viewed as fundamental fields of a Poisson-Nijenhuis structure on the infinite dimensional manifold of "potentials", a concept introduced in [12].

For all these aspects of Recursion Operators see also the monograph book [5] which contains an extensive bibliography for publications prior to 2008.

The Recursion Operators $\tilde{\Lambda}_\pm$ arise naturally when one tries to find the hierarchy of Lax pairs related to a particular auxiliary GZS linear problem. Assume this problem has the form $\tilde{L} = \mathrm{i}\partial_x - \lambda S$ where S is in the orbit of the element J_0 with no additional assumptions on S and we have that $\tilde{L} = \psi_0^{-1} L^0 \psi_0$ where L^0 is a GZS system with $J = J_0$ and ψ_0 is a solution to (11) for $\lambda = 0$. We have

$$\tilde{\Lambda}_\pm = \operatorname{Ad}(\psi_0^{-1}) \circ \Lambda_\pm \circ \operatorname{Ad}(\psi_0)$$

where $\Lambda_\pm$ are the Recursion Operators for L^0, see [16]. The explicit form of $\tilde{\Lambda}_\pm$ is

$$\tilde{\Lambda}_\pm(Z) = \mathrm{i}\operatorname{ad}_S^{-1}\pi_S \left\{ \partial_x Z + \frac{S_x}{12} \int_{\pm\infty}^{x} \langle Z, S_y \rangle \mathrm{d}y + \frac{S_{1x}}{4} \int_{\pm\infty}^{x} \langle Z, S_{1y} \rangle \mathrm{d}y \right\}$$

where $S_1 = S^2 - \frac{2}{3}\mathbf{1}$ and $S_{1x} = (S_1)_x$. GMV$_\varepsilon$ is a particular case of a $\mathfrak{sl}(3)$ problem so the operators $\tilde{\Lambda}_\pm$ are the Recursion Operators (or Generating Operators) for GMV$_\pm$ system and give the corresponding NLEEs. However, one must be a little more cautious here if one wants to obtain those NLEEs that are compatible with the reduction group. Indeed, the Lax pairs that obey the reductions give hierarchies of equations that have the form:

$$\operatorname{ad}_S^{-1}\partial_t S = \sum_{k=0}^{r} a_{2k}(\tilde{\Lambda}_\pm)^{2k}\operatorname{ad}_S^{-1}(S_x) + \sum_{k=1}^{m} a_{2k-1}(\tilde{\Lambda}_\pm)^{2k-1}\operatorname{ad}_S^{-1}(S_{1x}) \quad (18)$$

where a_i are some real constants. We shall not enter in more details here, see [18] for this, but one can see that when one considers the hierarchies of the NLEEs the next equation in the hierarchy is obtained not using $\tilde{\Lambda}_\pm$ but $\tilde{\Lambda}_\pm^2$. So one needs to understand what happens with the expansions that play role of generalized Fourier transform.

5 Spectral Theory of the Recursion Operators

The properties of the fundamental analytic solutions (FAS) of the GZS systems play a paramount role in the spectral theory of such systems. In fact, from the canonical FAS (denoted by $\chi^\pm$) in canonical gauge one immediately obtains FAS in the pole gauge $\tilde{\chi}^\pm$ (with the same analytic properties) [5]. In our case we have $\tilde{\chi}^\pm(x,\lambda) = \psi_0^{-1}\chi^\pm(x,\lambda)$. The superscripts $\pm$ mean that the corresponding solution is analytic in $\mathbb{C}_\pm$ (upper and lower half-plane). For these solutions one has

Theorem 2 *The FAS $\tilde{\chi}^\pm(x,\lambda)$ corresponding to the GMV$_\varepsilon$ system satisfy*[2]*:*

$$Q_\varepsilon(\tilde{\chi}^\pm(x,\lambda^*))^\dagger Q_\varepsilon = (\tilde{\chi}^\mp(x,\lambda))^{-1}, \qquad H\tilde{\chi}^\pm(x,\lambda)H = \tilde{\chi}^\mp(x,-\lambda)KH. \quad (19)$$

Further, one builds the so-called adjoint solutions (or generalized exponents) for the GZS systems:

- GZS system in canonical gauge: $\mathbf{e}_\alpha^\pm = \pi_0\chi^\pm E_\alpha(\chi^\pm)^{-1}$.
- GZS system in pole gauge: $\tilde{\mathbf{e}}_\alpha^\pm = \pi_S\tilde{\chi}^\pm E_\alpha(\tilde{\chi}^\pm)^{-1}$.

One sees that $\tilde{\mathbf{e}}_\alpha^\pm = \operatorname{Ad}(\psi_0^{-1})\mathbf{e}_\alpha^\pm$ and then the fact that they are eigenfunctions of $\tilde{\Lambda}_\pm$ and the completeness relations for them become immediate from the classical results for the Recursion Operators in canonical gauge. Indeed, first

$$\tilde{\Lambda}_-(\tilde{\mathbf{e}}_\alpha^+(x,\lambda)) = \lambda\tilde{\mathbf{e}}_\alpha^+(x,\lambda), \qquad \tilde{\Lambda}_-(\tilde{\mathbf{e}}_{-\alpha}^-(x,\lambda)) = \lambda\tilde{\mathbf{e}}_{-\alpha}^-(x,\lambda),$$
$$\tilde{\Lambda}_+(\tilde{\mathbf{e}}_{-\alpha}^+(x,\lambda)) = \lambda\tilde{\mathbf{e}}_{-\alpha}^+(x,\lambda), \qquad \tilde{\Lambda}_+(\tilde{\mathbf{e}}_\alpha^-(x,\lambda)) = \lambda\tilde{\mathbf{e}}_\alpha^-(x,\lambda)$$

[2]In [3, 4] FAS have mistakenly been claimed to satisfy $H\tilde{\chi}^\pm(x,\lambda)H = \tilde{\chi}^\mp(x,-\lambda)$.

and the completeness relations could be written into the following useful form [7]:

$$\delta(x-y)\tilde{P}_0 = \mathrm{DSC_p} + \frac{1}{2\pi}\int_{-\infty}^{\infty}\left[\sum_{\alpha\in\Delta_+}\tilde{\mathbf{e}}^+_{\alpha}(x,\lambda)\otimes\tilde{\mathbf{e}}^+_{-\alpha}(y,\lambda) - \tilde{\mathbf{e}}^-_{-\alpha}(x,\lambda)\otimes\tilde{\mathbf{e}}^-_{\alpha}(y,\lambda)\right]\mathrm{d}\lambda$$

where $\mathrm{DSC_p}$ is the discrete spectrum contribution. The second term is the continuous spectrum contribution which we denote by $\mathrm{CSC_p}$. Also, in the above

$$\tilde{P}_0 = \sum_{\alpha\in\Delta}\frac{1}{\alpha(J_0)}(\tilde{E}_{\alpha}\otimes\tilde{E}_{-\alpha}), \qquad \tilde{E}_{\alpha} = \mathrm{Ad}\,(\psi_0^{-1})E_{\alpha} = \psi_0^{-1}E_{\alpha}\psi_0,$$
$$\tilde{\mathbf{e}}^{\pm}_{\alpha} = \mathrm{Ad}\,(\psi_0^{-1})\mathbf{e}^{\pm}_{\alpha}\,.$$

For $\mathrm{DSC_p}$, assuming that one has N^+ poles λ_i^+, $1\leq i\leq N^+$ in the upper half-plane $\mathbb{C}_+$ and N^- poles λ_i^-, $1\leq i\leq N^-$ in the lower half-plane $\mathbb{C}_-$, we get

$$\begin{aligned}
&\mathrm{DSC}_p = \\
&-\mathrm{i}\sum_{\alpha\in\Delta_+}\sum_{k=1}^{N^+}\mathrm{Res}\,(\tilde{Q}^+_{\alpha}(x,y,\lambda);\lambda_k^+) - \mathrm{i}\sum_{\alpha\in\Delta_+}\sum_{k=1}^{N^-}\mathrm{Res}\,(\tilde{Q}^-_{-\alpha}(x,y,\lambda);\lambda_k^-),\\
&\tilde{Q}^+_{\alpha}(x,y,\lambda) = \tilde{\mathbf{e}}^+_{\alpha}(x,\lambda)\otimes\tilde{\mathbf{e}}^+_{-\alpha}(y,\lambda), \qquad \mathrm{Im}(\lambda) > 0,\\
&\tilde{Q}^-_{-\alpha}(x,y,\lambda) = \tilde{\mathbf{e}}^-_{-\alpha}(x,\lambda)\otimes\tilde{\mathbf{e}}^-_{\alpha}(y,\lambda), \qquad \mathrm{Im}(\lambda) < 0.
\end{aligned} \tag{20}$$

6 Λ-Operators and Reductions

Let us see now what are the implications of the reductions on the expansions over adjoint solutions. We start with the reduction defined by h. Since for the FAS we have the properties stated in Theorem 2, for $\beta\in\Delta$ we obtain

$$h(\tilde{\mathbf{e}}^{\pm}_{\beta}(x,\lambda)) = \tilde{\mathbf{e}}^{\mp}_{\mathscr{K}\beta}(x,-\lambda).$$

Changing the variable λ to $-\lambda$, taking into account that $\mathscr{K}$ maps the positive roots into the negative ones and vice versa, we obtain after some algebraic transformations

$$\mathrm{CSC_p} = \frac{A_h}{2\pi}\int_{-\infty}^{\infty}\left[\sum_{\alpha\in\Delta_+}\tilde{\mathbf{e}}^+_{\alpha}(x,\lambda)\otimes\tilde{\mathbf{e}}^+_{-\alpha}(y,\lambda) - \tilde{\mathbf{e}}^-_{-\alpha}(x,\lambda)\otimes\tilde{\mathbf{e}}^-_{\alpha}(y,\lambda)\right]\mathrm{d}\lambda$$

where

$$A_h = \frac{1}{2}(\mathrm{id} - h \otimes h). \tag{21}$$

Let us explain what means the presence of the "multiplier" A_h. For simplicity let us first assume we have only continuous spectrum.

Assume $\tilde{Z}(x)$ is such that $h(\tilde{Z}) = \tilde{Z}$ and let us make a contraction first to the right followed by integration over y and to the the left followed by integration over x. Then taking into account that h is an automorphism and the Killing form is invariant under automorphisms we get

$$\tilde{Z}(x) = \frac{1}{2\pi} \int_{-\infty}^{\infty} \left[\sum_{\alpha \in \Delta_+} \tilde{\mathbf{s}}_\alpha^\eta (x, \lambda) \mu_\alpha^\eta - \tilde{\mathbf{s}}_{-\alpha}^{-\eta}(x, \lambda)) \mu_\alpha^{-\eta} \right] \mathrm{d}\lambda$$

where $\eta = +$ ($\eta = -$) depending whether we contract to the left or to the right and

$$\begin{aligned}
\mu_\alpha^\eta &= \langle\langle \tilde{\mathbf{a}}_{-\alpha}^\eta, [S, \tilde{Z}] \rangle\rangle, \quad \mu_\alpha^{-\eta} = \langle\langle \tilde{\mathbf{a}}_\alpha^\eta, [S, \tilde{Z}] \rangle\rangle, \\
\tilde{\mathbf{s}}_{\pm\alpha}^\eta (x, \lambda) &= \frac{1}{2} \left(\tilde{\mathbf{e}}_{\pm\alpha}^\eta (x, \lambda) + h(\tilde{\mathbf{e}}_{\pm\alpha}^\eta (x, \lambda)) \right), \\
\tilde{\mathbf{a}}_{\pm\alpha}^\eta (x, \lambda) &= \frac{1}{2} \left(\tilde{\mathbf{e}}_{\pm\alpha}^\eta (x, \lambda) - h(\tilde{\mathbf{e}}_{\pm\alpha}^\eta (x, \lambda)) \right),
\end{aligned}$$

and for two functions $\tilde{Z}_1(x)$, $\tilde{Z}_2(x)$ with values in $\mathfrak{sl}(3)$ we used the notation

$$\langle\langle \tilde{Z}_1, \tilde{Z}_2 \rangle\rangle = \int_{-\infty}^{+\infty} \langle \tilde{Z}_1(x), \tilde{Z}_2(x) \rangle \mathrm{d}x.$$

If instead of $h(\tilde{Z}) = \tilde{Z}$ we assume that $h(\tilde{Z}) = -\tilde{Z}$ then in the same manner we shall obtain expansions over the functions $\tilde{\mathbf{a}}_\alpha^\eta$ and the coefficients are calculated via the functions $\tilde{\mathbf{s}}_\alpha^\eta$. Since $(\mathrm{id} \pm h)/2$ are in fact projectors onto the ± 1 eigenspaces of h

$$h(\tilde{\mathbf{s}}_{\pm\alpha}^\eta (x, \lambda)) = \tilde{\mathbf{s}}_{\pm\alpha}^\eta (x, \lambda), \qquad h(\tilde{\mathbf{a}}_{\pm\alpha}^\eta (x, \lambda)) = -\tilde{\mathbf{a}}_{\pm\alpha}^\eta (x, \lambda).$$

Thus, in case $h(\tilde{Z}) = \tilde{Z}$ or $h(\tilde{Z}) = -\tilde{Z}$ the expansions could be written in terms of new sets of adjoint solutions, $\tilde{\mathbf{s}}_{\pm\alpha}^\eta$ or $\tilde{\mathbf{a}}_{\pm\alpha}^\eta$ that reflect the symmetry of $\tilde{Z}$. For $\alpha \in \Delta_+$ one obtains that

$$\begin{aligned}
\tilde{\Lambda}_-(\tilde{\mathbf{s}}_\alpha^+ (x, \lambda)) &= \lambda \tilde{\mathbf{a}}_\alpha^+ (x, \lambda), \qquad & \tilde{\Lambda}_-(\tilde{\mathbf{s}}_{-\alpha}^- (x, \lambda)) &= \lambda \tilde{\mathbf{a}}_{-\alpha}^- (x, \lambda), \\
\tilde{\Lambda}_-(\tilde{\mathbf{a}}_\alpha^+ (x, \lambda)) &= \lambda \tilde{\mathbf{s}}_\alpha^+ (x, \lambda), \qquad & \tilde{\Lambda}_-(\tilde{\mathbf{a}}_{-\alpha}^- (x, \lambda)) &= \lambda \tilde{\mathbf{s}}_{-\alpha}^- (x, \lambda)
\end{aligned}$$

and also

$$\begin{aligned}
\tilde{\Lambda}_+(\tilde{\mathbf{s}}_{-\alpha}^+ (x, \lambda)) &= \lambda \tilde{\mathbf{a}}_{-\alpha}^+ (x, \lambda), \qquad & \tilde{\Lambda}_+(\tilde{\mathbf{s}}_\alpha^- (x, \lambda)) &= \lambda \tilde{\mathbf{a}}_\alpha^- (x, \lambda), \\
\tilde{\Lambda}_+(\tilde{\mathbf{a}}_{-\alpha}^+ (x, \lambda)) &= \lambda \tilde{\mathbf{s}}_{-\alpha}^+ (x, \lambda), \qquad & \tilde{\Lambda}_+(\tilde{\mathbf{a}}_\alpha^- (x, \lambda)) &= \lambda \tilde{\mathbf{s}}_\alpha^- (x, \lambda).
\end{aligned}$$

One sees that the functions in the expansions when we have some symmetry with respect to h are eigenfunctions for $\tilde{\Lambda}_{-}^{2}$ ($\tilde{\Lambda}_{+}^{2}$) with eigenvalue λ^2. This together with the fact that when recursively finding the coefficients for the Lax pairs one effectively uses $\tilde{\Lambda}_{+}^{2}$ leads to the interpretation that in case we have $\mathbb{Z}_2$ reduction defined by h the role of the generating operator is played by $\tilde{\Lambda}_{\pm}^{2}$.

All this happens because of the new form of the expansions, involving the "multiplier" $A_h = (\mathrm{id} + h \otimes h)/2$. The point is that the "multiplier" A_h has simple algebraic meaning:

Theorem 3 *The operator* $A_h = (id + h \otimes h)/2$ *(acting on* $\mathfrak{g} \otimes \mathfrak{g}$ *where* $\mathfrak{g} = \mathfrak{sl}(3, \mathbb{C})$*) is a projector onto the space*

$$V = (\tilde{\mathfrak{g}}^{[0]} \otimes (\tilde{\mathfrak{g}}^{[1]}) \oplus ((\tilde{\mathfrak{g}}^{[1]} \otimes (\tilde{\mathfrak{g}}^{[0]}).$$

Consequently, when for $B \in V$ one makes a contraction (from the right or from the left) with $[S, X]$ where X is in $\tilde{\mathfrak{g}}^{[s]}$, then $[S, X] \in \tilde{\mathfrak{g}}^{[s+1]}$ and $B\,[S, X] \in \tilde{\mathfrak{g}}^{[s]}$.

Let us consider now the discrete spectrum term more closely. For a GZS system in pole gauge in general position one has N^+ poles λ_i^+, $1 \le i \le N^+$ in the upper half-plane $\mathbb{C}_+$ and N^- poles λ_i^-, $1 \le i \le N^-$ in the lower half-plane $\mathbb{C}_-$. If we have reduction defined by h we see that we must have $N^+ = N^-$ since if $\tilde{\mathbf{e}}_\alpha^+(x, \lambda)$ has a pole of some order at $\lambda = \lambda_s^+$ in $\mathbb{C}_+$ then $\tilde{\mathbf{e}}_{\mathscr{K}\alpha}^-(x, \lambda)$ will have the same type of singularity at $-\lambda_s^+$ in $\mathbb{C}_-$. In order to simplify the notation we shall put $\lambda_s^+ = \lambda_s$, $\lambda_s^- = -\lambda_s$ and $N^+ = N^- = N$. Of course, in order to make concrete calculations one needs some assumption on the discrete spectrum. Assume that all the singularities are simple poles, let us consider the contribution from $\tilde{Q}_\beta(\lambda)$ for a fixed β and two poles: one pole $\lambda = \lambda_0$ located in $\mathbb{C}_+$ and one pole $\lambda = -\lambda_0$ located in $\mathbb{C}_-$. Then for $\beta \in \Delta$ in some discs around λ_0 and $-\lambda_0$ we have the Laurent expansions that hold uniformly on x:

$$\tilde{\mathbf{e}}_\beta^+(x, \lambda) = \frac{\tilde{A}_\beta^+(x)}{\lambda - \lambda_0} + \tilde{B}_\beta^+(x) + \tilde{C}_\beta^+(x)(\lambda - \lambda_0) + \cdots, \tag{22}$$

$$\tilde{\mathbf{e}}_{\mathscr{K}\beta}^-(x, \lambda) = \frac{\tilde{A}_{\mathscr{K}\beta}^-(x)}{\lambda + \lambda_0} + \tilde{B}_{\mathscr{K}\beta}^-(x) + \tilde{C}_{\mathscr{K}\beta}^-(x)(\lambda + \lambda_0) + \cdots \tag{23}$$

From the properties of the FAS we see that for $\beta \in \Delta$, $h(\tilde{\mathbf{e}}_\beta^\pm(x, \lambda)) = \tilde{\mathbf{e}}_{\mathscr{K}\beta}^\mp(x, -\lambda)$ and consequently

$$h\tilde{A}_\beta^+(x) = -\tilde{A}_{\mathscr{K}\beta}^-(x), \quad h\tilde{B}_\beta^+(x) = \tilde{B}_{\mathscr{K}\beta}^-(x), \quad h\tilde{C}_\beta^+(x) = -\tilde{C}_{\mathscr{K}\beta}^-(x). \tag{24}$$

For $\alpha \in \Delta_+$ the calculation gives

$$\mathrm{Res}\,(\tilde{Q}_\beta^+(x, y, \lambda);\ \lambda_0) = \tilde{A}_\beta^+(x) \otimes \tilde{B}_{-\beta}^+(y) + \tilde{B}_\beta^+(x) \otimes \tilde{A}_{-\beta}^+(y) \tag{25}$$

but since the singularities occur in pairs we can combine the contributions from λ_0 and $-\lambda_0$. After performing it, we put the formula for $\mathrm{DSC_p}$ into a form in which the poles in the upper and lower half-plane play equal role introducing the notation

$$\lambda_i = \lambda_i^+, \qquad \lambda_{i+N} = \lambda_i^- = -\lambda_i^+, \qquad 1 \leq i \leq N.$$

Then one could write

$$\mathrm{DSC_p} = -A_h \sum_{\alpha \in \Delta_+} \sum_{s=1}^{2N} \mathrm{i}\,\mathrm{Res}\,(\tilde{Q}_\alpha(x, y, \lambda)\,;\,\lambda_s). \qquad (26)$$

Now, making contractions to the right (left) by $[S, \tilde{Z}]$ and integrating one gets the discrete spectrum contribution to the expansion of a given function $\tilde{Z}$:

$$\mathrm{DSC_p}([S, \tilde{Z}]) = \qquad (27)$$
$$-2\mathrm{i} \sum_{\alpha \in \Delta_+} \sum_{k=1}^{N} \tilde{A}_{\varepsilon\alpha,k}^{[+;s]}(x)\langle\langle \tilde{B}_{-\varepsilon\alpha,k}^{[+;s+1]}, [S, \tilde{Z}]\rangle\rangle + \tilde{B}_{\varepsilon\alpha,k}^{[+;s]}(x)\langle\langle \tilde{A}_{-\varepsilon\alpha,k}^{[+;s+1]}, [S, \tilde{Z}]\rangle\rangle$$

for $\varepsilon = \pm 1$ (depending on what side we contracted). In the above

$$\tilde{A}_\beta^{[+;s]}(y) = \frac{1}{2}(\mathrm{id} + (-1)^s h)\tilde{A}_\beta^+(y), \qquad \tilde{B}_\beta^{[+;s]}(y) = \frac{1}{2}(\mathrm{id} + (-1)^s h)\tilde{B}_\beta^+(y) \qquad (28)$$

and s is understood modulo 2. The action of the Recursion Operators on the discrete spectrum is not hard to find. For for $\beta \in \Delta$ and the coefficients of the expansion about $\lambda = \lambda_0$ of $\tilde{\mathbf{e}}_\beta^+(x, \lambda)$ we get:

$$\tilde{\Lambda}_\pm \tilde{A}_\beta^+(x) = \lambda_0 \tilde{A}_\beta^+(x), \qquad \tilde{\Lambda}_\pm \tilde{B}_\beta^+(x) = \lambda_0 \tilde{B}_\beta^+(x) + \tilde{A}_\beta^+(x). \qquad (29)$$

Consider now the space $\tilde{V}_\beta^+$ spanned by the vectors $\tilde{A}_\beta^+$, $\tilde{B}_\beta^+$. Of course, we must have $\tilde{A}_\beta^+ \neq 0$, otherwise there is no singularity. One sees that also $\tilde{B}_\beta^+ \neq 0$. Next, one checks immediately that the above relations could be true only if $\tilde{A}_\beta^+$, $\tilde{B}_\beta^+$ are linearly independent, so $\tilde{V}_\beta^+$ has dimension 2 and the matrix of $\tilde{\Lambda}_\pm$ in the basis $\tilde{A}_\beta^+$, $\tilde{B}_\beta^+$ consists of 2×2 Jordan block having λ_0 on the diagonal.

The situation with the spaces $V_{\nu,\beta}^{[+;s]}$ spanned by the vectors $A_\beta^{[+;s]}$, $B_\beta^{[+;s]} \neq 0$ is very similar but slightly more complicated. Indeed, since $\tilde{\Lambda}_\pm \circ h = -h \circ \tilde{\Lambda}_\pm$ we obtain that

$$\tilde{\Lambda}_\pm \tilde{A}_\beta^{[+;s]}(x) = \lambda_0 \tilde{A}_\beta^{[+;s+1]}(x), \qquad (30)$$
$$\tilde{\Lambda}_\pm \tilde{B}_\beta^{[+;s]}(x) = \lambda_0 \tilde{B}_\beta^{[+;s+1]}(x) + \tilde{A}_\beta^{[+;s+1]}(x) \qquad (31)$$

and therefore

$$\tilde{\Lambda}_{\pm}^{2}\tilde{A}_{\beta}^{[+;s]}(x) = \lambda_0^2\tilde{A}_{\beta}^{[+;s]}(x), \tag{32}$$

$$\tilde{\Lambda}_{\pm}^{2}\tilde{B}_{\beta}^{[+;s]}(x) = \lambda_0^2\tilde{B}_{\beta}^{[+;s]}(x) + 2\lambda_0\tilde{A}_{\beta}^{[+;s]}(x). \tag{33}$$

We have the following options:

- $\tilde{A}_{\beta}^{[+;s]} \neq 0$. Then one sees that $\tilde{B}_{\beta}^{[+;s]} \neq 0$ and $\tilde{A}_{\beta}^{[+;s]}$, $\tilde{B}_{\beta}^{[+;s]}$ must be linearly independent, $\tilde{V}_{\beta}^{[+;s]}$ has dimension 2 and the matrix of $\tilde{\Lambda}_{\pm}^{2}$ in the basis $2\lambda_0\tilde{A}_{\beta}^{[+;s]}$, $\tilde{B}_{\beta}^{[+;s]}$ consists of 2×2 Jordan block having λ_0^2 on the diagonal.
- $\tilde{A}_{\beta}^{[+;s]} = 0$. Then if $\tilde{B}_{\beta}^{[+;s]} \neq 0$ the space $\tilde{V}_{\beta}^{[+;s]}$ is one dimensional and it is an eigenspace with eigenvalue λ_0^2.
- If $\tilde{A}_{\beta}^{[+;s]} = \tilde{B}_{\beta}^{[+;s]} = 0$ then $\tilde{V}_{\beta}^{[+;s]} = 0$.

In all the cases we see that for reduction defined by h the spaces $\tilde{V}_{\beta}^{[+;s]}$ are not invariant under the action of $\tilde{\Lambda}_{\pm}$ but are invariant under the action of $\tilde{\Lambda}_{\pm}^{2}$. This will happen, of course, when we consider the contribution from all the poles given by the expression (27).

We have the same effect from the reduction defined by the complex conjugation σ_ε. Both for the continuous and for the discrete spectrum we obtain

$$\mathrm{CSC_p} = A_{\sigma_\varepsilon}\mathrm{CSC_p}, \qquad \mathrm{DSC_p} = A_{\sigma_\varepsilon}\mathrm{DSC_p}$$

where

$$A_{\sigma_\varepsilon} = \frac{1}{2}(\mathrm{id} - \sigma_\varepsilon \otimes \sigma_\varepsilon). \tag{34}$$

Finally, if the reductions defined by h and σ_ε act simultaneously then

$$\mathrm{CSC_p} = A_h A_{\sigma_\varepsilon}\mathrm{CSC_p}, \qquad \mathrm{DSC_p} = A_h A_{\sigma_\varepsilon}\mathrm{DSC_p} \tag{35}$$

where A_h and A_{σ_ε} are as in (21) and (34). Note that A_h and A_{σ_ε} commute. Of course, in the case of the complex conjugation σ_ε the role of ''symmetric'' w.r.t. the action of h is taken by "real" functions with respect to σ_ε and the role of "anti-symmetric" w.r.t. the action of h is taken by "imaginary" functions with respect to σ_ε **etc**. However, both in the case of one $\mathbb{Z}_2$ reduction (h), and in the case of $\mathbb{Z}_2 \times \mathbb{Z}_2$ reduction (h and σ_ε) the role played previously by the operators $\tilde{\Lambda}_{\pm}$ is played by $\tilde{\Lambda}_{\pm}^{2}$ – the square of these operators.

7 Conclusion

We have already discussed that effectively the operators "shifting" the equations along the hierarchies of NLEEs (18) are $\tilde{\Lambda}_{\pm}^{2}$. We have showed that when one uses expansions over adjoint solutions to investigate these evolution equations, then according to the symmetry of the r.h.s. with respect to h and σ_ε the expansions modify depending on the symmetries of the functions we expand. So these expansions are over the eigenfunctions of $\tilde{\Lambda}^2_\pm$ and in the generalized Fourier expansions the role previously played by $\tilde{\Lambda}_\pm$ is played now by $\tilde{\Lambda}^2_\pm$.

Acknowledgements The work has been supported by the NRF incentive grant of South Africa and grant DN 02–5 of Bulgarian Fund "Scientific Research".

References

1. Ablowitz, M.J., Kaup, D.J., Newell, A.C., Segur, H.: The inverse scattering problem - Fourier analysis for nonlinear problems. Studies in Appl. Math. **53**, 249–315 (1974)
2. Gerdjikov, V.S.: Generalized Fourier transforms for the soliton equations. Gauge-covariant formulation. Inverse Problems **2**, 51–74 (1986)
3. Gerdjikov, V.S., Grahovski, G.G., Mikhailov, A.V., Valchev, T.I.: Polynomial bundles and generalized Fourier transforms for integrable equations on A.III-type symmetric spaces. Symmetry, Integrability and Geometry: Methods and Applications (SIGMA) **7**, 096 (2011)
4. Gerdjikov, V.S., Mikhailov, A.V., Valchev, T.I.: Reductions of integrable equations on A. III-symmetric spaces. J. Phys. A: Math. Theor. **43**, 434015 (2010)
5. Gerdjikov, V.S., Vilasi, G., Yanovski, A.B.: Integrable Hamiltonian Hierarchies – Spectral and Geometric Methods. Springer, Heidelberg (2008)
6. Gerdjikov, V.S., Yanovski, A.B.: Gauge-covariant theory of the generating operator. I. Commun. Math. Phys. **103**, 549–68 (1986)
7. Gerdjikov, V.S., Yanovski, A.B.: Completeness of the eigenfunctions for the Caudrey-Beals-Coifman system. J. Math. Phys. **35**, 3687–721 (1994)
8. Gerdjikov, V.S., Yanovski, A.B.: CBC systems with Mikhailov reductions by Coxeter automorphism: I. Spectral theory of the recursion operators. Studies in Appl. Maths. **1342**, 145–180 (2014)
9. Golubchik, I.Z., Sokolov, V.V.: Multicomponent generalization of the hierarchy of the Landau-Lifshitz equation. Theor. Math. Phys. **124**(1), 909–917 (2000)
10. Goto, M., Grosshans, F.: Semisimple Lie Algebras. Lecture Notes in Pure and Applied Mathematics **38**. M. Dekker Inc., New-York & Basel (1978)
11. Iliev, I.D., Khristov, E.Kh., Kirchev, K.P.: Spectral Methods in Soliton Equations. Pitman Monographs and Surveys in Pure and Applied Mathematics **73**. Wiley, New-York (1994)
12. Magri, F.: A simple model of the integrable Hamiltonian equations. J. Math. Phys. **19**, 1156–1162 (1978)
13. Mikhailov, A.V.: Reduction in the integrable systems. Reduction groups. Lett. JETF (Letts. Sov. J. Exper. Theor. Phys.) **32** 187–92 (1979)
14. Mikhailov, A.V.: The reduction problem and inverse scattering method. Physica **2D**, 73–117 (1981)
15. Yanovski, A.B.: Gauge-covariant approach to the theory of the generating operators for soliton equations. Ph.D. thesis, Joint Institute for Nuclear Research (JINR) 5–87–222 (1987)
16. Yanovski, A.B.: Generating operators for the generalized Zakharov-Shabat system and its gauge equivalent system in $\mathfrak{sl}(3, \mathbb{C})$ case. Preprint: Universität Leipzig, Naturwissenchaftlich Theoretisches Zentrum Report N20 http://cdsweb.cern.ch/record/256804/files/P00019754.pdf (1993)
17. Yanovski, A.B.: Gauge-covariant theory of the generating operators associated with linear problems of Caudrey-Beals-Coifman type in canonical and in pole gauge with and without reductions. In: Slavova, A. (ed.) Proc. BGSIAM'14, pp. 2–43, Sofia(2015)

18. Yanovski, A.B., Valchev, T.I.: Pseudo-Hermitian reduction of a generalized Heisenberg ferromagnet equation. I. Auxiliary system and fundamental properties. J. Nonl. Math. Phys. **25**(02), 324–350. arXiv:1709.09266v1 [nlin.SI] (2018)
19. Yanovski, A.B., Vilasi, G.: Geometry of the recursion operators for the GMV system. J. Nonl. Math. Phys. **19**, 1250023-1/18 (2012)
20. Yanovski, A.B., Vilasi G.: Geometric Theory of the recursion operators for the generalized Zakharov-Shabat system in pole gauge on the algebra $\mathfrak{sl}(n; \mathbb{C})$ with and without Reductions. SIGMA 087 (2012)
21. Zakharov, V.E., Takhtadjan, L.A.: Equivalence between nonlinear Schrödinger equation and Heisenberg ferromagnet equation. Theor. Math. Phys. (TMF) **38**, 26–35 (1979)

Printed in the United States
By Bookmasters